# Chlorophyll Biosynthesis and Technological Applications

Constantin A. Rebeiz

# Chlorophyll Biosynthesis and Technological Applications

Springer

Constantin A. Rebeiz
Rebeiz Foundation for Basic Research
Champaign, IL, USA

Additional material to this book can be downloaded from http://extras.springer.com

ISBN 978-94-007-7133-8 ISBN 978-94-007-7134-5 (eBook)
DOI 10.1007/978-94-007-7134-5
Springer Dordrecht Heidelberg New York London

Library of Congress Control Number: 2013951182



Printed on acid-free paper

Springer is part of Springer Science+Business Media (www.springer.com)

*This book is dedicated to*

*Govindjee**

*For his ground-breaking discoveries in oxygenic photosynthesis and his pursuit of excellence in promoting chloroplast and photosynthesis research.*

*And All Past and Future Recipients of the Rebeiz Foundation for Basic Research Awards (www.vlpbp.org).*

**Govindjee was born on October 24, 1932, in Allahabad, India. He obtained his B.Sc. and M.Sc. degrees from the Allahabad University, and his Ph.D. (under the pioneers of Photosynthesis: Robert Emerson and Eugene Rabinowitch) from the University of Illinois at Urbana—Champaign. He is a Fellow of the American Association of the Advancement of Science (AAAS); Fellow and Life Member of the National Academy of Sciences (India); Past President of the American Society for Photobiology (1980–1981); Honorary President of the 2004 International Photosynthesis Congress (Montreal, Canada); the first recipient of the Lifetime Achievement Award of the Rebeiz*

*Foundation for Basic Research (2006); recipient of the Communication Award of the International Society of Photosynthesis Research (ISPR), 2007; and the Liberal Arts and Sciences Lifetime Achievement Award of the University of Illinois at Urbana-Champaign, 2008. He uses only one name; there is even a hilarious poem on him: see: http://thelegendofberkley.blogspot.com/2009/01/govindjee.html*
*His 75th birthday and his 50-year research in photosynthesis was celebrated by a special issue of* Photosynthesis Research*, in 2 parts (Part A: Volume 93 (1–3), pp 1–244, 2007; Part B: Volume 94 (2–3), pp 153–466); it was edited by his past Ph.D. student Dr. Julian J. Eaton-Rye. A nice account on him is in: J.J. Eaton-Rye (2007) "Snapshots of the Govindjee lab from the late 1960s to the late 1990s, and beyond...".* Photosynthesis Research *94: 153–178. For further information on Govindjee, see his web page: http://www.life.illinois.edu/govindjee/*

# Preface

Heme and chlorophyll (Chl) are porphyrins. Porphyrins (also referred to as tetrapyrroles) are essential for life in the biosphere. Chlorophyll catalyzes the conversion of solar energy to chemical energy via the process of photosynthesis. Organic life in the biosphere is made possible by consumption of the chemical energy generated by photosynthesis.

Hemes are the prosthetic groups of cytochromes which are involved in electron transport during oxidative phosphorylation and photosynthetic phosphorylation which generate ATP and NADPH. The latter are essential for many cellular functions.

Chlorophyll on the other hand catalyzes the process of photosynthesis. Indeed, life in the biosphere depends on the process of photosynthesis which converts light energy, carbon dioxide and water into the chemical energy, required for the formation of food and fiber. Photosynthetic efficiency is controlled by extrinsic factors such as the availability of water, $CO_2$, inorganic nutrients, ambient temperature and the metabolic and developmental state of the plant, as well as by intrinsic factors (Lien and San Pietro 1975). The most important intrinsic factor is the efficiency of the photosynthetic electron transport system (PETS).

Conventional agriculture is one of the few human activities that have not undergone a revolution to join other activities such as overcoming gravity by flying, and landing on the moon, crossing underwater the polar cap, and communicating wirelessly over long distances via electromagnetic waves. We now feel that enough biochemical and molecular biological knowledge has accumulated to render this dream amenable to experimentation. We believe that the time has come to bioengineer chloroplasts capable of synthesizing a short chain carbohydrate such as glycerol at rates that approach the upper theoretical limits of photosynthesis [Rebeiz, C. A. (2010) Investigations of possible relationships between the chlorophyll biosynthetic pathway and the assembly of chlorophyll-protein complexes and photosynthetic efficiency. In: Rebeiz, C. A. Benning, C., Bohnert, H.J., Daniell, H., Hoober J. K., Lichtenthaler, H. K., Portis, A. R., and Tripathy, B. C. eds. The chloroplast: Basics and Applications. Springer. The Netherlands, pp 1–24]. In order to achieve this goal a thorough knowledge of the Chl biosynthetic pathway is needed along with knowledge in other domains (Rebeiz 2010). In this context, this monograph is devoted to an in-depth discussion of our present knowledge of the Chl biosynthetic pathway. The complexity and biochemical heterogeneity of the Chl biosynthetic pathway and the relationship of this complexity to the structural and biosynthetic complexity of photosynthetic membranes will be emphasized. We will also emphasize in historical perspective, key stages in our understanding of the Chl biosynthetic heterogeneity. The reader should keep in mind that a complex biosynthetic process is only fully understood when it becomes possible to reconstitute in vitro every step of the process. We are not yet at this stage of understanding of thylakoid membrane biogenesis. Considerable progress has been achieved, however, in the understanding of numerous facets of the Chl biosynthetic pathway, namely, (a) detection and identification of various major and minor metabolic intermediates, (b) precursor-product relationships between various intermediates, (c) structure and regulation of many enzymes of the pathway, and (d) the relationship of the Chl biosynthetic heterogeneity to the structural and functional heterogeneity of thylakoid membranes.

In addition, topics related to the development of Analytical techniques, Cell-free systems, Herbicides, Insecticides, and Cancericides are also discussed. Topics covered in the various chapters are discussed below.

Chapter 1: Some Major Steps in the Understanding of the Chemistry and Biochemistry of Chl

Chapter 2: Synopsis

Chapter 3: Development of Analytical and Preparatory Techniques

Chapter 4: Development of Cell-Free Systems

Chapter 5: Reactions Between δ-Aminolevulinic Acid and Protoporphyrin IX

Chapter 6: The Iron and Magnesium Branches of the Porphyrin Biosynthetic Pathway

Chapter 7: The Chl *a* Carboxylic Biosynthetic Routes: Reactions Between Mg-Protoporphyrin IX and Protochlorophyllide *a*

Chapter 8: The Chl *a* Carboxylic Biosynthetic Routes: Protochlorophyllide *a*

Chapter 9: The Chl *a* Carboxylic Biosynthetic Routes: (Photo)Conversion of Protochlorophyllides (Pchlides) *a* to Chlorophyllide (Chlide) *a*

Chapter 10: The Chl *a* Carboxylic Biosynthetic Routes: Conversion of Chlide *a* to Chl *a*

Chapter 11: The Fully Esterified Chlorophyll *a* Biosynthetic Routes: Reactions Between Mg-Protoporphyrin IX Diester and Chl *a*

Chapter 12: The Chlorophyll *b* Biosynthetic Pathway: Novel Metabolic Intermediates

Chapter 13: The Chl *b* Biosynthetic Pathway: Intermediary Metabolism

Chapter 14: Relationship of Chlorophyll Biosynthetic Heterogeneity to the Greening Group Affiliation of Plants

Chapter 15: Relationship of Chlorophyll Biosynthesis to the Assembly of Chlorophyll-Protein Complexes

Chapter 16: The Chlorophyll Biosynthetic Heterogeneity and Chloroplast Bioengineering

Chapter 17: Photodynamic Herbicides
Chapter 18: Porphyric Insecticides
Chapter 19: ALA-Dependent Cancericides
Appendix I: The Molecular Biology of Chlorophyll Biosynthetic Enzymes is discussed
Appendix II: The Molecular Biology of The Various Apoproteins of Pigment-Protein Complexed is discussed

Finally, we hope that this monograph will provide graduate students and leading research scientists involved in Chl and photosynthesis research and their technological fall-outs a useful data base for their research.

Champaign, IL, USA Constantin A. Rebeiz

# Contents

# Abbreviations

| | |
|---|---|
| 4VChlR | 4-Vinyl Chl *a* reductase |
| 4VCR | 4-Vinyl Chlide *a* reductase |
| 4VMPR | 4-Vinyl Mg-Proto reductase |
| 4VMPER | 4-Vinyl Mg-Proto monoester reductase |
| 4VPideR | 4-Vinyl Pchlide *a* reductase |
| 4VR | 4-Vinyl reductase |
| ALA | δ-Aminolevulinic acid |
| BSA | Bovine serum albumin |
| Chlide | Chlorophyllide |
| Chl | Chlorophyll |
| CP29 | Chl-Protein complex 29 |
| CP47 | Chl-Protein complex 47 |
| CP57 | Chl-Protein complex 57 |
| Copro III | Coproporphyrin III |
| Coprogen III | Coproporphyrinogen III |
| DDV-LDV | Dark Divinyl-Light Divinyl |
| DHGG | Dihydrogeranylgeraniol |
| DMV | Dark Monovinyl |
| DMV-LDV-LDMV | Dark Monovinyl-Light Divinyl-Light Dark Monovinyl |
| Dpy | α,α-dipyridyl |
| DV | Divinyl |
| DV Pchlide *a* | Divinyl Pchlide *a* |
| GA | Gibberellic acid |
| ER | Endoplasmic reticulum |
| GG | Geranylgeraniol |
| GGDP | Geranylgeraniol diphosphate |
| GSA | Glutamate semialdehyde |
| HAT | Hydroxyaminotetrahydropyranone |
| HDR | Hydroxy pyruvate reductase |
| HHGG | Hexahydro Geranylgeraniol |

| | |
|---|---|
| HMBL | Hydroxymethylbilane |
| LCFA | Long chain fatty alcohol |
| LDV | Light Divinyl |
| LDDV | Light–dark Divinyl |
| LDMV | Light–dark Monovinyl |
| LHC | Light harvesting Chl-protein complex |
| LHCI | Light harvesting Chl-protein Complex I |
| LHCI-730 | Light harvesting Chl-Protein complex 730 of PSI |
| LHCI-680 | Light harvesting Chl-Protein complex 680 of PSI |
| LHCII | Light harvesting Chl-Protein complex II |
| LW | Long wavelength |
| M-PBR | Mitochondrial peripheral-type benzodiazepine receptor |
| Mpe | Mg-Proto monoester |
| Mpde | Mg- Proto diester |
| Mpd(e) | Mg-Proto ester and/or diester |
| MV | Monovinyl |
| MV Pchlide | Monovinyl Pchlide |
| NMR | Nuclear magnetic resonance |
| nt-Pchlide *a* | Nontransformable Pchlide *a* |
| nt-SW Pchlide *a* | Nontransformable short wavelength Pchlide *a* |
| Oph | Orthophenanthroline |
| Pchl | Protochlorophyll |
| Pchl H | Protochlorophyll holochrome |
| Pchlide | Protochlorophyllide |
| Pchl(ide) | Pchlide and/or Pchlide ester |
| Pchlide E | Pchlide ester |
| PDT | Photoradiation therapy |
| PETS | Photoelectron transport system |
| PORA | Pchlide Oxidoreductase A |
| Proto | Protoporphyrin IX |
| Protogen | Protoporphyrinogen IX |
| Protox | Protoporphyrin IX oxygenase |
| PSI | Photosystem I |
| PSII | Photosystem II |
| PSU | Photosynthetic Unit |
| PT | Permeability transition |
| RC | Reaction center |
| SAM | S-Adenosyl methionine |
| SAMMT | S-Adenosyl methionine methyl transferase |
| SW | Short wavelength |
| SCR | Succinate cytochrome c reductase |
| THGG | Tetrahydrogeranylgeraniol |
| TDPH | Tetrapyrrole-dependent photodynamic herbicide |
| t | Transformable |

| | |
|---|---|
| t-LW Pchlide *a* H | Transformable long wavelength Pchlide *a* holochrome |
| t-Pchlide *a* | Transformable Pchlide *a* |
| t-SW Pchlide a H | Transformable short wavelength Pchlide *a* holochrome |
| Uro | Uroporphyrin |
| Urogen | Uroporphyrinogen |

# Author Biography

**Constantin A. Rebeiz (Tino)** was born on July 11, 1936, in Beirut, Lebanon, where his family has been since the fourth century AD. After 14 years of French schooling in a private French school, he joined the American University of Beirut in 1956. He graduated with distinction in 1959 with a BS degree in General Agricultural Sciences. In 1960, he obtained an M.S. degree with Julian Crane in Horticulture from the University of California at Davis (UC Davis). His M.S. thesis described the production of parthenocarpic cherries and peaches. In 1964, he obtained a Ph.D. degree in Plant Physiology from UC Davis. His Ph.D. thesis with Paul Castelfranco described the discovery of the extra-mitochondrial β-oxidation of long chain fatty acids. From 1965 to 1969 he headed the Department of Biological Sciences at the National Agricultural Research Institute in Lebanon. There, in 1968, he co-founded the Lebanese Association for the Advancement of Sciences. In 1972, he joined the University of Illinois at Urbana-Champaign (UIUC) as an Associate Professor of Plant Biochemical Physiology. In 1984 he received the John P Trebellas Endowment for Biotechnological Research and moved his Laboratory of Plant Biochemistry and Photobiology into a newly remodeled Laboratory space in the Agricultural Biotechnology Building; then in 1992 his laboratory moved into the new Edward R. Madigan Building. Tino retired as a Professor Emeritus in May, 2005. He is currently the President of the Rebeiz Foundation for Basic Research which is involved in the promotion of chloroplast research and bioengineering nationally and internationally. Tino's research in tetrapyrrole biochemistry and chemistry (see http://www.vlpbp.org/ selected research highlights) spans the fields of botany, plant physiology, preparative methodologies, analytical biochemistry, Biochemistry, Chemistry, the development of pesticides, biomedical research as well as chloroplast bioengineering. His tetrapyrrole biochemical work led to the discovery of novel Mg-porphyrins, protochlorophylls, chlorophylls, the discovery of the multi-branched chlorophyll biosynthetic pathway, and the development of a blue print for bioengineering chloroplasts with higher photosynthetic efficiencies. His pioneering work on the development of photodynamic pesticides has led to the discovery of δ-aminolevulinic-dependent photodynamic cancer treatment and skin keratoses that are used worldwide in the medical community. Many laboratories around the world are now developing δ-aminolevulinic-dependent non- invasive photodynamic cancer treatment strategies for several types of cancer. Among the awards received by Tino are the Siemens award (1957), The University of Illinois Funk award (1985), selection by *Science Digest* as one of America's outstanding innovators responsible for the 100 most technological achievements (1985), the University of Illinois College of Agriculture Senior Faculty award for excellence in research (1985), the US Presidential Green Chemistry Challenge Award (1999), and the American University of Beirut Scientific Achievement award during commemoration of the 50th Anniversary of the College of Agriculture and Food Sciences (2002).

# Chapter 1
# Some Major Steps in the Understanding of the Chemistry and Biochemistry of Chl

*Meaningful scientific discoveries are those that help humans achieve a better understanding of themselves, of their environment and of the universe at large, as well as those that contribute to the betterment of the human, spiritual, psychological and physical condition.*

(Constantin A. Rebeiz)

Since the turn of the nineteenth century, the green color of plants has attracted the attention of a wide spectrum of scientists. In this section an effort will be made to list chronologically important scientific discoveries that had a clear impact on our understanding of the structure and function of porphyrins and chlorophyll (Ikeuchi and Murakami 1982) in particular. The many different bacteriochlorophylls are not considered in this presentation.

## 1.1 Some Major Steps in the Understanding of the Chemical Structure of Porphyrins and Chl

In this historical section, emphasis is placed on important scientific discoveries that had a clear impact on the understanding of the chemical structure of porphyrins and Chl.

### *1.1.1 Pelletier, 1818*

In 1818, Pelletier and Caventou first used the word chlorophyll to describe the pigment complex responsible for the green color of leaves (Pelletier and Caventou 1818). Since then the green pigments of higher and lower plants has been called Chlorophyll.

C.A. Rebeiz, *Chlorophyll Biosynthesis and Technological Applications*,
DOI 10.1007/978-94-007-7134-5_1, © Springer Science+Business Media Dordrecht 2014

### *1.1.2 Verdiel, 1844*

In 1844 Verdiel, suggested a relationship between Chl and heme upon chemical conversion of Chl to a red pigment (Verdeil 1844). That was an important suggestion that was later on confirmed by structural analysis.

### *1.1.3 Fremy, 1860*

In 1860, Fremy partitioned the plastid pigments between a yellow ethereal solution containing carotenoids and an acidic aqueous solution of blue-green pheophytins and pheophorbides, which he called phyllocyanin (Fremy 1860). That marked the beginning of the chromatographic analysis of plant pigments.

### *1.1.4 Stokes, 1864*

In 1864 as a result of spectroscopic studies, Stokes suggested that even after extracting the yellow carotenoids, the Chl fraction consisted of two different, green, red-fluorescent substances. This important observation paved the way for the discovery of the Chl *a* and *b* (Stokes 1864).

### *1.1.5 Borodin, 1882*

In1882, Borodin formed Chl crystals (actually ethyl chlorophyllide) by the action of ethyl alcohol on leaves (Borodin 1882). Although Borodin is known worldwide as a musician, he was a carrier chemist in Russia.

### *1.1.6 Monteverde, 1893*

In 1893, Monteverde isolated Chl crystals and determined their spectroscopic properties (Monteverde 1893).

### *1.1.7 Nencki, 1896*

In 1896 Nencki established that porphyrins were made up of Pyrrole nuclei. He proposed that the similar chemical properties of hemin and chlorophyll denotes a common origin of plant and animal life and that comparison of similar compounds of flora and fauna provides insight into chemical and organismal evolution (Nencki 1896).

### *1.1.8 Twsett, 1906*

1906 Twsett separated blue Chl (i.e. Chl *a*) from yellow Chl (*i.e.* Chl *b*) using column chromatography, and called them chlorophylls *alpha* and *beta*, which later became chlorophylls *a* and *b* (Tswett 1906).

### *1.1.9 Willstatter and Hocheder, 1907*

In 1907, Willstatter and Hocheder discovered and named pheophytin (Chl that has lost the central Mg atom) (Willstatter and Asahina 1909). More recently this important molecule has been implicated in photosynthetic electron transport in Photosystem II.

### *1.1.10 Willstatter and Fritzsche, 1909*

In 1909, Willstatter and Fritzsche applied alkaline degradation to the study of the chemical structure of Chl.

### *1.1.11 Willstatter and Ashina, 1909*

In 1909, Willstatter and Asahina applied chromic acid oxidation to the study of the chemical structure of Chl (Willstatter and Asahina 1909). This technique was used later on by Shemin to determine the structure of Protoporphyrin IX.

### *1.1.12 Willstatter and Stoll, 1910*

In1910, Willstatter and Stoll discovered and named chlorophyllide (Chl without esterification with phytol at position 7 of the macrocycle), pheophorbide (chlorophyllide without the central Mg-atom), and chlorophyllase (the enzyme that de-esterifies Chl) (Willsttater and Stoll 1910).

### *1.1.13 Willstatter and Stoll, 1911*

In1911, Willstatter and Stoll discovered and named Phytol (the major esterifying alcohol of Chl at position 7 of the macrocycle), pheophorbide (chlorophyllide without the central Mg-atom), chlorophyllase (the enzyme that de-esterifies Chl) and allomerization (Willsttater and Stoll 1911). The products of allomerization vary with the circumstances, but they all have oxygen instead of hydrogen bonded to C-10 of the Chl molecule.

### *1.1.14 Kuster, 1913*

In 1913, Kuster proposed a correct formula for the ring system of porphyrins in which four pyrrole rings are linked together into a macrocycle by 4 methine bridges (Kuster 1913). He was violently criticized by Fischer, the father of contemporary porphyrin chemistry who believed that such a large structure would be highly unstable. Later Fischer accepted the structure proposed by Kuster when he achieved the total synthesis of protoheme.

### *1.1.15 Willstatter and Stoll, 1913*

In 1913, Willstatter and Stoll published a monograph that summarized most of the research findings of Willstatter and collaborators and ushered the modern era of the field of Chl chemistry (Willsttater and Stoll 1913). Willstatter got the Nobel Prize for his porphyrin and Chl work.

### *1.1.16 Fischer and Lowenberg, 1928*

In 1928, Fischer and Lowenberg established the structure of phytol (Fischer and Lowenberg 1928). Phytol is the major alcohol that esterifies the propionic acid residue of Chl at position 7 of the Chl macrocycle.

### *1.1.17 Fischer and Lowenberg, 1929*

In 1929, Fischer and Lowenberg synthesized phytol form pseudoionone (Fischer and Lowenberg 1929).

### *1.1.18 Noack and Kiessling, 1929, 1930*

In 1929 and 1930, Noack and Kiessling, initiated the study of the protochlorophyll chemistry of pumpkin seed coat (Noack and Kiessling 1929, 1930). These studies led to the erroneous assumption that protochlorophyll (i.e. esterified protochlorophyllide) is the immediate precursor of Chl.

### *1.1.19 Frank Capra, 1932*

In 1932, in the academy award winning movie, You Can't Take It With You, it was emphasized that the understanding of photosynthesis which is catalyzed by Chl could usher an era of vast abundance.

### *1.1.20 Noack and Schneider, 1933*

In 1933, Noack and Schneider, started the study of bacteriochlorophyll (Noack and Scneider 1933).

### *1.1.21 Fischer and Stern, 1935*

In 1935, Fischer and Stern, proposed a structure for Chl (Fischer and Stern 1935). The structural formula for Chl was correct, except for the position of the two extra hydrogens which are now recognized to be located on ring IV at positions 7 and 8 of the macrocycle.

### *1.1.22 Fischer and Orth, 1937*

In 1937, Fischer and Orth published the first part of a monograph about Chl chemistry that described the various research findings of Fischer and collaborators that resulted in our modern understanding of the structure of Chl and its degradation products (Fischer and Orth 1937).

### 1.1.23 Fischer and Lambrecht, 1937 and 1938

In 1937 and 1938, Fischer and Lambrecht established the chemical relationship of bacteriochlorophyll to Chl by the preparation of common derivatives (Fischer and Lambrecht 1937, 1938).

### 1.1.24 Fischer and Wenderoth, 1939

In 1939, Fischer and Wenderoth correctly assigned the two extra hydrogens of Chl to positions 7 and 8 on ring IV of the Chl macrocycle.

### 1.1.25 Fischer and Coworkers, 1939, 1940

In 1939, and 1940, Fischer and coworkers identified the protochlorophyll of pumpkin seed coat as a vinyl pheoporphyrin analog of chlorophyll (Fischer et al. 1939; Fischer and Oestreicher 1940).

### 1.1.26 Fischer and Stern, 1940

In 1940, Fischer and Stern published the second part of a monograph about Chl chemistry that described the various research findings of Fischer and collaborators that resulted in our modern understanding of the structure of Chl and its degradation products (Fischer and Stern 1940). Fischer won the Nobel Prize for his work on porphyrin and Chl structure.

### 1.1.27 Strain, 1942, 1943

In 1942, 1943 Strain and coworkers showed that chlorofucine, later known as Chl *c*, is not an artifact.

### 1.1.28 Manning and Strain, 1943

In 1943, Manning and Strain, discovered Chl *d* (Manning and Strain 1943).

### *1.1.29 Cookson and Rimington, 1954*

In 1954, Cookson and Rimington finalized their investigations of the chemical structure of porphobilinogen (Cookson and Rimington 1954). Porphobilinogen is the precursor of Uroporphyrinogen III, the first tetrapyrrole precursor of heme and Chl.

### *1.1.30 Holt and Morley, 1959*

In 1959, Holt and Morley determined the chemical structure of Chl *d* (Holt and Morley 1959).

### *1.1.31 Woodward and Coworkers, 1960*

In 1960, Woodward and coworkers reported the total synthesis of chlorin $e_6$ trimethyl ester, an important degradation product of chl. Since the reactions between chlorin $e_6$ trimethyl ester and Chl had been supposedly worked out earlier (in 1913) by Willstatter and coworkers, it was erroneously assumed that Woodward had achieved the total synthesis of Chl. In the process of their work the authors discovered the remarkable susceptibility of Chl to electrophilic attack. Woodward was Awarded the Nobel Prize for his work on Chl.

### *1.1.32 Closs and Coworkers, and Katz and Coworkers, 1963*

In 1963, Closs and coworkers and Katz and coworkers, applied infra-red and nuclear magnetic resonance techniques to the study of Chl and some of its derivatives (Closs et al. 1963; Katz et al. 1963).

### *1.1.33 Dougherty and Coworkers, 1970*

In 1970, Dougherty et al., determined the chemical structure of Chl *c* (Dougherty et al. 1970).

## 1.2 Some Major Steps in the Understanding of the Biochemistry of Porphyrin and Chl Formation

In this historical section emphasis is placed on important scientific discoveries that had a clear impact on the understanding of the structure and function of intermediates and end-products of the porphyrin and Chl biosynthetic pathways. Most of these discoveries will be discussed in details in various chapters of this monograph.

### *1.2.1 Granick, 1948a*

In 1948, Granick demonstrated the accumulation of divinyl (DV) protoporphyrin IX (Proto) in *Chlorella* mutants inhibited in their capability to form Chl. Since the algal cultures that accumulated Proto were inhibited in their capabilities of forming Chl, Granick proposed that in plants, DV Proto is a precursor of monovinyl (MV) Chl *a* (Granick 1948a). In this context, the term DV refers to tetrapyrroles that contain vinyl groups at positions 2 and 4 of the macrocycle, while the term MV refers to tetrapyrroles containing a vinyl group at position 2 and an ethyl group at position 4 of the macrocycle.

### *1.2.2 Granick, 1948b*

In 1948, Granick also demonstrated the accumulation of divinyl DV Mg-protoporphyrin IX (Mg-Proto) in X-ray *Chlorella* mutants inhibited in their capability to form Chl, and proposed that in plants, DV Mg-Proto is a precursor of MV Chl *a* (Granick 1948b).

### *1.2.3 Koski and Smith, 1948*

In 1948, Koski and Smith, purified protochlorophyllide (Pchlide *a*) which they mistook for Pchlide *a* phytyl ester [*i.e.* protochlorophyll (Pchl) *a*] and determined its spectral absorption properties (Koski and Smith 1948). That mistake was facilitated by the then erroneous notion that Pchl was the main immediate precursor of Chl.

### 1.2.4 Smith, 1948

In 1948, on the basis of the correspondence of the newly published absorbance spectrum of MV Pchlide *a* (mistaken for MV Pchl *a*), and the action spectrum of MV Chl *a* formation, smith proposed that MV Pchl *a* (in fact, MV Pchlide *a*) is the immediate precursor of MV Chl *a* (Smith 1948). That dogma was dropped later on (see below).

### 1.2.5 Muir and Neuberger, and Wittenberg and Shemin, 1949

In1949, Muir and Neuberger and Wittenberg and Shemin showed that one carbon atom and the nitrogen atom of each pyrrole ring of protoheme is derived from the alpha-atom and the associated nitrogen atom of glycine (Muir and Neuberger 1949).

### 1.2.6 Granick, 1950

In 1950 Granick demonstrated the accumulation of monovinyl (MV) Pchlide *a* in *Chlorella* mutants inhibited in their capability to form Chl, and proposed that in plants, Pchlide *a* is the immediate precursor of Pchlide *a* phytyl ester (i.e. Pchl *a*). Then Granick organized DV Proto, DV Mg-Proto, MV Pchlide *a*, MV Pchlide *a* phytyl ester, and MV Chl *a* by order of increasing chemical complexity into a paper chemistry, single branched, Chl *a* biosynthetic pathway that originated in DV Proto and ended in the formation of MV Chl *a* (Granick 1950).

### 1.2.7 Muir and Neuberger and Wittenberg and Shemin, 1950

In 1950, Muir and Neuberger and Wittenberg and Shemin showed that each of the four methine bridge carbon atoms of protoheme is derived from the alpha-carbon of glycine. Wittenberg and Shemin used chromic acid oxidation, a technique developed by Willstatter and Ashina in 1909 (see above).

### 1.2.8 Shemin and Wittenberg, 1951

In 1951, Shemin and Wittenberg concluded that all four pyrrole rings of protoheme arise from a common pyrrole precursor (Shemin and Wittenberg 1951).

### 1.2.9 Smith, 1952

In 1952, Smith, prepared the first Pchl *a*-apoprotein complex (actually containing a mixture of Pchlide *a* and Pchlide *a* phytyl ester) from etiolated barley, and determined its absorption maximum at 650 nm. Later the complex was named Pchl-holochrome by Smith and collaborators (Smith 1952).

### 1.2.10 Shemin and Kumin, 1952

In 1952, Shemin and Kumin demonstrated that the remaining carbon atoms as well as the side chains of protoheme are derived from succinate, which led Shemin to suggest that the carbon atoms of succinate enter porphyrin metabolism as succinyl-CoA (Shemin and Kumin 1952).

### 1.2.11 Westall, 1952

In 1952, Westall, crystallized porphobilinogen (PBG) from the urine of a patient with acute porphyria, and made pure PBG crystals available to other researchers (Westall 1952). The availability of pure PBG crystals helped determine its chemical structure (Westall 1952).

### 1.2.12 Shemin and Russell, 1953

In1953, Shemin and Russell proposed that glycine and succinate did not enter porphyrin metabolism as individual compounds but as a new compound δ-aminolevulinic acid (ALA) (Shemin and Russel 1953). That marked the involvement of ALA as the first building block of tetrapyrroles.

### 1.2.13 Della Rosa, 1953

In 1953, Della Rosa et al., demonstrated the incorporation of $^{14}$C-glycine and $^{14}$C-acetate into MV Chl *a*.

### 1.2.14 Bogorad and Granick, 1953a

In 1953, Bogorad and Granick described a *Chlorella* mutant capable of accumulating porphyrins with two, three, four, five, six, seven and eight carboxyl groups, and proposed that these porphyrins may be intermediates in the formation of DV Proto (Bogorad and Granick 1953a).

### 1.2.15 Bogorad and Granick, 1953b

In 1953, Bogorad and Granick demonstrated the conversion of exogenous PBG to DV Proto in frozen and thawed Chlorella cells, and proposed a single branched paper chemistry pathway that originated in glycine and succinate and ended with the formation of Proto via ALA, Uroporphyrin III (Uro), Coproporphyrin III (Copro), hematoporphyrin IX, and DV Proto (Bogorad and Granick 1953b).

### 1.2.16 Granick, 1954

In 1954, Granick, demonstrated the conversion of ALA to PBG and porphyrins by extracts of Chlorella cells, of spinach and chicken erythrocytes (Granick 1954).

### 1.2.17 Smith and Benitez

In 1954, Smith and Benitez, described the kinetics of Pchl *a* (actually mainly Pchlide *a*) photoconversion to chlorophyll(ide) [Chl(ide)] *a* in etiolated barley leaves. At the time Smith and Benitez believed that they were converting Pchl into Chl (Smith and Benitez 1954).

### 1.2.18 Neve and Labbe, 1956

In 1956, Neve and Labbe, recognized that the actual tetrapyrrole intermediates between PBG and DV proto are not porphyrins but reduced porphyrins, namely porphyrinogens, i.e. hexahydro porphyrins (Neve and Labbe 1956).

### 1.2.19 Smith and Kupke, 1956

In 1956, they extended their studies of Pchl-holochrome (Smith and Kupke 1956).

### 1.2.20 Goldberg, 1956

In 1956, Goldberg et al., described the insertion of ferrous iron into Proto by ferrochelatase (Goldberg et al. 1956).

### 1.2.21 Shibata, 1957

In 1957, Shibata described an opal glass technique for the determination of the spectral properties of intact leaves, and described a Chl-apoprotein spectral shift, the Shibata shift, during greening of etiolated tissues (Shibata 1957).

### 1.2.22 Wolff and Price, 1957

In 1957, Wolff and Price demonstrated that the immediate product of Pchlide *a* photoconversion is chlorophyllide (Chlide) *a*, which is then esterified to Chl *a* in the dark. They proposed that the main route of Chl formation proceeds from Pchlide to Chl *a* via Chlide *a* (Wolff and Price 1957).

### 1.2.23 Bogorad, 1958

In 1958, demonstrated the conversion of PBG to hexahydro Uro i.e. uroporphyrinogen III (Urogen III) in a wheat germ extract (Bogorad 1958).

### 1.2.24 Mauzerall and Granick, 1958

In 1958, demonstrated the conversion of Urogen III to coproporphyrinogen III (Coprogen III) in duck erythrocytes (Mauzerall and Granick 1958).

### 1.2.25 *Sano and Granick, 1961*

In 1961, Sano and Granick described the conversion of Coprogen III to DV Proto by beef liver mitochondria (Sano and Granick 1961; Sano 1966).

### 1.2.26 *Granick, 1961*

In 1961, Granick, demonstrated the accumulation of DV Mg-Proto monomethyl ester (Mpe) in *Chlorella* mutants inhibited in their capability to form chlorophyll (Chl), and proposed that in plants, DV Mpe is a precursor of MV Pchlide *a* (Granick 1961). In that same article Granick reported on the first usage of 2,2′-dipyridyl (Dpy) to induce the accumulation of Mpe in etiolated barley leaves.

### 1.2.27 *Tait and Gibson*

In1961, Tait and Gibson, demonstration the conversion of Mg-Proto to Mpe by *R. Spheroides* chromatophores (Tait and Gibson 1961).

### 1.2.28 *Gibson, 1963*

In 1963, Gibson et al., detected S-adenosylmethionine-Mg-Proto methyl transferase in *R. Spheroides* (Gibson et al. 1963).

### 1.2.29 *Jones, 1963*

In 1963 Jones, detected DV Pchlide *a* in R. Spheroides inhibited in their growth by 8-hydroxyquinoline and suggested that DV Pchlide *a* is a transient precursor of MV Pchlide *a* (Jones 1963).

### 1.2.30 *Jones, 1966*

In 1966, Jones, detected DV Pchlide *a* phytyl ester in the pumpkin inner seed coat and proposed its involvement in Chl *a* Biosynthesis (Jones 1966).

### 1.2.31 *Sironval, 1967*

In 1967, Sironval et al., proposed that the Pchl(ide) *a* holochrome acts as a shuttling photoenzyme that catalyzes the conversion of Pchlide *a* to Chlide *a* (Sironval et al. 1967).

### 1.2.32 *Shemin, 1968*

In 1968, Shemin, proposed a detailed mechanism of the mode of action of ALA dehydratase, the enzyme that converts 2 moles of ALA to PBG (Shemin 1968).

### 1.2.33 *Ellsworth and Aronoff, 1969*

In 1969, Ellsworth and Aronoff, detected intermediates between Mpe and Pchlide *a* involving putative DV and MV metal-free acrylic, hydroxy and keto tetrapyrrole derivatives in ultraviolet *Chlorella* mutants, and proposed that in plants, the formation of DV and MV Pchlide involves a β-oxidation sequence of the methyl propionate of DV and novel MV Mpe, at position 6 of the macrocycle. They also proposed a MV/DV biosynthetic loop that started at Mpe and finished at Pchlide (Ellsworth and Aronoff 1969).

### 1.2.34 *Rebeiz, 1970*

In 1970, Rebeiz et al., reported that kinetic analysis of the formation of the $^{14}$C-Pchl (ide) *a* pool, does not support the currently accepted notion that Pchlide *a* is the immediate precursor of Pchlide *a* phytyl ester. This observation constituted the first evidence of a potential Chl biosynthetic heterogeneity in plants (Rebeiz et al. 1970).

### 1.2.35 *Rebeiz and Castelfranco, 1971a*

In 1971, reported the total biosynthesis of $^{14}$C-Pchlide *a* and its phytyl ester from $^{14}$C-ALA in an in organello system from higher plants (Rebeiz and Castelfranco 1971a).

### 1.2.36 Rebeiz and Castelfranco, 1971b

In 1971, Rebeiz and Castelfranco also reported the total biosynthesis of $^{14}$C-Chl *a* and *b* from $^{14}$C-ALA in an in organello system from higher plants (Rebeiz and Castelfranco 1971b).

### 1.2.37 Gorchein, 1972

In 1972, Gorchein, demonstrated the conversion of exogenous Proto to Mpe in R. Spheroides, and the ATP requirement of the process (Gorchein 1972).

### 1.2.38 Griffiths, 1974

In 1974, Griffith, demonstrated that NADPH is the hydrogen donor for the photoreduction of Pchlide *a* to Chlide *a* (Griffiths 1974).

### 1.2.39 Beale and Castelfranco

In 1974, Beale and Castelfranco demonstrated that in green plants ALA is formed from glutamic acid. That led to the formulation of the C-5 biosynthetic pathway of ALA formation in green plants (Beale and Castelfranco 1974).

### 1.2.40 Rebeiz, 1975

In 1975, Rebeiz et al., illustrated the usefulness of Fluorescence spectroscopy by detecting the formation of several Mg-porphyrins during greening of etiolated tissues, and described their spectrofluorometric properties (Rebeiz et al. 1975b).

### 1.2.41 Poulson and Polglase, 1975

In 1975, Poulson and Polglase demonstrated that protoporphyrinogen IX oxidase (Protox for short) catalyzes the conversion of Protogen IX to Proto (Poulson and Polglase 1975).

### 1.2.42 *Smith and Rebeiz, 1977*

In 1977, Smith and Rebeiz, described the enzymic insertion of Mg into Proto, in higher plants in organello, to yield Mg-Proto (Smith and Rebeiz 1977). The formation of Mg-Proto was accompanied by the formation of Zn-Proto.

### 1.2.43 *Mattheis and Rebeiz, 1977a*

In 1977, Mattheis and Rebeiz, described the conversion of exogenous Proto to Pchlide *a*, in organello in higher plants (Mattheis and Rebeiz 1977a).

### 1.2.44 *Mattheis and Rebeiz, 1977b*

In 1977, Mattheis and Rebeiz also, described the conversion of exogenous Mpe to Pchlide *a*, in organello in higher plants (Mattheis and Rebeiz 1977b).

### 1.2.45 *Griffiths, 1978*

In 1978, Griffiths, proposed that the Pchl holochrome i.e. Pchlide *a* oxidoreductase, NADPH and Pchlide *a* form a photoactive ternary Pchlide *a* NADPH-enzyme complex with a red absorption maximum at 652 nm (Griffiths 1978).

### 1.2.46 *Belanger and Rebeiz, 1979*

In 1979, Belanger and Rebeiz, reported that the Pchlide *a* pool of etiolated tissues consisted of two components (probably MV and DV components) which are photoconvertible into two distinct Chlide *a* species (Belanger and Rebeiz 1979). This work heralded the discovery of the Chl biosynthetic heterogeneity and multiple Chl biosynthetic routes.

### 1.2.47 *Battersby and Jordan and Seerah, 1979*

In 1979, Battersby et al., and Jordan and Seerah, determined that 1-hydroxymethylbilane (HMBL) (also called preuroporphyrinogen) is the immediate precursor of Uroporphyrinogen III (Battersby et al. 1979; Jordan and Seehra 1979).

### 1.2.48 Apel, 1980

In 1980, Apel et al., described the purification of Pchlide *a* oxidoreductase (POR-A) from etiolated barley (Apel et al. 1980).

### 1.2.49 Pardo, 1980

In 1980, Pardo et al., confirmed that ATP was a mandatory cofactor for Mg-insertion into Proto and that higher concentration of added ATP eliminated the formation of Zn-Proto (Pardo et al. 1980).

### 1.2.50 Belanger and Rebeiz, 1980a

In 1980, Belanger and Rebeiz, described the detection of DV Pchlide *a* in higher plants (Belanger and Rebeiz 1980a).

### 1.2.51 Belanger and Rebeiz, 1980b

In 1980, Belanger and Rebeiz also, described the formation of DV Chlide *a* and DV Chl *a* in higher plants (Belanger and Rebeiz 1980b).

### 1.2.52 Schoch, 1980

In 1980, Schoch, wrapped up the demonstration that in etiolated tissues subjected to a light treatment followed by darkness, Chlide *a* is first esterified with geranylgeraniol (GG) to yield Chl a GG, which is reduced stepwise to Chl a dihydroGG (DHGG), tetrahyddroGG (THGG) and finally to hexahydroGG, i.e. phytylated Chl *a* (Schoch 1978).

### 1.2.53 Belanger and Rebeiz, 1980c

In 1980, Belanger and Rebeiz, also, described the detection of DV Pchlide *a* phytyl ester in etiolated higher plants (Belanger and Rebeiz 1980c).

### *1.2.54 McCarthy, 1981*

In 1981, McCarthy et al., described the detection of a fully esterified Mpe pool in etiolated higher plants treated with ALA and 2,2′-dipyridyl (McCarthy et al. 1981).

### *1.2.55 Rebeiz, 1981*

In 1981, Rebeiz et al., proposed a 4-branched Chl *a* biosynthetic pathway on the basis of the experimental evidence available at that time (Rebeiz et al. 1981).

### *1.2.56 Santel and Apel, 1981*

In 1981, Santel and Apel, demonstrated that during greening of etiolated tissues a rapid decline of POR-A is observed. After 6 h of continuous illumination, when the rate of Chl *a* accumulation is at its peak, only traces of the POR-A protein were detected (Santel and Apel 1981).

### *1.2.57 Bazzaz, 1981*

In 1981, Bazzaz, described a lethal maize mutant Nec 2, (ex-ON 2) which accumulates only DV Chl *a* and *b*.

### *1.2.58 Belanger and Rebeiz*

In 1982, Belanger and Rebeiz, detected the occurrence of MV Mg-Proto, MV Mpe and MV Mpe diester in higher plants (Belanger and Rebeiz 1982).

### *1.2.59 Belanger*

In 1982, Belanger et al., ascertained the chemical structure of DV Chlide *a* (Belanger et al. 1982).

### 1.2.60 Duggan and Rebeiz, 1982a

In 1982, Duggan and Rebeiz, described the induction of massive accumulation of DV Chlide *a* in greening tissues (Duggan and Rebeiz 1982a).

### 1.2.61 Duggan and Rebeiz, 1982b

In 1982, Duggan and Rebeiz, also detected [4-vinyl] chlorophyllide *a* reductase (4VCR) activity in higher plants (Duggan and Rebeiz 1982b).

### 1.2.62 Duggan and Rebeiz, 1982c

In 1982, Duggan and Rebeiz, also detected the occurrence of Chlide *b* in higher plants (Duggan and Rebeiz 1982c).

### 1.2.63 McCarthy, 1982

In 1982, McCarthy et al., demonstrated that Pchlide *a* and Pchlide *a* phytyl ester are formed via two distinct biosynthetic routes in higher plants (McCarthy et al. 1982).

### 1.2.64 Rebeiz, 1983

In1983, Rebeiz et al., proposed a 6-branched Chl *a* biosynthetic pathway on the basis of the experimental evidence available at that time (Rebeiz et al. 1983).

### 1.2.65 Wu and Rebeiz, 1984

In 1984, Wu and Rebeiz, ascertained the chemical structure of DV Pchlide *a*, and DV Chlide *a*, by nuclear magnetic resonance (NMR) spectroscopy.

### *1.2.66 Daniell and Rebeiz, 1984*

In 1984, Daniell and Rebeiz, demonstrated that direct esterification of endogenous Chlide *a* with exogenous phytol in the presence of added ATP, and Mg was also observed, in etiochloroplasts, which led to the proposal that depending on the stage of plastid development, the conversion of Chlide *a* to Chl *a* may follow different biosynthetic routes having different substrate and cofactor requirements (Daniell and Rebeiz 1984).

### *1.2.67 Wu and Rebeiz, 1985*

In 1985, Wu and Rebeiz, ascertained the chemical structure of DV Chl *b*, by NMR spectroscopy.

### *1.2.68 Carey and Rebeiz, 1985*

In 1985, Carey and Rebeiz discovered the DV and MV greening group affiliation of plants (Carey and Rebeiz 1985).

### *1.2.69 Tripathy and Rebeiz, 1986*

In 1986, Tripathy and Rebeiz, demonstrated precursor-product relationships among the various MV and DV monocarboxylic routes of the proposed multibranched Chl *a* biosynthetic pathway (Tripathy and Rebeiz 1986).

### *1.2.70 Wu and Rebeiz, 1988*

In 1988, Wu and Rebeiz, ascertained the chemical structure of 10-OH-Chl *a* lactone by NMR spectroscopy. This short wavelength Chl may play a role in PSII reaction centers (Wu and Rebeiz 1988).

### 1.2.71 Tripathy and Rebeiz, 1988

In 1988, Tripathy and Rebeiz, demonstrated that only part of the MV Pchlide *a* in DMV-LDV-LDMV plant species such as barley, can arise by vinyl reduction of DV Pchlide *a*, the rest of the MV Pchlide *a* pool, is formed via an independent route (Tripathy and Rebeiz 1988).

### 1.2.72 Chisholm, 1988

In 1988, Chisholm, reported that the major Chls in some prochlorophytes are DV Chl *a* and *b*.

### 1.2.73 Walker, 1988

In1988, Walker et al., described the conversion of beta-OH and beta-keto methyl propionate Pchlide *a* to Pchlide *a* (Walker et al. 1988).

### 1.2.74 Shedbalkar, 1991

In 1991, Shedbalkar et al., detected the occurrence of MV Pchlide *b* in higher plants (Shedbalkar et al. 1991).

### 1.2.75 Parham and Rebeiz, 1992

In 1992, Parham and Rebeiz, determined that NADPH is a mandatory cofactor for VCR activity (Parham and Rebeiz 1992).

### 1.2.76 Porra, 1993

In 1993, Porra et al., reported that mass spectra of [7-hydroxymethyl]-chlorophyll *b* extracted from leaves greened in the presence of either $^{18}O_2$ or $H_2{}^{18}O_2$ revealed that $^{18}O$ was incorporated only from molecular oxygen into the 7-formyl group of Chl *b* (Porra et al. 1993).

### 1.2.77 Armstrong, 1995

In 1995, Armstrong et al., demonstrated that in *Arabidopsis thaliana* and Barley, two different genes PorA and PorB (with about 75 % homology) code for two different protochlorophyllide oxidoreductases, namely POR-A and POR-B (Armstrong et al. 1995).

### 1.2.78 Jensen, 1995

In 1995, Jensen et al., expressed the three Mg-Proto chelatase genes (*chlI*, *chlD*, and *chlH*) in *E. coli*, and demonstrated that the three cognate proteins are required for activity (Jensen et al. 1995).

### 1.2.79 Kim and Rebeiz, 1996

In 1996, Kim and Rebeiz, detected [4-vinyl] Mg-Proto reductase (4VMPR) in green plants (Kim and Rebeiz 1996).

### 1.2.80 Abd El Mageed

In 1997, Abd El Mageed et al., discovered the LD-MV and LD-DV and LD-MV greening groups of plants (Abd-El-Mageed et al. 1997).

### 1.2.81 Adra and Rebeiz, 1998

In 1998, Adra and Rebeiz, discovery the occurrence of transients DV Chl *a* and [4-vinyl] Chl reductase (4VChlR) in plants (Adra and Rebeiz 1998).

### 1.2.82 Rebeiz, 1999

In 1999, Rebeiz et al., proposed an integrated multibranched Chl *a*/*b* biosynthetic pathway in plants (Rebeiz et al. 1999).

### *1.2.83 Kolossov and Rebeiz, 2003*

In 2003, Kolossov and Rebeiz discovered energy transfer in plastids between Chl-protein complexes and nascent tetrapyrroles (Kolossov et al. 2003).

### *1.2.84 Rebeiz, 2003*

In 2003, Rebeiz et al., formulated models for the assembly of Chl-protein complexes (Rebeiz et al. 2003).

### *1.2.85 Rebeiz, 2004*

In 2004, Rebeiz et al. proposed an experimental approach for increasing the photosynthetic efficiency of green plants (Rebeiz et al. 2004).

### *1.2.86 Kolossov and Rebeiz*

In 2005, Kolossov and Rebeiz discovered ALA esterases in higher plants.

## 1.3 Some Major Steps in the Development of Tetrapyrrole Analytical Techniques

In this historical section, emphasis is placed on the development of analytical techniques that had a clear impact on the understanding of the structure and function of intermediates and end-products of the porphyrin and Chl biosynthetic pathways.

### *1.3.1 Jope and O'Brion, 1945*

In 1945, Jope and O'Brion determined the spectral absorption and fluorescence properties of Copro I and III (Jope and O'Brion 1945).

### 1.3.2 Rimington and Sveinsson, 1950

In 1950, Rimington and Sveinsson developed a spectrophotometric method for the determination of uroporphyrin (Uro) (Rimington and Sveinsson 1950).

### 1.3.3 Koski, 1950

In 1950, Koski, derived simultaneous equations for the determination of Pchl (ide) and Chl *a* and *b* by absorbance spectroscopy, in unsegregated plant extracts (Koski 1950).

### 1.3.4 Nicolas and Rimington, 1951

In 1951, Nicolas and Rimington completed the development of a paper chromatographic technique for the separation of porphyrins with 2–8 carboxylic groups.

### 1.3.5 Dresel and Falk, 1956

In 1956, Dresel and Falk developed a method for the separation of Uro, Copro and Proto, by solvent partition (Dresel and Falk 1956).

### 1.3.6 Seliskar, 1966

In 1966, Seliskar developed a thin-layer chromatographic technique for the separation of fully esterified and monocarboxylic tetrapyrroles.

### 1.3.7 Rebeiz, 1975

In 1975, Rebeiz et al. developed spectrofluorometric techniques for the detection and quantitative determination of Proto and Pchl(ide) *a* formation in greening tissues (Rebeiz et al. 1975a).

### 1.3.8 Smith and Rebeiz, 1977

In 1977, Smith and Rebeiz developed spectrofluorometric techniques for the detection of Mg-porphyrins in the presence of Zn-porphyrins (Smith and Rebeiz 1977).

### 1.3.9 Bazzaz and Rebeiz, 1977

1979, Bazzaz and Rebeiz developed very sensitive analytical techniques for the quantitative determination of Chls and pheophytin and Chlides and Pheophorbides by spectrofluorometry in unsegregated mixtures (Bazzaz and Rebeiz 1979).

### 1.3.10 Daniell and Rebeiz, 1982a

In 1982, Daniell and Rebeiz developed an organello system capable of the massive conversion of ALA to Pchlide *a* (Daniell and Rebeiz 1982a).

### 1.3.11 Daniell and Rebeiz, 1982b

In 1982, Daniell and Rebeiz, described an organello system capable of the massive conversion of ALA to Chl *a* in the light (Daniell and Rebeiz 1982b).

### 1.3.12 Rebeiz et al., 1984

In 1984, Rebeiz et al., described the coupling of Chl *a* accumulation to thylakoid assembly in organello (Rebeiz et al. 1984a).

### 1.3.13 Tripathy and Rebeiz

In 1985, Tripathy and Rebeiz developed spectrofluorometric equations for the quantitative determination of MV and DV Mg-porphyrins and MV and DV Pchl (ides) at 77 K (Tripathy and Rebeiz 1985).

### *1.3.14 Rebeiz et al., 1988*

In 1988, Rebeiz et al. developed equations for the determination of precursor-product relationships, in vivo, and in vitro (Rebeiz et al. 1988b).

### *1.3.15 Wu et al., 1989*

In 1989, Wu et al. developed spectrofluorometric equations for the quantitative determination of DV and MV chlorophyll(ide) *a* and *b* at 77 K (Wu et al. 1989).

### *1.3.16 Shedbalkar and Rebeiz, 1992*

In1992, Shedbalkar and Rebeiz, determined the molar extinction coefficients of DV Chl *a* and *b* and their pheophytins (Shedbalkar and Rebeiz 1992).

### *1.3.17 Parham and Rebeiz, 1995*

In 1995, Parham and Rebeiz, developed an assay for [4-vinyl] Chlide *a* reductase using exogenous DV Chlide *a* (Parham and Rebeiz 1995).

### *1.3.18 Ioannides et al., 1997*

In 1997, Ioannides et al. developed spectrofluorometric equations for the quantitative determination of MV Pchlide *b* (Ioannides et al. 1997).

### *1.3.19 Kopetz et al., 2004*

In 2004, Kopetz et al. developed analytical tools for probing the relationship between Chl biosynthesis and the topography of photosynthetic membranes (Kopetz et al. 2004).

## 1.4 Some Major Steps in the Development of Tetrapyrrole-Dependent Photobiotechnologies

In this historical section, emphasis is placed on the development of tetrapyrrole-dependent photobiotechnologies.

### *1.4.1 Lipson et al., 1961*

In 1961, Lipson et al. synthesized hematoporphyrin derivative (HPD), the first practical photodynamic therapy drug used in the treatment of cancer (Lipson et al. 1961).

### *1.4.2 Rebeiz, et al., 1984*

In 1984, Rebeiz et al., described the concept and phenomenology of tetrapyrrole-dependent photodynamic herbicides (Rebeiz et al. 1984b).

### *1.4.3 Pottier et al., 1986*

In 1986, Pottier et al., used delta-aminolevulinic acid (ALA)-induced Proto in the treatment of cancerous tumors (Pottier et al. 1986).

### *1.4.4 Rebeiz et al., 1988*

In 1988, Rebeiz et al. described the concept and phenomenology of tetrapyrrole-dependent photodynamic insecticides (Rebeiz et al. 1988a).

### *1.4.5 Matringe et al., 1989*

1989: Matringe et al., demonstrated the inhibition of protoporphyrinogen IX oxidase by diphenyl ether herbicides.

## References


Abd-El-Mageed HA, El Sahhar KF, Robertson KR et al (1997) Chloroplast biogenesis 77. Two novel monovinyl and divinyl light–dark greening groups of plants and their relationship to the chlorophyll *a* biosynthetic heterogeneity of green plants. Photochem Photobiol 66:89–96

Adra AN, Rebeiz CA (1998) Chloroplast biogenesis 81. Transient formation of divinyl chlorophyll *a* following a 2.5 ms light flash treatment of etiolated cucumber cotyledons. Photochem Photobiol 68:852–856

Apel K, Santel HJ, Redlinger TE et al (1980) The protochlorophyll ide holochrome of barley (Hordeum vulgare L.). Isolation and characterization of the NADPH: protochlorophyllide oxidoreductase. Eur J Biochem 111:251–258

Armstrong GA, Runge S, Frick G et al (1995) Identification of NADPH: protochlorophyllide oxidoreductases A and B: a branched pathway for light-dependent chlorophyll biosynthesis in *Arabidopsis thaliana*. Plant Physiol 108:1505–1517

Battersby AR, Fookes CJR, Matcham GWJ et al (1979) Order of assembly of the four pyrrole rings during the biosynthesis of the natural porphyrins. J Chem Soc Chem Commun:539–541

Bazzaz MB, Rebeiz CA (1979) Chloroplast culture V. Spectrofluorometric determination of chlorophyll(ide) *a* and *b* and pheophytin (or pheophorbide) *a* and *b* in unsegregated pigment mixtures. Photochem Photobiol 30:709–721

Beale SI, Castelfranco PA (1974) The biosynthesis of δ-aminolevulinic acid in higher plants. II. Formation of $^{14}$C-δ-aminolevulinic acid from labelled precursors in greening plant tissues. Plant Physiol 53:297–296

Belanger FC, Rebeiz CA (1979) Chloroplast biogenesis XXVII. Detection of novel chlorophyll and chlorophyll precursors in higher plants. Biochem Biophys Res Commun 88:365–472

Belanger FC, Rebeiz CA (1980a) Chloroplast biogenesis 30. Chlorophyll(ide) (E459F675) and Chlorophyll(ide) (E449F675) the first detectable products of divinyl and monovinyl protochlorophyll photoreduction. Plant Sci Lett 18:343–350

Belanger FC, Rebeiz CA (1980b) Chloroplast biogenesis. Detection of divinyl protochlorophyllide in higher plants. J Biol Chem 255:1266–1272

Belanger FC, Rebeiz CA (1980c) Chloroplast biogenesis. Detection of divinyl protochlorophyllide ester in higher plants. Biochemistry 19:4875–4883

Belanger FC, Rebeiz CA (1982) Chloroplast biogenesis: detection of monovinyl magnesium protoporphyrin monoester and other monovinyl magnesium porphyrins in higher plants. J Biol Chem 257:1360–1371

Belanger FC, Dugan JX, Rebeiz CA (1982) Chloroplast biogenesis: identification of chlorophyllide *a* (E458F674) as a divinyl chlorophyllide *a*. J Biol Chem 257:4849–4858

Bogorad L (1958) The enzymic synthesis of porphyrins from porphobilinogen. I. Uroporphyrinogen I. J Biol Chem 233:501–509

Bogorad L, Granick S (1953a) Protoporphyrin precursors produced by a *Chlorella* mutant. J Biol Chem 202:793–800

Bogorad L, Granick S (1953b) The enzymatic synthesis of porphyrins from porphobilinogen. Proc Natl Acad Sci 39:1176–1188

Borodin A (1882) Botan Ztg 40:608

Carey EE, Rebeiz CA (1985) Chloroplast biogenesis 49. Difference among angiosperms in the biosynthesis and accumulation of monovinyl and divinyl protochlorophyllide during photoperiodic greening. Plant Physiol 79:1–6

Closs GL, Katz JJ, Pennington FC et al (1963) Nuclear magnetic resonance spectra and molecular association of chlorophyll *a* and *b*, methyl chlorophyllides, pheophytins, and methyl pheophorbides. J Am Chem Soc 85:3809–3821

Cookson GH, Rimington C (1954) Porphobilinogen. Biochem J 57:476–484

Daniell H, Rebeiz CA (1982a) Chloroplast culture IX. Chlorophyll(ide) *a* biosynthesis in vitro at rates higher than in vivo. Biochem Biophys Res Commun 106:466–470

Daniell H, Rebeiz CA (1982b) Chloroplast culture VIII. A new effect of kinetin in enhancing the synthesis and accumulation of protochlorophyllide in vitro. Biochem Biophys Res Commun 104:837–843

Daniell H, Rebeiz CA (1984) Bioengineering of photosynthetic membranes: requirement of magnesium for the conversion of chlorophyllide a to chlorophyll a during the greening of etiochloroplasts in vitro. Biotechnol Bioeng 26:481–487
Dougherty RC, Strain HH, Svec WA et al (1970) The role of chlorophyll *b* in photosynthesis: hypothesis. J Am Chem Soc 92:2826
Dresel EIB, Falk JE (1956) Studies on the biosynthesis of blood pigments. 2. Haem and porphyrin formation in intact erythrocytes. Biochem J 63:72–79
Duggan JX, Rebeiz CA (1982a) Chloroplast biogenesis 38. Quantitative detection of a chlorophyllide *b* pool in higher plants. Biochim Biophys Acta 714:248–260
Duggan JX, Rebeiz CA (1982b) Chloroplast biogenesis 42. Conversion of DV chlorophyllide *a* to monovinyl chlorophyllide *a* in vivo and in vitro. Plant Sci Lett 27:137–145
Duggan JX, Rebeiz CA (1982c) Chloroplast biogenesis 37: induction of chlorophyllide *a* (E459F675) accumulation in higher plants. Plant Sci Lett 24:27–37
Ellsworth RK, Aronoff S (1969) Investigations of the biogenesis of chlorophyll *a*. IV. Isolation and partial characterization of some biosynthetic intermediates between Mg-protoporphine IX monomethyl ester and Mg-vinylpheoporphine $a_5$, obtained from *Chlorella* mutants. Arch Biochem Biophys 130:374–383
Fischer H, Lambrecht R (1937) Z Physiol Chem 249:1
Fischer H, Lambrecht R (1938) Z Physiol Chem 253:1
Fischer FG, Lowenberg K (1928) Ann Chem 464:469
Fischer FG, Lowenberg K (1929) Ann Chem 475:183
Fischer H, Oestreicher A (1940) Protochlorophyll precursor of chlorophyll. Z Physiol Chem 262:243
Fischer H, Orth H (1937) Die Chimie des Pyrrols. Akad. Verlagsges, Leipzig
Fischer H, Stern A (1935) Ann Chem 520:88
Fischer H, Stern A (1940) Die Chimie des Pyrroles. Kademische Verlagsgesellschaft
Fischer H, Mittenzwei H, Oestreicher A (1939) Z Physiol Chem 257:IV
Fremy E (1860) Compt Rend 50:405
Gibson KD, Neuberger A, Tait GH (1963) Studies on the biosynthesis of porphyrins and bacteriochlorophyll by Rhodopseudomonas spheroides. S-adenosylmethionine-magnesium protoporphyrin methyltransferase. Biochem J 88:325–334
Goldberg A, Ashenbrucker M, Cartwright GE et al (1956) Studies on the biosynthesis of heme in vitro by avian erythrocytes. Blood 11:821–833
Gorchein A (1972) Magnesium protoporphyrin chelatase activity in Rhodopseudomonas spheroides. Studies with whole cells. Biochem J 127:97–106
Granick S (1948a) Protoporphyrin 9 as a precursor of chlorophyll. J Biol Chem 172:717–727
Granick S (1948b) Magnesium protoporphyrin as a precursor of chlorophyll in *Chlorella*. J Biol Chem 175:333–342
Granick S (1950) Magnesium vinyl pheoporphyrin $a_5$, another intermediate in the biological synthesis of chlorophyll. J Biol Chem 183:713–730
Granick S (1954) Enzymatic conversion of delta-aminolevulinic acid to porphobilinogen. Science 120:1105–1106
Granick S (1961) Magnesium protoporphyrin monoester and protoporphyrin monomethyl ester in chlorophyll biosynthesis. J Biol Chem 236:1168–1172
Griffiths WT (1974) Source of reducing equivalents for the in vitro synthesis of chlorophyll from protochlorophyll. FEBS Lett 46:301–304
Griffiths WT (1978) Reconstitution of chlorophyllide formation by isolated etioplast membranes. Biochem J 174:681–692
Holt AS, Morley HV (1959) Proposed structure of chlorophyll *d*. Can J Chem 37:507
Ikeuchi M, Murakami S (1982) Measurement and identification of NADPH: protochlorophyll ide oxidoreductase solubilized with Triton-X-100 from etioplast membranes of squash cotyledons. Plant Cell Physiol 23:1089–1099
Ioannides IM, Shedbalkar VP, Rebeiz CA (1997) Quantitative determination of 2-monovinyl protochlorophyll(ide) *b* by spectrofluorometry. Anal Biochem 249:241–244

Jensen PE, Gibson LCD, Henningsen KW et al (1995) Expression of the *chlI*, *chlD* and *chlH* genes from the cyanobacterium *Synechocystis* PCC6803 in *Escherichia coli* and demonstration that the three cognate proteins are required for magnesium-protoporphyrin chelatase activity. J Biol Chem 271(28):1662–1667
Jones OTG (1963) The inhibition of bacteriochlorophyll biosynthesis in *Rhodopseudomonas spheroides* by 8-hydroxyquinoline. Biochem J 88:335–343
Jones OTG (1966) A protein-protochlorophyll complex obtained from inner seed coats of *Cucurbita pepo*. Biochem J 101:153–160
Jope EM, O'Brion JR (1945) Spectral absorption properties and fluorescence of coproporphyrin I and III and the melting point of their tetramethyl esters. Biochem J 39:239–245
Jordan PM, Seehra JS (1979) The biosynthesis of uroporphyrinogen III: order of assembly of the four porphobilinogen molecules in the formation of the tetrapyrrole ring. FEBS Lett 104:364–366
Katz JJ, Closs GL, Pennington FC et al (1963) Infrared spectra of methyl chlorophyllides and pheophytins in various solvents. J Am Chem Soc 85:3801
Kim JS, Rebeiz CA (1996) Origin of the chlorophyll *a* biosynthetic heterogeneity in higher plants. J Biochem Mol Biol 29:327–334
Kolossov VL, Kopetz KJ, Rebeiz CA (2003) Chloroplast biogenesis 87: evidence of resonance excitation energy transfer between tetrapyrrole intermediates of the chlorophyll biosynthetic pathway and chlorophyll *a*. Photochem Photobiol 78:184–196
Kopetz KJ, Kolossov VL, Rebeiz CA (2004) Chloroplast biogenesis 89: development of analytical tools for probing the biosynthetic topography of photosynthetic membranes by determination of resonance excitation energy transfer distances separating metabolic tetrapyrrole donors from chlorophyll *a* acceptors. Anal Biochem 329:207–219
Koski VM (1950) Chlorophyll formation in seedlings of *Zea mays* L. Arch Biochem 29:339–343
Koski VM, Smith JHC (1948) The isolation and spectral absorption properties of protochlorophyll from etiolated barley seedlings. J Am Chem Soc 70:3558–3562
Kuster W (1913) Z Physiol Chem 82:463–483
Lipson RL, Baldes EJ, Olsen AM (1961) Hematoporphyrin-derivative fluorescence in malignant neoplasms. J Natl Cancer Inst 26:1–11
Manning WM, Strain HH (1943) Chlorophyll *d*, a green pigment of red algae. J Biol Chem 151:1
Mattheis JR, Rebeiz CA (1977a) Chloroplast biogenesis. Net synthesis of protochlorophyllide from magnesium protoporphyrin monoester by developing chloroplasts. J Biol Chem 252:4022–4024
Mattheis JR, Rebeiz CA (1977b) Chloroplast biogenesis. Net synthesis of protochlorophyllide from protoporphyrin IX by developing chloroplasts. J Biol Chem 252:8347–8349
Mauzerall D, Granick S (1958) Porphyrin biosynthesis in erythrocytes. III. Uroporphyrinogen and its decarboxylation. J Biol Chem 232:1141–1162
McCarthy SA, Belanger FC, Rebeiz CA (1981) Chloroplast biogenesis: detection of a magnesium protoporphyrin diester pool in plants. Biochemistry 20:5080–5087
McCarthy SA, Mattheis JR, Rebeiz CA (1982) Chloroplast biogenesis: biosynthesis of protochlorophyll(ide) via the acidic and fully esterified biosynthetic branches in higher plants. Biochemistry 21:242–247
Monteverde NA (1893) Acta Horti Petrolitani 13:148
Muir HM, Neuberger A (1949) The biogenesis of porphyrins. The distribution of $^{15}$N in the ring system. Biochem J 45:163
Nencki M (1896) Ber Deut Chem Ges 29:2877
Neve RA, Labbe RF (1956) Reduced uroporphyrinogen III in the biosynthesis of heme. J Am Chem Soc 78:691–692
Noack K, Kiessling W (1929) Z Physiol Chem 182:13
Noack K, Kiessling W (1930) Z Physiol Chem 182:97
Noack K, Scneider E (1933) Naturwisswnchaften 21:835
Pardo AD, Chereskin BM, Castelfranco PA et al (1980) ATP requirement for Mg chelatase in developing chloroplasts. Plant Physiol 65:956–960
Parham R, Rebeiz CA (1992) Chloroplast biogenesis: [4-vinyl] chlorophyllide *a* reductase is a divinyl chlorophyllide *a*-specific NADPH-dependent enzyme. Biochemistry 31:8460–8464

Parham R, Rebeiz CA (1995) Chloroplast biogenesis 72: a [4-vinyl] chlorophyllide *a* reductase assay using divinyl chlorophyllide *a* as an exogenous substrate. Anal Biochem 231:164–169
Pelletier PJ, Caventou JB (1818) Ann Chim et Phys 9:194–196
Porra RJ, Schafer W, Cmiel E et al (1993) Derivation of the formyl-group oxygen of chlorophyll *b* from molecular oxygen in greening leaves of higher plants (*Zea mays*). FEBS 323:31–34
Pottier RH, Chow YFA, Laplante JP et al (1986) Non invasive technique for obtaining fluorescence excitation and emission spectra in vivo. Photochem Photobiol 44:679–687
Poulson R, Polglase WJ (1975) The enzymic conversion of protoporphyrinogen IX to protoporphyrin IX. Protoporphyrinogen oxidase activity in mitochondrial extracts of Saccharomyces cerevisiae. J Biol Chem 250:1269–1274
Rebeiz CA, Castelfranco P (1971a) Chlorophyll biosynthesis in a cell-free system from higher plants. Plant Physiol 47:33–37
Rebeiz CA, Castelfranco P (1971b) Protochlorophyll biosynthesis in a cell-free system from higher plants. Plant Physiol 47:24–32
Rebeiz CA, Yaghi M, Abou Haidar M et al (1970) Protochlorophyll biosynthesis in cucumber (Cucumis sativus, L) cotyledons. Plant Physiol 46:57–63
Rebeiz CA, Mattheis JR, Smith BB et al (1975a) Chloroplast biogenesis. Biosynthesis and accumulation of protochlorophyll by isolated etioplasts and developing chloroplasts. Arch Biochem Biophys 171:549–567
Rebeiz CA, Mattheis JR, Smith BB et al (1975b) Chloroplast biogenesis. Biosynthesis and accumulation of Mg-protoporphyrin IX monoester and longer wavelength metalloporphyrins by greening cotyledons. Arch Biochem Biophys 166:446–465
Rebeiz CA, Belanger FC, McCarthy SA et al (1981) Biosynthesis and accumulation of novel chlorophyll *a* and *b* chromophoric species in green plants. In: Akoyounoglou G (ed) Proceedings of the 5th international congress on photosynthesis, vol V. International Science Services, Jerusalem, pp 197–192
Rebeiz CA, Wu SM, Kuhadje M et al (1983) Chlorophyll *a* biosynthetic routes and chlorophyll *a* chemical heterogeneity. Mol Cell Biochem 58:97–125
Rebeiz CA, Montazer-Zouhoor A, Daniell H (1984a) Chloroplast culture X: thylakoid assembly in vitro. Isr J Bot 33:225–235
Rebeiz CA, Montazer-Zouhoor A, Hopen HJ et al (1984b) Photodynamic herbicides: 1. Concept and phenomenology. Enyme Microbiol Technol 6:390–401
Rebeiz CA, Juvik JA, Rebeiz CC (1988a) Porphyric insecticides 1. Concept and phenomenology. Pestic Biochem Physiol 30:11–27
Rebeiz CA, Tripathy BC, Mayasich JM (1988b) Chloroplast biogenesis 61: kinetic analysis of precursor-product relationships in complex biosynthetic pathways. J Theor Biol 133:319–326
Rebeiz CA, Ioannides IM, Kolossov V et al (1999) Chloroplast biogenesis 80. Proposal of a unified multibranched chlorophyll *a*/*b* biosynthetic pathway. Photosynthetica 36:117–128
Rebeiz CA, Kolossov VL, Briskin D et al (2003) Chloroplast biogenesis: chlorophyll biosynthetic heterogeneity, multiple biosynthetic routes and biological spin-offs. In: Nalwa HS (ed) Handbook of photochemistry and photobiology, vol 4. American Scientific Publishers, Los Angeles, pp 183–248
Rebeiz CA, Kolossov VL, Kopetz KK (2004) Chloroplast bioengineering: photosynthetic efficiency, modulation of the photosynthetic unit size, and the agriculture of the future. In: Nelson DW (ed) Agricultural applications in green chemistry, vol 887. American Chemical Society, Washington, DC, pp 81–105
Rimington C, Sveinsson SL (1950) The spectrophotometric determination of uroporphyrin. Scand J Clin Lab Invest 2:209–216
Sano S (1966) 2,4-Bis-(B-hydroxypropionic acid) deuteroporphyrinogen IX, a possible intermediate between coproporphyrinogen III and Protoporphyrin IX. J Biol Chem 241:5276–5283
Sano S, Granick S (1961) Mitochondrial coproporphyrinogen oxidase and protoporphyrin formation. J Biol Chem 236:1173–1180
Santel HJ, Apel K (1981) The protochlorophyll ide Holochrome of Barley (*Hordeum vulgare* L.). The effect of light on the NADPH: protochlorophyll ide oxidoreductase. Eur J Biochem 120:95–103

Schoch S (1978) The esterification of chlorphyllide a in greening bean leaves. Z Naturforsch 33 c:712–714
Shedbalkar VP, Rebeiz CA (1992) Chloroplast biogenesis: determination of the molar extinction coefficients of divinyl chlorophyll *a* and *b* and their pheophytins. Anal Biochem 207:261–266
Shedbalkar VP, Ioannides IM, Rebeiz CA (1991) Chloroplast biogenesis. Detection of monovinyl protochlorophyll(ide) *b* in plants. J Biol Chem 266:17151–17157
Shemin D (1968) Mechanism and control of pyrrole synthesis. In: Goodwin GT (ed) Porphyrins and related products. Academic, New York, pp 75–89
Shemin D, Kumin S (1952) The preparation of S-succinyl coenzyme A. J Biol Chem 198:827
Shemin D, Russel CS (1953) J Am Chem Soc 76:4873
Shemin D, Wittenberg J (1951) Location in protoporphyrin of the carbon atoms derived from the alpha-carbon of glycine. J Biol Chem 192:315
Shibata K (1957) Spectroscopic studies on chlorophyll formation in intact leaves. J Biochem 44:147–172
Sironval C, Kuyper Y, Michel JM et al (1967) The primary photoact in the conversion of protochlorophyllide into chlorophyllide. Stud Biophys 5:43–50
Smith JHC (1948) Protochlorophyll, precursor of chlorophyll. Arch Biochem 19:449–454
Smith JHC (1952) Yearb Carneg Inst 51:151
Smith JHC, Benitez A (1954) The effect of temperature on the conversion of protochlorophyll to chlorophyll a in etiolated barley leaves. Plant Physiol 29:135–143
Smith JHC, Kupke DW (1956) Some properties of extracted protochlorophyll holochrome. Nature 178:751–752
Smith BB, Rebeiz CA (1977) Spectrofluorometric determination of Mg-protoporphyrin monoester and longer wavelength metalloporphyrins in the presence of Zn-protoporphyrin. Photochem Photobiol 26:527–532
Stokes GG (1864) Proc R Soc 13:144
Tait GH, Gibson HD (1961) The enzymic formation of magnesium protoporphyrin monomethyl ester. Biochim Biophys Acta 52:614–616
Tripathy BC, Rebeiz CA (1985) Chloroplast biogenesis. Quantitative determination of monovinyl and divinyl Mg-protoporphyrins and protochlorophyll(ides) by spectrofluorometry. Anal Biochem 149:43–61
Tripathy BC, Rebeiz CA (1986) Chloroplast biogenesis. Demonstration of the monovinyl and divinyl monocarboxylic routes of chlorophyll biosynthesis in higher plants. J Biol Chem 261:13556–13564
Tripathy BC, Rebeiz CA (1988) Chloroplast biogenesis 60. Conversion of divinyl protochlorophyllide to monovinyl protochlorophyllide in green(ing) barley, a dark monovinyl/light divinyl plant species. Plant Physiol 87:89–94
Tswett M (1906) Ber Deut Bot Ges 24:384
Verdeil F (1844) J Prakt Chem 33:478
Walker CJ, Mansfield KE, Rezzano IN et al (1988) The magnesium-protoporphyrin IX (oxidative) cyclase system. Studies of the mechanism and specificity of the reaction sequence. Biochem J 255:685–692
Westall RG (1952) Isolation of porphobilinogen from the urine of a patient with acute porphyria. Nature 170:614–616
Willstatter R, Asahina Y (1909) Ann Chem 373:227
Willsttater R, Stoll A (1910) Ann Chem 378:18
Willsttater R, Stoll A (1911) Ann Chem 387:317
Willsttater R, Stoll A (1913) Untersuchungen uber Chlorophyll. Springer, Berlin
Wolff JB, Price L (1957) Terminal steps of chlorophyll *a* biosynthesis in higher plants. Arch Biochem Biophys 72:293–301
Wu SM, Rebeiz CA (1988) Chloroplast biogenesis. Molecular structure of short wavelength chlorophyll a (E432 F662). Phytochemistry 27:353–356
Wu SM, Mayasich JM, Rebeiz CA (1989) Chloroplast biogenesis: quantitative determination of monovinyl and divinyl chlorophyll(ide) *a* and *b* by spectrofluorometry. Anal Biochem 178:294–300

# Chapter 2
# Synopsis

*A journey of 10,000 miles starts with the first mile (Confucius).*

## 2.1 Prologue

I was born on July 11, 1936 at 5:00 a.m. in the hospital of my uncle, Nicolas Rebeiz, a famous Lebanese general surgeon. It was a Saturday. My dad named me Constantin, nick named Tino, at birth, after the uncle who raised him. My grandfather had eloped with his mistress Anastasia, to Venezuela, at the turn of the twentieth century and left my dad and his wife in the care of his brother Constantin (Fig. 2.1).

I was told that when I was 3 day-old, Alice the head nurse in the Rebeiz Hospital was carrying me in her arms on her daily tour of the patients. When she entered the room of a Moroccan astrologer, he told her to get closer, looked at my face, asked her when I was exactly born, and told her "tell his folks that if they are not well enough to educate him, he is going to become a great criminal mind. But if he gets educated he will become a well-known scientist". That event has puzzled me, since I was told that story 71 years ago. Indeed, my family was mainly engaged in business. The closest thing to science in the family was Constantin Rebeiz, the Famous Lebanese MD who raised my father, and Nicolas Rebeiz, the surgeon, in whose Hospital I was born. Nicolas had married Marcelle Rebeiz, the only child of Constantin Rebeiz, and the first cousin of my father.

## 2.2 From the Lycee Francais of Beirut Lebanon to the American University of Beirut

After 13 year of French schooling in a French private school (Fig. 2.2), The Lycee Francais of Beirut Lebanon, and after passing successfully the mandatory French first and second part baccalaureate exams, I decided to attend the newly founded

C.A. Rebeiz, *Chlorophyll Biosynthesis and Technological Applications*,
DOI 10.1007/978-94-007-7134-5_2, © Springer Science+Business Media Dordrecht 2014

**Fig. 2.1** Tino at age one, frowning as if the weight of the whole world rested on his shoulders

**Fig. 2.2** Summer 1943. Tino beginning French schooling at the Lycee Francais

College of Agricultural Sciences at the American University of Beirut (AUB), as my father had acquired a 100-acre fruit farm, in the central Bekaa valley after retiring from his combo business. Since I knew very little English, I enrolled in an intensive summer English course at AUB, and in October 1956, I started the agricultural curriculum at AUB while toting a French-English dictionary (Fig. 2.3). In July 1959, I graduated with distinction from AUB with a BS in General Agricultural Sciences.

**Fig. 2.3** Spring 1958, Tino at age 22 on a trip to Tripoli Lebanon with fellow AUB Students. From *Left* to *right*: Tino, Bahram Bahmanyar an Iranian Student and Fayez Kasawinah who later became President of the University of Yarmouk in Jordan

## 2.3 From AUB to the University of California at Davis, California

After graduation from AUB, and in order to further my education, I decided to join the MS curriculum in the Department of Pomology at the University of California at Davis (UC. Davis) (Fig. 2.4). In August 1959 I choose Professor Julian Crane, a well-known horticulturist, as a thesis adviser. By June 1960 I graduated with an MS in Pomology as we made history by developing seedless (parthenocarpic) peaches and cherries (Crane and Rebeiz 1961; Rebeiz and Crane 1961).

That summer I was offered a teaching assistantship (TA) in the Department of Botany to help Professor Elliot Weir with the Laboratory of Botany I, an introductory botanical course. At the same time, I Joined the Ph.D. curriculum in that department. A year later the TA assistantship was converted into a research assistantship (RA).

At that time a young postdoctoral trainee by the name of Joe Key, joined the Department of Botany at UC Davis (UCD), and I was allowed to start my Ph.D. research under his supervision. Joe was studying the effect of auxin on the incorporation of $^{14}$C-Pyrimidines into the nucleic acids of cucumber hypocotyls. What impressed me most at that time was the rapidity with which the discarded excised cucumber cotyledons turned green, usually within a few hours after excision. Soon thereafter Joe left the UC department of Botany and accepted an

**Fig. 2.4** Fall 1959 Tino, far *right*, at UCD in Reid Brook's fruit morphology class in the Department of Pomology, at UC Davis

assistant professorship at Purdue University. I stayed behind at UC Davis and in the spring of 1960, joined the laboratory of a newly recruited assistant professor by the name of Paul Castelfranco who was sharing the laboratory with Joe Key. Paul had obtained his Ph.D. with Paul Stumph at UC Berkeley, and worked on intermediary metabolism of the α–oxidation of long chain fatty acids in peanut cotyledons. He wanted me to solubilize the various membrane-bound enzymes involved in the pathway.

Then in the fall of 1961, I met a new graduate student from Berkeley California, Carole Conness, who Joined UCD to work on her Master Degree in the Botany Department with Professor John Tucker. In the spring of 1962 we got engaged (Fig. 2.4) and got married in August 1962 (Fig. 2.5).

After 2 years of work trying to solubilize the α–oxidation enzymes, very little progress was achieved. Then I made an observation, when Paul was on sabbatical in Milan Italy, in the laboratory of Giorgio Forti that led to the discovery of the extra-mitochondrial β–oxidation pathway. After about a year and a half of clean ups, my Ph.D. thesis was defended successfully and was published in three parts in Plant Physiology (Rebeiz and Castelfranco 1964; Rebeiz et al. 1965a, b).

## 2.4 From UC Davis Back to Beirut Lebanon

Nineteen sixty four was a very good year for us at UC Davis. Carole obtained her MS degree in Botany and I successfully defended my Ph.D. thesis after having spent 4 wonderful years in Paul Castelfranco's Laboratory, where years of open scientific and philosophical discussions, in English and French with Paul, had shaped my

**Fig. 2.5** Carole and Tino announcing their engagement at Elliott's Weir Garden Party in the spring of 1962. From *Left* to *right*: Elliott Weir, Tino Rebeiz, and Carole Conness

research acumen. By that time I had accepted a postdoctoral position at the French CNRS in Paris with Jules Bove'. Bove' had previously spent a sabbatical in Paul Stumph laboratory. By that time Carole and I had a 1 year old son, Paul. Carole, Paul and I headed to Paris France, on the Christophoro-Colombo, a transatlantic cruise ship. On the way, we discovered that Carole was pregnant with our second child and that the Postdoctoral stipend at the French CNRS was so meager that there was no way that a family of four could survive on it in Paris. So we decided to continue to Lebanon, my native country, via a Mediterranean cruise ship, the Lusitania, and try our luck in Lebanon with the optimism of a 28 year-old new Ph.D. and a beautiful 25 years-old wife.

## 2.5 Joining the Lebanese National Research Institute at Tal El-Amara

In Lebanon, my dad immediately handed me the keys to a beautiful fourth floor roof-garden flat which was supposed to be our home in Beirut as long as we stayed in Lebanon. Once in Beirut, Carole and I started thinking about our future. It soon became apparent that my interests were in continuing my research career rather than following in the Family business traditions.

At that time Lebanon being an underdeveloped country, had poorly-established research traditions. The only extensive research in the area was being carried out in neighboring Israel and to a much lesser extent at the American University of Beirut. It was very fortunate however that about that time the Lebanese administration was developing an interest in new research initiatives. Thus, because of family connections, and the Lebanese way of doing things, I was offered the job of head of the Department of Soil Sciences at the Lebanese National Research Institute which at that time was teaming with French LORSTOM, soil scientists. It was agreed that once I got the soil and leaf analyses routines going, I would devote my attention to the development of a research-oriented Department of Biological

**Fig 2.6** Spring 1965. Carole and Tino dancing in a Beirut night club

Sciences. To that effect I was awarded an excellent research budget that allowed me to comfortably initiate these undertakings.

Thus in February, 1965, I started my new Job at the Lebanese National Research Institute of Tal-El-Amara, located in the biblical central Bekaa valley where my dad had his fruit farm. As I was working out the soil and leaf analysis routines, I started thinking about initiating my own research program. I did not want to continue in lipid metabolism and compete with the laboratories of Paul Castelfranco and Paul Stumph, who by the way was one of the examiners on my Ph.D. oral exam.

I remember very vividly that day in April 1965 when sitting in my first floor office at Tal-El-Amara, in the central Lebanese Bekaa valley, and looking out the picture window, I was struck by the beauty of the Mediterranean spring colors and particularly by the explosion of green color surrounding me. I thought that if there was so much green in nature it had to be important. Since I knew very little about the biochemistry of plant pigments or chlorophyll, I decided to consult the pigment literature at the Library of the American University of Beirut.

By that time, Carole and I had established a regular weekly routine. We stayed in a newly built rented house in the medieval town of Zahle, 6 miles away from the Research Institute, and worked at the Institute Monday through Thursday. After work on Thursday we drove for about 90 min to our house in Beirut where I spent Friday and Saturday at the well-stocked AUB Medical Library. In the meanwhile Carole visited with family and friends. Night life was wonderful and Beirut lived up to its reputation as the Paris of the Middle East (Fig. 2.6).

## 2.6 Research in Lebanon

At the AUB medical library a search of the Chemical Abstracts, netted the latest review on the chlorophyll biosynthetic pathway authored by Smith and French (1963). Upon examining the review, I realized that this excellent piece of work was short on hard facts and long on hypotheses.

The Reactions between porphobilinogen (PBG) and protoporphyrin IX which are common to the heme and chlorophyll pathways were well covered, backed by the pioneering work of David Shemin, Lawrence Bogorad and Sam Granick (Chapter III). However the reactions between Proto and Chl were very tentative. It relied mainly on the pioneering work of Sam Granick with X-ray Chlorella mutants (Granick 1948a, b, 1950). Since cell free systems for these reactions were not available, no precursor-product relationships had been established, and the pathway proposed by Granick was a paper pathway backed only by incomplete in vivo data. With these facts in mind I concluded that the field of Chl biosynthesis was still a virgin field ripe for potential discoveries.

## *2.6.1 Chlorophyll and Carotenoid Research at Tel-el-Amara*

### 2.6.1.1 In Vivo Experimentation

The picture of excised cucumber cotyledons turning green rapidly under laboratory light within hours after excision was still very vivid in my mind. Therefore, I thought that etiolated (dark-grown) excised cucumber cotyledons would be a good system to study Chl biosynthesis as they rapidly turned green under illumination. A search of the literature revealed that Withrow et al. (1955) and Hemming Virgin (1960) had studied the greening of etiolated bean and wheat seedlings upon illumination. I therefore decided to repeat some of their work using excised etiolated cucumber cotyledons. The finished work was published in the Journal of the National Research Institute entitled Magon (Rebeiz 1967). The journal Magon was named after a Carthaginian scientist who published a treatise about agriculture and veterinary medicine in 22 volumes, about 140 B.C.

Since I knew little about the pigment composition of etiolated cucumber cotyledons, I decided to investigate the pigment profile of these cotyledons before undertaking systematic studies of Chl biosynthesis. The pigment composition of the etiolated cotyledons consisted mainly of carotenoids and was published in Magon in two articles (Rebeiz 1968a, b).

### 2.6.1.2 In Vitro Experimentation

When the above work was completed I reasoned that the best way of moving the field of Chl biosynthesis forward was to develop cell-free systems capable of Chl biosynthesis in vitro. I reckoned that such systems would allow a stepwise investigation of the Chl biosynthetic pathway and establish in vitro, the missing precursor-product relationships. The research was started by growing in darkness cucumber cotyledons, then excising the etiolated cotyledons and exposing them to white light for 3 h to trigger the greening process. The greening cotyledons were rapidly homogenized, the homogenate was filtered through cheese cloth and the filtrate was exposed to white light for a few minutes. It was conjectured that since the greening process was proceeding very rapidly in the excised cotyledons, I may be able to detect some carry-over Chl biosynthesis in the filtrate exposed to light for a brief period to time. After a few minutes of illumination the filtrate was extracted with aqueous acetone and the amount of Chl in the aqueous acetone was evaluated before and after exposure of the filtrate to light, with a very sensitive Unicam null-point spectrophotometer. In order to increase the sensitivity of the procedure, spectrophotometric cells with 10 cm path length were used. Mixed results were obtained and I decided to shift to the more sensitive usage of $^{14}$C –ALA as a substrate for the Chl biosynthetic work.

### *2.6.2 Establishment of the Joint Master of Sciences Research Program at Tal-el-Amara and the Faculty of Pedagogy of the Lebanese University*

In 1967, I met a very enthusiastic Professor of Chemistry by the name of Elie Trad, at the Faculty of Pedagogy of the Lebanese University. Dr. Trad had gotten his doctorate in Tcheckoslovakia. Soon thereafter Professor Trad became Dean of the Sciences Department at the faculty of Pedagogy and visited my laboratory at Tal-El-Amara. He invited me to teach part-time, an introductory biochemistry course at the Faculty of Pedagogy, on my regular week-end visits to Beirut. He also invited Carole to teach introductory plant physiology. Very soon we realized that the classes were made up of very bright students that deserved exposure to graduate work. I discussed with Dean Trad the possibility of initiating a Master of Sciences Program administered jointly by my Laboratory in Tal-El-Amara and the Faculty of Pedagogy of the Lebanese University. We started the program by recruiting four bright students who did their research in my laboratory and did their course work at the Faculty of Pedagogy. The students started by shuttling between my laboratory in the Bekaa valley for their research and the Faculty of Pedagogy, in Beirut, for their course-work. Then in early 1968 my whole laboratory was moved to the suburb of Beirut where the National Research Institute had built a new research center in the Fanar district. That was very convenient for everybody. Carole and I moved back to our house in Beirut, and the students shuttled between my 5,000 square feet Laboratory in Fanar, and the Faculty of Pedagogy, a few miles away.

With the new laboratory came new equipment, namely a refrigerated high speed centrifuge, an ultracentrifuge, a sophisticated Beckman gas chromatograph, a Hewlett Packard radio chromatogram scanner, a Beckmann 100 S liquid scintillation counter, a Beckman recording double beam spectrophotometer, and all the needed accessory equipment. Soon my facility was transformed into a state-of-the-art laboratory. With the new equipment, it became possible to finally perform sophisticated experiments involving in vivo and in vitro studies of the Chl biosynthetic pathway using $^{14}$ C–ALA as substrate. Other pigment and lipid research was also initiated. I started working jointly with the students on four different projects. With Mounir

Abou-Haidar, we studied the incorporation of $^{14}$C –ALA into metalloporphyrins in vitro. With Moustapha Yaghi we studied the incorporation of $^{14}$C –ALA into protochlorophyllide (Pchlide) and its ester Pchlide ester in vivo in order to try to demonstrate a precursor-product relationship between the two protochlorophylls. Both Pchlide and its ester are precursors of Chl. With Antoine Chamai we studied the anthocyanins of the Starking Delicious apple variety grown in Lebanon. With Georges Saliba, we studied the lipid profile of cucumber cotyledons. After obtaining their MS degress, Mounir Abou Haidar, Moustapha Yaghi and Antoine Chamai continued their studies in France and obtained their doctorate degrees there.

### 2.6.3 Foundation of the Lebanese Association for the Advancement of Sciences

In 1968 I discussed with a young French-trained inorganic chemist by the name of Emile Samaha, the creation of a Lebanese Association for the Advancement of Sciences (LAAS), patterned after the US American Association for the Advancement of Sciences (AAAS). Emile had just returned to Lebanon from a postdoctoral training in the Chemistry department at UC Berkeley in California. We got together a group of Lebanese academics who became the founders of the association, namely from the American University of Beirut, professors: Samir Thabet an inorganic chemist, Charles Abou-Chaar, a marine Biologist, and Jamal Karam-Harfouche, a public health scientist, and from the Lebanese University, Professors Rafic Eido, a physicist, and Hafez Khobeisi and Mounir Abou-Hajal two mathematicians.

We drafted a constitution for the association that was approved by the Lebanese Government. Then we created a yearly scientific symposium where research done in Lebanon and the Arab World was presented. The first Symposium was held in August 1969. Unfortunately by that time I had returned to the United States. However my students presented their research findings at the symposium.

### 2.6.4 The Winds of War

After creation of the Lebanese Association for the Advancement of Sciences, Emile Samaha headed a study, sponsored by LAAS about the social conditions in Lebanon. Our investigation was helped by a French scientist, Marcel Piganiol, CEO of Pyrex-Sovirell, who was then an advisor to Charles Helou, President of the Lebanese Republic. The study came to the conclusion that some kind of revolution was brewing and as a consequence I started thinking of leaving Lebanon, since there was very little hope that anything would be done to correct the many problems facing the country.

## 2.7 From Lebanon Back to UC Davis

By the beginning of 1969, things were getting pretty bad at the National Research Institute. My Research budget was cut drastically, and the research working conditions left much to be desired. Essentially there was no set up infrastructure for the protection of the right and status of research scientists, and we failed to convince government officials of creating one. Therefore Carole and I decided that the best we could do for our immediate family was to get back and settle in the United States. Then in June 1969 Paul Castelfranco offered me a 3 months summer stay in his lab at UCD. Therefore, Carole, my son Paul, my daughter Natalie who was born in 1965, and I left everything behind and headed to California suit cases filled with our cloths and unfinished manuscripts. Once at UCD I started putting the last touches on unfinished work that was started in my Lab at Tal-El-Amara. My employment was extended by Paul Castelfranco and the UCD Botany department. Soon thereafter I resigned from the Lebanese National Research Institute. In 1970 our first two manuscripts were submitted for Publication in Plant Physiology and were very well received (Rebeiz et al. 1970a, b). The first evidence of Chl biosynthesis in vitro was observed in my laboratory at Tel-Al-Amara in 1967. Once in Davis I finished the research by working out the cofactor requirement for the incorporation of $^{14}$C –ALA into Chl. Two manuscripts were submitted to Plant Physiology. They appeared back to back out in Jan 1971 (Rebeiz and Castelfranco 1971a, b). Martin Gibbs, editor-in-chief of Plant physiology congratulated us and wrote "You have achieved what others had tried and failed". Paul Stumph in whose lab I used the chromatogram scanner needed for the work, congratulated me earlier by shaking my hand and telling me "Tino you have gotten one of the last large molecules out of the way".

## 2.8 From UC Davis to Fresno State College

Early In 1971, Julian Crane, my MS thesis adviser, invited me to join the department of Pomology at UCD on a visiting basis, and I started looking for permanent employment in the US. In the spring of 1972, while doing research at UCD, I accepted a part time job at Sacramento State, College, which later became Cal State Sacramento. There, I taught a course and a lab on the use of radiochemical techniques. Also in the spring of 1971, I interviewed at the University of Illinois, in Urbana-Champaign (UI) for a position in the department of Horticulture. However due to budget restraints the position was momentarily suspended. Then late in the spring of 1971, I interviewed for an assistant professorship position in the Biology Department at Fresno State College, which later became Cal State Fresno. I accepted the offer for an assistant Professor step four, which is one step below an associate professor. In August 1971 Carole, Paul, Natalie and I moved to Fresno, Cal. I started the 1971 fall semester by teaching two courses and trying to setup a laboratory to initiate some kind of research. Then soon thereafter I received a phone call from Charlie Birkeland, the Head of the Department of Horticulture at the University of Illinois offering me the professorial position at the University of Illinois. That position had been occupied by Dan McCollum, a tomato breeder. His wife Ashti McCollum is the lady who won the Supreme Court case against allowing prayers release time in public schools. Charlie Birkeland gave me the choice of an assistant professorship at a higher salary, or an associate professorship at a lower salary. For security purposes, I choose the latter.

## 2.9 From Fresno State College to the University of Illinois at Urbana Champaign

When I joined the Department of Horticulture at the UI, on February 1, 1972, as an Associate Professor of Plant Physiology, I shared a laboratory with Walter Splittstoesser. Walter and I had met earlier at UCD, where after getting his Ph.D. degree with Harry Beevers at Purdue University, he had postdoctoral training in amino acid metabolism in the laboratory of Mendel Mazelis at UCD. Mendel had graduated earlier from Paul Stumpf's Laboratory in Berkeley. Later on, I learned that Walter Splittstoesser was on the search committee that assessed my credentials for the UI position.

### *2.9.1 Demonstration of Precursor Product Relationships During Chlorophyll Biosynthesis*

Even though the total biosynthesis of Chl was achieved in vitro, the biosynthetic steps from Proto to Chl were based on a paper chemistry pathway proposed in 1950 by Sam Granick (1950). The pathway was based on the detection and partial identification of various tetrapyrroles in *Chlorella* mutants, that were organized into a paper pathway by order of increasing structural complexity (Granick 1948a, b, 1950). Since at that time no cell free systems beyond the biosynthesis of Proto were available, the proposed pathway was not subjected to the rigors of testing for precursor-product relationships.

After thinking of ways to test Granick's pathway via precursor-product relationships during Chl Biosynthesis, I came to the conclusion that this task could not be accomplished with the use of the cell-free Chl biosynthesis system developed earlier with the use of $^{14}$C-putative precursors. Indeed, with the use of $^{14}$C-anabolic precursors, investigators have to rely mainly on chromatographic mobility to identify the $^{14}$C-products. I decided that this methodology was not rigorous enough for a pathway as complex as the Chl biosynthetic pathway. I reckoned that what was needed was the development of new analytical tools that permitted the viewing and determination of the actual chemical structures of the generated end-products. It also required the development of cell-free systems that produced enough end-products to allow rigorous chemical structure determinations.

#### 2.9.1.1 Choice of Analytical Spectroscopic Technique for Room Temperature Spectroscopy

After a few trials using absorption spectroscopy, it became obvious that absorption spectrophotometry was not suitable as a major tool for achieving our goals. The method lacked the required sensitivity and was not rigorous enough for the task on hands. Then I remembered that while at UC Davis, I was impressed by the fluorescence spectroscopic techniques used by Eloise Tappel to study the formation of malonylaldehydes in his Vitamin E metabolic studies. I therefore investigated the market for the availability of recording fluorescence spectrofluorometers and was very impressed by what I learned about their sensitivity and their two-window functionality. Indeed, the great sensitivity of these instruments and the possibility of varying the excitation wavelength (one window) or the emission wavelength (second window) imparted a considerable flexibility and usefulness to this technique. Thus when my first tetrapyrrole proposal submitted to the UI Research Board in 1972 was funded, I immediately purchased a Perkin-Elmer MPF-3 recording spectrofluorometer, which at that time was a state-of-the-art instrument. Consequently, I started using fluorescence spectrofluorometry in my Chl biosynthesis studies.

Usually at the end of an incubation period, I made an 80 % Acetone extract of the incubation mixture, subjected it to fluorescence analysis, and looked for fluorescing metabolic intermediates. However there was so much Chl in the extract that its fluorescence masked everything else. I then started extracting the 80 % acetone extract with hexane and started monitoring the fluorescence of the hexane extract for possible metabolic intermediates. That effort also failed because the Chl passed into the hexane and its fluorescence masked everything else.

Then one day I noticed that after extraction with hexane, the residue what was left behind was clear, and slightly yellowish. That residue was usually dumped. I felt that this hexane-extracted acetone residue (HEAR) contained so few metabolites that it was not worth looking at. Then one day I decided for the fun of it, to look at the fluorescence of the HEAR. To my great surprise, the fluorescence spectrofluorometer kept on recording large peak after peak, the identity of which was unknown. I then realized that (a) all the metabolic products of the incubation were hydrophilic enough to pass in the HEAR that we usually dumped, and (b) that although the HEAR looked devoid of any metabolites, I had greatly underestimated the sensitivity of fluorescence spectroscopy. After this unexpected discovery, I purchased nearly every available tetrapyrrole I could find on the market and ran their emission and excitation spectra in order to build a database of porphyrin fluorescence that would help me identify the fluorescence peaks being generated during in vitro incubations.

#### 2.9.1.2 Development of Room Temperature Analytical Fluorescence Techniques

After having built a voluminous data base of known tetrapyrroles fluorescence emission and excitation peaks at room temperature, I was able to identify all the Chl metabolic intermediates detected at room temperature during the Chl biosynthetic studies. Usually the metabolic intermediates that were formed and passed into the HEAR at room temperature depended upon the incubation conditions and the plastid fraction that was used. In most cases the accumulated metabolites consisted of uroporphyrin, (Uro) coproporphyrin (Copro), Proto, Mg-porphyrins, protochlorophyllide (Pchlide) and chlorophyllide (Chlide). These metabolites were detected by their emission and excitation peaks. However since several peaks occurred in every recorded spectrum, there was a considerable fluorescence band overlap. In order to quantitate the amount of every detected tetrapyrrole by reference to standard calibration curves, the various fluorescence band overlaps had to be deconvoluted and computed out. Thus several simultaneous fluorescence equations were derived that allowed the deconvolution of the various fluorescence bands with great precision (Rebeiz et al. 1975). These techniques will be referred to throughout this monograph.

#### 2.9.1.3 Development of Cell-Free Systems Capable of the Net Synthesis of Chlorophyll Biosynthetic Metabolic Intermediates

Armed with sensitive fluorescence techniques and quantitative measuring capabilities we proceeded with the development of cell-free systems that formed net metabolic tetrapyrrole intermediates in vitro (Rebeiz et al. 1982). That effort spanned many years of research and culminated with the development of cell-free systems capable of very high rates of Pchlide and Chl biosynthesis in vitro (Daniell and Rebeiz 1982). These cell-free systems will be referred to throughout this monograph.

#### 2.9.1.4 Demonstration of Precursor-Product Relationships During Chlorophyll Biosynthesis

By 1975 we felt confident that with the newly developed techniques it would be possible to start systematic investigations of possible precursor product relationships during chlorophyll biosynthesis. It was thus possible to detect the insertion of Mg into Proto (Smith and Rebeiz 1977) and the conversion of exogenous Proto, Mg proto and its monoester into Pchlide (Mattheis and Rebeiz 1977b, c). Although these results confirmed the paper pathway proposed by Granick in 1950 (Granick 1950), we could not demonstrate the conversion of exogenous Pchlide into its phytylated analog, Pchlide ester as was commonly believed (Mattheis and Rebeiz 1977a). These precursor-product relationships will be discussed in various chapters of this monograph.

## 2.10 Discovery of the Chlorophyll Biosynthetic Heterogeneity

In 1970 an in vivo kinetic study of the biosynthetic relationship between Pchlide (Fig. 2.7, 3b) and its esterified analog, Pchlide ester, failed to demonstrate a precursor-product relationship between these two protochlorophylls (Pchls) (Rebeiz et al. 1970b) as was commonly believed.

When the in vitro studies mentioned above failed to detect a direct precursor-product relationship between these two Pchls (Fig. 2.7), we became suspicious that the Chl biosynthetic pathway was more complex than proposed by Sam Granick. Therefore we kept on the lookout for such evidence as we proceeded with our Chl biosynthetic studies.

### *2.10.1 Discovery of Novel Tetrapyrrole Intermediates*

After finishing the initial precursor-product studies mentioned above, we focused our attention on studying the various Pchl-protein complexes, also known as Pchl-holochromes (PchlHs) in various green plants. A postdoctoral trainee, by the name of Charley Cohen, started working on the problem. Charley had just graduated from Jerry Schiff's laboratory at Brandeis University. He started by looking at the emission and excitation spectra at 77 K of PchlHs extracted from various green plants grown under different conditions (Cohen and Rebeiz 1978, 1981). During one of our regular meetings I suggested to Charley that he should extract some of the PchlHs with 80 % acetone and transfer the extracted pigments to ether in order to establish standard 77 K emission and excitation fluorescence spectra in ether for the purpose of adding this data to our tetrapyrrole database. That morning Charley came running into my office and informed me that he was seeing strange looking recorded spectra. I looked at the recorded 77 K Pchl spectra in ether and immediately realized that we were upon something exciting related to the suspected putative Chl biosynthetic heterogeneity.

I asked Charley to continue his PchlH work and put a newly arrived Ph.D. graduate student by the name of Faith Belanger on the Pchl extracts problem.

#### 2.10.1.1 Discovery of the Ubiquitous Occurrence of Divinyl Protochlorophyllide Occurrence in Higher Plants

In 1963, Jones reported that cultures of *Rhodopseudomonas spheroides* grown in the presence of 8-hydroxyquinoline accumulated a novel Pchl that he identified as 2–4 divinyl Pchlide (DV Pchlide) (Fig. 2.7, 3a) (Jones 1963). He proposed that DV Pchlide was a transient intermediate in the formation of Chl that was rapidly converted to conventional Pchlide (Fig. 2.7, 3b) [i.e. Monovinyl Pchlide (MV-Pchlide)] by

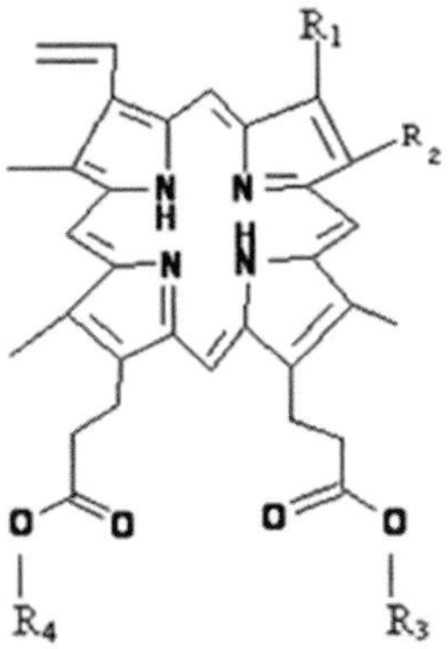


**1. Protoporphyrin IX**

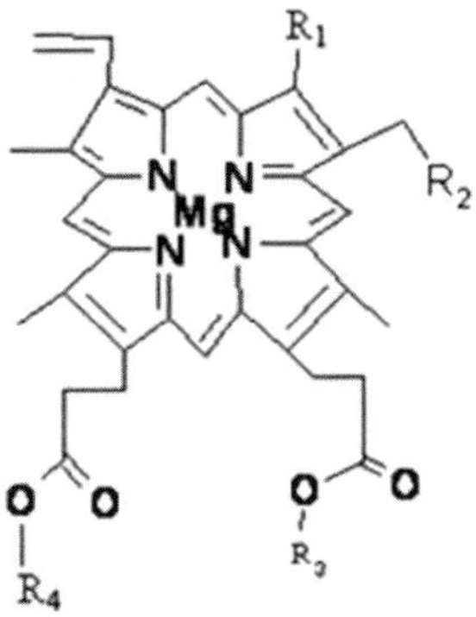


**2. Mg-Protoporphyrin IX**

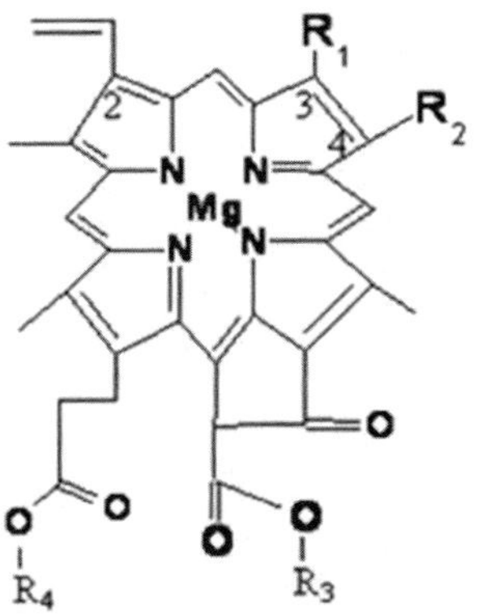


**3. Protochlorophyllide *a***

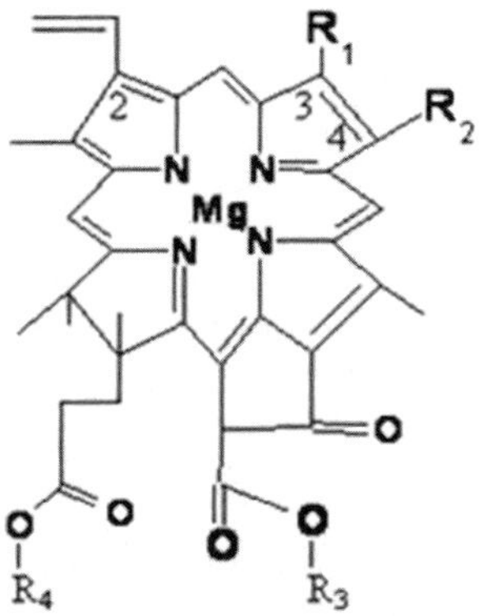


**4. Chlorophyllide *a***

| | $R_1$ | $R_2$ | $R_3$ | $R_4$ | Compound |
|---|---|---|---|---|---|
| 1 | Methyl | Vinyl | H | H | DV Proto |
| 2a | Methyl | Vinyl | H | H | DV Mg-Proto |
| 2b | Methyl | Ethyl | H | H | MV Mg-Proto |
| 2c | Methyl | Vinyl | Methyl | H | DV Mpe |
| 2d | Methyl | Ethyl | Methyl | H | MV Mpe |
| 3a | Methyl | Vinyl | Methyl | H | DV Pchlide *a* |
| 3b | Methyl | Ethyl | Methyl | H | MV Pchlide *a* |
| 4a | Methyl | Vinyl | Methyl | H | DV Chlide *a* |
| 4b | Methyl | Ethyl | Methyl | H | MV Chlide *a* |
| 4c | Methyl | Vinyl | Methyl | Phytol | DV Chl*a* |
| 4d | Methyl | Ethyl | Methyl | Phytol | MV Chl*a* |

**Fig. 2.7** Chemical structure of some common tetrapyrroles

reduction of the vinyl group to ethyl at position 4 of the macrocycle. In other words he assumed that under normal etiolation or greening conditions, DV-Pchlide did not accumulate, and was not detectable in green plants. By suspecting that the strange Pchl fluorescence profile observed in plant extracts in ether at 77 K was due to the presence of DV-Pchlide, we were proposing that contrary to previous assumptions this Pchl did indeed accumulate ubiquitously in green plants.

To check the above hypothesis, we obtained samples of *R spheroides* DV-Pchlide from June Lascelles who was then at UCLA, and recorded its fluorescence emission and excitation spectra in ether at 77 K. The recorded emission and excitation peaks of standard DV Pchlide were identical to those observed in our plant extracts (Belanger and Rebeiz 1979). In 1980, the chemical structure of the newly detected DV Pchlide was ascertained by chemical derivatization (Belanger and Rebeiz 1980c) and in 1984, its chemical structure was confirmed by NMR spectroscopy and mass spectroscopy (Wu and Rebeiz 1985). The metabolism of MV and DV Pchlide *a* will be discussed in Chaps. 7–9.

#### 2.10.1.2 Discovery of Divinyl Chlorophyllide Occurrence in Green Plants

To our surprise most of the accumulated DV Pchlide was found to be convertible into Chl (Fig. 2.7, 4d) (Belanger and Rebeiz 1979). Soon thereafter the accumulated DV Pchlide was shown to be convertible into putative DV Chlide *a* (Fig. 2.7, 4a) (Belanger and Rebeiz 1980b). The chemical structure of the newly formed DV Chlide *a* was ascertained by chemical derivatization (Belanger et al. 1982) and in 1984, its chemical structure was confirmed by NMR and mass spectroscopy (Wu and Rebeiz 1984). These observations, in addition to the suspected Chl biosynthetic heterogeneity involved in the biosynthesis of Pchlide ester suggested the existence of multiple biosynthetic routes during the biosynthesis of Chl. These routes will be fully discussed in various chapters.

#### 2.10.1.3 Discovery of the Occurrence of Divinyl Protochlorophyllide Ester in Green Plants

In 1939 and 1940: Fischer and coworkers (Fischer and Oestricher 1940; Fishcher et al. 1939) identified the protochlorophyll of pumpkin seed coat as a vinyl pheoporphyrin analog of chlorophyll (Fig. 2.8, II). Since this esterified MV Pchlide differed from Chl by the absence of only two hydrogen atoms at the 7–8 position of the tetrapyrrole macrocycle (Fig. 2.7, 4d) they proposed that this esterified Pchl was the immediate precursor of Chl. This notion was adopted by Granick and became dogma until Wolfe and Price demonstrated that in green plants, the immediate precursor of Chl was unesterified Chlide (Fig. 2.7, 4a) (Wolff and Price 1957). After 1957 Pchlide ester became a tetrapyrrole with an unknown function and was no longer considered to be an intermediate in Chl biosynthesis. Then in 1966, Jones reported that in addition to the esterified Pchl, i.e. MV Pchlide

Fig. 2.8 DV and MV Pchlide E

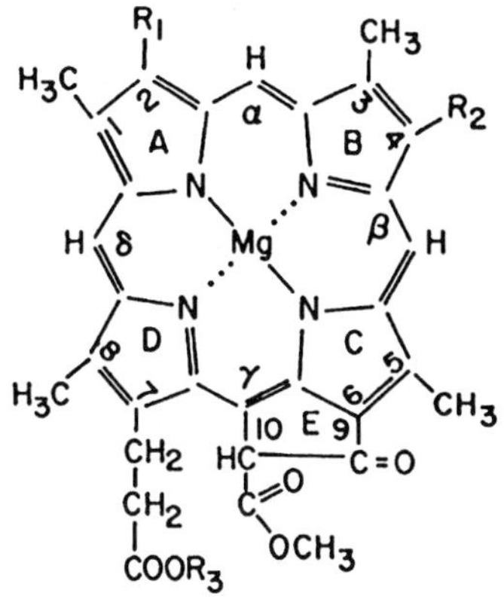


**Protochlorophyllide *a* Ester (Pchlide *a* E)**

| | R1 | R2 | R3 | Compound |
|---|---|---|---|---|
| I | Vinyl | Vinyl | LCFAH | DV Pchlide *a* E |
| II | Vinyl | Ethyl | LCFAH | MV Pchlide *a* E |

ester, he reported the occurrence of another esterified Pchl namely DV Pchlide ester (Fig. 2.8, I) in the inner pumpkin seed coat (Jones 1966). In 1969, this finding was confirmed by Houssier and Sauer (1969). Since then, it was assumed that only MV Pchlide ester (Fig. 2.8, II) normally occurred in green plants, except in some particular cases such as the inner cucurbid seed coat.

Then in 1980, Belanger and Rebeiz reported that in many cases, in addition to MV Pchlide ester, small quantities of DV Pchlide ester (Fig. 2.8, I) occurred routinely in green plants (Belanger and Rebeiz 1980a). The function of these Pchls in Chl Biosynthesis will be discussed in Chap. 11.

#### 2.10.1.4 Discovery of Divinyl Chlorophyll *a* Occurrence in Green Plants

With the discovery of the occurrence of DV Pchlide and DV Chlide, it was conjectured that it was a matter of time before the occurrence of DV Chl *a* (Fig. 2.7, 4c) would be reported in green plants. Indeed, in 1980 Belanger and Rebeiz reported that upon a 47 ms illumination of etiolated cucumber cotyledons containing MV and DV Pchlide *a* as well as Pchlide ester *a*, the appearance of DV Chlide *a* and DV Chl *a* was noted along with the MV analogs (Belanger and Rebeiz 1980b; Wu and Rebeiz 1984). Also in 1981 and 1982, Bazzaz reported the detection of DV Chl *a* accumulation in a lethal corn mutant (Bazzaz 1981; Bazzaz et al. 1982). Then in 1983 Gieskes and Kraay (1983) reported the occurrence of a Chl in the phytoplankton of various oceans, that had slightly different spectroscopic properties than conventional Chl *a*. That Chl had the spectroscopic properties of the DV Chl *a* that we described in 1980 (Belanger and Rebeiz 1980b). In future publications, it was acknowledged to be DV Chl *a* by various authors (Chisholm et al. 1988, 1992; Goerike and Repeta 1992). These observations were very important as they pointed out that DV Chl *a* and *b* [see below, and (Wu and Rebeiz 1985)] were as abundant if not more abundant than their MV analogs in nature. The biosynthetic routes that involve these Chls will be discussed in Chaps. 7, 9, 10 and 11.

#### 2.10.1.5 Discovery of Monovinyl Mg Porphyrins in Green Plants

Until 1969, it was commonly believed that the only porphyrins and Mg-porphyrins that accumulated beyond Coproporphyrinogen III were DV porphyrins (Granick 1948a, b, 1961). Then in 1969, through the use of mass, visible and infrared spectrometry, Ellsworth and Aronoff proposed the existence of MV Mg-porphyrin intermediates between DV MPE and MV Pchlide in the *AJ* and *BE Chlorella* mutants (Ellsworth and Aronoff 1969).

Ellsworth and Aronoff proposed a Chl biosynthetic loop that started with MV and DV MPE and finished with MV and DV Pchlide. This work, which was difficult to read, was poorly received by the scientific community, and the detection of MV intermediates was ascribed to the vagaries of mass spectroscopy. Later on, the ubiquitous occurrence of MV Mg porphyrins in green plants was unambiguously described by Belanger and Rebeiz (1982). The chemical structure of MV MPE (Fig. 2.7, 2d) was ascertained by chemical derivatization coupled to 77 K spectrofluoromety.

Also the ubiquitous occurrence of MV Mg-Proto (Fig. 2.7, 2b) was detected. Earlier McCarthy and Rebeiz had described the occurrence of another MV Mg-Proto namely a fully esterified Mg-protoporphyrin monoester (MPE Diester) pool in higher plants (McCarthy et al. 1981). This pool too was shown to consist of MV and DV components (Fig. 2.9). The involvement of MV Mg-porphyrins in Chl biosynthetic routes will be discussed in Chap. 7.

#### 2.10.1.6 Discovery of Monovinyl Protochlorophyllide *b* Occurrence in Green Plants

Up to 1991, it was commonly believed that the only Pchls that occurred in green plants were Pchls *a*.

Then in 1991, Shedblakar et al. detected the occurrence of MV Pchlide *b* in higher plants (Shedbalkar et al. 1991) (Fig. 2.10). The chemical structure of the MV Pchlide *b* was ascertained by chemical derivatization coupled to 77 K spectrofluorometry as well as by NMR and fast atom bombardment mass spectroscopy. Since 1991 Pchl *a* and *b* have been distinguished from one another by an *a* or *b* designation. The involvement of MV Pchlide *b* in Chl biosynthetic routes will be discussed in Chaps. 12 and 13.

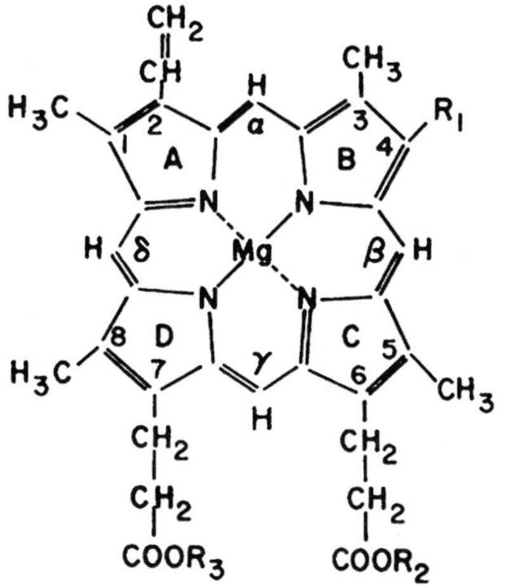


**Mg-Protoporphyrin IX diester (Mpde)**

| | R1 | R2 | R3 | Compound |
|---|---|---|---|---|
| I | Vinyl | Methyl Propionate | Long chain fatty alcohol | DV Mpde |
| II | Ethyl | Methyl Propionate | Long chain fatty alcohol | MV Mpde |

**Fig. 2.9** MV and DV Mg-Proto diesters

**Fig. 2.10** *MV Pchlide b*

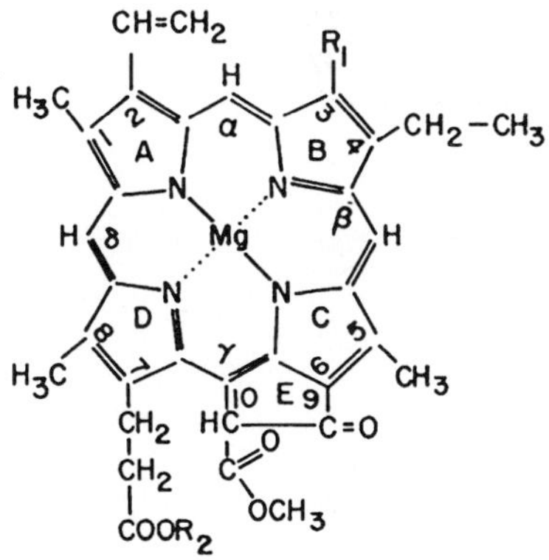


**Protochlorophyllide *b* (Pchlide *b*)**

| R1 | R2 | Compound |
|---|---|---|
| Formyl | H | MV Pchlide *b* |

Fig. 2.11 DV and MV Chlide *b*

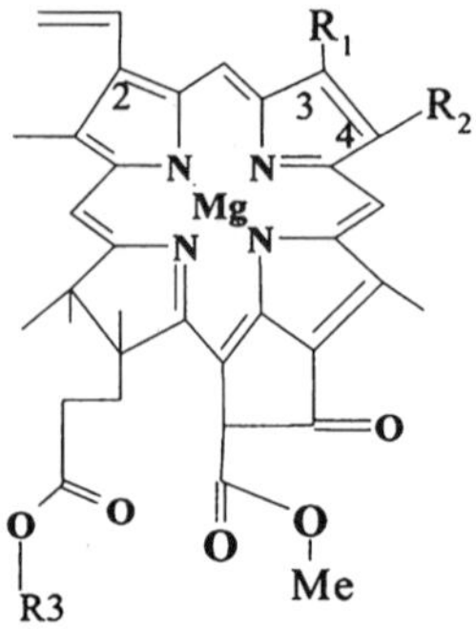


**Chlorophyllide *b* (Chlide *b*)**

| | R1 | R2 | R3 | Compound |
|---|---|---|---|---|
| I | Formyl | Vinyl | H | DV Chlide *b* |
| II | Formyl | Ethyl | H | MV Chlide *b* |

#### 2.10.1.7 Discovery of Monovinyl Chlorophyllide *b* Occurrence in Green Plants

Over the years, the origin of Chl *b* had been the subject of many debates. Shlyk championed the idea that Chl *b* was formed from newly formed Chl *a* (Shlyk 1971). On the other hand, Oelze-Karow and Mohr, suggested that Chl *b* was formed from Chlide *a* (Oelze-Karow and Mohr 1978).

Then in 1981 and 1982, Duggan and Rebeiz, reported the detection of a MV Chlide *b* pool in green plants (Duggan and Rebeiz 1981, 1982c) (Fig. 2.11, II). The MV Chlide *b* chemical structure was ascertained by chemical derivatization coupled to 77 K spectrofluorometry. The ubiquitous detection of MV Chlide *b* in green plants indicated that the biosynthesis of MV Chl *b* most probably involved MV Chlide *b*. The involvement of MV Chlide *b* in the biosynthesis of MV Chl *b* will be discussed in Chaps. 12 and 13.

#### 2.10.1.8 Discovery of Divinyl Chlorophyll *b* Occurrence in the Corn Nec 2 Mutant

The unambiguous occurrence of DV Chl *b* in the Corn Nec 2 mutant was reported by Brereton et al. and was characterized by mass spectroscopy (Brereton et al. 1983). It was characterized by Wu and Rebeiz (1985) by NMR and mass spectroscopy. The metabolism of DV Chl *b* will be discussed in Chaps. 12 and 13.

**Fig. 2.12** DV and MV Chl *b*

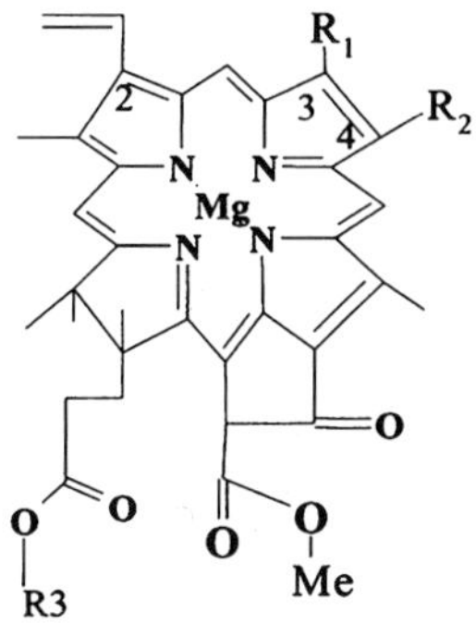


**Chlorophyll *b* (Chl *b*)**

| | R1 | R2 | R3 | Compound |
|---|---|---|---|---|
| I | Formyl | Vinyl | Phytol | DV Chl *b* |
| II | Formyl | Ethyl | Phytol | MV Chl *b* |

## *2.10.2 Discovery of Novel Chl Biosynthetic Routes*

Chlorophyll biosynthetic heterogeneity refers either (a) to spatial biosynthetic heterogeneity, (b) to chemical biosynthetic heterogeneity, or (c) to a combination of spatial and chemical biosynthetic heterogeneities (Rebeiz 2010). Spatial biosynthetic heterogeneity refers to the biosynthesis of an anabolic tetrapyrrole or end product by identical sets of enzymes, at several different locations of the thylakoid membranes. On the other hand, chemical biosynthetic heterogeneity refers to the biosynthesis of an anabolic tetrapyrrole or end product at several different locations of the thylakoid membranes, via different biosynthetic routes, each involving at least one different enzyme.

The chemical heterogeneity of Chl resides mainly in the MV or DV substitutions at positions 2 and 4 of the Chl macrocycle (Figs. 2.7, 4c, 4d and 2.12). It also involves esterification with different long chain fatty alcohols of the propionic acid residue at position 7 of the macrocycle, and substitution of a lactone ring for a cyclopentanone ring at positions 5 and 6 of the macrocycle (Wu and Rebeiz 1988). This chemical heterogeneity is catalyzed by various Chl biosynthetic routes and involves various enzymes. One family of enzymes, the vinyl reductase enzyme family plays a prominent role in this process. All these issues will be discussed in Chaps. 5, 6, 7, 8, 9, 10, 11, 12, and 13. A Brief introduction to the various 4-vinyl reductases is given below.

#### 2.10.2.1 Discovery of DV Mg-Proto Vinyl Reductase

4-vinyl Mg-protoporphyrin IX reductase (VMPR) catalyzes the reduction of the vinyl group of Mg-Proto to ethyl, at position 4 of the Macrocycle (Fig. 2.7, 2a, 2b).

It was first reported by Kim and Rebeiz (1996). Later on Kolossov, solubilized the enzyme from Barley etiochloroplasts (Kolossov and Rebeiz 2010). The involvement of VMPR in Chl biosynthesis is discussed in Chap. 7.

#### 2.10.2.2 Discovery of DV Mg-Proto Monoester Vinyl Reductase

4-vinyl MPE reductase (VMPER) catalyzes the reduction of the vinyl group of MPE to ethyl, at position 4 of the Macrocycle (Fig. 2.7, 2c, 2d).

In 1973, Ellsworth and Hsing reported the reduction of DV MPE to MV MPE by a soluble NADH-dependent enzyme in etiolated wheat homogenates (Ellsworth and Hsing 1973). However to our knowledge, no one, including ourselves has been able to duplicate this work.

The unambiguous detection of 4VMPER was first reported and solubilized by Kolossov and Rebeiz (2010). The involvement of VMPER in Chl biosynthesis is discussed in Chap. 7.

#### 2.10.2.3 Discovery of DV Pchlide *a* Vinyl Reductase

4-vinyl Pchlide reductase (VPideR) catalyzes the reduction of the vinyl group of DV Pchlide *a* to ethyl, at position 4 of the Macrocycle (Fig. 2.7, 3a, 3b).

The detection of VPideR was first reported by Tripathy and Rebeiz (1988).

It was solubilized by Kolossov and Rebeiz (2010). The VPideR gene has been detected in green sulfur bacteria (Gomez Maqueo Chew and Bryant 2007). The involvement of VMPER in Chl biosynthesis is discussed in Chaps. 8 and 9.

#### 2.10.2.4 Discovery of DV Chlorophyllide *a* Vinyl Reductase

4-vinyl Chlide reductase (VChlideR) catalyzes the reduction of the vinyl group of DV Chlide *a* to ethyl, at position 4 of the Macrocycle (Fig. 2.7, 4a, 4b).

VChlideR was the first vinyl reductase to be detected. I was first detected by Duggan and Rebeiz in 1982 and shown to be a very fast acting enzyme (Duggan and Rebeiz 1982a, b). It was shown to have an absolute requirement for NADPH, and was solubilized and partially purified by Parham and Rebeiz (1992, 1995).

It was studied further by Kolossov and Rebeiz (2010). The VChlideR gene has been first detected in *Arabidopsis* (Nagata et al. 2005). The involvement of VChlideR in Chl biosynthesis is discussed in Chap. 10.

#### 2.10.2.5 Discovery of DV Chlorophyll *a* Vinyl Reductase

4-vinyl Chl *a* reductase (VChlaR) catalyzes the reduction of the vinyl group of DV Chl *a* to ethyl, at position 4 of the Macrocycle (Fig. 2.7, 4c, 4d).

It was unambiguously detected by Adra and Rebeiz, when they showed the transient vinyl reduction of DV Chl *a* to MV Chl *a* in etiolated cucumber cotyledons following a 2.5 ms light flash treatment (Adra and Rebeiz 1998). Very recently the VChlaR gene has been detected in rice (Wang et al. 2010). The involvement of VChlideR in Chl biosynthesis is discussed in Chap. 10.

### 2.10.3 Proposal of the Multibranched Chlorophyll Biosynthetic Pathway

The discovery of novel tetrapyrrole intermediates and enzymes involved in Chl biosynthesis via various biosynthetic routes led to a gradual modification of the Chl biosynthetic pathway.

The first modification was proposed in 1971 after the lack of precursor-product relationship between Pchlide and Pchlide ester was observed and after the total biosynthesis of Pchls was achieved in organello (Rebeiz and Castelfranco 1971a). It depicted the pathway as splitting in two branches starting at the level of MPE. One branch led to the formation of Pchlide *a* while the other led to the formation of Pchlide E *a*.

Then after the discovery of the DV/MV heterogeneity, the Chl biosynthetic pathway was considered to consist of 4 Chl biosynthetic routes starting at DV and MV Mg-proto and leading to the formation of multiple Chl a spectroscopic species (Rebeiz et al. 1981). However, the origin of the biosynthetic heterogeneity was erroneously assigned to the level of the DV/MV Proto pool.

In 1983, in order to accommodate the discovery of the MV Mg-proto, MV MPE and MV MPE ester pools the Chl biosynthetic pathway was proposed to consist of six Chl biosynthetic routes originating in DV/MV Proto(gen) pools (Rebeiz et al. 1983).

Then in 1999 with the incorporation of the roles of MV Pchlide *b* and DV and MV Chlide *b* in Chl biosynthesis and the assignment of the origin of the Chl biosynthetic heterogeneity to the Mg-Proto pool (Kim and Rebeiz 1996), the Chl biosynthetic pathway was considered to consist of ten biosynthetic routes leading to the formation of MV and DV Chl *a* and *b* (Rebeiz et al. 1999).

With the incorporation of newly discovered additional data, in 1983, the Chl biosynthetic pathway was proposed to consist of 12 carboxylic and 2 fully esterified routes (Rebeiz et al. 2003). Finally in its latest version, the chlorophyll biosynthetic pathway is considered to consist of eight carboxylic routes in DV plant species (Kolossov and Rebeiz 2010), of 7 carboxylic routes in MV plant species (Kolossov and Rebeiz 2010), and of three fully esterified routes (Kolossov et al. 2003). The latest version of the Chl biosynthetic pathway will be discussed at length in various chapters of this monograph.

## 2.10.4 Discovery of the Greening Group Affiliation of Plants

With the belief in a uniform, single-branched Chl biosynthetic pathway (Granick 1950), there was no need to even contemplate the existence of plants greening differently from one another, depending on their taxonomical affiliation. However, soon after the discovery of the DV-MV Chl biosynthetic heterogeneity, and after a brief survey of the plant kingdom, it became apparent that plants used different MV or DV Chl biosynthetic routes to make Chl (Abd-El-Mageed et al. 1997; Carey and Rebeiz 1985; Carey et al. 1985; Ioannides et al. 1994). The greening group affiliation of plants will be discussed at length in Chap. 14.

## 2.10.5 Discovery of Photodynamic Herbicides

In 1982, after having researched the chemistry and biochemistry of the greening process for 18 years, it was felt that enough was known about this important biological phenomenon to translate it into biotechnological developments. In looking for a handle on the problem we opted for the development of photodynamic herbicides. That decision was prompted by two considerations: (a) the size and importance of the herbicide industry, and (b) the interesting photosensitizing properties of tetrapyrroles, which came to our attention.

Tetrapyrrole-dependent photodynamic herbicides (TDPH) consist of compounds that force green plants to accumulate undesirable amounts of metabolic intermediates of the Chl and heme biosynthetic pathways, namely tetrapyrroles. In the light the accumulated tetrapyrroles photosensitize the formation of singlet oxygen which kills the treated plants by oxidation of their cellular membranes. Tetrapyrrole-dependent photodynamic herbicides usually consist of a 5-carbon amino acid, δ-aminolevulinic acid (ALA), the precursor of all tetrapyrroles in plant and animal cells, and one of several chemicals referred to as modulators. δ-Aminolevulinic acid and the modulators act in concert. The amino acid serves as a building block of tetrapyrrole accumulation, while the modulator alters quantitatively and qualitatively the pattern of tetrapyrrole accumulation (Rebeiz et al. 1988b). In the light, tetrapyrroles are excited to the singlet state. It is believed that the excited, singlet tetrapyrroles can readily be converted to the triplet state via intersystem crossing. Since, in the ground state, oxygen exists in the triplet state, triplet-triplet energy transfer can readily take place between the excited triplet tetrapyrroles and the ground state triplet oxygen. As a consequence of this energy transfer, oxygen is excited to the singlet state. Being a very powerful oxidant, singlet oxygen oxidizes the unsaturated fatty acids of the lipoprotein membranes which are converted to hydroperoxides. The latter in turn produce free radicals which attack the unsaturated membrane lipoproteins thus setting in motion a greatly damaging free-radical chain reaction. The reaction stops when most of the cell membranes have been destroyed, causing plant death (Rebeiz et al. 1984).

The tetrapyrrole-dependent connotation of this herbicidal system is meant to differentiate between this class of photodynamic herbicides from other light activated herbicides such as paraquat which are not dependent on tetrapyrrole metabolism for herbicidal activity. During the past 20 years, the scope of TDPH research has expanded considerably, as some established herbicides which act via the TDPH phenomenon have been discovered.

The discovery of photodynamic herbicides and their development will be discussed at length in Chap. 17.

### *2.10.6 Discovery of Porphyric Insecticides*

The discovery of porphyric insecticides (Rebeiz et al. 1988a) was built upon the discovery and development of photodynamic herbicides (see above). In this case however the tetrapyrrole which accumulated upon incubation of insects with ALA and modulators was identified as Proto, the precursor of heme in insects and animals, and of Chl in plants. Proto in turn generates damaging singlet oxygen upon light exposure as was observed in plants. The discovery and development of porphyric insecticides will be described in Chap. 18.

### *2.10.7 Discovery of Photodynamic Cancericides*

Following in the footsteps of the photodynamic herbicide and porphyric insecticide technologies, the porphyrin-inducing properties of delta-aminolevulinic acid (ALA) have been adapted for the photodynamic destruction of cancer cells (Kennedy et al. 1990). Topical administration ALA to various skin lesions has, in particular, been very successful in clinical trials. For instance Kennedy et al. (1990) treated successfully basal cell carcinomas, superficial squamous cell carcinomas, and actinic keratoses, with a response rate of 90 % for basal cell carcinomas. Overall, topical, oral, or systemic administration of ALA and subsequent photodynamic therapy has been successful in a variety of tumor models, including amelanotic melanomas (Abels et al. 1994), pancreatic cancer and colon tumors, and breast cancer (Rebeiz et al. 1996). Description of the development of ALA-dependent photodynamic cancericides will be discussed in Chap. 19.

### *2.10.8 Chloroplast Bioengineering*

The notion of chloroplast bioengineering stems from the observation (a) that conventional agriculture is one of the few human activities that have not undergone a revolution to join other activities such as defying gravity by flying and landing on the moon, crossing underwater the polar cap, and communicating wirelessly over long distances via electromagnetic waves, and (b) that enough biochemical and molecular biological knowledge has accumulated to render this dream amenable to experimentation. We believe that the time has come to bioengineer chloroplasts capable of synthesizing a short chain carbohydrate such as glycerol at rates that approach the upper theoretical limits of photosynthesis (Rebeiz et al. 2004).

Chloroplast bioengineering will be discussed in Chap. 16.

## 2.11 Retirement and the Creation of the Rebeiz Foundation for Basic Research

In 2005 I decided that after 33 years at the University of Illinois it was time to retire and devote my attention to other undertakings. Thus on June 30, 2005, I officially retired from the University of Illinois and devoted my attention to the creation of the Rebeiz Foundation for Basic Research (RFFBR) (www.vlpbp.org) (Fig. 2.13). My laboratory, the Laboratory of Plant Biochemistry and Photobiology was immediately converted to the Virtual Laboratory of Plant Biochemistry and Photobiology (VLPBP). In Addition to its dedication to the promotion of research dealing with the greening process, the VLPBP website was expanded to become the voice

**Fig. 2.13** Headquarters of the Rebeiz Foundation for Basic Research, in Champaign, Illinois

of the RFFBR. The Foundation became a clearing house for the promotion of Chloroplast Research and Bioengineering, nationally and internationally. At the present, the foundation delivers annual prizes for the best research papers in the fields of chloroplast biochemistry and molecular biology, as well as one yearly Life-Time Achievement Award for a qualified chloroplast scientist. It also supports selected chloroplast symposia.

At present the RFFBR is run by Constantin A. Rebeiz, and 11 international scientists, namely:

Constantin A. Rebeiz, (crebeiz@uiuc.edu), President, Rebeiz Foundation for Basic Research.

Christoph Benning, (benning@msu.edu), Michigan State University.

Hans Bohnert, (bohnert@life.uiuc.edu), University of Illinois.

Donald Bryant, Pennsylvania State University (dab@psu.edu).

Henry Daniell, (daniell@mail.ucf.edu), Central Florida University.

Govindjee, University of Illinois at Urbana-Champaign for the light reactions of photosynthesis (gov@life.uiuc.edu).

William Lucas, University of California at Davis (wjlucas@ucdavis.edu).

Archie Portis, (arportis@uiuc.edu), University of Illinois.

Harald Paulsen, (paulsen@uni-mainz.de) Institut fur Allgemaine Botanik, Johannes Gutenberg University.

Carole Rebeiz, (ccrebeiz@aol.com), Secretary, Rebeiz Foundation for Basic Research.

Baishnab Tripathy, (bctripathy@mail.jnu.ac.in), Nehru University.

## 2.12 Epilogue: The Static and the Dynamic

It has been my experience that all phenomena can be conveniently classified as dynamic or static phenomena. Dynamic phenomena encompass the present and immediate future, and consist of our ever changing daily actions. Research is a dynamic phenomenon as research scientists carry their daily research and try to build a scientific legacy. What is a bandwagon at the present time may become ordinary in a few years as other bandwagons come into being. Therefore by its nature dynamic undertakings such as research have a built in transient characteristic.

On the other hands, static phenomena such as significant past research discoveries belong into the realm of history. They have a more permanent influence when it is realized that historically significant discoveries such as those of Maxwell, Pasteur, Darwin, Einstein and others keep influencing the course of humanity, on and on.

I am hoping that by catching a time capsule of significant static events in Chl biosynthesis we can enlighten the way of future scientists.

## Added Note

On January 3, 2013, Xiaojian Dend and his colleagues, confirmed the existence of multiple 4-vinyl reductases in an article that appeared in Plant Physiology:

Pingrong Wang, Chummei Wan, Zhengjun Xu, Pingyu Wang, Wenming Wang, Changhui Sun, Xiaozhi Ma, Yunhua Xiao, Jianqing Zhu, Xiaoling Gao, and Xiaojian Deng (2013). One divinyl reductase reduces the 8-vinyl groups in various intermediates of chlorophyll biosynthesis in a given higher plant species, but the isozyme differs between species. Plant Physiol. 161: 521–534.

## References

Abd-El-Mageed HA, El Sahhar KF, Robertson KR et al (1997) Chloroplast biogenesis 77. Two novel monovinyl and divinyl light–dark greening groups of plants and their relationship to the chlorophyll *a* biosynthetic heterogeneity of green plants. Photochem Photobiol 66:89–96

Abels C, Heil P, Dellian M et al (1994) In vivo kinetics and spectra of 5-aminolevulinic acid-induced fluorescence in an amelanotic melanoma of the hamster. Br J Cancer 70:826–833

Adra AN, Rebeiz CA (1998) Chloroplast biogenesis 81. Transient formation of divinyl chlorophyll *a* following a 2.5 ms light flash treatment of etiolated cucumber cotyledons. Photochem Photobiol 68:852–856

Bazzaz MB (1981) New chlorophyll chromophores isolated from a chlorophyll deficient mutant of maize. Photobiochem Photobiophys 2:199–207

Bazzaz MB, Bradley CV, Brereton RG (1982) 4-Vinyl-4-desethyl chlorophyll *a*: characterization of a new naturally occurring chlorophyll using fast atom bombardment field desorption and "in beam" impact mass spectroscopy. Tetrahedron Lett 23:1211–1214

Belanger FC, Rebeiz CA (1979) Chloroplast biogenesis XXVII. Detection of novel chlorophyll and chlorophyll precursors in higher plants. Biochem Biophys Res Commun 88:365–472

Belanger FC, Rebeiz CA (1980a) Chloroplast biogenesis. Detection of divinyl protochlorophyllide ester in higher plants. Biochemistry 19:4875–4883

Belanger FC, Rebeiz CA (1980b) Chloroplast biogenesis 30. Chlorophyll(ide) (E459F675) and chlorophyll(ide) (E449F675) the first detectable products of divinyl and monovinyl protochlorophyll photoreduction. Plant Sci Lett 18:343–350

Belanger FC, Rebeiz CA (1980c) Chloroplast biogenesis. Detection of divinyl protochlorophyllide in higher plants. J Biol Chem 255:1266–1272

Belanger FC, Rebeiz CA (1982) Chloroplast biogenesis: detection of monovinyl magnesium protoporphyrin monoester and other monovinyl magnesium porphyrins in higher plants. J Biol Chem 257:1360–1371

Belanger FC, Dugan JX, Rebeiz CA (1982) Chloroplastbiogenesis: identification of chlorophyllide *a* (E458F674) as a divinyl chlorophyllide *a*. J Biol Chem 257:4849–4858

Brereton RG, Bazzaz MB, Santikarn S et al (1983) Positive and negative ion fast atom bombardment mass spectrometric studies on chlorophylls: structure of 4-vinyl-4-desethyl chlorophyll *b*. Tetrahedron Lett 24:5775–5778

Carey EE, Rebeiz CA (1985) Chloroplast biogenesis 49. Difference among angiosperms in the biosynthesis and accumulation of monovinyl and divinyl protochlorophyllide during photoperiodic greening. Plant Physiol 79:1–6

Carey EE, Tripathy BC, Rebeiz CA (1985) Chloroplast biogenesis 51. Modulation of monovinyl and divinyl protochlorophyllide biosynthesis by light and darkness in vitro. Plant Physiol 79:1059–1063

Chisholm S, Olson RJ, Zettler ER et al (1988) A novel free-living prochlorophyte abundant in the oceanic euphotic zone. Nature 334:340–343

Chisholm SW, Frankel S, Goerike R et al (1992) *Prochlorococcus marinus* nov. gen. sp.: an oxyphototrophic marine prokaryote containing divinyl chlorophyll *a* and *b*. Arch Microbiol 157:297–300

Cohen CE, Rebeiz CA (1978) Chloroplast biogenesis 22. Contribution of short wavelength and long wavelength protochlorophyll species to the greening of higher plants. Plant Physiol 61:824–829

Cohen CE, Rebeiz CA (1981) Chloroplast biogenesis 34. Spectrofluorometric characterization *in situ* of the protochlorophyll species in etiolated tissues of higher plants. Plant Physiol 67:98–103

Crane JC, Rebeiz CA (1961) Gibberellin-induced parthenocarpy in the J. H. Hale peach and the probable cause of "button" production. Proc Am Soc Hortic Sci 78:111–118

Daniell H, Rebeiz CA (1982) Chloroplast culture VIII. A new effect of kinetin in enhancing the synthesis and accumulation of protochlorophyllide in vitro. Biochem Biophys Res Commun 104:837–843

Duggan JX, Rebeiz CA (1981) Detection of a naturally occurring chlorophyllide *b* pool in higher plants. Plant Physiol 67(suppl):267

Duggan JX, Rebeiz CA (1982a) Chloroplast biogenesis 42. Conversion of DV chlorophyllide a to monovinyl chlorophyllide a in vivo and in vitro. Plant Sci Lett 27:137–145

Duggan JX, Rebeiz CA (1982b) Chloroplast biogenesis 37: induction of chlorophyllide *a* (E459F675) accumulation in higher plants. Plant Sci Lett 24:27–37

Duggan JX, Rebeiz CA (1982c) Chloroplast biogenesis 38. Quantitative detection of a chlorophyllide b pool in higher plants. Biochim Biophys Acta 714:248–260

Ellsworth RK, Aronoff S (1969) Investigations of the biogenesis of chlorophyll *a*. IV. Isolation and partial characterization of some biosynthetic intermediates between Mg-protoporphine IX monomethyl ester and Mg-vinylpheoporphine $a_5$, obtained from *Chlorella* mutants. Arch Biochem Biophys 130:374–383

Ellsworth RK, Hsing AS (1973) The reduction of vinyl-side chains of Mg-protoporphyrin IX monomethyl ester in vitro. Photosynthetica 313:119–129

Fischer H, Oestricher A (1940) Protochlorophyll precursor of chlorophyll. Z Physiol Chem 262:243

Fishcher H, Mittenzwi H, Oestricher A (1939) Z Physiol Chem 257:IV

Gieskes WW, Kraay GW (1983) Unknown chlorophyll *a* derivatives in the North Sea and the tropical Atlantic Ocean revealed by HPLC analysis. Limnol Oceanogr 28(4):757–766

Goerike R, Repeta D (1992) The pigments of *Prochlorococcus marinus*. The presence of divinyl-chlorophyll *a* and *b* in a marine prochlorophyte. Limnol Oceanogr 37:425–433

Gomez Maqueo Chew A, Bryant DA (2007) Characterization of a plant-like protochlorophyllide *a* divinyl reductase in green sulfur bacteria. J Biol Chem 282(5):2967–2975

Granick S (1948a) Protoporphyrin 9 as a precursor of chlorophyll. J Biol Chem 172:717–727

Granick S (1948b) Magnesium protoporphyrin as a precursor of chlorophyll in *Chlorella*. J Biol Chem 175:333–342

Granick S (1950) Magnesium vinyl pheoporphyrin $a_{5,}$ another intermediate in the biological synthesis of chlorophyll. J Biol Chem 183:713–730

Granick S (1961) Magnesium protoporphyrin monoester and protoporphyrin monomethyl ester in chlorophyll biosynthesis. J Biol Chem 236:1168–1172

Houssier C, Sauer K (1969) Optical properties of the protochlorophyll pigments II. Electronic absorption, fluorescence and circular dichroism spectra. Biochim Biophys Acta 172:492–502

Ioannides IM, Fasoula DM, Robertson RK (1994) An evolutionary study of chlorophyll biosynthetic heterogeneity in green plants. Biochem Syst Ecol 22:211–220

Jones OTG (1963) Magnesium 2,4-divinyl phaeoporphyrin $a_5$ monomethyl ester, a protochlorophyll-like pigment produced by *Rhodopseudomonas spheroides*. Biochem J 89:182–189

Jones OTG (1966) A protein-protochlorophyll complex obtained from inner seed coats of *Cucurbita pepo*. Biochem J 101:153–160

Kennedy J, Pottier RH, Pross DC (1990) Photodynamic therapy with endogenous protoporphyrin IX. Basic principles and present clinical experience. J Photochem Photobiol B Biol 6:143–148

Kim JS, Rebeiz CA (1996) Origin of the chlorophyll *a* biosynthetic heterogeneity in higher plants. J Biochem Mol Biol 29:327–334

Kolossov VL, Rebeiz CA (2010) Evidence for various 4-vinyl reductase activities in higher plants. In: Rebeiz CA, Benning C, Bohnertet HJ et al (eds) The chloroplast: basics and applications. Springer, Dordrecht, pp 25–38

Kolossov VL, Kopetz KJ, Rebeiz CA (2003) Chloroplast biogenesis 87: evidence of resonance excitation energy transfer between tetrapyrrole intermediates of the chlorophyll biosynthetic pathway and chlorophyll *a*. Photochem Photobiol 78:184–196

Mattheis JR, Rebeiz CA (1977a) Chloroplast biogenesis XVII. Metabolism of protochlorophyllide and protochlorophyllide ester in developing chloroplasts. Arch Biochem Biophys 184:189–196

Mattheis JR, Rebeiz CA (1977b) Chloroplast biogenesis. Net synthesis of protochlorophyllide from magnesium protoporphyrin monoester by developing chloroplasts. J Biol Chem 252:4022–4024

Mattheis JR, Rebeiz CA (1977c) Chloroplast biogenesis. Net synthesis of protochlorophyllide from protoporphyrin IX by developing chloroplasts. J Biol Chem 252:8347–8349

McCarthy SA, Belanger FC, Rebeiz CA (1981) Chloroplast biogenesis: detection of a magnesium protoporphyrin diester pool in plants. Biochemistry 20:5080–5087

Nagata N, Tanaka R, Satoh S et al (2005) Identification of a vinyl reductase gene for chlorophyll synthesis in *Arabidopsis thaliana* and implications for the evolution of prochlorococcus species. Plant Cell 17:233–240

Oelze-Karow H, Mohr H (1978) Control of chlorophyll b biosynthesis by phytochrome. Photochem Photobiol 27:189–193

Parham R, Rebeiz CA (1992) Chloroplast biogenesis: [4-vinyl] chlorophyllide *a* reductase is a divinyl chlorophyllide *a*-specific NADPH-dependent enzyme. Biochemistry 31:8460–8464

Parham R, Rebeiz CA (1995) Chloroplast biogenesis 72: a [4-vinyl] chlorophyllide *a* reductase assay using divinyl chlorophyllide *a* as an exogenous substrate. Anal Biochem 231:164–169

Rebeiz CA (1967) Studies on chlorophyll biosynthesis in etiolated excised cotyledons of germinating cucumber at different stages of seedling development. Magon Serie Scientifique 13:1–21

Rebeiz CA (1968a) Dark and light carotenoids accumulation in etiolated and greening cucumber cotyledons. Magon Serie Scientifique 23:1–10

Rebeiz CA (1968b) The chloroplast pigments of etiolated and greening cucumber cotyledons. Magon Serie Scientifique 21:1–25

Rebeiz CA (2010) Investigations of possible relationships between the chlorophyll biosynthetic pathway and the assembly of chlorophyll-protein complexes and photosynthetic efficiency. In: Rebeiz CAB, Bohnert C, Daniell HJ, Hoober H, Lichtenthaler JK, Portis HK (eds) The chloroplast: basics and applications. Springer, Dordrecht, pp 1–24

Rebeiz CA, Castelfranco PA (1964) An extra-mitochondrial enzyme system from peanuts catalyzing the ß-oxidation of fatty acids. Plant Physiol 39:932–938

Rebeiz CA, Castelfranco P (1971a) Protochlorophyll biosynthesis in a cell-free system from higher plants. Plant Physiol 47:24–32

Rebeiz CA, Castelfranco P (1971b) Chlorophyll biosynthesis in a cell-free system from higher plants. Plant Physiol 47:33–37

Rebeiz CA, Crane JC (1961) Growth regulator-induced parthenocarpy in the Bing cherry. Proc Am Soc Hortic Sci 78:69–75

Rebeiz CA, Castelfranco PA, Breidenbach RW (1965a) Activation and oxidation of acetic acid 1-14C by cell-free homogenates of germinating peanuts cotyledons. Plant Physiol 49:286–289

Rebeiz CA, Castelfranco PA, Engelbrecht AH (1965b) Fractionation and properties of a extra-mitochondrial enzyme system from peanuts catalyzing the B-oxidation of palmitic acid. Plant Physiol 40:281–286

Rebeiz CA, Abou Haidar M, Yaghi M et al (1970a) Porphyrin biosynthesis in cell-free homogenates from higher plants. Plant Physiol 46:543–549

Rebeiz CA, Yaghi M, Abou Haidar M et al (1970b) Protochlorophyll biosynthesis in cucumber (Cucumis sativus, L.) cotyledons. Plant Physiol 46:57–63

Rebeiz CA, Mattheis JR, Smith BB et al (1975) Chloroplast biogenesis. Biosynthesis and accumulation of protochlorophyll by isolated etioplasts and developing chloroplasts. Arch Biochem Biophys 171:549–567

Rebeiz CA, Belanger FC, McCarty SA et al (1981) Biosynthesis and accumulation of novel chlorophyll *a* and *b* chromophoric species in green plants. In: Akoyunoglou G (ed) Photosynthesis V. Chloroplast development. Balaban International Services, Philadelphia, pp 197–212

Rebeiz CA, Daniell H, Mattheis JR (1982) Chloroplast bioengineering: the greening of chloroplasts in vitro. In: Scott CD, Scott CD (eds) Biotechnology bioengineering symposium, John Wiley, New York, vol 12, pp 414–439

Rebeiz CA, Wu SM, Kuhadje M et al (1983) Chlorophyll *a* biosynthetic routes and chlorophyll *a* chemical heterogeneity. Mol Cell Biochem 58:97–125

Rebeiz CA, Montazer-Zouhoor A, Hopen HJ et al (1984) Photodynamic herbicides: 1. Concept and phenomenology. Enyme Microbiol Technol 6:390–401

Rebeiz CA, Juvik JA, Rebeiz CC (1988a) Porphyric insecticides 1. Concept and phenomenology. Pestic Biochem Physiol 30:11–27

Rebeiz CA, Montazer-Zouhoor A, Mayasich JM et al (1988b) Photodynamic herbicides. Recent developments and molecular basis of selectivity. Crit Rev Plant Sci 6:385–434

Rebeiz N, Arkins S, Rebeiz CA et al (1996) Induction of tumor necrosis by delta-aminolevulinic acid and 1,10-phenanthroline. Cancer Res 56:339–344

Rebeiz CA, Ioannides IM, Kolossov V et al (1999) Chloroplast biogenesis 80. Proposal of a unified multibranched chlorophyll *a*/*b* biosynthetic pathway. Photosynthetica 36:117–128

Rebeiz CA, Kolossov VL, Briskin D et al (2003) Chloroplast biogenesis: chlorophyll biosynthetic heterogeneity, multiple biosynthetic routes and biological spin-offs. In: Nalwa HS (ed) Handbook of photochemistry and photobiology, vol 4. American Scientific Publishers, Los Angeles, pp 183–248

Rebeiz CA, Kolossov VL, Kopetz KK (2004) Chloroplast bioengineering: photosynthetic efficiency, modulation of the photosynthetic unit size, and the agriculture of the future. In: Nelson DW (ed) Agricultural applications in green chemistry, vol 887. American Chemical Society, Washington, DC, pp 81–105

Shedbalkar VP, Ioannides IM, Rebeiz CA (1991) Chloroplast biogenesis. Detection of monovinyl protochlorophyll(ide) *b* in plants. J Biol Chem 266:17151–17157

Shlyk AA (1971) Biosynthesis of chlorophyll b. Annu Rev Plant Physiol 22:169–184

Smith JHC, French CS (1963) The major accessory pigment in photosynthesis. Annu Rev Plant Physiol 14:181–224

Smith BB, Rebeiz CA (1977) Chloroplast biogenesis: detection of Mg-protoporphyrin chelatase in vitro. Arch Biochem Biophys 180:178–185

Tripathy BC, Rebeiz CA (1988) Chloroplast biogenesis 60. Conversion of divinyl protochlorophyllide to monovinyl protochlorophyllide in green(ing) barley, a dark monovinyl/light divinyl plant species. Plant Physiol 87:89–94

Virgin HI (1960) Pigment transformations in leaves of wheat after irradiation. Physiol Plant 13:155–164

Wang P, Gao J, Chunmei W et al (2010) Divinyl chlorophyll(ide) *a* can be converted to Monovinyl chlorophyl(lide) *a* by a divinyl reductase in rice. Plant Physiol 153:994–1003

Withrow RB, Wolff JB, Price L (1955) Elimination of the lag phase of chlorophyll synthesis in dark-grown bean leaves by a pretreatment with low irradiances of monochromatic energy. Plant Physiol 31:XIII

Wolff JB, Price L (1957) Terminal steps of chlorophyll *a* biosynthesis in higher plants. Arch Biochem Biophys 72:293–301

Wu SM, Rebeiz CA (1984) Chloroplast biogenesis 45: molecular structure of protochlorophyllide (E443 F625) and of chlorophyllide *a* (E458 F674). Tetrahydron 40(4):659–664
Wu SM, Rebeiz CA (1985) Chloroplast biogenesis. Molecular structure of chlorophyll *b* (E489 F666). J Biol Chem 260:3632–3634
Wu SM, Rebeiz CA (1988) Chloroplast biogenesis. Molecular structure of short wavelength chlorophyll a (E432 F662). Phytochemistry 27:353–356

# Chapter 3
# Development of Analytical and Preparatory Techniques

*Integrity without knowledge is weak and useless, and knowledge without integrity is dangerous and dreadful (Samuel Johnson)*

## 3.1 Prologue

The successful investigations of the chlorophyll biosynthetic pathway would not have been possible without the development of new analytical and preparatory techniques. These techniques form the object of this chapter.

Prior to 1975 practically all investigations involving metabolic tetrapyrroles used absorption spectrophotometric techniques. Then in the early 1970s it came to our attention that metabolic tetrapyrroles were fluorescent. A comparison of absorption and fluorescence techniques revealed that fluorescence techniques were about a 100 fold more sensitive and were more flexible than absorption techniques. Indeed while absorption spectroscopy is a one-window technique fluorescence is a two windows technique. This means that with fluorescence techniques it is possible to record sensitive fluorescence emission spectra by varying the emission wavelength while keeping the excitation wavelength constant. Or one can record equally sensitive excitation spectra by varying the excitation wavelength and keeping the emission wavelength constant.

## 3.2 Determination of Metabolic Tetrapyrroles by Room Temperature Spectrofluorometry

After the usage of $^{14}C$-δ-aminolevulinic acid (ALA) to achieve the total biosynthesis of protochlorophyllide *a* (Pchlide *a*), and Chl *a* and *b* in organello (Rebeiz and Castelfranco 1971a, b). The need arose to achieve the net synthesis of Chl,

C.A. Rebeiz, *Chlorophyll Biosynthesis and Technological Applications*, 65
DOI 10.1007/978-94-007-7134-5_3, © Springer Science+Business Media Dordrecht 2014

protochlorophyll (Pchl) and tetrapyrrole intermediates in organello. In order to achieve this goal it became mandatory to develop sensitive spectroscopic techniques to monitor the products of in organello incubations at room temperature. These techniques are described below.

### *3.2.1 Calculation of Protochlorophyllide Ester by Fluorescence Spectroscopy at Room Temperature*

Upon extraction of the chlorophyll (Chl) in hexane, protochlorophyllide *a* ester (Pchlide a E) is extracted along with the Chl (see later sections below, dealing with preparative techniques). Its determination by fluorescence spectroscopy at room temperature is described below.

Let *E* refer to the Soret excitation maximum of a fluorescent compound and *F* to the fluorescence amplitude of the short-wavelength emission maximum of that same compound. Then let (E440F631) and (E440 F670) represent the respective fluorescence amplitudes of a mixture of Pchlide *a* E and Chl *a* + *b* at 631 and 670 nm respectively the fluorescence being elicited by a 440-nm excitation of the hexane extract. In hexane, a mixture of Chl *a* + *b*, where the ratio of Chla/Chl *b* is about 2.6–3.0, usually exhibits an emission maximum at about 670 nm.

The following simultaneous equations can be written for the Pchlide E, Chl mixture

$$\begin{aligned}(\text{E440 F670}) = {} & \text{Pchlide a E (E440 F631)} + \text{Chl (E440 F633)} \\ & + \text{Pchlide (E400 F631)},\end{aligned} \tag{3.1}$$

and,

$$(\text{E440 F670}) = k_1 \text{ Pchlide a E (E440 F631)} + k_2 \text{ Chl (E440 F631)} \tag{3.2}$$

where

$$\begin{aligned}k_1 &= \text{Pchlide a E (E440 F631)/Proto (E400 F670)} \\ k_2 &= \text{Chl (E440 F670)/Chl (E440 F631)}\end{aligned} \tag{3.3}$$

According to the above terminology, Pchlide E (E440 F631) = fluorescence amplitude of Pchlide E at 631 nm, which is elicited by a 440-nm excitation of the hexane extract. Chl (E440 F631) = fluorescence emission amplitude of Chl *a* + *b* at 631 nm, which is elicited by a 440-nm excitation of the hexane extract; etc. By solving Eqs. (3.1) and (3.2) for Pchlide E (E440 F631) the following equation is obtained

$$\text{Pchlide a E (E440 F631)} = [(\text{E440 F631}) - ((\text{E440 F670})/k_2)](1/\text{K}) \tag{3.4}$$

where

$$K = 1 - (k_1/k_2) \tag{3.5}$$

Constants $k_l$ and $k_2$ were calculated according to Eq. (3.3). The fluorescence amplitudes used in Eq. (3.3) were derived from the emission spectra of standard Pchlide E and of a mixture of standard Chl *a* + *b* dissolved separately in acetone-extracted hexane (see below, the sections on preparatory techniques). The ratio of Chl *a* to Chi *b* in the mixture was about 2.6. Under our instrumental conditions the mean of five determinations performed on different concentrations of standard Pchlide E and Chl were 0.144 for $k_l$ and 145 for $k_2$. The average value for *K* calculated according to Eq. (3.5) was 0.999 (Rebeiz et al. 1975a).

## 3.2.2 *Calculation of Protochlorophyllide a by Fluorescence Spectroscopy at Room Temperature*

In developing analytical preparatory techniques for monitoring tetrapyrrole metabolism in organello and in vitro it was realized that the best way to monitor Pchlide *a* was in the hexane-extracted acetone fraction (HEAR). In Addition to Pchlide *a* that fraction contained chlorophyllide *a* (Chlide *a*), Chlide *b*, protoporphyrin IX (Proto) as well as coproporphyrin (Copro).

The spectrofluorometric correction for these tetrapyrroles in the HEAR fraction while measuring Pchlide *a* is described below.

### 3.2.2.1 Correction for the Chlorophyllide *a* + *b* Fluorescence at 640 nm

Let E and F be defined as above and let (E′440 F′676) and (E′440 F′640) represent the fluorescence amplitudes at 676 and 640 nm, respectively, of a mixture of several pigments containing Chl and Chlide [Chl(ide)] *a* + *b* and which are elicited by a 440-nm excitation.

Then let Chl(ide) (E440 F676) and Chl(ide) (E440 F640) represent the Chl (ide) a + b fluorescence amplitudes at 676 and 640 nm, respectively, upon excitation at 440 nm. Finally let (E440 F640) represent the Chl(ide)-free fluorescence amplitude at 640 nm of the pigment mixture containing Chl(ide) upon excitation at 440 nm.

The following equation can then be written:

$$(\mathrm{E'440\ F640}) = (\mathrm{E440\ F640}) + k(\mathrm{E440\ F676}) \tag{3.6}$$

where

$$k = \mathrm{Chl(ide)\ (E440\ F640)/Chl(ide)\ (E440\ F676)} = 0.015 \tag{3.7}$$

The value of $k$ was the mean of eight different determinations performed on eight different concentrations of Chl $a + b$ dissolved in hexane-extracted acetone. Upon substituting for $k$ in Eq. (3.6), Eq. (3.8) is obtained:

$$(\mathrm{E}'440\ \mathrm{F}'640) = (\mathrm{E}440\ \mathrm{F}640) + 0.015\ \mathrm{Chl(ide)}\ (\mathrm{E}440\ \mathrm{F}676) \tag{3.8}$$

In a mixture containing several pigments in addition to Chl(ide) and where the contribution of the other pigment components to the Chl(ide) fluorescence at 676 nm is negligible, the following equation can be written

$$\mathrm{Chl(ide)}\ (\mathrm{E}440\ \mathrm{F}676) = (\mathrm{E}'440\ \mathrm{F}'676) \tag{3.9}$$

Substituting Eq. (3.9) into Eq. (3.8) gives:

$$(\mathrm{E}'440\mathrm{F}'640) = (\mathrm{E}440\mathrm{F}640) + 0.015\ (\mathrm{E}'440\ \mathrm{F}'676) \tag{3.10}$$

and

$$(\mathrm{E}440\ \mathrm{F}640) = (\mathrm{E}'440\ \mathrm{F}'640) - 0.015\ (\mathrm{E}'440\ \mathrm{F}'676) \tag{3.11}$$

In the extracts of irradiated etioplasts that are still in the lag phase of Chl accumulation, the contribution of Chl fluorescence to Pchlide fluorescence at 640 nm is negligible, therefore:

$$k = \mathrm{Chl(ide)}\ (\mathrm{E}440\ \mathrm{F}6640)/\mathrm{Chl(ide)}\ (\mathrm{E}440\ \mathrm{F}676) = 0$$

and Eq. (3.11) becomes

$$(\mathrm{E}440\ \mathrm{F}640) = \mathrm{E}'440\ \mathrm{F}'640) \tag{3.12}$$

In the hexane-extracted acetone fractions of developing chloroplasts excited at 440 nm, the contribution of Copro, Proto and Pchlide to Chli(de) fluorescence at 676 nm is also negligible. This is caused by the small amounts of Copro + Proto + Pchlide encountered in the extracts as well as by the very low fluorescence yield of these tetrapyrroles at 676 nm. This was checked further by preparing mixtures of Copro + Proto + Pchlide in hexane-extracted acetone that contained the same concentration of tetrapyrroles as the extracts of the plastids. Excitation at 440 nm did not elicit any fluorescence at 676 nm. It was therefore concluded that the use of Eq. (3.12) for calculating the Chl(ide)free fluorescence at 640 nm (E440 F640) of the hexane-extracted acetone fractions of developing chloroplasts is valid.

#### 3.2.2.2 Determination of the Chlorophyll(ide)-Free Amplitudes of Extracts at 633 and 622 nm

The Chl(ide)-free fluorescence amplitudes at 633 and 622 nm [(E400 F633) and (E400 F622)] of the hexane-extracted acetone fractions were obtained directly from the emission spectra of the extracts after excitation at 400 nm. Indeed no correction was necessary for the contribution of Chl(ide) fluorescence at these wavelengths since the Chl(ide) $a + b$ of the extracts exhibited a totally negligible fluorescence at 633 and 622 nm (Rebeiz et al. 1975b).

#### 3.2.2.3 Calculation of Protochlorophyllide *a* After Excitation at 640 nm

Let (E440 F640), (E400 F633) and (E400F622) represent the Chl(ide)-free fluorescence amplitudes of a mixture of Chl(ide), Pchlide, Proto and Copro at 640, 633 and 622 nm, respectively, which are elicited by 440- and 400-nm excitations, of the HEAR fractions. The Chl(ide)- free fluorescence amplitudes can then be treated as emanating from mixtures containing only Copro, Proto and Pchlide and lacking Chl (ide). (E440 F640) is then calculated according to Eq. (3.11) for developing chloroplasts and from Eq. (3.12) for etioplasts. (E400 F633) and (E400 F622) are obtained directly from the emission spectra of the hexane-extracted acetone fractions as described above. For such extracts the following simultaneous equation can then be written:

$$\begin{aligned}(\text{E440 F640}) = & \text{Pchlide (E440 F640)} \\ & + \text{Proto (E440 F640)} + \text{Copro (E440 F640)},\end{aligned} \tag{3.13}$$

and

$$\begin{aligned}(\text{E400 F633}) = & k'_1 \text{ Pchlide (E440 F640)} \\ & + k'_2 \text{ Proto (E440 F640)} \\ & + k'_3 \text{ Copro (E440 F640)},\end{aligned} \tag{3.14}$$

and

$$\begin{aligned}(\text{E400 F622}) = & k'_4 \text{ Pchlide (E440 F640)} \\ & + k'_5 \text{ Proto (E440 F640)} \\ & + k'_6 \text{ Copro (E440 F640)},\end{aligned} \tag{3.15}$$

where

$$\begin{aligned} k'_1 &= \text{Pchlide } a \text{ (E400 F633)/Pchlide } a \text{ (E440 F640)} \\ k'_2 &= \text{Proto (E400 F633)/Proto (E440 F640)} \\ k'_3 &= \text{Copro (E400 F633)/Copro (E440 F640)} \\ k'_4 &= \text{Pchlide } a \text{ (E400 F622)/Pchlide } a \text{ (E440 F640)} \\ k'_5 &= \text{Proto (E400 F622)/Proto (E440 F640)} \\ k'_6 &= \text{Copro (E400 F622)/Copro (E440 F640)} \end{aligned} \tag{3.16}$$

According to the above terminology:Pchlide (E440 F640) = fluorescence emission amplitude of Pchlide at 640 nm which is elicited by a 440-nm excitation of the hexane-extracted acetone fraction. Proto (E440 F640) = fluorescence emission amplitude of Proto at 640 nm which is elicited by a 440-nm excitation of the hexane-extracted acetone fraction; etc. By solving Eqs. (3.13), (3.14) and (3.15) for Pchlide (E440, F640) the following equation is obtained:

$$\text{Pchlide (E440 F640)} = [\text{(E440 F640)} - C'_1 \text{ (E400 F633)} - C'_2 \text{ (E400 F622)}]/C'_3 \tag{3.17}$$

Where (E440 F640) is calculated from Eqs. (3.11) or (3.12) and were

$$\begin{aligned} C'_1 &= 1/k'_2 - K'_2.K'_4/K'_3 \\ C'_2 &= K'_2/K'_3 \\ C'_3 &= K'_1 + K'_2.K'_5/K'_3 \end{aligned} \tag{3.18}$$

and

$$\begin{aligned} K'_1 &= 1 - k'_1/k'_2 \\ K'_2 &= 1 - k'_3/k'_2 \\ K'_3 &= k'_6 - k'_5.k'_3/k'_2 \\ K'_4 &= k'_5/k'_2 \\ K'_5 &= k'_5.k'_1/k'_2 - k'_4 \end{aligned} \tag{3.19}$$

The values of the $k'$ constants were calculated according to Eq. (3.16). The fluorescence amplitudes used in Eq. (3.16) were derived from the emission spectra of standard Pchlide, Proto and Copro dissolved separately in acetone/$H_2O$/ 0.1NNH4O(9:2:1, v/v) that was previously extracted with hexane (Rebeiz et al. 1975a). The emission spectra of the standards were elicited by two successive excitations at 440 and 400 nm. The $k'$ values used for the calculations of the $K'$ and $C$ constants were the mean of at least five determinations performed on five different concentrations of standard Pchlide, Proto, and Copro. The $K'$ constants were calculated from Eq. (3.19) and the C′ constants from Eq. (3.18). Under our instrumental conditions the values obtained were 0.03 for $C_1$, 0.0 for $C_2'$, and 0.99 for $C_3'$.

### 3.2.3 *Development of Fluorescence Equations for the Determination of Protoporphyrin IX by Room Temperature Spectrofluorometry*

Prior to 1975, Protoporphyrin IX (Proto) was monitored by absorption spectrophotometry (Granick 1948). Below is described the determination of Proto in a misture of metabolic tetrapyrroles by fluorescence spectroscopy.

Let *E* refer to the Soret excitation maximum of Proto and *F* to the fluorescence amplitude of its short-wavelength emission maximum.

Then let (E400 F622), (E400 F633), and (E440F640) represent the chlorophyllide (Chlide) -free fluorescence amplitudes of a mixture of Chl(ide), Proto, coproporphyrin (Copro) and protochlorophyllide a (Pchlide a) at 622, 633, and 640 nm, respectively, which are elicited by 400- and 440-nm excitations of the hexane-extracted acetone fraction. The following simultaneous equation can then be written:

$$\begin{aligned}(\text{E400 F633}) = {} & \text{Proto (E400 F633)} + \text{Copro (E400 F633)} \\ & + \text{Pchlide (E400 F633)},\end{aligned} \tag{3.20}$$

$$\begin{aligned}(\text{E400 F622}) = {} & k_1'' \text{ Proto (E400 F633)} + k_2'' \text{ Copro (400 F633)} \\ & + k_3'' \text{ Pchlide (400 F633)}\end{aligned} \tag{3.21}$$

$$\begin{aligned}(\text{E440 F640}) = {} & k_4'' \text{ Proto (E400 F633)} + k_5'' \text{Copro (E400 F633)} \\ & + k_6'' \text{ Pchlide (E400 F633)},\end{aligned} \tag{3.22}$$

where

$$\begin{aligned}
k_1'' &= \text{Proto (E400 F622)/Proto (E400 F633)} \\
k_2'' &= \text{Copro (E400 F622)/Proto (E400 F633)} \\
k_3'' &= \text{Pchlide (E400 F622)/Pchlide (E400 F633)} \\
k_4'' &= \text{Proto (E440 F640)/Proto (E400 F633)} \\
k_5'' &= \text{Copro (E400 F640)/Copro (E400 F633)} \\
k_6'' &= \text{Pchlide (E440 F640)/Pchlide (E400 F633)}
\end{aligned} \tag{3.23}$$

According to the above terminology: Proto (E400 F633) = fluorescence emission amplitude of Proto at 633 nm which is elicited by a 400-nm excitation of the hexane-extracted acetone fraction. Copro (E400F633) = fluorescence emission amplitude of Copro at 633 nm, which is elicited by a 400-nm excitation of the hexane-extracted acetone fraction; etc. By solving Eqs. (3.20), (3.21) and (3.22) for Proto (E400 F633) the following equation is obtained:

$$\text{Proto (E400 F633)} = \text{(E400 F633)} - C_1'' \text{ (E400 F622)} - C_2'' \text{ (E440 F640)}/C_3'' \quad (3.24)$$

where (E440 F640) is calculated from Eqs. (3.11) or (3.12), and where

$$\begin{aligned} C_1'' &= 1/k_2'' - K_2''.K_4''/K_3'' \\ C_2'' &= K_2''/K_3'' \\ C_3'' &= K_1'' + K_2''.K_5''/K_3'' \end{aligned} \quad (3.25)$$

and

$$\begin{aligned} K_1'' &= 1 - k_1''/k_2'' \\ K_2'' &= 1 - k_3''/k_2'' \\ K_3'' &= k_6'' - k_5''.k_3''/k_2'' \\ K_4'' &= k_5''/k_2'' \\ K_5'' &= k_5''.k_1''/k_2'' - k_4'' \end{aligned} \quad (3.26)$$

The values of the k″ constants were calculated according to Eq. (3.23). The fluorescence amplitudes used in Eq. (3.23) were derived from the emission spectra of standard Proto, Copro, and Pchlide dissolved separately in acetone/$H_2O$/0.1 N $NH_4OH$ (9:2:1, vv) that was previously extracted twice with hexane (Rebeiz et al. 1975b). The emission spectra of the standards were elicited by two successive excitations at 400 and 440 nm. The k″ values used for the calculations of the $K''$ and C″ constants were the mean of at least five determinations performed on five different concentrations of standard Proto, Copro, and Pchlide. The $K''$ constants were calculated from Eq. (3.26) and the C″ constants from Eq. (3.25). Under our instrumental conditions the values obtained were 0.25 for $C_1''$, 0.24 for $C_2''$, and 0.95 for $C_3''$.

## 3.3 Spectrofluorometric Determination of Mg-Protoporphyrin Monoester and Longer Wavelength Metalloporphyrins in the Presence of Zn-Protoporphyrin IX at Room Temperature

When incubation for the biosynthesis of Mg-porphyrins are carried out in organello or in vitro without adding enough ATP to the incubation mixture, the biosynthesis of Mg-porphyrins is accompanied by the formation of Zn-Proto. Spectrofluorometric equations were developed for the calculation of Mg-porphyrins in the presence of Zn-Proto as described below (Smith and Rebeiz 1977b).

### 3.3.1 Calculation of the Fluorescence Integral Between 592 and 620 nm Which Is Contributed Solely by Mg-Porphyrins in Mixtures Containing Zn-Proto

Let E refer to the Soret excitation of a fluorescent compound. Then let E420 $\int_{570}^{592} f(\lambda)d\lambda$ and E420 $\int_{592}^{620} f(\lambda)d\lambda$ represent the respective fluorescence integrals between 570 and 592 nm and between 592 and 620 nm for a mixture of Zn-Proto and Mg-porphyrins, which are elicited by a 420 nm excitation of the hexane-extracted acetone solution (Smith and Rebeiz 1977a). The following simultaneous equations can then be written for the fluorescent mixture:

$$\mathrm{E420}\int_{592}^{620} f(\lambda)d(\lambda)$$
$$= \text{Mg-Porphyrins}\left(\mathrm{E420}\int_{592}^{620} f(\lambda)d\lambda\right) + \text{Zn-Proto E420}\left(\int_{592}^{620} f(\lambda)d\lambda\right) \tag{3.27}$$

and

$$E420\int_{570}^{592} f(\lambda)d\lambda$$
$$= (\mathrm{k}_1)\ \text{Mg-porphyrins}\left(\mathrm{E420}\int_{592}^{620} f(\lambda)d\lambda\right) + \mathrm{k}_2\,\text{Zn-Proto}\left(\mathrm{E420}(\int_{592}^{620} f(\lambda)d\lambda\right)$$

where:

$$\mathrm{k}_1 = \text{Mg-Porphyrins (E420}\int_{570}^{592} f(\lambda)d\lambda/\text{Mg-Porphyrins (E420}\int_{592}^{620} f(\lambda)d(\lambda) \tag{3.28}$$

and,

$$\mathrm{k}_2 = \text{Zn-Proto (E420}\int_{570}^{592} f(\lambda)d\lambda/\text{Zn-Proto (E420}\int_{592}^{620} f(\lambda)d(\lambda)$$

Where:

$$\mathrm{k}_1 = \text{Mg-Porphyrins (E420}\int_{570}^{592} f(\lambda)d\lambda/\text{Mg-Porphyrins (E420}\int_{592}^{620} f(\lambda)d\lambda) \tag{3.29}$$

and,

$$k_2 = \text{Zn-Proto } (E420 \int_{570}^{592} f(\lambda)d\lambda)/\text{Zn-Proto } (E420 \int_{592}^{620} f(\lambda)d\lambda)$$

According to the above terminology: Mg-Porphyrins (E420 $\int_{592}^{620} f(\lambda)d\lambda$) = the fluorescence integral of the Mg-Porphyrins between 592 and 620 nm which was elicited by a 420 nm excitation of the hexane-extracted acetone solution. Zn-Proto (E420 $\int_{592}^{620} f(\lambda)d\lambda$) = the fluorescence integral of Zn-Proto between 592 and 620 nm which was elicited by a 420 nm excitation of the hexane-extracted acetone solution. Zn-Proto (E420 $\int_{570}^{592} f(\lambda)d\lambda$) = the fluorescence integral of Zn-Proto between 570 and 592 nm which was elicited by a 420 nm excitation of the hexane-extracted acetone solution, etc. By solving Eqs. (3.27) and (3.28) for Mg-Porphyrins (E420 $\int_{592}^{620} f(\lambda)d\lambda$), the following equation is obtained:

$$\text{Mg-Porphyrins}\left(E420 \int_{592}^{620} f(\lambda)d\lambda\right) = \left[E420 \int_{592}^{620} f(\lambda)d\lambda) - (E420 \int_{570}^{592} f(\lambda)d\lambda/k_2\right] 1/K \tag{3.30}$$

Where:

$$K = 1 - k_1/k_e \tag{3.31}$$

$k_1$and $k_2$ were calculated according to Eq. (3.29). The fluorescence integrals used in Eq. (3.29) were determined by planimetry from the emission spectra of standard MPE-equivalent and Zn-Prot in hexane-extracted acetone as described in (Smith and Rebeiz 1977a).The value of $k_1$ was 0.12 and was the mean of 51 different determinations performed on various concentrations of standard Mg-Porphyrins. The value of $k_2$ was 1.22 and was the mean of 19 different determinations performed on various concentrations of Zn-Proto. The value of *K*, calculated according to Eq. (3.31) was 0.90 (Smith and Rebeiz 1977a). By substituting the values for $k_2$ and *K* in Eq. (3.30) the following equation was obtained:

$$\text{Mg-Porphyrins } (E420 \int_{592}^{620} f(\lambda)d\lambda) = (1.11)(E420 \int_{592}^{620} f(\lambda)d\lambda) - (0.91)\,(E420 \int_{570}^{592} f(\lambda)d\lambda) \tag{3.32}$$

### 3.3.2 Validation of Equation (3.32)

In order to determine the reliability of Eq. (3.32), the fluorescence emission spectra of various concentrations of standard Mg-Porphyrins, and Zn-Proto were recorded before and after mixing with one another. The various mixtures were prepared in the same proportions as encountered in organello experiments (i.e. the mixtures contained 20–100 % Mg-Porphyrins). The Mg-porphyrins fluorescence integral between 592 and 620 nm, i.e. Mg-Porphyrins (E420 $\int_{592}^{620} f(\lambda)d\lambda$) was determined by planimetry from the emission spectrum of the Mg-porphyrins before mixing with Zn-Proto. After mixing with Zn-Proto the fluorescence integral of the Mg-Porphyrins between 592 and 620 nm was calculated from the emission spectrum of the mixture according to Eq. (3.32). The actual and calculated Mg-Porphyrins Mg-Porphyrins (E420 $\int_{592}^{620} f(\lambda)d\lambda$) values were compared and both the percent error and the discrepancy were determined (Smith and Rebeiz 1977a). In addition, the mean percent error ± the standard deviation of the mean percent error and the mean discrepancy ± the standard error of the mean discrepancy were also determined for the following single component and double component solutions: (a) Mg-Porphyrins, (b) Zn-Proto and (c) Mg-Porphyrins + Zn-Proto (Smith and Rebeiz 1977a).

The mean discrepancies for all errors ranged from 0 to –16 % (Smith and Rebeiz 1977a).

## 3.4 Determination of Chlorophyll and Chlorophyllide *a* [Chl(ide *a*)] in the Presence of Chl(ide) *b* and Pheophytin and Pheophorbide [Pheo(bide)] *a* and *b* Spectrofluorometry at Room Temperature

The use of absorption spectrophotometry to monitor the amount of Chl *a* and *b*, which is extractable in organic solvents, is satisfactory when the ratio Chl *a* to Chl *b* is less than 6.0 (Bazzaz and Rebeiz 1979). This technique is not suitable, however, for detecting and determining small amounts of Chl degradation products in the presence of large amounts of Chl. Since it now appears that the formation of small amounts of chlorophyllides (Chlides) (Chl that has lost its phytol) and pheophorbides (Pheobides) (Chlides hat have lost their Mg) signals the beginning of chloroplast degradation, before the disappearance of Chl becomes evident, (Bazzaz and Rebeiz 1978, 1979) it has become essential to develop sensitive techniques for the early detection of these degradation products with minimum processing of the chloroplast extract. This precaution was dictated by the lability of the Chl extracted in organic solvents. Indeed, it has been our experience that Chl can easily generate chemical artifacts during chromatography, which interfere with the detection of small amounts of Chlides and Pheobides. As a consequence sensitives pectrofluorometric techniques for the early detection and quantitative determination of picomole quantities of Chlide *a* and *b* and Pheobide *a* and *b* were developed. These techniques were based on the observation that although Chlide *a* exhibited similar red emission and absorption maxima as Pheobide *a* and Chlide *b* exhibited similar red emission and absorption maxima as Pheobide *b*, these four tetrapyrroles exhibited distinct Soret excitation maxima (Bazzaz and Rebeiz 1979). Therefore this observation was used, to derive four simultaneous equations which permitted the determination of picomole amounts of Chlide *a*, Chlide *b*, Pheobide *a*, and Pheobide *b* in mixtures of these tetrapyrroles without prior chromatographic segregation. The derivation and testing of these equations is described below.

## 3.4.1 Calculation of Chl(ide) *a*

The calculation of Chl(ide) *a* in various pigment mixtures is described below.

### 3.4.1.1 Calculation of Chl(ide) *a* in a Mixture Containing Chl(ide) *a* and *b* and Pheo(bide) *a* and *b*

Let E refer to the Soret excitation maximum of a fluorescent compound and F to the fluorescence emission maximum of that same compound. Then let (F674 E433), (F674E412), (F660 E460) and (F660 E438) represent the respective fluorescence excitation amplitudes at 433, 412, 460 and 438 nm, of a mixture of Chl(ide) *a*, Pheobide) *a*, Chl(ide) *b* and Pheo(bide) *b* which were recorded either at 674 or at 660 nm. The Soret excitation amplitude contributed solely by Chl(ide) *a* [i.e. Chl (ide) *a* (F674 E433)] may then be calculated with a formula that can be derived from appropriate simultaneous equations. Since all the tetrapyrroles in the mixture contribute to the Soret excitation amplitudes between 412 and 460 nm. The following equations can be written:

$$\begin{aligned}(\text{F674 E433}) = {} & \text{Chl(ide)}a\ (\text{F674 E433}) + \text{Pheo(bide)}a\ (\text{F674 E433}) \\ & + \text{Chl(ide)}b\ (\text{F674 E433}) + \text{Pheo(bide)}b\ (\text{F674 E433})\end{aligned}$$

$$\begin{aligned}(\text{F674 E4I2}) = {} & k_1\text{Chl(ide)}a\ (\text{F674 E433}) + k_2\text{Pheo(bide)a}\ (\text{F674 E433}) \\ & + k_3\text{Chl(ide)b}\ (\text{F674 E433}) + k_4\text{Pheo(bide)b}\ (\text{F674 E433})\end{aligned}$$

$$\begin{aligned}(\text{F660 E460}) = {} & \text{k}_5\text{Chl(ide)a}\ (\text{F674 E433}) + \text{k}_6\text{Pheo(bide)}a\ (\text{F674 E433}) \\ & + k_7\text{Chl(ide)}b\ (\text{F674 E433}) + k_8\text{Pheo(bide)}b\ (\text{F674 E433})\end{aligned}$$

$$\begin{aligned}(\text{F660 E438}) = {} & k_9\text{Chl(ide)}a\ (\text{F674 E433}) + k_{10}\text{Pheo(bide)}a\ (\text{F674 E433}) \\ & + k_{11}\text{Chl(ide)}b\ (\text{F674 E433}) + k_{12}\text{Pheo(bide)}b\ (\text{F674 E433})\end{aligned}$$

where

$$k_1 = \text{Chl(ide)}a\ (\text{F674 E412})/\text{Chl(ide)}a\ (\text{F674 E433})$$
$$k_2 = \text{Pheo(bide)}a\ (\text{F674 E412})/\text{Pheo(bide)a}\ (\text{F674 E433})$$
$$k_3 = \text{Chl(ide)}b\ (\text{F674 E412})/\text{Chl(ide)}b\ (\text{F674 E433})$$
$$k_4 = \text{Pheo(bide)}b\ (\text{F674 E412})/\text{Pheo(bide)}b\ (\text{F674 E433})$$
$$k_5 = \text{Chl(ide)}a\ (\text{F660 E460})/\text{Chl(ide)}a\ (\text{F674 E433})$$
$$k_6 = \text{Pheo(bide)}a\ (\text{F660 E460})/\text{Pheo(bide)}a\ (\text{F674 E433})$$
$$k_7 = \text{Chl(ide)}b\ (\text{F660 E460})/\text{Chl(ide)}b\ (\text{F674 E433})$$
$$k_8 = \text{Pheo(bide)}b\ (\text{F660 E460})/\text{Pheo(bide)}b\ (\text{F674 E433})$$
$$k_9 = \text{Chl(ide)}a(\text{F660 E438})/\text{Chl(ide)}a(\text{F674 E433})$$
$$k_{10} = \text{Pheo(bide)}a\ (\text{F660 E438})/\text{Pheo(bide)}a\ (\text{F674 E433})$$
$$k_{11} = \text{Chl(ide)}b\ (\text{F660 E438})/\text{ChI(ide)}b\ (\text{F674 E433})$$
$$k_{12} = \text{Pheo(bide)}b\ (\text{F660 E438}/\text{Pheo(bide)}b\ (\text{F674 E433})$$

The numerical values of $k_1$–$k_{12}$ are reported in Table 3.1 which is displayed below.

The $k$ values were calculated with the above k equations, from fluorescence excitation spectra recorded at 674 and 660 nm, respectively, on purified samples of Chl *a*, Chl *b*, Pheo *a* and Pheo *b*. Every $k$ value reported in Table 3.1 is the mean of 15–25 different determinations. By substituting the numerical values of $k_l$–$k_{l2}$ into the four simultaneous equations described above and by solving for Chl(ide)*a* (F674 E433). The following equation was obtained:

$$\text{Chl(ide)}a\ (\text{F674 E433}) = 1.248\ (\text{F674 E433}) - 0.145\ (\text{F674 E412}) - 0.068\ (\text{F660 E460}) - 0.313\ (\text{F660 E438}) \quad (3.33')$$

#### 3.4.1.2 Determination of the Reliability of Eq. (3.33′) and Determination of Its $X_1$ and $Sx_1$ Correction Factors

The reliability of Eq. (3.33′) was tested via the following equation:

$$\text{Chl(ide)}a\ (\text{F674 E433}) = \left[A_1 + (A_1)(X_1)/100\right] \pm (A_1)(S_{X1})/100 \quad (3.33)$$

Where:

$A_1$ = Eq. (3.33′)

$[(A_1)(X_1)/100]$ = are relative concentration-correction factors, which vanish as the X values approach zero. The $X_1$ correction values are listed in Table 3.2 and their derivation and usage is explained below. Finally, the last term in Eq. (3.33), i.e. $(A_1)$

**Table 3.1** Numerical values of the constants utilized in solving the various simultaneous equations

| Equation | Constant | Mean value ± standard deviation |
|---|---|---|
| (1′) | $k_1$ | 0.88 ± 0.02 |
| | $k_2$ | 8.05 ± 0.83 |
| | $k_3$ | 0.42 ± 0.06 |
| | $k_4$ | 0.53 ± 0.01 |
| | $k_5$ | 0.02 ± 0.01 |
| | $k_6$ | 000 |
| | $k_7$ | 5.91 ± 0.58 |
| | $k_8$ | 0.33 ± 0.05 |
| | $k_9$ | 0.38 ± 0.02 |
| | $k_{10}$ | 0.26 ± 0.08 |
| | $k_{11}$ | 2.51 ± 0.17 |
| | $k_{12}$ | 3.66 ± 0.3 |
| (2′) | $k_{13}$ | 1.13 ± 0.03 |
| | $k_{14}$ | 0.18 ± 0.07 |
| | $k_{15}$ | 2.47 ± 0.31 |
| | $k_{16}$ | 1.90 ± 0.14 |
| | $k_{17}$ | 0.00 |
| | $k_{18}$ | 0.00 |
| | $k_{19}$ | 13.87 ± 2.01 |
| | $k_{20}$ | 0.66 ± 0.11 |
| | $k_{21}$ | 0.42 ± 0.03 |
| | $k_{22}$ | 0.00 |
| | $k_{23}$ | 5.95 ± 0.77 |
| | $k_{24}$ | 7.09 ± 0.79 |
| (3′) | $k_{25}$ | 19.39 ± 3.78 |
| | $k_{26}$ | 1.41 ± 0.38 |
| | $k_{27}$ | 0.43 ± 0.02 |
| | $k_{28}$ | 10.85 ± 1.0 |
| | $k_{29}$ | 52.37 ± 10.67 |
| | $k_{30}$ | 2.82 ± 0.73 |
| | $k_{31}$ | 0.18 ± 0.02 |
| | $k_{32}$ | 2.97 ± 0.46 |
| | $k_{33}$ | 46.54 ± 9.45 |
| | $k_{34}$ | 17.9 ± 1.73 |
| | $k_{35}$ | 0.06 ± 0.02 |
| | $k_{36}$ | 1.58 ± 0.31 |
| (4′) | $k_{37}$ | 0.00 |
| | $k_{38}$ | 0.00 |
| | $k_{39}$ | 2.36 ± 0.15 |
| | $k_{40}$ | 0.09 ± 0.01 |
| | $k_{41}$ | 2.73 ± 0.18 |
| | $k_{42}$ | 2.82 ± 0.32 |
| | $k_{43}$ | 0.38 ± 0.06 |
| | $k_{44}$ | 0.27 ± 0.02 |
| | $k_{45}$ | 2.45 ± 0.00 |
| | $k_{46}$ | 29.54 ± 6.63 |

(continued)

**Table 3.1** (continued)

| Equation | Constant | Mean value ± standard deviation |
|---|---|---|
| | $k_{47}$ | 0.16 ± 0.04 |
| | $k_{48}$ | 0.14 ± 0.02 |

The constants were calculated from fluorescence excitation spectra recorded at 674 and 660 nm, respectively, on purified samples of Chl *a*, Chl *b*, Pheo *a* and Pheo *b*. Every constant reported in the Table is the mean of 15–25 different determinations (Adapted from Bazzaz and Rebeiz 1979)

$(Sx_1)/100$ refers to the standard deviation ($Sx_1$) of the determinations expressed as a percentage of the pigment content. The $Sx_1$ values are also listed in Tables 3.2 and their derivation is discussed in the ensuing sections.

Because of the partial overlap of the Soret excitation bands of Chl(ide) *a*, Pheo (bide) a, Chl(ide) *b* and Pheo(bide) *b* (Bazzaz and Rebeiz 1979), it was expected that the uncertainty in determining the Soret excitation amplitude contributed solely by Chl(ide) *a* in a mixture of the four tetrapyrroles would depend on the relative concentration of the pigments in the mixture. For example, if the relative concentration of Chl(ide) *a* in the mixture was high, its determination would be much more reliable than if its relative concentration was low. Thus, it was necessary to test the reliability of Eq. (3.33′) with mixtures of tetrapyrroles containing various proportions of Chl(ide) *a*. However, due to the infinite number of mixtures having different relative concentrations that may be tested, only three ranges of concentrations were investigated. In every case it was made certain that the pigments in the mixture responded linearly to different dilutions. In these mixtures, the relative proportion of Chl(ide) *a* was about 25, 10, and 5 % of the total tetrapyrrole content in the mixture. The relative concentration of the other three tetrapyrroles in the mixtures were always adjusted to equal proportions.

Thus, the reliability of Eq. (3.33′) in calculating $A_1$ i.e.in calculating the Soret excitation amplitude at 433 nm, contributed solely by Chl(ide) *a* in a mixture of Chl (ide) *a*, Pheo(bide) *a*, Chl(ide) *b*, and Pheo(bide) *b*, was tested as follows: 1 ml of HEAR solution containing 198 pmol of Chl *a* was freshly prepared, by dilution from a stock solution of Chl *a* which was monitored by absorption spectrophotometry, and its Soret excitation amplitude in relative fluorescence units, was determined from the Soret excitation spectrum which was recorded at the emission maximum of Chl *a* at 674 nm. These experimentally determined fluorescence excitation amplitudes are reported in Table 3.2 (second row, column 6). A 1 ml HEAR solution containing 27 % (198 pmol) of Chl *a*, 26 %′ (195 pmol), 24 % (178 pmol) and 23 % (170 pmol) of Chl *b*, Pheo *a* and Pheo *b*, respectively, were prepared as described elsewhere (Bazzaz and Rebeiz 1979).Then two Soret excitation spectra recorded on the mixture at an emission wavelength of 674 nm and 660 nm, respectively. Three to four more identical mixtures were prepared and their Soret excitation spectra were also recorded at 674 and 660 nm, respectively. The Soret excitation amplitudes contributed solely by the 198 pmol of Chl *a* in the Five duplicate mixtures of the four tetrapyrroles was then calculated from the Soret excitation spectra with the use of Eq. (3.33′). These values, in arbitrary fluorescence

**Table 3.2** Determination of the reliability of Eq. (3.33′) which was used in the calculation of the Soret excitation amplitude of Chl *a* at E 433 nm in mixtures containing Chl *a*, Chl *b*, Pheo *a* and Pheo *b*

| | Pigments present | | | | Soret excitation fluorescence units at 433 nm | | | |
|---|---|---|---|---|---|---|---|---|
| Amount of pigments | Chl *a* | Chl *b* | Pheo *a* | Pheo *b* | Added | Calculated | Percent error between amount added and amount calculated | Mean percent error ($X_1$) ± standard deviation ($Sx_1$) of the mean per cent error |
| Mixture 1: Picomoles per m/ | 175.0 | 0.0 | 0.0 | 0.0 | 54.1 | 54.4 | +0.6 % | 0.7 % ± 0.6 % |
| Percentage of total pigment | 100.0 % | 0.0 | 0.0 | 0.0 | 53.4 | 53.3 | −0.2 % | |
| | | | | | 55.5 | 56.1 | +1.1 % | |
| | | | | | 54.4 | 55.1 | +1.3 % | |
| | | | | | 57.0 | 57.5 | +0.9 % | |
| Mixture 2: Picomoles per m/ | 198 | 195 | 178 | 170 | 61.6 | 64.3 | +4.4 % | 2.2 % ± 2.8 % |
| Percentage of total pigment | 27 % | 26 % | 24 % | 23 % | 61.6 | 63.2 | +2.6 % | |
| | | | | | 61.6 | 63.9 | +3.7 % | |
| | | | | | 61.6 | 59.9 | −2.8 % | |
| | | | | | 61.6 | 63.4 | +2.9 % | |
| Mixture 3: Picomoles per m/ | 90 | 320 | 240 | 280 | 26.6 | 28.4 | +6.8 % | 13.6 % ± 6.4 % |
| Percentage of total pigment | 10 % | 34 % | 26 % | 30 % | 26.6 | 31.0 | +16.5 % | |
| | | | | | 26.6 | 30.9 | +16.2 % | |
| | | | | | 26.6 | 32.3 | +21.4 % | |
| Mixture 4: Picomoles per m/ | 170 | 820 | 900 | 900 | 52.5 | 60.2 | +14.7 % | 23.9 % ± 11.5 % |
| Percentage of total pigment | 6 % | 29 % | 32 % | 32 % | 52.5 | 72.0 | +37.1 % | |
| | | | | | 52.5 | 68.2 | +29.9 % | |
| | | | | | 52.5 | 59.8 | +13.9 % | |

The Soret excitation spectra were recorded at emission maxima of 674 and 660 nm respectively units, are also reported in Table 3.2 (column 7). The percent error between the actual and calculated Soret excitation amplitudes for Chl *a* in the mixtures were calculated and reported in column 8 of Table 3.2. The mean per cent error (i.e. $X_1$) and the standard deviation of the mean per cent error (i.e.$Sx_1$) were then calculated and reported in column 9 of Table 3.2. The same procedure was applied for mixtures of tetrapyrroles containing only 10 and 6 % of Chl *a*, respectively, and equal amounts of the other three tetrapyrroles. The $X_1$ and $Sx_1$ values were calculated and also reported in column 9 of Table 3.2. Also reported in Table 3.2 (Mixture 1) is the data on the reliability of determining the Soret excitation amplitude of pure solutions of Chl *a* in HEA, in the absence of other added tetrapyrroles.

Altogether the data reported in Table 3.2 indicated that both the mean per cent error ($X_1$) between the amount of pigment added and the amount determined by calculation and the standard deviation of these determinations (Sx,) increased as the proportion of Chl *a* in the mixture was decreased. The systematic variation in the mean per cent error was corrected for in Eq. (3.33) by the $[(A_1)(X_1)/100$factor. It is obvious that this correction becomes negligible for pure solutions of Chl *a* ($X_1 = 0.7$ %) and for mixtures in which the Chl *a* relative concentration amounted to 27 % or more of the total tetrapyrrole content ($X_1 = 2.2$ %). The correction becomes significant, however, for mixtures containing between 10 and 6 % of Chl(ide) *a* (Table 3.2). In all cases the standard deviation ($Sx_1$) of the mean per cent error was reasonable and ranged from ±0.6 % to ±11.5 % of the $A_1$calculated values (Table 3.2).

It is therefore recommended that for precise calculations with Eq. (3.1), the proper values for $X_1$, and $Sx_1$, should be calculated by interpolation from Table 3.2. For less rigorous calculations an overall average value of +10 % for Xx and ±5 % for Sx, may be used.

## 3.4.2 Calculation of Pheo(phorbide) a

### 3.4.2.1 Calculation of Pheophytin and Pheophorbide *a* [Pheo (Phorbide)] *a* in Various Pigment Mixtures

The equation for calculating pheo(phorbide) *a*(F674 E4121] was derived from the following simultaneous equations:

$$\begin{aligned}(\text{F674 E412}) = &\ \text{Chl(ide)}a\ (\text{F674 E412}) + \text{Pheo(bide)}a\ (\text{F674 E412})\\ &+ \text{Chl(ide)}b\ (\text{F674 E412}) + \text{Pheo(bide)}b\ (\text{F674 E412})\end{aligned}$$

$$\begin{aligned}(\text{F674 E433}) = &+ k_{13}\text{Chl(ide)}a\ (\text{F674 E412}) + k_{14}\text{Pheo(bide)}a\ (\text{F674 E412})\\ &+ k_{15}\text{Chl(ide)}b\ (\text{F674 E412}) + k_{16}\text{Pheo(bide)}b\ (\text{F674 E412})\end{aligned}$$

$$\text{F660 E460}) = +k_{17}\text{Chl(ide)}a\ (\text{F674 E412}) + k_{18}\text{Pheo(bide)}a\ (\text{F674 E412}) + k_{19}\text{Chl(ide)}b\ (\text{F674 E412}) + k_{20}\text{Pheo(bide)}b\ (\text{F674 E412})$$

$$\text{F660 E438}) = +k_{21}\text{Chl(ide)}a\ (\text{F674 E412}) + k_{22}\text{Pheo(bide)}a\ (\text{F674 E412}) + k_{23}\text{Chl(ide)}b\ (\text{F674 E412}) + k_{24}\text{Pheo(bide)}b\ (\text{F674 E412})$$

Where

$$k_{13} = \text{Chl(ide)}a\ (\text{F674 E433})/\text{Chl(ide)}a\ (\text{F674 E412})$$
$$k_{14} = \text{Pheo(bide)}a\ (\text{F674 E433})/\text{Pheo(bide)}a\ (\text{F674 E412})$$
$$k_{15} = \text{Chl(ide)}b\ (\text{F674 E433})/\text{Chl(ide)}b\ (\text{F674 E412})$$
$$k_{16} = \text{Pheo(bide)b}\ (\text{F674 E433})/\text{Pheo(bide)b}\ (\text{F674 E412})$$
$$k_{17} = \text{Chl(ide)}a\ (\text{F660 E460})/\text{Chl(ide)}a\ (\text{F674 E412})$$
$$k_{18} = \text{Pheo(bide)}a\ (\text{F660 E460})/\text{Pheo(bide)}a\ (\text{F674 E412})$$
$$k_{19} = \text{Chl(ide)}b\ (\text{F660 E460})/\text{Chl(ide)}b\ (\text{F674 E412})$$
$$k_{20} = \text{Pheo(bide)}b\ (\text{F660 E460})/\text{Pheo(bide)}b\ (\text{F674 E412})$$
$$k_{21} = \text{Chl(ide)a}\ (\text{F660 E438})/\text{Chl(ide)a}\ (\text{F674 E412})$$
$$k_{22} = \text{Pheo(bide)}a\ (\text{F660 E438})/\text{Pheo(bide)}a\ (\text{F674 E412})'$$
$$k_{23} = \text{Chl(ide)}b\ (\text{F660 E438})/\text{Chl(ide)}b\ (\text{F674 E412})$$
$$k_{24} = \text{Pheo(bide)b}\ (\text{F660 E438})/\text{Pheo(bide)b}\ (\text{F674 E412})$$

The numerical values of $k_{13}$–$k_{24}$ are reported in Table 3.1.They were calculated with the above $k$ equations from the same Soret excitation spectra used to calculate $k_1$–$k_{12}$. Every $k$ value reported in Table 3.1 is the mean of 15–25 different determinations.

By substituting the numerical values of $k_{13}$–$k_{24}$ into the four simultaneous equations described above and by solving for Pheo(bide) *a* (F674 E412) the following equation was obtained:

$$\text{Pheo(bide)}a\ (\text{F674 E412}) = 1.198(\text{F674 E412}) - 1.100(\text{F674 E433}) + 0.057(\text{F660 E460}) + 0.110(\text{F660 E438}) \tag{3.34'}$$

#### 3.4.2.2 Determination of the Reliability of Eq. (3.34′) and Determination of Its $X_1$ and $Sx_1$ Correction Factors

The reliability of Eq. (3.34′) was tested via the following equation:

$$\text{Pheo(bide)}a\ (\text{F674 E433}) = [\text{A}_2 + (\text{A}_2)(\text{X}_2)/100]\ (\text{A}_2)(\text{S}_{\text{X2}})/100 \qquad (3.34)$$

Where:

$A_2$ = Eq. (3.34′)

[($A_2$)($X_2$)/100] = are relative concentration-correction factors, which vanish as the X values approach zero. The $X_2$ correction values are listed in Table 3.3 and their derivation and usage is as for Eq. (3.33). Finally, the last term in Eqs. (3.34), i.e. ($A_2$) ($Sx_2$)/100 refers to the standard deviation ($Sx_2$) of the determinations expressed as a percentage of the pigment content. The $Sx_2$ values are also listed in Tables 3.3 and their derivation is as discussed for Eq. (3.33).

The determination of the $X_2$ and $Sx_2$ correction values for various relative concentrations of Pheo(bide) *a* are reported in Table 3.3. The mean per cent error ($X_2$) and standard deviation ($Sx_2$) ranged from 0.0 + 0.8 % for pure solutions of Pheo *a* to −16.8 % ± 2.6 % for mixtures containing 11 % of Pheo a, and −12.9 % ± 6.7 % for mixtures containing 5 % of Pheo *a*.

Based on the results of Table 3.3, it is therefore recommended that for precise calculations with Eq. (3.2), the proper values for $X_2$ and $Sx_2$ should be calculated by interpolation from Table 3.3. For less rigorous calculations an overall average value of −8.5 % for $X_2$ ± 3.7 % for $Sx_2$ may be utilized.

## 3.4.3 Calculation of Chloropyll(ide) b

### 3.4.3.1 Calculation of Chlorophyll and Chlorophyll(ide) *b* [Chl(ide)]*b* in Various Pigment Mixtures

The equation for calculating Chl(ide) *b*(F660 E460] was derived from the following simultaneous equations:

$$\begin{aligned}(\text{F660 E4460}) = {} & \text{Chl(ide)}a\ (\text{F660 E460}) + \text{Pheo(bide)}a\ (\text{F66604 E460}) \\ & + \text{Chl(ide)}b\ (\text{F660 E460}) + \text{Pheo(bide)}b\ (\text{F660 E460})\end{aligned}$$

$$\begin{aligned}\text{F660 E438}) = {} & + k_{25}\text{Chl(ide)}a\ (\text{F660 E460}) + k_{26}\text{Pheo(bide)}a\ (\text{F660 E460}) \\ & + k_{27}\text{Chl(ide)}b\ (\text{F660 E460}) + k_{28}\text{Pheo(bide)}b\ (\text{F660 E460})\end{aligned}$$

$$\begin{aligned}\text{F674 E433}) = {} & + k_{29}\,\text{Chl(ide)}a\ (\text{F660 E460}) + k_{30}\,\text{Pheo(bide)}a\ (\text{F660 E460}) \\ & + k_{31}\text{Chl(ide)}b\ (\text{F660 E460}) + k_{32}\text{Pheo (bide)}b\ (\text{F660 E460})\end{aligned}$$

$$\begin{aligned}\text{F674 E412}) = {} & + k_{33}\text{Chl(ide)}a\ (\text{F660 E60}) + k_{34}\text{Pheo(bide)}a\ (\text{F660 E460}) \\ & + k_{35}\text{Chl(ide)}b\ (\text{F660 E460}) + k_{36}\text{Pheo(bide)}b\ (\text{F660 E460})\end{aligned}$$

**Table 3.3** Determination of the reliability of Eq. (3.34′) which was used in the calculation of the Soret excitation amplitude of Pheo(bide) *a* at E 412 nm in mixtures containing Chl *a*, Chl *b*, Pheo *a* and Pheo *b*

| | Pigments present | | | | Soret excitation fluorescence units at 412 nm | | | |
|---|---|---|---|---|---|---|---|---|
| Amount of pigments | Chl *a* | Chl *b* | Pheo *a* | Pheo *b* | Added | Calculated | Percent error between amount added and amount calculated (%) | Mean percent error ($X_2$) ± standard deviation ($Sx_2$) of the mean percent error |
| Mixture 5: Picomoles per m/ | 0 | 0 | 275 | 0 | 75.9 | 76.6 | +1.0 | 0.0 % ± 0.8 % |
| | | | | | 77.2 | 76.8 | −0.5 | |
| | | | | | 73.6 | 74.0 | +0.5 | |
| Percentage of total pigment | 0 | 0 | 100 % | 0 | 76.2 | 75.5 | −0.9 | |
| | | | | | 76.2 | 76.1 | −0.1 | |
| Mixture 6: Picomoles per m/ | 198 | 195 | 178 | 170 | 54.8 | 56.4 | +2.9 | −4.4 % ± 4.6 % |
| | | | | | 54.8 | 50.9 | −7.1 | |
| Percentage of total pigment | 27 % | 26 % | 24 % | 23 % | 54.8 | 49.9 | −9.1 | |
| | | | | | 54.8 | 51.9 | −5.3 | |
| | | | | | 54.8 | 53.0 | −3.2 | |
| Mixture 7: Picomoles per m/ | 820 | 820 | 314 | 900 | 87.6 | 72.4 | −17.3 | −16.8 % ± 2.6 % |
| | | | | | 87.6 | 70.5 | −19.5 | |
| Percentage of total pigment | 29 % | 29 % | 11 % | 32 % | 87.6 | 72.6 | −17.1 | |
| | | | | | 87.6 | 75.9 | −13.3 | |
| Mixture 8: Picomoles per m/ | 790 | 859 | 116 | 770 | 31.3 | 28.0 | −10.5 | −12.9 % ± 6.7 % |
| | | | | | 31.3 | 24.3 | −22.4 | |
| Percentage of total pigment | 31 % | 34 % | 5 % | 30 % | 31.3 | 29.2 | −6.7 | |
| | | | | | 31.3 | 27.5 | −12.1 | |

The Soret excitation spectra were recorded at an emission maximum of 674 and 660 nm respectively

Where

$$K_{25} = \text{Chl(ide)}a\ (\text{F660 E438})/\text{Chl(ide)}a\ (\text{F660 E460})$$
$$K_{26} = \text{Pheo(bide)}a\ (\text{F660 E438})/\text{Pheo(bide)}a\ (\text{F660 E460})$$
$$K_{27} = \text{Chl(ide)}b\ (\text{F660 E438})/\text{Chl(ide)}b\ (\text{F660 E4I2})$$
$$K_{28} = \text{Pheo(bide) b(F660 E438)/Pheo(bide)b (F660 E460)}$$
$$K_{29} = \text{Chl(ide)}a\ (\text{F674 E433})/\text{Chl(ide)}a\ (\text{F660 E460})$$
$$K_{30} = \text{Pheo(bide)}a\ (\text{F674 E433})/\text{Pheo(bide)}a\ (\text{F660 E460})$$
$$K_{31} = \text{Chl(ide)}b\ (\text{F674 E433})/\text{Chl(ide)}b\ (\text{F660 E460}$$
$$K_{32} = \text{Pheo(bide)}b\ (\text{F674 E433})/\text{Pheo(bide)}b\ (\text{F660 E460})$$
$$K_{33} = \text{Chl(ide)a (F674 E412)/Chl(ide)a (F60 E460)}$$
$$K_{34} = \text{Pheo(bide)}a\ (\text{F674 E412})/\text{Pheo(bide)}a\ (\text{F660 E460})$$
$$K_{35} = \text{Chl(ide)}b\ (\text{F674 E412})/\text{Chl(ide)}b\ (\text{F660 E460})$$
$$K_{36} = \text{Pheo(bide)b (F674 E412)/Pheo(bide)b (F660 E460)}$$

The numerical values of $k_{25}$–$k_{36}$ are reported in Table 3.4.They were calculated with the above *k* equations from the same Soret excitation spectra used to calculate $k_1$–$k_{12}$.

Every *k* value reported in Table 3.4 is the mean of 15–25 different determinations.

By substituting the numerical values of $k_{25}$–$k_{36}$ into the four simultaneous equations described above and by solving for Chl(ide)*b*(F660 E460) the following equation was obtained:

$$\begin{aligned}\text{Chl(ide)}b\ (\text{F660 E460}) = {} & 0.067(\text{F674 E433}) + 1.028(\text{F660 E460}) \\ & + 0.057(\text{F660 E460}) - 0.086(\text{F660 E438}) \\ & - 0.061(\text{F674 E412}) \end{aligned} \qquad (3.35')$$

#### 3.4.3.2 Determination of the Reliability of Eq. (3.35′) and Determination of Its $X_1$ and $Sx_1$ Correction Factors

The reliability of Eq. (3.35′) was tested via the following equation:

$$\text{Chl(ide)}b\ (\text{F660 E460}) = [A_3 + (A_3)(X_3)/100]\ (A_3)(S_{X3})/100 \qquad (3.35)$$

Where:

$A_3$ = Eq. (3.35)

$[(A_3)(X_3)/100]$ = are relative concentration-correction factors, which vanish as the X values approach zero. The $X_3$ correction values are listed in Table 3.4 and their derivation and usage is as for Eq. (3.33). Finally, the last term in Eqs. (3.35), i.e. $(A_3)$ $(Sx_3)/100$ refers to the standard deviation $(Sx_3)$ of the determinations expressed as a percentage of the pigment content. The $Sx_3$ values are also listed in Tables 3.4 and their derivation is as discussed for Eq. (3.33).

The determination of the $X_3$ and $Sx_3$ correction values for various relative concentrations Chl(ide) B are reported in Table 3.4. The mean per cent error $(X_3)$

**Table 3.4** Determination of the reliability of Eq. (3.35′) which was used in the calculation of the Soret excitation amplitude of Ch(lide)) *b* at E 460 nm in mixtures containing Chl *a*, Chl *b*, Pheo *a* and Pheo *b*

| | Pigments present | | | | Soret excitation fluorescence units at 460 nm | | | |
|---|---|---|---|---|---|---|---|---|
| Amount of pigments | Chl *a* | Chl *b* | Pheo *a* | Pheo *b* | Added | Calculated | Percent error between amount added and amount calculated (%) | Mean percent error ($X_2$) ± standard deviation ($Sx_2$) of the mean percent error |
| Mixture 9: picomoles per m/ | 0 | 675 | 0 | 0 | 90.8 | 90.5 | −0.3 | −0.3 % ± 0.1 % |
| | | | | | 89.1 | 88.8 | −0.3 | |
| Percentage of total pigment | 0 | 100 % | 0 | 0 | 88.4 | 88.1 | −0.3 | |
| | | | | | 86.1 | 85.7 | −0.5 | |
| | | | | | 89.1 | 88.9 | −0.2 | |
| Mixture 10: Picomoles per m/ | 198 | 195 | 178 | 170 | 25.6 | 22.7 | −11.3 | −10.0 % ± 2.6 % |
| | | | | | 25.6 | 22.4 | −12.5 | |
| Percentage of total pigment | 27 % | 26 % | 24 % | 23 % | 25.6 | 22.9 | −14.1 | |
| | | | | | 25.6 | 23.5 | −8.2 | |
| | | | | | 25.6 | 23.5 | −8.2 | |
| Mixture 11: Picomoles per m/ | 240 | 121 | 243 | 279 | 16.0 | 12.0 | −24.4 | |
| | | | | | 16.0 | 12.2 | −23.8 | |
| Percentage of total pigment | 27 % | 14 % | 28 % | 32 % | 16.0 | 12.8 | −20.0 | −23.3 % ± 2.2 % |
| | | | | | 16.0 | 12.0 | −25.0 | |
| Mixture 12: Picomoles per m/ | 693 | 160 | 815 | 880 | 21.1 | 13.9 | −34.1 | |
| | | | | | 21.1 | 13.5 | −36.0 | |
| Percentage of total pigment | 27 % | 6 % | 32 % | 35 % | 21.1 | 12.0 | −43.1 | −38.9 % ± 3.9 % |
| | | | | | 21.1 | 12.2 | −42.2 | |
| | | | | | 21.1 | 12.8 | −39.3 | |

The Soret excitation spectra were recorded at a emission maxima of 674 and 660 nm respectively and standard deviation ($Sx_3$) ranged from −0.3 % ± 0.1 % for pure solutions of Chl *b* to −38.9 % ± 3.9 % for mixtures containing 6 % of Chl *b*.

Based on the results of Table 3.4, it is there forem recommended that for precise calculations with Eq. (3.35), the proper values for $X_3$ and $Sx_3$ should be calculated by interpolation from Table 3.4. For less rigorous calculations an overall average value of −18.1 % for $X_3$ ± 2.2 % for $Sx_3$ may be utilized.

## 3.4.4 Calculation of Pheo(phorbide) b

### 3.4.4.1 Calculation of Pheophytin and Pheophorbide *b* [Pheo(phorbide)*b*] in Various Pigment Mixtures

The equation for calculating pheo(phorbide) *b*(F660 E4381] was derived from the following simultaneous equations:

$$\begin{aligned}(\mathrm{F660\ E438}) = {} & \mathrm{Chl(ide)}a\,(\mathrm{F660\ E4438}) + \mathrm{Pheo(bide)}a\,(\mathrm{F660\ E438}) \\ & + \mathrm{Chl(ide)}b\,(\mathrm{F660E438}) + \mathrm{Pheo(bide)}b\,(\mathrm{F660\ E438})\end{aligned}$$

$$\begin{aligned}(\mathrm{F660\ E460}) = {} & + k_{37}\mathrm{Chl(ide)}a\,(\mathrm{F660\ E438}) + k_{38}\mathrm{Pheo(bide)}a\,(\mathrm{F660\ E438}) \\ & + k_{39}\mathrm{Chl(ide)}\,b\,(\mathrm{F660\ E438}) + k_{40}\,\mathrm{Pheo(bide)}b\,(\mathrm{F660\ E438})\end{aligned}$$

$$\begin{aligned}\mathrm{F674\ E433}) = {} & k_{41}\mathrm{Chl(ide)}a\,(\mathrm{F660\ E438}) + k_{42}\mathrm{Pheo(bide)}a\,(\mathrm{F660\ E438}) \\ & + k_{43}\,\mathrm{Chl(ide)}b\,(\mathrm{F660\ E438}) + k_{44}\mathrm{Pheo(bide)}b\,(\mathrm{F660\ E438})\end{aligned}$$

$$\begin{aligned}\mathrm{F674\ E412}) = {} & + k_{45}\mathrm{Chl(ide)}a\,(\mathrm{F660\ E438}) + k_{46}\mathrm{Pheo(bide)}a\,(\mathrm{F660\ E438}) \\ & + k_{47}\mathrm{Chl(ide)}\,b\,(\mathrm{F660\ E438}) + k_{48}\mathrm{Pheo(bide)}b\,(\mathrm{F660\ E438})\end{aligned}$$

Where

$$\begin{aligned}
K_{37} &= \mathrm{Chl(ide)}a\,(\mathrm{F660\ E460})/\mathrm{Chl(ide)}a\,(\mathrm{F660\ E438}) \\
K_{38} &= \mathrm{Pheo(bide)}a\,(\mathrm{F660\ E460})/\mathrm{Pheo(bide)}a\,(\mathrm{F660\ E438}) \\
K_{39} &= \mathrm{Chl(ide)}b\ (\mathrm{F660\ E460})/\mathrm{Chl(ide)}b(\mathrm{F660\ E438}) \\
K_{40} &= \mathrm{Pheo(bide)b(F660\ E460)/Pheo(bide)b\ (F660\ E438)} \\
K_{41} &= \ \mathrm{Chl(ide)}a\,(\mathrm{F674\ E433})/\mathrm{Chl(ide)}a\,(\mathrm{F660\ E438}) \\
K_{42} &= \ \mathrm{Pheo(bide)}a\,(\mathrm{F674\ E433})/\mathrm{Pheo\ (bide)}a\,(\mathrm{F660\ E4438}) \\
K_{43} &= \ \mathrm{Chl(ide)}b\,(\mathrm{F674\ E433})/\mathrm{Chl(ide)}b\,(\mathrm{F660\ E438} \\
K_{44} &= \mathrm{Pheo(bide)}b\,(\mathrm{F674\ E412})/\mathrm{Pheo(bide)}b\,(\mathrm{F660\ E438}) \\
K_{45} &= \mathrm{Chl(ide)a\ \ (F674\ E412)/Chl(ide)a\,(F660\ E438)} \\
K_{46} &= \ \mathrm{Pheo(bide)}a\,(\mathrm{F674\ E412})/\mathrm{Pheo(bide)}a\,(\mathrm{F660\ E438}) \\
K_{47} &= \mathrm{Chl(ide)}b\ (\mathrm{F674\ E412})/\mathrm{Chl(ide)}b\,(\mathrm{F660\ E438}) \\
K_{48} &= \mathrm{Pheo(bide)b\ \ (F674\ E412)/Pheo(bide)b\ \ (F660\ E438)}
\end{aligned}$$

The numerical values of $k_{37}$–$k_{48}$ are reported in Table 3.1.They were calculated with the above $k$ equations from the same Soret excitation spectra used to calculate $k_1$–$k_{12}$.

Every $k$ value reported in Table 3.1 is the mean of 15–25 different determinations.

By substituting the numerical values of $k_{37}$–$k_{48}$ into the four simultaneous equations described above and by solving for Pheo(bide) $b$(F660 E438) the following equation was obtained:

$$\begin{aligned}\text{Pheo(bide)}b\,(\text{F660 E412}) =& 0.001(\text{F674 E412}) + 1.152(\text{F660 E438}) \\ & - 0.423(\text{F674 E433}) - 0.420(\text{F660 E460}) \end{aligned} \quad (3.36')$$

#### 3.4.4.2 Determination of the Reliability of Eq. (3.36′) and Determination of Its $X_1$ and $Sx_1$ Correction Factors

The reliability of Eq. (3.36′) was tested via the following equation:

$$\text{Pheo(bide)}b\ (\text{F660 E438}) = [A_4 + (A_4)(X_2)/100]\ (A_4)(S_{X2})/100 \quad (3.36)$$

Where:

$A_4$ = Eq. (3.36′)

$[(A_4)(X_4)/100]$ = are relative concentration-correction factors, which vanish as the X values approach zero. The $X_4$ correction values are listed in Table 3.5 and their derivation and usage is as for Eq. (3.33). Finally, the last term in Eqs. (3.36), i.e. $(A_4)$ $(Sx_4)/100$ refers to the standard deviation ($Sx_4$) of the determinations expressed as a percentage of the pigment content.

The $Sx_2$ values are also listed in Tables 3.5 and their derivation is as discussed for Eq. (3.33).

The determination of the $X_4$ and $Sx_4$ correction values for various relative concentrations of Pheo(bide) $a$ are reported in Table 3.5. The mean per cent error ($X_4$) and standard deviation ($Sx_2$) ranged from $-1.1\ \% + 0.6\ \%$ for pure solutions of Pheo $b$ to $-14.0\ \% \pm 22.0\ \%$ for mixtures containing 14 % of Pheo$b$. The performance of Eq. (4′) broke down for mixtures containing 6 % or less of Pheo(bide) $b$ and its use is not recommended for these low concentrations of Pheo(bide) $b$.

Based on the results of Table 3.5, it is however recommended that for precise calculations with Eq. (3.4), the proper values for $X_4$ and $Sx_4$ should be calculated by interpolation from Table 3.5.

For less rigorous calculations an overall average value of $-9.5\ \%$ for $X_4 \pm 9.7\ \%$ for $Sx_4$ may be utilized.

**Table 3.5** Determination of the reliability of Eq. (3.36′) which was used in the calculation of the Soret excitation amplitude of Pheo(bide) *b* at E 438 nm in mixtures containing Chl *a*, Chl *b*, Pheo *a* and Pheo *b*

| | Pigments present | | | | Soret excitation fluorescence units at 460 nm | | | |
|---|---|---|---|---|---|---|---|---|
| Amount of pigments | Chl *a* | Chl *b* | Pheo *a* | Pheo *b* | Added | Calculated | Percent error between amount added and amount calculated (%) | Mean percent error ($X_4$) ± standard deviation ($Sx_4$) of the mean per cent error |
| Mixture 13: Picomoles per m/ | 0 | 0 | 0 | 1940 | 94.0 | 93.7 | −0.3 | −1.1 % ± 0.6 % |
| | | | | | 90.0 | 89.2 | −1.0 | |
| Percentage of total pigment | 0 | 0 | 0 | 100 % | 89.1 | 88.1 | −1.1 | |
| | | | | | 88.8 | 87.6 | −1.3 | |
| | | | | | 87.8 | 86.1 | −1.9 | |
| Mixture 14: Picomoles per m/ | 198 | 195 | 178 | 170 | 8.5 | 6.9 | −18.8 | |
| | | | | | 8.5 | 7.6 | −10.6 | −13.5 % ± 6.4 % |
| Percentage of total pigment | 27 % | 26 % | 24 % | 23 % | 8.5 | 8.0 | −5.9 | |
| | | | | | 8.5 | 6.9 | −18.8 | |
| Mixture 15: Picomoles per m/ | 220 | 250 | 277 | 120 | 5.9 | 6.4 | +8.5 | |
| | | | | | 5.9 | 3.6 | −39.0 | |
| Percentage of total pigment | 25 % | 29 % | 32 % | 14 % | 5.9 | 4.4 | −25.4 | −14.0 % ± 22.0 % |
| | | | | | 5.9 | 5.9 | 0.0 | |
| Mixture 16: Picomoles per m/ | 700 | 750 | 810 | 140 | 6.8 | 3.1 | −54.4 | |
| | | | | | 6.8 | 0.1 | −98.5 | |
| Percentage of total pigment | 29 % | 31 % | 34 % | 6 % | 6.8 | 0.9 | −86.7 | −133.5 % ± 78.7 % |
| | | | | | 6.8 | −9.8 | −244.1 | |
| | | | | | 6.8 | −5.7 | −183.8 | |

The Soret excitation spectra were recorded at an emission maximum of 674 and 660 nm respectively

## 3.5 Quantitative Determination of Monovinyl (MV) and Divinyl (DV) Mg-Protoporphyrins (Mg-Protos) by Spectrofluorometry at 77 K

At 77°K electronic spectroscopic methods of analysis are not suited for direct quantitative determinations. Indeed, at low temperatures solutions do not freeze evenly and generate macroscopically non-homogeneous glasses. As a consequence, low-temperature glasses prepared from the same fluorescent solution exhibit wide variations in the magnitude of their fluorescence emission and excitation signals, which in turn depend on the condition of the glass. However, it has been our experience that, in frozen samples of a solution containing MV and DV tetrapyrroles, the ratio of MV to DV fluorescence signals is independent of the condition of the glass. This is probably due to the random microscopic distribution of the MV and DV molecules before and after freezing. It was therefore conjectured that if one determines, by 293°K spectrofluorometry, the total amount of MV + DV tetrapyrroles in a mixture of the two compounds, and determines the MV/DV ratio in the mixture by 77°K spectrofluorometry, then the calculation of the amounts of MV and DV tetrapyrroles in the mixture reduces to simple arithmetic.

The methodology for determining small amounts of tetrapyrroles at 293°K has already been described in Sects. 3.2 and 3.3 (see above). Therefore, what remains to be done is to develop the methodology for determining by 77°K spectrofluorometry the ratio of MV and DV components in various tetrapyrrole mixtures. This in turn reduces to deriving equations that permit the determination, at 77°K, of the net fluorescence signal generated by any MV tetrapyrrole in a MV + DV tetrapyrrole mixture, and of that generated by its DV analog. The ratio of MV to DV fluorescence signals is then computed and that ratio can be readily converted to a ratio of MV/DV tetrapyrrole concentrations by reference to a standard calibration curve. The latter would relate various MV/DV tetrapyrrole concentration ratios to the ratio of their net MV and DV fluorescence signals. The amount of MV and DV tetrapyrroles in the sample is then computed (a) from the total tetrapyrrole concentration which has been determined from the sample at 293°K and (b) from the authentic MV/DV tetrapyrrole ratio for that particular sample, which has been determined from the 77°K spectra.

The derivation of generalized equations for calculating net MV and DV tetrapyrrole fluorescence signals from low-temperature fluorescence spectra, in the absence and presence of other interfering MV and DV tetrapyrroles, is described below.

Next, the quantitative determinations of MV and DV Mg-Protos and MV and DV Pchl(ides) in mixtures of MV and DV Mg-tetrapyrroles, similar to those encountered in extracts of etiolated and greening tissues will then be described.

### 3.5.1 General Equations for the Determination of Net Monovinyl and Divinyl Fluorescence Signals in the Absence of Interference by Other Monovinyl and Divinyl Signals

We have reported in this chapter that when two fluorescent compounds, *X* and *Y*, having different but overlapping fluorescence excitation and emission properties, occur together in a particular sample, the net fluorescence signals generated by compound *X* can be separated from the fluorescence signals generated by compound *Y* by calculation via two unknown simultaneous equations (see above) and (Tripathy and Rebeiz 1985).

For example, let *a* and *b* represent any two fluorescence excitation or emission wavelengths of compound *X*. Likewise let *c* and *d* represent any fluorescence excitation or emission wavelengths of compound *Y*. We have demonstrated earlier in this chapter that the deconvoluted net fluorescence signals, *X(Ea Fb)* and *Y (EC Fd),* generated by compounds X and *Y,* respectively, at the designated wavelengths *a, b, c, d,* can be determined as described by Eq. (3.4), from

$$X(E_a\, F_b) = [(F_a\, F_b) - (E_c\, F_d)/k_2]\ (l/K_1) \tag{3.37}$$

and

$$Y(E_C\, F_d) = [(\mathrm{F_c}\, F_d) - (E_a\, F_b)/k_4]\ (l/K_2) \tag{3.38}$$

where (*E*a *F*b) and (Ec *F*d) represent the fluorescence excitation (*E*) and emission (*F*) amplitudes of the *X* + *Y* mixture, at the *a, b,* and *c, d* wavelengths, respectively, and where

$$\begin{aligned} K_1 &= 1 - (k_1/k_2) \\ K_2 &= 1 - (k_3/k_4) \end{aligned} \tag{3.39}$$

and

$$\begin{aligned} k_1 &= X(E_c\, F_d)/X(E_a\, F_b); k_2 = Y(E_c\, F_d)/Y(E_a\, F_b) \\ k_3 &= Y(E_a\, F_b)/Y(E_a\, F_d); k_4 = X(E_a\, F_b)/X(E_a\, F_d) \end{aligned} \tag{3.40}$$

In what follows we will illustrate the use of these generalized equations for the calculation of the net fluorescence excitation signals of mixtures of MV and DV MPE and of mixtures of MV and DV Pchl(ides).

### 3.5.2 *Calculation of the Net Fluorescence Amplitudes at the Monovinyl and Divinyl Soret Excitation Maxima of Monovinyl and Divinyl Mg-Protoporphyrins in a Mixture of the Two Tetrapyrroles*

The various involved procedures will be discussed below.

#### 3.5.2.1 Choice of the Excitation and Emission Wavelengths That Give the Best Distinction Between Monovinyl and Divinyl signals in a Mixture of Monovinyl and Divinyl Mg-Protoporphyrins

Dicarboxylic, monocarboxylic and fully esterified Mg-Protos, i.e., Mg-Proto, Mpe, and Mg-Proto diester, exhibit nearly identical fluorescence emission and excitation properties. Therefore the equations derived for any MV and DV Mg-Protos pair are valid for the other two MV and DV Mg-Proto pairs.

The adaptation of Eqs. (3.37) and (3.38) to the deconvolution of the MV and DV Mg-Proto signals consists essentially in choosing appropriate *a, c* excitation wavelengths and *b, d* emission wavelengths for Eqs. (3.37) and (3.38). An appropriate choice is one that gives the most precise determination of the net MV and DV Mg-Proto Soret excitation signals. Indeed we have previously demonstrated that MV and DV Mg-Protos exhibit readily distinguishable fluorescence excitation maxima in glasses of diethyl ether at 77°K (Belanger and Rebeiz 1982).Under these conditions, the DV and MV Mg-Proto species occur mainly (95 %) in the pentacoordinated state (Belanger and Rebeiz 1984; Rebeiz and Belanger 1984). The rest occurs in the hexacoordinated state (Belanger and Rebeiz 1984; Rebeiz and Belanger 1984). The pentacoordinated DV Mg-Protos species exhibit a Soret excitation maximum at 424–425 nm and a fluorescence emission maximum at 591 nm (Belanger and Rebeiz 1982). Thus their designation as DV Mg-Protos (E424 F591) On the other hand, the pentacoordinated MV Mg-Proto species exhibit Soret excitation and fluorescence emission maxima at 417 and 589 nm, respectively (Belanger and Rebeiz 1982). Thus their designation as MV Mg-Protos (E417 F589) (Belanger and Rebeiz 1982). An examination of the aforementioned DV and MV Mg-Proto fluorescence parameters ties readily indicated that the most convenient way of differentiating between the MV and DV Mg-Proto signals is by their Soret excitation maxima, which are about 7 nm apart in ether at 77°K. Furthermore, as a consequence of the overlap of the DV and MV Mg-Proto fluorescence bands, we have observed that the best resolution of the MV and DV Soret excitation signals in mixtures of MV and DV Mg-Protos is achieved as follows: (a) by recording the DV excitation spectrum at the emission maximum of DV Mg-Proto, i.e., at 591 nm and (b) by recording the MV excitation spectrum at the short wavelength tail of the emission band, between 587 and 584 nm. In what follow the adaptation of Eqs. (3.37) and (3.38) to the determination of the net fluorescence amplitudes at the Soret excitation maxima of pentacoordinated DV Mg-Proto, and MV Mg-Proto, from two excitation spectra will be described. One of the two spectra is recorded at an emission wavelength of 591 nm, in order to optimize the detection of the DV Mg-Proto Soret excitation maximum at 424–425 nm. The other spectrum is recorded at an emission wavelength of 587 nm, in order to optimize the detection of the MV Mg-Proto Soret excitation maximum at 417 nm.

#### 3.5.2.2 Calculation of the Net Fluorescence Amplitudes at the Monovinyl and Divinyl Soret Excitation Maxima of Monovinyl and Divinyl Mg-Protoporphyrins in a Mixture of the Two Tetrapyrroles

Let the 424-nm Soret excitation amplitude of the MV + DV Mg-Proto mixture, which is recorded at an emission wavelength of 591 nm, be referred to as (E424 F591). Likewise let the 417-nm Soret excitation amplitude of the MV + DV Mg-Proto mixture, which is recorded at an emission wavelength of 587 nm, be referred to as (E417 F587). In addition, let *X and Y* in Eqs. (3.37) and (3.38) represent DV Mg-Proto and MV Mg-Proto, respectively. By substituting (E424 F591) for ($E_a\,F_b$) and (E417F 587) for ($E_c\,Fd$) in Eqs. (3.37) and (3.38), the latters transform into

$$\text{DV Mg}-\text{Proto (E424 F591)} = [(\text{E424 F591}) - (\text{E417 F587})/k_2]\ (1/K_1) \quad (3.41)$$

$$\text{MV Mg}-\text{Proto (E417 F587)} = [(\text{E417 F587}) - (\text{E424 F591 })/k4]\ (1/K2) \quad (3.42)$$

where $K_1$ and $K_2$ are as defined by Eq. (3.39)and Where

$$\begin{aligned} k_1 &= \text{DV Mg Proto (E417F 587)/DV Mg Proto (E424 F591)} \\ k_2 &= \text{MV Mg Proto (E417 F587)/MVMg Proto (E424 F591)} \\ k_3 &= \text{MV Mg Proto (E424 F591)/MVMg Proto (E417 F587)} \\ k_4 &= \text{DV Mg Proto (E424 F591)/DVMg Proto (E417 F587)} \end{aligned} \quad (3.43)$$

in this context, DV Mg-Proto (E417 F587) refers to the magnitude of the Soret excitation maximum of authentic DV Mg-Proto at 417 nm, when the Soret excitation spectrum is recorded at an emission wavelength of 587 nm, etc. . . .

The mean *k* and *K* values of five determinations for the DV and MV Mg-Proto pair amounted to: $k_1 = 0.130 \pm 0.012$; $k_2 = 9.550 \pm 0.450$; $k_3 = 0.103 \pm 0.005$; $k_4 = 7.750 \pm 0.670$; $K_1 = K_2 = 0.987$. By substituting for the values of $k2$, $k_4$, $K_1$

and $K_2$ in Eqs. (3.41) and (3.42) and by further reduction of the substituted equations, the latter assume the following form:

$$\text{DV MPE (E424 F591)} = 1.014\ \text{(E424 F591)} - 0.106\ \text{(E417 F587)} \quad (3.44)$$

and

$$\text{MV MPE (E417 F587)} = 1.013\ \text{(E417 F587)} - 0.131\text{(E424 F591)} \quad (3.45)$$

#### 3.5.2.3 Calculation of Small Proportions of MV Mg-Protoporphyrins in the Presence of Much Larger Proportions of DV Mg-Protoporphyrins and in the Absence of Interference by Other Tetrapyrroles

To detect very small proportions of MV Mg-Protos (<5 %) in the presence of much larger proportions of DV Mg-Protos (>95 %), the shortwavelength excitation spectrum of the $X + Y$ mixture is best recorded at an emission wavelength of 584 nm instead of at 587 nm. In this case the (E417 F587) Soret excitation amplitudes in Eqs. (3.41), (3.42), (3.43), (3.44) and (3.45) are replaced by (E417F 584) Soret excitation amplitudes. The latter further optimize the detection of very small proportions of MV Mg- Protos. As a consequence Eqs. (3.41) and (3.42) are transformed to:

$$\text{DV MPE (E424 F591)} = 1.010\ \text{(E424 F591)} - 0.439\ \text{(E417 F584)} \quad (3.46)$$

and

$$\text{MV MPE (E417 F584)} = 1.010\ \text{(E417 F584)} - 0.023\ \text{(E424 F591)} \quad (3.47)$$

#### 3.5.2.4 Conversion of the MV and DV Mg Protoporphyrins Soret Excitation Ratios to MV and DV Mg-Protoporphyrins Concentrations

As was just pointed out, Eqs. (3.44) and (3.45)or (3.46) and (3.47) allow the calculation of the net Soret excitation amplitudes of the MV and DV Mg-Protos components in mixtures of these two tetrapyrroles. Although the absolute values of the two Soret excitation magnitudes depend on the condition of the frozen sample (vide supra), the relative amplitudes of the MV and DV Mg-Proto signals, for any one sample, are independent of the frozen condition of that sample. This is because the excitation spectra which are needed for solving Eqs. (3.44) (3.45), (3.46), (3.47) and (3.47) are recorded on the same frozen sample. As a consequence the ratio of the MV to DV Soret excitation amplitudes which can be readily computed for any given sample, from the relative amplitudes of the net MV and DV Soret excitation signals (vide supra), is in turn independent of the frozen condition of the sample.

The ratio of the MV net Soret excitation amplitude at 417 nm to the DV net Soret excitation amplitude at 424 nm is not necessarily identical, to the authentic ratio of the amounts of MV to DV Mg-Protos in the mixture. This is because the MV and DV Soret excitation magnitudes of the MV and DV components depend (a) on the fluorescence quantum yield of the MV and DV MgProto components, which may not be necessarily identical, and (b) on the fluorescence emission wavelength at which the MV and DV excitation spectra are recorded.

However, once the MV to DV Mg-Proto net Soret excitation ratio has been determined it can be readily converted to an authentic ratio of MV to DV Mg-Proto concentrations, by reference to a standard calibration curve. The latter is constructed (a) by mixing authentic MV Mg-Proto and DV Mg-Proto in known proportions, (b) by recording the two required 77°K excitation spectra on every mixture, (c) by calculating the ratio of the net MV to the net DV Soret excitation amplitudes for every mixture with the help of Eqs. (3.44) and (3.45) or (3.46) and (3.47), and finally (d) by plotting the authentic MV/DV Mg-Proto concentration ratios, on the abscissa for example, against the calculated MV/DV ratio of the net Soret excitation amplitudes on the ordinate. Under our instrumental conditions, such a plot of authentic concentration ratios of MV/DV MPE against the corresponding net Soret excitation ratios of MV MPE (E417/587)/DV MPE (E424/591), yielded a straight curve that passed through the origin and that exhibited a slope of 1.12 and a coefficient of correlation, r, of 0.998. The concentration of MV MPE and DV MPE in the mixture was then calculated from the total amount of MV + DV MPE, which can be readily determined at 293°K for any MV + DV MPE mixture (see Sect. 3.5.2.1), and from the slope of the calibration curve (vide infra).

The reliability of Eqs. (3.44) and (3.45) in determining the proportions of MV and DV Mg Proto Monoester in mixtures of the two tetrapyrroles amounted to $-0.8 \pm 6$ % and to $1.85 \pm 4.8$ %, respectively (Table 3.6).

#### 3.5.2.5 Sample Calculation of the Amounts of MV Mg-Proto Monoester and DV Mg-Proto Monoester Present in a Mixture of the Two Tetrapyrrole

1. The sample of MV and DV MPE, under consideration, was free of Zn-Proto contamination. The total amount of MV + DV Mg-Proto monoester in the sample was determined at 293°K in hexane-extracted acetone from the area under the fluorescence emission band between 580 and 610 nm, as previously described (Rebeiz et al. 1975a; Smith and Rebeiz 1977a). It amounted to 20 pmol of MV + DV Mpe/ml of solution.

**Table 3.6** Determination of the reliability of Eqs. (3.44) and (3.45) used to calculate excitation amplitudes of DV and MV Mpe at 424 and 417 respectively

| Amount of MPE added (pmol/ml) | | Amount of MPE calcd (pmol/ml) | | Percentage error between amount of MPE added and calcd (%) | | Mean percentage error ± SD | |
|---|---|---|---|---|---|---|---|
| MV MPE | DV MPE | MV MPE | DV MPE | MV MPE | DV MPE | MV MPE | DV MPE |
| 2 | 18 | 2.1 | 17.9 | 5.00 | −0.55 | −0.8 % ± 6 % | 1.85 % ± 4.8 % |
| 4 | 16 | 3.3 | 16.7 | −17.51 | 4.3 | | |
| 6 | 14 | 6.8 | 13.2 | 13.3 | −5.7 | | |
| 8 | 12 | 7.5 | 12.5 | −6.25 | 4.00 | | |
| 10 | 10 | 10.2 | 9.8 | 2.00 | −2.00 | | |
| 12 | 8 | 12 | 8 | 0 | 0 | | |
| 14 | 6 | 13.6 | 6.4 | −2.85 | 6.80 | | |
| 16 | 4 | 16 | 4 | 0 | 0 | | |
| 18 | 2 | 17.8 | 2.2 | −1.00 | 10.00 | | |

Note: The excitation spectra were recorded at emission maxima of 591 and 587 nm. Other experimental details areas under materials and methods

2. The MV + DV MPE were next transferred to diethyl ether and two Soret excitation spectra were recorded on the mixture at77°K. One spectrum was recorded at the emission maximum of the DV Mpe component at 591 nm and the other was recorded on the shortwave length tail of the MV Mpe emission band at 587 nm.
3. The nondeconvoluted Soret excitation amplitudes at 417 and 424 nm were then determined in relative fluorescence units, from the (E417 F587) and (E424 F591) Soret excitation spectra. The (E417) and (E424) relative Soret excitation amplitudes amounted to 9.39 and 6.06 fluorescence units, respectively.
4. The deconvoluted net Soret excitation amplitudes of the DV and MV Mpe components were next calculated from the aforementioned (E417) and (E424) Soret excitation amplitudes with the help of Eqs. (3.44) and (3.45), respectively. The deconvoluted DV MPE (E424) net Soret excitation amplitude amounted to 5.06 fluorescence units. That of MV Mpe (E417) amounted to 8.60 relative fluorescence units.
5. The apparent ratio of the MV (E417)to DV (E424) net Soret excitation amplitudes of the mixture amounted to 8.60/5.06 = 1.69.
6. The true ratio, $R$, of the MV MPE to DV Mpe concentrations in the mixture was calculated from the apparent MV (E417)/DV (E424) ratio of 1.69 and from the inverse of the slope (1/1.12) of the calibration curve as follows: $R = (1.69)(1/1.12) = (1.51/1)$. This, in turn, corresponded to a MV MPE percentage of $(1.51/1 + 1.51) \times 100 = 60\,\%$, and to a DV MPE percentage of $100{-}60 = 40\,\%$.
7. Since the sample contained a total amount of 20 pmol/ml of MV + DV MPE, the concentration of MV MPE amounted to$(20/100) \times 60 = 12$ pmol/ml. That of DV Mpe amounted to $20{-}12 = 8$ pmol/ml

## 3.6 Quantitative Determination of Monovinyl (MV) and Divinyl (DV) Protochlorophyllides (Pchlides) by Spectrofluorometry at 77 K

The determinations of DV and MV Pchlides in the absence and presence of other tetrapyrroles are described below.

### *3.6.1 Generalized Equations for the Determination of the Net Monovinyl and Divinyl Fluorescence Signals of a Particular Tetrapyrrole Pair in the Presence of a Third Interfering Tetrapyrrole*

Such a situation arises when the determination of the amount of MV and DV Pchl (ides) is attempted in the presence of large amounts of Mg-Protos. This condition is encountered under certain in vivo (Rebeiz et al. 1984b) and in vitro (Daniell and Rebeiz 1984; Rebeiz et al. 1984a) conditions. The problem arises because Mg-Protos exhibit a pronounced and broad vibrational emission band, which in diethyl ether at 77°K extends from about 618 to 660 nm (Belanger and Rebeiz 1982). As a consequence, when Pchl(ide) Soret excitation spectra are recorded at the emission maximum of Pchlide, i.e., at 625 nm or at a shorter wavelength, at 618 nm for example, the Pchl(ide) Soret excitation bands overlap with the DV Mg-Protos Soret excitation band between 430 and 450 nm(Belanger and Rebeiz 1982). Therefore in order to calculate the net Soret excitation amplitudes of MV and DV Pchl(ide), the contribution of the DV Mg-Protos Soret excitation amplitudes to the Pchlide Soret excitation bands must be eliminated. The reverse is not true, however, since the Mg-Protos Soret excitation bands, which are recorded at an emission wavelength between 591 and 584 nm, do not overlap with those of MV and DV Pchl(ides). This is because MV and DV Pchl(ides) do not exhibit any significant fluorescence between 584 and 591 nm.

The deconvolution of the net fluorescence signals Ea and Fb or Ec and Fd of two different compounds *X* and *Y*, from the fluorescence signals, Eg and Fh, of a third interfering compound, Z, has already been discussed above (see Sect. 3.2). It was then pointed out that the net fluorescence signals generated by any one of the three aforementioned compounds can be separated from the fluorescence signals generated by the other two compounds with the use of three unknown simultaneous equations. The latter are of the form

$$X\ (E_a\ F_b) = (E_a\ F_b) - Q_1\ \ (Ec\ F_d) - Q2\ (E_g\ F_h)/Q_3 \qquad (3.48)$$

and

$$Y\ (E_cF_d) = (E_cF_d) - Q_4\ (E_a\ \ F_b) - Q_5\ (E_g\ F_h)/Q_6 \tag{3.49}$$

Where

$$\begin{aligned} &Q_1 = 1/k_{10} - K_6K_8/K_7; && Q_2 = K_6/K_7; \\ &Q_3 = K_5 + K_6K_9/K_7; && Q_4 = 1/k_{16} - K_{11}K_{13}/K_{12} \\ &Q_5 = K_{11}/K_{12}; && Q_6 = K_{10} + K_{11}K_{14}/K_{12} \end{aligned} \tag{3.50}$$

and

$$\begin{aligned} &K_5 = \ 1 - k_9/k_{10}; && K_6 = 1 - k_{11}/k_{10}; \\ &K_7 = k_{14} - k_{13}k_{11}/k_{10}; && K_8 = k_{13}/k_{10} \\ &K_9 = (k_{13}k_9/k) - k_{12}; && K_{10} = 1 - k_{15}/k_{16}; \\ &K_{11} = 1 - k_{17}/k_{16}; && K_{12} = k_{20} - k_{19}k_{17}/k_{16}; \\ &K_{13} = \ k_{19}/k_{16}; && K_{14} = (k_{19}k_{15}/k_{16}) - k_{18} \end{aligned} \tag{3.51}$$

and,

$$\begin{aligned} &k_9 = X(E_cF_d)/X(E_aF_b); && k_{10} = Y(E_cF_d)/Y(E_aF_b); \\ &k_{11} = Z(E_cF_d)/Z(E_aF_b); && k_{12} = X(E_gF_h)/X(E_aF_b); \\ &k_{13} = Y(E_gF_h)/Y(E_aF_b); && k_{14} = Z(E_gF_h)/Z(E_aF_b); \\ &k_{15} = Y(E_aF_b)/Y(E_cF_d); && k_{16} = X(E_aF_b)/X(E_cF_d); \\ &k_{17} = Z(E_aF_b)/Z(E_cF_d); && k_{18} = Y(E_gF_h)/Y(E_cF_d); \\ &k_{19} = X(E_gF_h)/X(E_cF_d); && k_{20} = Z(E_gF_h)/Z(E_cF_d); \end{aligned} \tag{3.52}$$

In this context, (Ea *Fb*, *(Ec Fd)*, and *(Eg Fh)* represent the fluorescence excitation or fluorescence emission amplitudes of the *X*, *Y*, *Z* mixture, at the $E_a$ or *Fb*, *Ec* or *Fd*, and *Eg* or *Fh* wavelengths, respectively

### *3.6.2 Calculation of the Amounts of MV and DV Protochlorophyll(ides) in a Mixture of These Two Compounds, and in the Absence of Interference by Other Tetrapyrroles*

This case can also be solved by appropriate adaptation of Eqs. (3.37) and (3.38). The difference between the Pchl(ide) and the Mg-Proto equations lies in an appropriate choice of the excitation and emission wavelengths, which in turn influences the values of the $k2$, $k_4$, $K_1$, and $K_2$ constants in Eqs. (3.37) and (3.38). As expected, the choice of the appropriate Pchl(ide) *a, c* excitation wavelengths and of the *b, d* emission wavelengths is dictated by the fluorescence emission and excitation characteristics of MV and DV Pchl(ide) in ether at 77°K. At this temperature (a) 90 % of the Pchl(ide) pool occurs in the pentacoordinated state, while the remaining 10 % occurs in the hexacoordinated state (Belanger and Rebeiz 1984), (b) the B (0–0) Soret excitation electronic transitions of the pentacoordinated MV and DV Pchl(ide) are very clearly split into shorter wavelength By (0–0) and longer wavelength Bx (0–0) components (Rebeiz and Lascelles 1982), (c) although pentacoordinated MV Pchl(ide) (E437/F625) and pentacoordinated DV Pchl(ide) (E443/F625) possess very similar fluorescence emission maxima at 624 and625 nm, respectively, they exhibit different By (0–0) Soret excitation maxima, E, at 437and 443 nm respectively. The corresponding Bx (0–0) transitions of pentacoordinated MV and DV Pchl(ide) exhibit excitation maxima at 443 and 451 nm respectively (13). As a consequence of these observations and because of the forementioned Soret excitation overlap between the pentacoordinated MV Pchl(ide) Bx (0–0) and the DV By (0–0) transitions at 443 nm the following strategy was adopted. It was decided to discriminate between the pentacoordinated MV and DV Pchl(ides) in Eqs. (3.37) and (3.38), via the MV Pchlide By (0–0) Soret excitation maximum at 437 nm and via the DV Pchlide Bx (0–0) Soret excitation maximum at 451 nm, respectively.

Therefore in Eqs. (3.37) and (3.38) $X$ was considered to represent the deconvoluted net Soret excitation amplitude of pentacoordinated MV Pchl(ide) By (0–0) (E437/F625) at437 nm. Likewise $Y$ was considered to represent the Soret excitation amplitude of pentacoordinated DV Phclide Bx (0–0) ($E451F625$) at 451 nm. Therefore in this context ($EaFb$) refers to ($E437F625$) and represents the Soret excitation amplitude at 437 nm, when the excitation spectrum of a MV + DV Pchl(ide) mixture is recorded at an emission wavelength of 625 nm. Likewise ($Ec\ Fd$) refers to ($E451F625$) and represents the Soret excitation amplitude of the MV + DV Pchl(ide) mixture at 451 nm, when the excitation spectrum of the mixture is recorded at an emission wavelength of 625 nm.

With the forementioned assignment, Eqs. (3.37) and (3.38) transform into:

$$\text{MV Pchl(ide) (E437 F625)} = [\text{(E437 F625)} - \text{(E451 F625)}/k_6]\ (1/\text{K}_3) \quad (3.53)$$

$$\text{DV Pchl(ide) (E451F625)} = [\text{(E451 F625)} - \text{(E437 F625)}/k8]\ (1/K_4) \quad (3.54)$$

and Eq. (3.40) becomes

$$\begin{aligned}
k_5 &= \text{MV Pchl(ide) (E451F625)/MV Pchl(ide) (E437F625)} \\
k_6 &= \text{DV Pchl(ide) (E451F625)/DV Pchl(ide) (E437F625)} \\
k_7 &= \text{DV Pchl(ide) (E437 F625)/DV Pchl(ide) (E451 F625)} \\
k_8 &= \text{MV Pchl(ide) (E437F625)/MV Pchl(ide) (E451F625)}
\end{aligned} \quad (3.55)$$

In this context, MV Pchl(ide) (E451F625) refers to the magnitude of the Soret excitation maximum of authentic MV Pchl(ide) at 451 nm, when its Soret excitation spectrum is recorded at an emission wavelength of 625 nm, etc. . . .

The mean "*k*" values of five determinations for the MV and DV Pchl(ide) pair amounted to $k_5 = 0.063 \pm 0.004$; $k_6 = 1.100 \pm 0.07$; $k_7 = 0.902 \pm 0.030$; $k_8 = 15.68 \pm 1.000$. The "*K*" values were calculated from Eqs. (3.39) and amounted to $K_3 = 0.943$ and $K_4 = 0.942$. By substituting for the values of $k_6$, $k_8$, $K_3$ and $K_4$ in Eqs. (3.53) and (3.54) and by further reduction of the substituted equations, the latter transform into

$$\text{MV Pchl(ide) (E437F625)} = 1.060\ \text{(E437F625)} - 0.964\ \text{(E451F625)} \quad (3.55a)$$

$$\text{DV Pchl(ide) (E451F625)} = 1.061\ \text{(E451F625)} - 0.068\ \text{(E437F625)} \quad (3.56)$$

### 3.6.3 *Calculation of Small Proportions of MV Protochlorophyll(ide) in the Presence of Much Larger Proportions of DV Protochlorophyll(ide) in the Absence of Interference by Other Tetrapyrroles*

To calculate small proportions of MV Pchl(ide) (<5 %) in the presence of larger proportions of DV Pchl(ide) (>95 %), the Soret excitation spectrum of the MV and DV Pchl(ide) mixture is best recorded at 618 nm. This emission wavelength is at the short-wavelength tail of the MV Pchl(ide) emission band. Because of slight differences in the overlap of the MV and DV Pchl(ide) emission bands, the detectability of small amounts of MV Pchlide in the presence of much larger amounts of DV Pchlide at this emission wavelength is improved considerably. Under these circumstances, the (*E*437*F*625) and (E451F625) Soret excitation amplitudes in Eqs. (3.53), (3.54) and (3.55), are replaced by (E436F618) and (E450F618) Soret excitation values. As a consequence, Eqs. (3.53) and (3.54) are transformed into

$$\text{MV Pchl(ide) (E436F618)} = 1.023\ \text{(E436F618)} - 0.843\ \text{(E450F618)} \quad (3.57)$$

$$\text{DV Pchl(ide) (E450F618)} = 1.023\ \text{(E450F618)} - 0.029\ \text{(E436F618)} \quad (3.58)$$

**Table 3.7** Determination of the reliability of equations 3.57 and 3.58 used to calculate the excitation amplitude of MV Pchl(ide) and DV Pchl(ide) at 437 and 451 nm, respectively

| Amount of Pchlide added (pmol/ml) | | Amount of Pchlide (pmol/ml) | | Percentage error between amount of Pchlide added and calcd (%) | | Mean percentage error ± SD | |
|---|---|---|---|---|---|---|---|
| MV Pchlide | DV Pchlide | MV Pchlide | DV Pchlide | MV Pchlide | DV Pchlide | MV Pchlide | DV Pchlide |
| 4 | 16 | 4.2 | 15.8 | −5 | 1.25 | −2.24% ± 3.97% | 0.87 ± 6.9% |
| 10 | 10 | 9.6 | 10.4 | −4 | 4 | | |
| 12 | 8 | 11.1 | 8.9 | −7.5 | 11.25 | | |
| 14 | 6 | 14.1 | 5.9 | 0.7 | −1.6 | | |
| 16 | 4 | 16.4 | 3.6 | 2.5 | −10 | | |
| 18 | 2 | 18 | 2 | 0 | 0 | | |

*Note:* The excitation spectra were recorded at an emission of maximum of 625 nm. Other experimental details are as in materials and methods

### 3.6.4 Calculation of the Amounts of Monovinyl and Divinyl Protochlorophyll(ides) in the Presence of DV Mg-Protos in a Mixture of the Three Tetrapyrroles

This situation can be resolved by appropriate adaptation of Eqs. (3.48)] and (3.49). In this case there is no reason to change the wavelength assignment of ($E_a\ F_b$) and ($E_cF_d$) which was adopted in Eqs. (3.53) and (3.54) for MV Pchl(ides) and DV Pchl (ides), respectively. As may be recalled, in these equations, $X\ (E_a\ F_b)$ referred to the deconvoluted net Soret excitation amplitude of the pentacoordinated MV Pchl(ides) By (0–0) (E437F625) transition at 437 nm and $Y\ (E_c\ F_d)$ referred to that of the DV Pchl(ides) Bx (0–0) (E451F625) transition at 451 nm. Therefore, in this context, ($E_aF_b$) refers to (E437F625) and represents the Soret excitation amplitude at 437 nm, when the excitation spectrum of the tetrapyrrole mixture containing MV + DV Pchl(ides) and DV Mpe is recorded at an emission wavelength of 625 nm. Likewise ($E_c\ F_d$) represents (*E*451*F*625) and refers in the same *F*625-nm excitation spectrum to the Soret excitation amplitude of the mixture at 451 nm. On the other hand the (*Eg Fh*) terms in Eqs. (3.48), (3.49), (3.50), (3.51) and (3.52) refer to (*E*424*F*625) and represent the Soret excitation amplitude at 424 nm in the *F*625-nm excitation spectrum (Table 3.7).

With the aforementioned assignments, Eqs. (3.48) and (3.49) transform into:

$$\text{MV Pchl(ides) (E437F625)} = [(\text{E437F625}) - Q_1\ (\text{E451F625}) - Q2\ (\text{E424F625})]/Q_3 \quad (3.59)$$

$$\text{DV Pchl(ides) (E451F625)} = [(\text{E451F625}) - Q4\ (\text{E437F625}) - Q_5(\text{E424 F625})]/Q_6 \quad (3.60)$$

Likewise Eqs. (3.52) transform into:

$$k_9 = \text{MV Pchl(ide) (E451 F625)/MV Pchl(ide) (E437 F625)}$$
$$k_{10} = \text{DV Pchl(ide) (E451 F625)/DV Pchl(ide) (E437 F625)}$$
$$k_{11} = \text{DV Mpe (E451 F625)/DV Mpe (E437 F625)}$$
$$k_{12} = \text{MV Pchl(ide) (E424 F625)/MV Pchl(ide) (E437 F625)}$$
$$k_{13} = \text{DV Pchl(ide) (E424 F625)/DV Pchl(ide) (E437 F625)}$$
$$k_{14} = \text{DV Mpe (E437 F625)/DV Mpe (E437 F625)}$$
$$k_{15} = \text{DV Pchl(ide) (E437 F625)/DV Pchl(ide) (E451 F625)}$$
$$k_{16} = \text{MV Pchl(ide) (E437 F625)/MV Pchl(ide) (E451 F625)}$$
$$k_{17} = \text{DV Mpe (E437 F625)/DV Mpe (E451 F625)}$$
$$k_{18} = \text{DV Pchl(ide) (E424 F625)/DV Pchl(ide) (E451 F625)}$$
$$k_{19} = \text{MV Pchl(ide) (E424 F625)/MV Pchl(ide) (E451 F625)}$$
$$k_{20} = \text{DV Mpe (E437 F625)/DV Mpe (E451 F625)}$$

According to the above terminology, MV Pchl(ides) (E451F625) is as defined for Eqs. (3.55). Divinyl MPE (E451F625) refers to the amplitude of the Soret excitation maximum of authentic DV Mpe at 451 nm, when its Soret excitation spectrum is recorded at an emission wavelength of 625 nm, etc.. The mean "*k*" values of five determinations for authentic MV and DV Pchlide and for authentic DV Mpe are reported in Table 3.8. Likewise the $K_5$ to $K_{14}$ values as calculated from Eqs. (3.51) are also reported in Table 3.8. The $Q_1$ to $Q_6$ values were calculated from Eqs. (3.50) and are also reported in Table 3.8. By substituting for the $Q_1$ to $Q_6$ values in Eqs. (3.59) and (3.60) and by further reduction of the substituted equations, the latter transform into:

$$\text{MV Pchl(ides) (E437F625)} = 1.093\ \text{(E437 F625)} - 0.624\ \text{(E451 F625)} - 0.217\ \text{(E424 F625)} \quad (3.60a)$$

$$\text{DV Pchl(ides) (E451 F625)} = 1.070\ \text{(E451 F625)} - 0.020\ \text{(E437 F625)} - 0.163\ \text{(E424 F625)} \quad (3.61)$$

The MV Pchl(ides) (E437)/(F625) and DV Pchl(ides) E451 F625 net Soret excitation ratios, which are calculated from Eqs. (3.60) and (3.61), are converted to authentic ratios of MV to DV Pchl(ides) concentrations by reference to a standard calibration curve. The latter is constructed exactly as described for Eqs. (3.55) and (3.56). However, in this particular case the various mixtures of authentic MV Pchlide + DV Pchlide were adjusted to a total of 40 pmol/ml. Furthermore, in all the tetrapyrrole mixtures, the ratio of DV Mpe + DV Pchlides was maintained at a molar ratio of 1:1. The curve which related the authentic MV/DV Pchlide concentration ratios (on the abscissa) to the calculated MV/DV

**Table 3.8** Numerical values of constants used in solving 3.21–3.26

| Constant determined from excitation spectra acc. to Eq. (3.26) | Mean value of constant ± SD | Constant calcd acc. to Eqs. (3.21) and (3.22) | Value of constant |
|---|---|---|---|
| $k_9$ | 0.060 ± 0.006 | $K_5$ | 0.961 |
| $k_{10}$ | 1.555 ± 0.074 | $K_6$ | 0.659 |
| $k_{11}$ | 0.530 ± 0.037 | $K_7$ | 3.326 |
| $K_{12}$ | 0.255 ± 0.002 | $K_8$ | 0.348 |
| $K_{13}$ | 0.541 ± 0.034 | $K_9$ | −0.234 |
| $K_{14}$ | 3.510 ± 0.113 | $K_{10}$ | 0.962 |
| $K_{15}$ | 0.644 ± 0.029 | $K_{11}$ | 0.888 |
| $K_{16}$ | 16.98 ± 1.580 | $K_{12}$ | 5.805 |
| $K_{17}$ | 1.896 ± 0.130 | $K_{13}$ | 0.260 |
| $K_{18}$ | 0.348 ± 0.022 | $K_{14}$ | −0.180 |
| $K_{19}$ | 4.430 ± 0.410 | $Q_1$ | 0.571 |
| $K_{20}$ | 6.300 ± 0.326 | $Q_2$ | 0.198 |
| | | $Q_3$ | 0.914 |
| | | $Q_4$ | 0.019 |
| | | $Q_5$ | 0.153 |
| | | $Q_6$ | 0.934 |

Note: The constants were calculated from fluorescence excitation spectra recorded at 625 samples of DV MPE, MV Pchlide, and DV Pchlide. Every constant reported in the table is the mean of five to seven different determinations

Pchlide ratios of the net Soret excitation amplitudes (on the ordinate), was a straight line that passed through the origin. It exhibited a slope of 0.900 and an index of correlation of 0.998.

The reliability of Eqs. (3.60) and (3.61) in determining the proportions of MV and DV Pchlides in the presence of DV Mpe amounted to 1.24 ± 3.7 % and to 0.32 ± 1.9 %, respectively. Equations (3.60) and (3.61) gave the same calculated values for MV and DV Pchlide in the presence and in the absence of various proportions of DV Mpe.

### 3.6.5 *Calculation of Small Proportions of Monovinyl Protochlorophyll(ide) in the Presence of Much Larger Proportions of Divinyl Protochlorophyll(ide) and in the Presence of Divinyl Mg-Protoporphyrins*

To calculate small proportions of MV Pchl(ides) (<5 %) in the presence of larger proportions of DV Pchl(ides) (>95 %) and in the presence of DV Mpe, the Soret excitation spectrum of the tetrapyrrole mixtures is recorded at an emission wavelength of 618 nm as was described for Eqs. (3.57) and (3.58). As a consequence the (*E*437*F*625), (E451F625), and (E424F625) Soret excitation amplitudes in Eqs. (3.24) and (3.25) are replaced by (E436 F618), (E450 F618), and (E424 F618) Soret excitation amplitudes. This in turn transforms Eqs. (3.59) and (3.60) into:

$$\text{MV Pchl(ide) (E436 F618)} = 1.046\ \text{(E436 F618)} - 0.829\ \text{(E450F/618)} - 0.099\ \text{(E424 F618)} \quad (3.62)$$

$$\text{DV Pchl(ide) (E450 F618)} = 1.033\ \text{(E450 F618)} - 0.01\ \text{(E436 F618)} - 0.079\ (\text{E424}\ F\text{618}) \quad (3.63)$$

The calibration curve which relates authentic MV/DV Pchlide concentration ratios (abscissa) to the calculated MV/DV Pchlide ratios of the net Soret excitation amplitudes (ordinate) was a straight line that passed through the origin. It exhibited a slope of 2.100 and a correlation coefficient of 0.990.

### 3.6.6 Sample Calculation of the Amount of Monovinyl and Divinyl Protochlorophyllides in a Tetrapyrrole Mixture Containing Divinyl Mg-Protoporphyrin Monoester

1. Authentic MV Pchlide, DV Pchlide, and DV Mpe were dissolved together in hexane-extracted acetone. The total amount of MV + DV Pchlide and of Mpe in the solution was determined by spectrofluorometry at 293°K (see Sects. 3.2 and 3.3) and amounted to 40 pmol of MV + DV Pchlides and 40 pmol of DV Mpe/ml of solution.
2. The tetrapyrrole mixture was transferred to diethyl ether, and one Soret excitation spectrum was recorded on the mixture at77°K, at an emission wavelength of 625 nm.
3. The non-deconvoluted Soret excitation amplitudes at (E451), (E437), and (E424) nm were the next determined, in relative fluorescence units from the *F*625 nm excitation spectrum. The Soret excitation amplitudes at 451, 437, and 424 nm of the tetrapyrrole mixture amounted to 3.53, 8.02, and 4.03 relative fluorescence units, respectively.
4. The deconvoluted net Soret excitation amplitudes of MV and DV Pchlide were calculated from the aforementioned (E451), (E437), and (E424) Soret excitation amplitudes with the help of Eqs. (3.60) and (3.61). The deconvoluted MV Pchlide (*E*437) net Soret excitation amplitude amounted to 5.69 relative fluorescence units. That of DV Pchlide (*E*451) amounted to 2.96 relative fluorescence units.
5. The apparent ratio of the MV (E437)to DV (E451) net Soret excitation amplitudes amounted to 5.69/2.96 = 1.92/1.
6. The true ratio of the MV Pchlide to DV Pchlide concentrations, *R*, was calculated from the apparent MV (E437)/DV (E451)ratio of 1.92 and from the inverse of the

slope of the calibration curve (1/0.900) as $R = (1.92 \times 1/0.900) = 2.13/1$. This, in turn, corresponded to a MV Pchlide percentage of $(2.13/1 + 2.13)(100) = 68\,\%$.
7. Since the sample contained a total of 40 pmol/ml of MV + DV Pchlide, the concentration of MV Pchlide amounted to $(40/100)(68) = 27.2$ pmol/ml. That of DV Pchlide amounted to $40–27.2 = 12.8$ pmol/ml.

## 3.7 Quantitative Determination of Monovinyl and Divinyl Chlorophyll(ides) [Chli(des)]*a* and *b* by Spectrofluorometry at 77 K

The same experimental strategy was adopted for the determination of the amount of DV and MV Chl(ide) *a* and *b*, in mixtures of these compounds, from room temperature and 77°K spectrofluorometric analysis as was used for the analysis of MV and DV [Pchl(ides)] described in Sect. 3.5. This involved the following:

(a) Determination of the amount of DV + MV Chl(ide) *a* and *b* at room temperature, either by room temperature spectrofluorometry (Sect. 3.4) and/or 77°K spectrofluorometry (Sect. 3.6). Since chlorophylls and chlorophyllides exhibit identical electronic spectroscopic properties at room temperature and at 77°K, the same equations used for the determination of Chl *a* or *b* are also valid for the determination of Chlide *a* or *b*.
(b) Selection of appropriate fluorescence wavelengths for the best discrimination between DV and MV Chl(ide) *a* and between DV and MV Chl(ide) *b* at 77°K.
(c) Adaptation of previously derived, general purpose simultaneous equations, for the calculation of the net fluorescence signals of DV and MV Chl(ide) *a* and *b* (Sect. 3.6).
(d) Calculation of the true 77°K MV/DV fluorescence ratios for Chl(ide) *a* and *b*.
(e) Calculation of the individual amounts of DV and MV Chl(ide) *a* and *b* from the total amount of DV + MV Chl(ide) a/or and from the respective MV/DV fluorescence ratios of Chl(ide) *a* or *b*.

### 3.7.1 *Choice of Excitation and Emission Wavelengths that Give the Best Distinction Between the Monovinyl a Divinyl Signals in Mixtures of Monovinyl and Divinyl Chlorophyll(ide) a and b*

The selection of appropriate wavelengths for the calculation of net DV and MV fluorescence signals from simultaneous equations was essentially based on two criteria, namely (a) reasonably pronounced fluorescence signals with a high signal-to-noise ratio, and (b) minimal contribution of the DV signal to the MV signal and vice versa.

In ether at 77°K, MV and DV Chl(ide) *a* exhibit similar fluorescence emission bands with maxima at 673–674 nm (Wu et al. 1989). Fortunately the Soret excitation bands are very different. While MV Chl(ide) *a* exhibits a Soret excitation maximum at 446–447 nm, DV Chl(ide) *a* exhibits a Soret excitation maximum at458–459 nm(Wu et al. 1989). Furthermore, at these wavelengths, the contribution of the DV Chl(ide) *a* Soret signal to the MV Chl(ide) *a* Soret signal (about 1/3 at 447 nm) and the contribution of the MV Chl(ide) *a* signal to the DV Chl(ide) *a* signal (about 1/14 at 458 nm) were suitable for efficient correction with simultaneous equations. It was therefore considered that the Soret excitation maxima of MV and DV Chl(ide) *a* at 447 and 458 nm, respectively, should be most suitable for the best signal deconvolution possible in mixtures of the two compounds.

The choice of appropriate fluorescence wavelengths or the mathematical Deconvolution of the DV Chl(ide) *b* signal from the MV Chl(ide) *b* signal proved to be more involved. At 77°K, MV and DV Chl(ide) *b* exhibited emission maxima at 659 and 666 nm, respectively (Wu et al. 1989). Their Soret excitation profile was more complicated, however. Monovinyl Chl(ide) *b* exhibited a major excitation maximum at 475 nm and a pronounced shoulder at 486–489 nm (Wu et al. 1989). Such a splitting of Soret excitation bands into By(0–0) and Bx(0–0) components is rather common in asymmetrical tetrapyrroles and was also observed in DV and MV Pchlides (Tripathy and Rebeiz 1985). Likewise, DV Chl(ide) *b* exhibited a split Soret excitation band with a maximum at 489–490 nm and a shoulder at 498–500 nm (Wu et al. 1989). Since the contribution of the DV Chl(ide) *b* signal to the MV signal amounted to 1/3.5 at 475 nm, while the contribution of the MV Chl (ide) *b* signal to the DV Chl(ide) *b* signal amounted to 1/6.6 at 498 nm, the two aforementioned wavelengths were judged to be most suitable for the mathematical deconvolution of the DV from the MV Chl(ide) *b* signals.

### 3.7.2 Calculation of the Net Fluorescence Amplitudes at 447 and 458 nm of Monovinyl and Divinyl Chlorophyll(ide) a, Respectively, in a Mixture of the Two Compounds

Let *a* and *b* represent any two fluorescence excitation(*a*) and emission (*b*) wavelengths of fluorescent compound X. Likewise, let *c* and *d* represent any two fluorescence excitation (*c*) and emission (*d*) wavelengths of fluorescent compound *Y*. It has been demonstrated above (Sect. 3.5.1) and elsewhere (Rebeiz et al. 1975a) that the deconvoluted net fluorescence signals, $X(E_aF_b)$ and $Y(E_cF_d)$, generated by compound *X* and Y, respectively, at the designated wavelengths *a, b, c,* and *d,* can be determined from the equations described in Sect. 3.5.1, namely

$$X(E_a\ F_b) = [(F_a\ F_b) - (E_c\ F_d)/k2]\ \ (l/K1) \tag{3.37}$$

and

$$Y(E_C\ F_d) = [(F_c\ F_d) - (E_a\ F_b)/k4]\ \ (l/K2) \tag{3.38}$$

where $(E_a\ F_b)$ and $(E_c\ F_d)$ represent the fluorescence excitation ($E$) and emission ($F$) amplitudes of the $X + Y$ mixture, at the $a, b$, and $c, d$ wavelengths, respectively, and where

$$\begin{aligned} K_1 &= 1 - (k_1/k_2) \\ K_2 &= 1 - (k_3/k_4) \end{aligned} \tag{3.39}$$

and

$$\begin{aligned} k_1 = X(E_c\ F_d)/X(E_a\ F_b);\quad & k_2 = Y(E_c\ F_d)/Y(E_a\ F_b) \\ k_3 = Y(E_a\ F_b)/Y\ (E_a\ F_d);\quad & k_4 = X(E_a\ F_b)/X(E_c\ F_d) \end{aligned} \tag{3.40}$$

Adaptation of general Eqs. (3.37) and (3.38) to the deconvolution of the DV and MV Chl *a* Soret excitation signals is essentially achieved by substituting appropriate DV and MV Chl(ide) *a* values in Eqs. (3.37) and (3.38) (Table 3.9).

Let the 447-nm Soret excitation amplitude of the DV + MV Chl(ide) *a* mixture, which is recorded at an emission wavelength of 674 nm [i.e., the emission maximum of DV and MV Chl(ide) *a*], be referred to as $(E_{447}F_{674})$. Likewise, let the458-nm Soret excitation amplitude of the DV + MV Chl mixture, which is also recorded at 674 nm, be referred to as $(E_{458}F_{674})$. Also let X and Y in Eqs. (3.37) and (3.38) represent DV and MV Chl(ide) *a*, respectively. By substitution of $(E_{458}F_{674})$ for $(E_aF_b)$, $(E_{447}F_{474})$ for $(E_cF_d)$, DV Chl(ide) *a* for X, and MV Chl(ide) *a* for Y, Eqs. (3.37) and (3.38) transform into:

$$\text{DV Chl(ide) } a\ \ (\text{E458 F674}) = [(\text{E458 F674}) - (\text{E447 F674})/k_2]\ \ (1/K_1) \tag{3.64}$$

$$\text{MV Chl(ide) } a\ \ (\text{E447 F674}) = [(\text{E447 F674}) - (\text{E458 F674})/k_4]\ \ (1/K_2) \tag{3.65}$$

where $K_1$ and $K_2$ are as defined by Eq. (3.39) and Where:

$$\begin{aligned} k_1 &= \text{DV Chl(ide) } a \text{ (E447 F674)/DV Chl(ide) } a \text{ (E458 F674)} \\ k_2 &= \text{MV Chl(ide) } a \text{ (E447 F674)/MV Chl(ide) } a \text{ (E458 F674)} \\ k_3 &= \text{MV Chl(ide) } a \text{ (E458 F674)/MV Chl(ide) } a \text{ (E447 F674)} \\ k_4 &= \text{DV Chl(ide) } a \text{ (E458 F674)/DV Chl(ide) } a \text{ (E447 F674)} \end{aligned} \tag{3.66}$$

in this context, DV Chl(ide) *a* (E458 F674) refers to the magnitude of the Soret excitation maximum of authentic DV Chl(ide) *a* at 458 nm when the excitation spectrum is recorded at an emission wavelength of 674 nm, etc.

**Table 3.9** Determination of reliability of equations 3.62 and 3.63 used to calculate excitation amplitudes of MV Pchl9ide) and DV Pchl(ide) at 437 and 451 nm, respectively

| | Amount of Pchlide added (pmol/ml) | | Amount of Pchlide calcd (pmol/ml) | | Percentage error between amount of pchlide added and calcd (%) | | Mean percentage error n SD | |
|---|---|---|---|---|---|---|---|---|
| Amount of DV MPE added (pmol/ml) | MV Pchlide | DV Pchlide | DV Pchlide | MV Pchlide | MV Pchlide | DV Pchlide | MV Pchlide | DV Pchlide |
| 40 | 4 | 36 | 4 | 36 | 0 | 0 | 1.24 % ± 3.7 % | 0.32 % ± 1.9 % |
| 40 | 8 | 32 | 8.7 | 31.3 | 8.75 | −2.2 | | |
| 40 | 20 | 20 | 20 | 20 | 0 | | | |
| 40 | 24 | 16 | 23.9 | 16.1 | −0.41 | 0.4 | | |
| 40 | 28 | 12 | 28 | 12 | 0 | | | |
| 40 | 32 | 8 | 31.7 | 8.3 | −0.9 | 3.75 | | |

Note: The excitation spectra were recorded at an emission maximum of 625 nm. Other experimental details are as under materials and methods

The mean $k$ and $K$ values of 20 determinations for DV and MV Chl(ide) *a* amounted *to* $k_1 = 0.3675 \pm 0.0066$; $k_2 = 18.3010 \pm 0.8119$; $k3 = 0.0548 \pm 0.0025$; $k_4 = 2.7220 \pm 0.0516$; $K_1 = 0.9799$; $K_2 = 0.9799$. By substitution for the values of $k_1$, $k_4$, $K_1$ and $K_2$ in Eqs. (3.64) and (3.65) and by further reduction of the substituted equations, the latter assume the form:

$$\text{DV Chl(ide) } a \ \ (E458\ F674) = 1.0205 \ \ (E458\ F674) - 0.0557 \ \ (E447\ F674) \tag{3.67}$$

and

$$\text{MV Chl(ide) } a \ \ (E447\ F674) = 1.0205 \ \ (E447\ F674) - 0.3749 \ \ (E458\ F674) \tag{3.68}$$

### *3.7.3 Conversion of the DV and MV Chlorophyll(ide) a Soret Excitation Ratios to DV and MV Chlorophyll(ide) a Concentrations*

As was pointed out in Sect. 3.5.2.4, the ratio of the net Soret excitation amplitudes must be converted to an authentic concentration ratio before actual MV and DV amounts can be calculated (Tripathy and Rebeiz 1985). This is achieved by reference to a standard calibration curve. For MV and DV Chl(ide) *a*, such a calibration curve can be constructed as follows:

(a) by mixing authentic MV and DV Chl(ide)*a* in known proportions,
(b) by recording the required77°K spectra on every mixture,
(c) by calculating the ratio of the net MV to the net DV Soret excitation amplitudes for every mixture, with the help of Eqs. (3.67) and (3.68), and finally
(d) by plotting the authentic MV/DV Chl(ide) *a* ratios, on the abscissa for example, against the experimental MV/DV ratio of the net Soret excitation amplitudes on the ordinate.

Under our instrumental conditions, for experimental Soret excitation ratios equal to or smaller than 2.5 such a plot yielded a straight line with an intercept of 0.3150, a slope of 0.5405, and a correlation coefficient of 0.9958. At experimental Soret excitation ratios larger than 2.5, the curve exhibited an ordinate intercept of 0.4248, a slope of 0.4675, and a correlation coefficient of 0.9971. The concentration of MV and DV Chl(ide) *a* in the mixture was then calculated from the total amount of MV + DV Chl(ide) *a* and from the slopes and intercepts of the calibration curve as illustrated below. The reliability of Eqs. (3.67) and (3.68) in determining the proportions of MV and DV Chl(ide) *a* in mixtures of the two tetrapyrroles amounted to $-0.27 \pm 2.69$ and $2.69 \pm 5.95$ %, respectively (Table 3.10).

**Table 3.10** Determination of the reliability of Eqs. (3.44) and (3.45) used to calculate excitation amplitudes of MV and DV Chl(ide) *a* at 447 and 458 nm respectively

| Amount of CHL *a* added (pmol/ml) | | Amount of CHL *a* calcd (pmol/ml) | | Percentage error between amount of CHL *a* added and calcd (%) | | Mean percentage error ± SD | |
|---|---|---|---|---|---|---|---|
| MV | DV | MV | DV | MV | DV | MV | DV |
| 13.10 | 30.54 | 12.26 | 31.38 | −6.41 | 2.75 | −0.27 ± 2.69 | 2.69 ± 5.95 |
| 13.10 | 15.27 | 13.87 | 14.50 | 5.88 | −5.40 | | |
| 26.20 | 15.27 | 26.42 | 15.05 | 0.84 | −1.44 | | |
| 52.40 | 15.27 | 51.95 | 15.72 | −0.86 | 2.95 | | |
| 40.00 | 10.00 | 40.13 | 9.87 | 0.33 | −1.30 | | |
| 52.40 | 7.64 | 51.64 | 8.40 | −1.45 | 9.95 | | |
| 40.00 | 5.00 | 39.95 | 5.05 | −0.13 | 1.00 | | |
| 78.60 | 7.64 | 77.82 | 8.42 | −0.99 | 11.02 | | |
| 104.80 | 7.64 | 105.16 | 7.28 | 0.34 | −4.71 | | |
| 100.00 | 5.00 | 99.92 | 5.08 | −0.08 | 1.60 | | |
| 150.00 | 5.00 | 149.34 | 5.66 | −0.44 | 13.20 | | |

Note: The excitation spectra were recorded at an emission wavelength of 674 nm

### 3.7.4 Sample Calculation of the Amounts of MV and DV Chldde a in a Mixture of the Two Tetrapyrroles

1. The total amount of MV + DV Chl(ide) *a* in the mixture was determined by spectrophotometry at 293°K (Bazzaz and Rebeiz 1979). It amounted to 86.24 pmol of MV + DV Chl(ide) *a*/ml of solution.
2. The MV + DV Chl(ide) *a* were transferred to diethyl ether, and a Soret excitation spectrum was recorded at 77°K, at an emission wavelength of 674 nm.
3. The non deconvoluted Soret excitation amplitudes at 447 and 458 nm which were contributed by MV + DV Chl(ide) *a* were determined in relative fluorescence units from the 77°K excitation spectrum. They amounted to 3.21 and 0.7874 fluorescence units, respectively.
4. The deconvoluted net Soret excitation amplitudes of the MV and DV Chl(ide) *a* components were calculated from the non deconvoluted Soret excitation amplitudes at 447 and 458 nm with Eqs. (3.67) and (3.68). They amounted to 2.981 and 0.6247 net fluorescence units, respectively. This in turn corresponded to an experimental net Soret excitation ratio of 4.7719 (i.e., 2.981/0.6247 = 4.7719).
5. The true ratio, *R*, of the MV to DV Chl(ide) *a* concentrations was calculated from the experimental ratio of 4.7719, from the slope (0.4675) and from the ordinate intercept (0.4248) of the calibration curve as follows:

   $R = (4.7719 - 0.4248)/0.4675 = 9.2986/1$. This in turn corresponded to a MV Chl(ide) *a* percentage of $(9.2986/1 + 9.2986) \times 100 = 90.29$ % and to a DV Chl(ide) *a* percentage of 100–90.29 = 9.71 %.

6. Since the sample contained a total amount of 86.2 pmol/ml of MV + DV Chl(ide) *a*, the concentration of MV Chl(ide) *a* amounted to (86.24/100) × 90.29 = 77.87 pmol/ml. That of DV Chl(ide) *a* amounted to 86.2−77.87 = 8.37 pmol/ml.

### 3.7.5 Calculation of the Net Fluorescence Amplitudes at 475 and 498 nm of MV and DV Chlorophyll(ide) b Respectively in Mixtures of the Two Compounds

Calculation of the net Soret excitation amplitudes of MV and DV Chl(ide) *b* in mixtures of these two tetrapyrroles was also achieved by adaptations of Eqs. (3.37) and (3.38).

As was mentioned earlier the optimum wavelengths for this analysis were determined to be in the Soret region, at 475 and 498 nm for MV and DV Chl(ide) b, respectively. Adaptation of Eqs. (3.37) and (3.38) was achieved via the following assignments:

X = deconvoluted net Soret excitation amplitude of MV Chl(ide) *b* (E475 F660) at 475 nm;

Y = deconvoluted net Soret excitation amplitude of DV Chl(ide) *b* (E498 F666) at 498 nm;

$(E_aF_b)$ = (E475 F660) = Soret excitation amplitude of the MV + DV mixture at 475 nm, when the excitation spectrum is recorded at an emission wavelength of 660 nm;

$(E_C\ F_D)$ = *(E498 F666 )* = Soret excitation amplitude of the MV + DV mixture at 498 nm, when the excitation spectrum is recorded at an emission wavelength of 666 nm.

By substitution of the above values for X, Y, $(E_a\ F_b)$, and $(E_c\ F_d)$ in Eqs. (3.37) and (3.38) the latter transform into:

$$\text{DV Chl(ide)}\,a\ (\text{E498 F666}) = [(\text{E475 F660}) - (\text{E475 F660})/k_6]\ (1/K_3) \tag{3.69}$$

$$\text{MV Chl(ide)}b\ (\text{E475 F660}) = [(\text{E475 F660}) - (\text{E498 F666})/k_8]\ (1/K_4) \tag{3.70}$$

where $K_3$ and $K_4$ are as defined by Eq. (3.39) and Where:

$$\begin{aligned} k_5 &= \text{DV Chl(ide)}b\ (\text{E475 F660})/\text{DV Chl(ide)}b\ (\text{E498 F666}) \\ k_6 &= \text{MV Chl(ide)}b\ (\text{E475 F660})/\text{MV Chl(ide)}b\ (\text{E498 F666}) \\ k_7 &= \text{MV Chl(ide)}b\ (\text{E498 F666})/\text{MV Chl(ide)}b\ (\text{E475 F660}) \\ k_8 &= \text{DV Chl(ide)}b\ (\text{E498 F666})/\text{DV Chl(ide)}b\ (\text{E475 F660}) \end{aligned} \tag{3.71}$$

in this context DV Chl(ide) *b* (E498 F666) refers to the magnitude of the Soret excitation maximum of authentic DV Chl(ide) *b* at 498 nm, when the excitation spectrum is recorded at an emission wavelength of 666 nm, etc.. . .

The mean $k$ and $K$ values of 20 determinations for DV and MV Chl(ide) *b* amounted to $k_5 = 0.4345 \pm 0.0144$; $k_6 = 23.9631 \pm 1.4382$; $k_7 = 0.0419 \pm 0.0026$; $k_8 = 2.3038 \pm 0.0763$; $K_3 = 0.9819$; $K_4 = 0.9818$. By substitution for the values of $k_6$, $k_8$, $K_3$, and $K_4$, Eqs. (3.69) and (3.70) transform into:

$$\text{DV Chl(ide)}b\ (\text{E498 F666}) = 1.0184\,(\text{E498 F666}) - 0.0425\,(\text{E475 F666}) \tag{3.72}$$

and

$$\text{MV Chl(ide)}b\ (\text{E475 F660}) = 1.0185\ (\text{E475 F666}) - 0.4421\ (\text{E498 }F666). \tag{3.73}$$

### 3.7.6 Conversion of the DV and MV Chl(ide) b Soret Excitation Ratios into DV and MV Chl(ide) b Concentrations

Conversion of the net Soret excitation ratios to authentic concentration ratios was achieved exactly as was described for DV and MV Chl(ide) *a* by reference to a standard calibration curve. The calibration curve was constructed by:

(a) mixing authentic MV and DV Chl(ide) *b* in known proportions,
(b) by recording the required 77°K spectra on every mixture,
(c) by calculating the ratio of the experimental, net MV/net DV Soret
(d) excitation amplitudes for every mixture, with the help of Eqs. (3.72) and (3.73), and finally,
(e) by plotting the authentic MV/DV Chl(ide) *b* ratios, on the abscissa, against the experimental MV/DV ratio of the net Soret excitation amplitudes on the ordinate.

Under our instrumental conditions, for experimental Soret excitation ratios equal to or smaller than 10.50, such a plot yielded a straight line with an intercept of 0.3806, a slope of 3.4392, and a correlation coefficient of 0.9974. At experimental Soret excitation ratios greater than 10.50, the curve exhibited an ordinate intercept of 3.1081, a slope of 2.0138, and a correlation coefficient of 0.9974.

The reliability of Eqs. (3.72) and (3.73) in determining the proportion of MV and DV Chl(ide) *b* in mixtures of the two tetrapyrroles amounted to 0.63 ± 5.79 % and −0.75 ± 5.79 %, respectively as displayed below in Table 3.11.

**Table 3.11** Determination of the reliability of Eqs. (3.44) and (3.45) used to calculate excitation amplitudes of MV and DV Chl(ide) *a* at 447 and 458 nm respectively

| Amount of CHL *a* added (pmol/ml) | | Amount of CHL *a* calcd (pmol/ml) | | Percentage error between amount of CHL *a* added and calcd (%) | | Mean percentage error ± SD | |
|---|---|---|---|---|---|---|---|
| MV | DV | MV | DV | MV | DV | MV | DV |
| 20.00 | 50.00 | 19.32 | 50.68 | −3.40 | 1.36 | −0.63 ± 1.78 | −0.75 ± 5.79 |
| 40.00 | 50.00 | 39.86 | 50.14 | −0.35 | 0.28 | | |
| 30.00 | 30.00 | 31.25 | 28.75 | 4.17 | −4.17 | | |
| 40.00 | 20.00 | 41.84 | 18.16 | 4.60 | −9.20 | | |
| 60.00 | 30.00 | 59.69 | 30.31 | −0.52 | 1.30 | | |
| 60.00 | 20.00 | 61.05 | 18.95 | 1.75 | −5.25 | | |
| 50.00 | 10.00 | 50.04 | 9.96 | 0.08 | −0.40 | | |
| 80.00 | 10.00 | 80.14 | 9.86 | 0.18 | −1.40 | | |
| 100.00 | 10.00 | 101.16 | 8.84 | 1.16 | −11.60 | | |
| 75.00 | 5.00 | 74.40 | 5.60 | 0.80 | 12.00 | | |
| 100.00 | 5.00 | 99.96 | 5.04 | −0.44 | 0.80 | | |
| 125.00 | 5.00 | 124.65 | 5.35 | −0.28 | 7.00 | | |
| 150.00 | 5.00 | 150.01 | 4.99 | −0.01 | −0.20 | | |

Note: The excitation spectra were recorded at an emission wavelength of 674 nm

## 3.8 Quantitative Determination of Monovinyl Protochlorophyllide *b* by Spectrofluorometry at 77 K

During analysis of Pchl(ide) *b*, two different situations are encountered in:

(a) Etiolated tissues with relatively high concentrations of Pchl(ide) *a* and,
(b) Greening or green tissues with high concentrations of Chl(ide) *a* and *b*.

Since Pchlide *b* and its phytyl ester exhibit identical electronic spectroscopic properties, including identical spectrofluorometric properties (Shedbalkar et al. 1991), the derived equations could be used either for the quantitative determination of Pchlide *b* or its phytyl ester.

### *3.8.1 Determination of the Amount of 2-MV Pchl(ide) b in the Presence of Pchl(ide) a, Using 293 and 77°K Spectrofluorometric Analysis: Overall Strategy*

The procedure involves:

(a) determination of the amount of 2-MV Pch(lide) *a* in a 2-MV Pchl(ide) *a* + 2-MV Pchl(ide) *b* mixture by spectrofluorometry, at room temperature as described elsewhere as described in Sect. 3.2

(b) selection of appropriate fluorescence wavelengths for the best discrimination between the 2-MV Pchl(ide) *b* and 2-MV Pchl(ide) *a* fluorescence signals,
(c) adaptation of previously derived, general purpose simultaneous equations, for the calculation of the net fluorescence signal generated by 2-MV Pchl(ide) *a* and *b* at 77°K (3),
(d) calculation of the molar ratio of 2-MV Pchl(ide) *a*/2-MV Pchl(ide) *b* at 77°K, from the net fluorescence signals,
(e) calculation of the amount of 2-MV Pchl(ide) *b* from the total amount of 2-MV Pchl(ide) *a*, which is determined at 293°K and from the calculated 2-MV Pchl (ide) *a*/2-MV Pchl(ide) *b* molar ratio.

#### 3.8.1.1 Selection of Appropriate Wavelengths for the Calculation of the Net 2-MV Pchl(ide) *a* and *b* Fluorescence Signals in a Mixture of the Two Compounds

Selection of appropriate wavelengths was based on two criteria: (a) reasonably pronounced fluorescence signals with a high signal to noise ratio and (b) minimal contribution of any other fluorescence signal to the 2-MV Pchl(ide) *a* and *b* signals at the selected wavelengths. On the basis of the aforementioned criteria, fluorescence emission wavelengths at 635 and 643 nm, elicited by excitation at 440 and 463 nm respectively were selected for the calculation of the net 2-MV Pchl(ide) *b* and *a* fluorescence signals in a mixture of both compounds. Fluorescence emission spectra of 2-MV Pchl(ide) *a* and *b* mixtures in diethyl ether at 77°K, elicited by excitation at 463 nm [the Soret excitation maximum of 2-MVPchl(ide) *b* in ether at 77°K], exhibit a pronounced 2-MV Pchl(ide) *b* fluorescence emission spectrum with a maximum at 643 nm, while 2-MV Pchlide *a*, exhibits much weaker emission signals. On the other hand, fluorescence emission spectra elicited by excitation at 440 nm [close to the Soret excitation maximum of 2-MV Pchl(ide) *a* in diethyl ether at 77°K], exhibit a pronounced 2-MV Pchl(ide) *a* fluorescence emission spectrum with a maximum at 625 nm, and a less pronounced Pchl(ide) *b* emission band between at 635 and 648 nm [see Chap. 12 and (Shedbalkar et al. 1991)].

#### 3.8.1.2 Calculation of the 2-MV Pchl(ide) a/2-MV Pchl(ide) b Molar Ratio in a Mixture of Both Compounds

The procedure involved:

(a) Calculation of the net fluorescence emission amplitudes at 635 and 643 nm of 2 MV Pchl(ide) *a* and *b* respectively in a mixture of both compounds, and
(b) Calculation of the 2-MV Pchl(ide) *a*/2-MV Pchl(ide) *b* molar ratio from the net fluorescence amplitudes.

Let the 635 nm emission amplitude of the 2-MV Pchl(ide) *a* and *b* mixture which is elicited by excitation 440 nm, be referred to as (E440 F635). Likewise let the 643 nm emission amplitude of the 2-MV Pchl(ide) *a* and *b* mixture which is elicited by excitation at 463 nm, be referred to as (E463 F643). Also, let X in Eq. (3.37) represent 2-MV Pchl(ide) *b* and Y (4) in Eq. (3.38) represent 2-MV Pchl(ide) *a* as depicted below:

$$X(E_a\ F_b) = [(F_a\ F_b) - (E_c\ F_d)/k_2]\quad (l/K_1) \tag{3.37}$$

and

$$Y(E_C\ F_d) = [(\mathrm{F_c}\ F_d) - (E_a\ F_b)/k_4]\quad (l/K_2) \tag{3.38}$$

By substituting (E463 F643) for $(E_a\ F_b)$, (E440 F635) for $(E_c\ F_d)$, 2-MV Pchl (ide) *b* for *X* and 2 MV Pchl(ide) *a* for *Y*, Eqs. (3.37) and (3.38) transform into:

$$\text{2-MV Pchl(ide)}b\ (\text{E463 F643}) = [(\text{E463 F643}) - (\text{E440 F635})/k_2]\,(1/K_1) \tag{3.74}$$

$$\text{2-MV Pchl(ide)}a\ (\text{E440 F635}) = [(\text{E440 F635}) - (\text{E463 F643})/k_4]\,(1/K_2) \tag{3.75}$$

where as reported elsewhere in Sect. 3.5, $K_1$, $K_2$ and $k_1$, $k_2$, $k_3$, $k_4$ are as defined by Eqs. (3.39) and (3.40)

$$\begin{aligned} K_1 &= 1 - (k_1/k_2) \\ K_2 &= 1 - (k_3/k_4) \end{aligned} \tag{3.39}$$

$$\begin{aligned} k_1 &= X(E_c\ F_d)/X(E_a\ F_b); \quad & k_2 &= Y(E_c\ F_d)/Y(E_a\ F_b) \\ k_3 &= Y(E_a\ F_b)/Y(E_c\ F_d); \quad & k_4 &= X(E_a\ F_b)/X(E_c\ F_d) \end{aligned} \tag{3.40}$$

By substitution for *X*, *Y*, $(E_aF_b)$, and $(E_cF_d)$, Eq. (3.40) transforms into:

$$k_1 = \text{2-MV Pchl(ide)}b\ (\text{E440 F635})/\text{2-MV Pchl(ide)}b\ (\text{E463 F643})$$

$$k_2 = \text{2-MV Pchl(ide)}a\ (\text{E440 F635})/\text{2-MV Pchl(ide)}a\ (\text{E463 F643})$$

$$k_3 = \text{2-MV Pchl(ide)}a\ (\text{E463 F643})/\text{2-MV Pchl(ide)}a\ (\text{E440 F635})$$

$$k_4 = \text{2-MV Pchl(ide)}b\ (\text{E463 F643})/\text{2-MV Pchl(ide)}b\ (\text{E440 F635})$$

The means of 12 determinations amounted to $k_1 = 0.07902 \pm 0.005$; $k_2 = 1.915 \pm 0.111$; $k_3 = 0.524 \pm 0.032$; $k_4 = 12.72 \pm 0.727$; $K_1 = 0.959$; $K_2 = 0.959$. By substitution of the above constants into Eqs. (3.74) and (3.75), the latter transform into:

$$\text{2-MV Pchl(ide)}b\ (\text{E463 F643}) = 1.04\ (\text{E463 F643}) - 0.54\ (\text{E440 F635}) \quad (3.76)$$

$$\text{2-MV Pchl(ide)}a\ (\text{E440 F635}) = 1.04\ (\text{E440 F635}) - 0.08\ (\text{E463 F643}) \quad (3.77)$$

The calculated ratio of the net fluorescence amplitudes of 2-MV Pchl(ide) *b* (E463 F643)/2-MV Pchl(ide) *a* (E440 F635) was next converted to an authentic molar ratio of 2-MV Pchl(ide) *b*/2-MV Pchl(ide) *a* prior to the calculation of the actual amount of 2-MV Pchl(ide) *b*. This was achieved by reference to a standard calibration curve that was constructed as follows:

(a) mixing authentic 2-MV Pchl(ide) *b* and 2-MV Pchl(ide) *a* in known proportions,
(b) recording the required 77°K emission spectra in diethyl ether for every mixture by excitation at E440 and E463 nm,
(c) calculating the net fluorescence amplitudes of 2-MV Pchl(ide) *b* (E463 F6443) and 2-MV Pchl(ide) *a* (E440 F635) with Eqs. (3.76) and (3.77),
(d) calculating the apparent 2-MV Pchl(ide) *b* (E463 F6443)/2-MV Pchl(ide) *a* (E440 F635) emission ratio, and
(e) plotting the 2-MV Pchl(ide) *a*/2-MV Pchl(ide) *b* molar ratios, on the abscissa, against the apparent 2-MV Pchl(ide) *a* (E440 F635)/Pchl(ide) *b* (E463 F6443) emission ratio on the ordinate, or *vice versa*. The linear relationship between the 2-MV Pchl(ide) *a*/2-MV Pchl(ide) *b* molar ratios (Y), and the apparent Pchl (ide) *a* (E440 F635)/Pchl(ide) *b* (E463 F6443) emission ratio (X) obeyed the following equation:

$$\mathrm{Y} = 1.827\ \mathrm{X} + 0.55\ (\mathrm{n} = 12, \mathrm{R}^2 = 0.976) \quad (3.78)$$

#### 3.8.1.3 Determination of the Amount of 2-MV Pchl(ide) b in a Mixture of 2-MV Pchl(ide) b and 2-MV Pchl(ide) a

The amount of 2-MV Pchl(ide) *b* in the presence of 2-MV Pchl(ide) *a* was determined from the total amount of Pchl(ide) *a* which was calculated from 293°K fluorescence emission spectra as described elsewhere in Sect. 3.2 and from the

**Table 3.12** Reliability of Eqs. (3.76) and (3.77) in the determination of the amount of 2-MV Pch (lide) *b* in the presence of 2-MV Pchl(ide) *a*

| Amount of 2-MV Pchl (ide) added error ± SD (pmol/ml) | | Amount of 2-MVPchl (ide) *b* calculated (pmol/ml) | Percentage error between amounts added and calculated (%) | Mean percentage error ± SD (% ± %) |
|---|---|---|---|---|
| Pchl(ide) *a* | Pchl(ide) *b* | | | |
| 131.0 | 121.9 | 131.0 | −7.5 | 1.0 ± 6.8 |
| 131.0 | 61.0 | 56.4 | 7.5 | |
| 170.3 | 122.0 | 118.0 | 3.3 | |
| 131.0 | 36.6 | 31.8 | 13.1 | |
| 103.0 | 78.4 | 77.3 | 1.4 | |
| 103.6 | 52.3 | 56.9 | −8.8 | |
| 103.6 | 104.5 | 99.6 | 4.7 | |
| 133.2 | 117.6 | 122.7 | −4.3 | |
| 281.2 | 156.8 | 154.1 | 1.7 | |
| 296.0 | 182.9 | 185.7 | 1.5 | |

2-MV Pchl(ide) *a*/2-MV Pchl(ide) *b* molar ratio which was calculated as described above. The reliability of calculating the amount of 2-MV Pchl(ide) *b* in the presence of 2-MV Pchl(ide) *a* amounted to 1.0 ± 6.8 % (Table 3.12).

#### 3.8.1.4 Sample Calculation of the Amount of 2-MV Pchl(ide) *b* in a Mixture of 2-MV Pchl(ide) *b* and 2-MV Pchl(ide) *a*

1. An ether solution containing 199.0 pmol of 2-MV Pchl(ide) *a* and 402.4 pmol of Pchl(ide) *b* per ml was prepared.
2. Two fluorescence emission spectra elicited by excitation at 435 and 463 nm were recorded on the ether solution at 77°K.
3. The net fluorescence emission amplitudes at 635 and 643 nm respectively, for the 2-MV Pchl(ide) *a* and 2-MV Pchl(ide) *b* components were calculated using Eqs. (3.76) and (3.77). The apparent (E440 F635)/(E43 F643) fluorescence ratio amounted to 0.0731/0.3397 = 0.2152.
4. The authentic molar ratio, of 2-MV Pchl(ide) *a*/2-MV Pchl(ide) *b* was calculated from the apparent (E440 F635)/(E43 F643) fluorescence ratio with Eq. (3.78). It amounted to 0.4829.
5. The calculated amount of 2MV-Pchl(ide) *b* in the mixture (412.0 pmol/ml) was determined from the relation:

$$\text{2-MV Pchl(ide) } a/\text{2-MV Pchl(ide) } b = 0.4829$$

where 2-MV Pchl(ide) *a*, as determined by 293°K fluorescence emission, amounted to 402.4 pmol/ml.

### 3.8.2 *Determination of the Amount of 2-MV Pchl(ide) b in the Presence of 2-MV Chl(ide) a and b, Using Room Temperature and 77°K Spectrofluorometric Analysis: Overall Strategy*

The procedure involved:

(a) determination of the amount of 2-MV Chl(ide) *b* in the mixture by spectrofluorometry, at room temperature as described in Sect. 3.4 and elsewhere (Bazzaz and Rebeiz 1979),
(b) selection of appropriate fluorescence wavelengths for the best discrimination between the 2-MV Pchl(ide) *b* and 2-MV Chl(ide) *b* fluorescence signals,
(c) adaptation of the previously derived, general purpose simultaneous equations, for the calculation of the net fluorescence signal generated by 2-MV Pchl(ide) *b* and 2-MV Chl(ide) *b* in diethyl ether at 77°K,
(d) calculation of the molar ratio of 2-MV Pchl(ide) *b*/2-MV Chl(ide) *b* from the net fluorescence signals,
(e) calculation of the amount of 2-MV Pchl(ide) *b* from the total amount of 2-MV Chl(ide) *b*, which is determined by spectrofluorometry at 293°K and from the 2-MV Pchl(ide) *b*/2-MV Chl(ide) *b* molar ratio.

#### 3.8.2.1 Selection of Appropriate Wavelengths for the Calculation of the Net 2-MV Pchl(ide) b Fluorescence Signal in the Presence of 2-MV Chl(ide) a and b

Fluorescence excitation wavelengths at 463 and 455 nm, were selected from excitation spectra recorded at emission wavelengths of 643 and 660 nm respectively, in diethyl ether at 77°K, for the calculation of the net 2-MV Pchl(ide) *b* and 2-MV Chl(ide) *b* fluorescence excitation signals.

At these wavelengths, in diethyl ether at 77°K, contribution of the 2-MV Chl(ide) *a* fluorescence excitation is nil, since at emission wavelengths of 643 and 660 nm, Chl (ide) *a* fluorescence emission is negligible (Belanger et al. 1982). In other words, under these conditions, a mixture of 2-MV Pchl(ide) *b* and 2-MV Chl(ide) *a* and *b*, behaves like a dual mixture of 2-MV Pchl(ide) *b* and 2-MV Chl(ide) *b*. In excitation spectra recorded at an emission wavelength of 643 nm in diethyl ether at 77°K [the emission maximum of 2-MV Pchl(ide) *b* in ether at 77°K], the 2-MV Pchl(ide) *b* component exhibits a pronounced Soret excitation band with a maximum at 463 nm, while 2-MV Chl(ide) *b* exhibits a weaker excitation signal (Ioannides et al. 1997). On the other hand, in excitation spectra recorded at an emission wavelength of 660 nm in diethyl ether at 77°K [the emission maximum of 2-MV Chl(ide) *b* in diethyl ether at 77°K], the 2-MV Chl *b* component exhibits a Soret excitation maximum at 475 nm and a measurable Soret excitation signal at 455 nm (Ioannides et al. 1997).

#### 3.8.2.2 Calculation of the 2-MV Pchl(ide) *b*/2-MV Chl(ide) *b* Molar Ratio in a Mixture of Both Compounds

This procedure involved:

(a) calculation of the net fluorescence excitation amplitudes at 455 and 463 nm of 2 MV Chl(ide) *b* and 2-MV Pchl(ide) *b* respectively in a mixture of both compounds, and
(b) calculation of the 2-MV Pchl(ide) *b*/2-MV Chl(ide) *b* molar ratio of the mixture from the calculated net fluorescence signals.

Let the 463 nm excitation amplitude of the 2-MV Pchl(ide) *b* and 2-MV Chl(ide) *b* mixture that is recorded at an emission wavelength of 643 nm in diethyl ether at 77°K be referred to as (E463 F660). Likewise let the 455 nm excitation amplitude of the same mixture which is recorded at an emission wavelength of 660 nm, be referred to as (E455 F660). Also, let *X* (4) in Eq. (3.37) represent 2-MV Pchl(ide) *b* and *Y* in Eq. (3.38) represent 2-MV Chl(ide) *b*. By substituting (E463 F643) for ($E_a$ $F_b$), (E455 F660) for ($E_c$ $F_d$), 2-MV Pchl(ide) *b* for *X* and 2-MV Chl(ide) *b* for *Y*, Eqs. (3.37) and (3.38) transform into:

$$\text{2-MV Pchl(ide) } b \text{ (E463 F643)} = [\text{(E463 F643)} - \text{(E455 F660)}/\text{k2}]\,(1/\text{K1}) \tag{3.79}$$

$$\text{2-MV Chl(ide) } b \text{ (E455 F660)} = [\text{(E455 F660)} - \text{(E463 F643)}/\text{k4}]\,(1/\text{K2}) \tag{3.80}$$

Where $K_1$, $K_2$ and $k_1$, $k_2$, $k_3$, $k_4$ are as defined by Eqs. (3.39) and (3.40). By substitution for *X*, *Y*, *($E_aF_b$)*, and *($E_cF_d$)*, Eq. (3.40) transforms into:

$$k_1 = \text{2-MV Pchl(ide)}b \text{ (E455 F660)}/\text{2-MV Pchl(ide)}b \text{ (E463 F643)}$$
$$k_2 = \text{2-MV Chl(ide)}b \text{ (E455 F660)}/\text{2-MV Chl(ide)}b \text{ (E463 F643)}$$
$$k_3 = \text{2-MV Chl(ide)}b \text{ (E463 F643)}/\text{2-MV Chl(ide)}b \text{ (E455 F660)}$$
$$k_4 = \text{2-MV Pchl(ide)}b \text{ (E463 F643)}/\text{2-MV Pchl(ide)}b \text{ (E455 F660)}$$

The mean of 12 determinations amounted to $k_1 = 0.0442 \pm 0.0052$; $k_2 = 19.7869 \pm 1.6540$; $k_3 = 0.0509 \pm 0.0043$; $k_4 = 22.8827 \pm 2.4596$; $K_1 = 0.9978$; $K_2 = 0.9978$. By substitution of the above constants into Eqs. (3.79) and (3.80), the latter transform into:

$$\text{2-MV Pchl(ide)}b \text{ (E463 F643)} = 1.0022 \text{ (E463 F643)} - 0.0507 \text{ (E440 F635)} \tag{3.81}$$

$$\text{2-MV Chl(ide)}b \text{ (E455 F660)} = 1.0022 \text{ (E455 F660)} - 0.0438 \text{ (E463 F643)} \tag{3.82}$$

The calculated ratio of the net fluorescence amplitudes of 2-MV Pchl(ide) *b* (E463 F643)/2-MV Chl(ide) *b* (E455 F660) was next converted to an authentic molar ratio of 2-MV Pchl(ide) *b*/2-MV Chl(ide) *b* prior to the calculation of the actual amount of 2-MV Pchl(ide) *b*. This was achieved by reference to a standard calibration curve that was constructed as follows:

(a) mixing diethyl ether solutions of authentic 2-MV Pchl(ide) *b* and 2-MV Chl (ide) *b* in known proportions,
(b) recording the required Soret excitation spectra at emission wavelength of 643 and 660 nm at 77°K for every mixture,
(c) calculating the net fluorescence excitation amplitudes of Pchl(ide) *b* (E463 F6443) at 463 nm, and Chl(ide) *b* (E455 F660) at 455 nm with Eqs. (3.81) and (3.82),
(d) calculating the apparent Chl(ide) *b* (E455 F660)/Pchl(ide) *b* (E463 F6443) excitation ratio, and
(e) plotting the authentic 2-MV Chl(ide) *b*/2-MV Pchl(ide) *b* molar ratios, on the ordinate, against the apparent Chl(ide) *b* (E455 F660)/Pchl(ide) *b* (E463 F6443) excitation ratio on the abscissa, or *vice versa*. The linear relationship between the 2-MV Chl(ide) *b*/2-MV Pchl(ide) *b* molar ratios (Y), plotted on the ordinate, against the apparent 2-MV Chl(ide) *a* (E455 F660)/2-MV Pchl(ide) *b* (E463 F6443) excitation ratio (X) plotted on the abscissa obeyed the following equation:

$$Y = 1.493\ X + 1.003\ (n = 10, R^2 = 0.9869) \tag{3.83}$$

#### 3.8.2.3 Determination of the Amount of 2-MV Pchl(ide) *b* in a Mixture of 2-MV Pchl(ide) b and 2-MV Chl(ide) *b*

The amount of 2-MV Pchl(ide) *b* in the presence of 2-MV Chl(ide) *b* was determined from the total amount of Chl(ide) *b* which was calculated from 293°K fluorescence excitation spectra as described in Sect. 3.4 and elsewhere (Bazzaz and Rebeiz 1979) (10), and from the 2-MV Chl(ide) *b*/2-MV Pchl(ide) *b* molar ratio which was calculated as described above. The reliability of calculating the amount of 2-MV Pchl(ide) *b* in the presence of 2-MV Chl(ide) *b* amounted to −2.7 ± 10.3 % (Table 3.13).

#### 3.8.2.4 Sample Calculation of the Amount of 2-MV Pchl(ide) b in a Mixture of 2-MV Pchl(ide) b and 2-MV Chl(ide)b

1. An ether solution containing 161.4 pmol of 2-MV Pchl(ide) *b* and 410.0 pmol of Chl(ide) *b* per ml was prepared.

**Table 3.13** Reliability of Eqs. (3.81), (3.82) and (3.83) in the determination of the amount of 2-MV Pchl(ide) *b* in the presence of 2-MV Chl(ide) *b*

| Amount of pigment added (pmol/ml) | | Amount of 2-MV Pchl(ide) *b* calculated (pmol/ml) | Percentage error between amounts added and calculated (% ) | Mean percentage error ± SD (% ± %) (% ± %) |
|---|---|---|---|---|
| Chl(ide) *b* | Pchl(ide) *b* | | | |
| 482.4 | 96.9 | 85.3 | −12.0 | −2.7 ± 10.3 |
| 265.3 | 161.4 | 143.9 | −10.8 | |
| 289.4 | 161.4 | 133 | −17.6 | |
| 337.7 | 161.4 | 186.8 | 15.7 | |
| 385.9 | 161.4 | 172.7 | 7.0 | |
| 410.0 | 161.4 | 153.5 | −4.9 | |
| 434.12 | 161.4 | 154.1 | −4.5 | |
| 458.2 | 161.4 | 164.8 | 2.1 | |
| 530.6 | 161.4 | 162.5 | 0.7 | |

2. Two fluorescence excitation spectra were recorded at 77°K, at emission wavelengths of 660 and 643 nm.
3. The net fluorescence excitation amplitudes at 455 and 463 nm respectively, for the 2-MV Chl(ide) *b* and 2-MV Pchl(ide) *b* components were calculated using Eqs. (3.81) and (3.82). The apparent (E455 F6660)/(E463 F643) fluorescence excitation ratio amounted to: 0.4210/0.3849 = 1.0938.
4. The molar ratio, of 2-MV Chl(ide) *b*/2-MV Pchl(ide) *b* was calculated from the apparent (E455 F660)/(E463 F643) fluorescence excitation ratio with Eq. (3.83). It amounted to 2.6707.
5. The calculated amount of 2MV-Pchl(ide) *b* in the mixture (153.5 pmol/ml) was determined from the relation:

$$\text{2-MV Chl(ide) } b/\text{2-MV Pchl(ide) } b = 2.6707$$

where 2-MV Chl(ide) *b*, as determined by fluorescence excitation at 293°K amounted to 410.0 pmol/ml

## 3.9 Kinetic Analysis of Precursor-Product Relationships in Complex Biosynthetic Pathways

The discovery of multiple chlorophyll biosynthetic routes in plants (Chaps. 1, 2, 3, 4, 5, 6, 7, 8, 9, 10, 11, 12 and 13) has highlighted the need for appropriate analytical tools that enable the determination of precursor-product relationships among intermediates of irreversible but interconnected biosynthetic routes (Rebeiz et al. 1988). In an attempt to resolve this question, we have developed kinetic equations that address this issue.

### 3.9.1 Modeling Strategy

The modeling strategy consisted in deriving equations that enabled the calculation of expected radiolabel incorporation from a precursor P into a compound B, when "B" is formed exclusively from an immediate precursor "A", via pathway I (Fig. 3.1).

It was conjectured that a comparison of expected radiolabel incorporations into "B", that are derived by calculation, with experimentally determined incorporations, should reveal, whether compound B is formed exclusively from compound A or not. For example, if compound B is formed exclusively from compound A, then within the range of experimental error, the theoretical (i.e. calculated) and experimental radiolabel incorporations into "B" should be identical or reasonably similar (Pathway I in Fig. 3.1). On the other hand, if compound B is not formed from compound A (pathway III, in Fig. 3.1) or is found to be partially formed from "A" via pathway II (Fig. 3.1), then the calculated and experimental radiolabel incorporations into "B"" should be different. In that latter case, another equation is derived that enables the evaluation of the partial contribution of compound A to the formation of compound B via pathway II (Fig. 3.1). The derivations of these equations are described below.

**B. Relationships Between the Specific Radioactivity of Compound A and Label Incorporation into Compound B, if Compound B is formed Exclusively from Compound A During Any Time Interval $t_1$–$t_2$**

Let a radiolabelled precursor P and radiolabeled products A and B be related by the precursor-product relationship depicted in Fig. 3.1a. Furthermore, let:

$Y_{A1}, Y_{A2}, Y_{A3}\ldots$ represent the specific radioactivity of compound A at the end of time intervals $t_0$–$t_1$, $t_1$–$t_2$, and $t_2$–$t_3$ respectively.

$Q_{B1}, Q_{B2}, Q_{B3}\ldots$ represent the amount of radiolabel incorporated into compound B by the end of time intervals

$$t_0 - t_1,\ t_1 - t_2,\ \text{and}\ t_2 - t_3 \tag{3.84}$$

$\Delta_{B1}, \Delta_{B2}, \Delta_{B3}$ represent the amount of compound B synthesized by the end of time intervals $t_0$–$t_1$, $t_1$–$t_2$, and $t_2$–$t_3$.

If the formation and accumulation of radio labeled compound B from radio labeled compound A during any time interval $t_1$ to $t_2$, is a linear function of time, then the accumulation of compound B can be described by the function

$$B = at \pm b, \tag{3.85}$$

where "$a$" is the rate constant and "$b$" the intercept on the ordinate axis. The rate of change of $B$ with respect to time is then given by

$$\mathrm{d}B/\mathrm{d}t = a \tag{3.86}$$

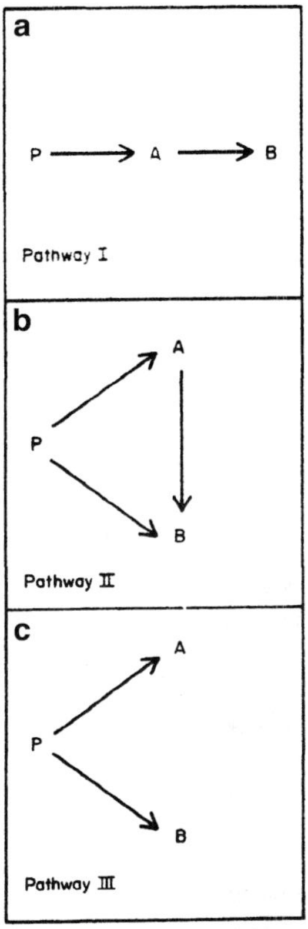


**Fig. 3.1** The three possible irreversible precursor-product relationships between two precursors, P, A, and one end product, B (Adapted from Rebeiz et al. 1988)

If compound $B$ is formed exclusively from A, the increase in $B$ during a small time interval dt is given by

$$\mathrm{d}B = a.\mathrm{d}t. \tag{3.87}$$

Furthermore, since it is assumed that compound $B$ is formed exclusively from $A$, the radioactivity $d_{qb}$ that accumulates in compound $B$ during time $dt$ is given by

$$dq_B = Y_A\ dB \tag{3.88}$$

where $Y_A$ is the specific radioactivity of compound $A$ during the small time interval $dt$, and $dB$ is the increase in $B$ during time interval d$t$.

By substituting Eq. (3.87) for d$B$ in Eq. (3.88), the latter transforms into

$$\mathrm{d}q_B = (Y_A)a.dt. \tag{3.89}$$

During a [14]C-incubation, $Y_A$ changes from a value $Y_{A1}$, at time t1 to a value $Y_{A2}$ at time $t_2$. If during time interval $t_1$ to $t_2$ $Y_A$ is a linear function of time, then at any time within that interval

$$Y_A = ct + Y_{A1} \tag{3.90}$$

where $c$ is the rate constant.

This type of linear specific radioactivity kinetics is not unusual for metabolic pathways where the substrate is rapidly converted to end products with very little accumulation of other intermediates in between (Tripathy and Rebeiz 1988). Even if in some cases, absolute linearity between any two points is not strictly obeyed, it may be reasonably approximated by a linear relationship by shortening the time interval between the two points. By substituting Eq. (3.90) for $Y_A$ in Eq. (3.89), the latter transforms into

$$dq_B = (ct + Y_{A1})a.dt. \tag{3.91}$$

For the time interval $t_1$ to $t_2$, the total radioactivity ($Q_{B2}$) that accumulates incompound $B$ is given by

$$Q_{B2} = \int_{t_1}^{t_2} (ct + YA1)a.dt \tag{3.92}$$

Which can be rewritten

$$Q_{B2} = a\int_{t1}^{t2} (YA1 + ct)dt \tag{3.93}$$

and which can be integrated and arranged (Rebeiz et al. 1988) to yield

$$Q_{B2} = Y_{A1}\ .\ at_2 + (at_2\ .\ ct_2)/2 - Y.\ at_1 - (at_1\ .\ ct_1)/2 \tag{3.94}$$

It follows from Eq. (3.85) that for the time interval $t_1$ to $t_2$ and for $\Delta B_2 = B_2 – B_1$, where $B_2$ = amount of compound B at time $t_2$ and $B_1$ = amount of compound $B$ at time $t_1$,

$$\Delta B_2 = at_2 - at_1 \tag{3.95}$$

Since for any time interval $t_1$ to $t_2$, or $t_3$ to $t_4$, etc., the starting time $t_1$, or $t_3$ can be taken as zero, Eq. (3.95) reduces to

$$\Delta B_2 = at_2 \tag{3.96}$$

By incorporating Eq. (3.96) into Eq. (3.94), and by taking into account that under these conditions $t_I = 0$ Eq. (3.94) transforms into

$$Q_{B2} = Y_{A1} \, . \, \Delta B_2 + \Delta B_2 \, . \, ct_2/2 \tag{3.97}$$

From Eq. (3.90) at times $t_I$–$t_2$ etc....

$$ct_2 = Y_{A2} - Y_{A1} \tag{3.98}$$

By substituting Eq. (3.98) into Eq. (3.97), the latter transforms into

$$\begin{aligned} \mathrm{Q}_{B2} &= Y_{A1} \, . \, \Delta B_2 + \Delta B_2 \; (Y_{A2\_}Y_{A1})/2 \\ &= Y_{A1} \, . \, \Delta B_2 + (Y_{A2}\Delta B_2)/2 - (Y_{A1} \, . \, \Delta B_2)/2 \end{aligned} \tag{3.99}$$

or

$$Q_{B2} = (Y_{A1} \, . \, \Delta B_2)/2 + Y_{A2} \, . \, \Delta B_2/2 \tag{3.100}$$

Equation (3.100) can be rewritten as

$$Q_{B2} = (Y_{A1} + Y_{A2})/2(\Delta B_2) \tag{3.101}$$

transforms into

Where:

$Q_{B2}$ = amount of radiolabel incorporated into compound B by the end of time interval $t_I$–$t_2$.

$Y_{AI}, Y_{A2}$ = specific radioactivity of compound $A$ by the end of the time intervals $t_O$–$t_I$ and $t_I$–$t_2$, respectively.

$\Delta B_2$ = amount of $B$ synthesized by the end of time interval $t_O$–$t_I$.

### 3.9.2 The Special Case of Time Interval $t_O$–$t_I$

During time interval $t_O$–$t_I$, $t_O$ corresponds to the beginning of incubation with radioactive precursor $P$. For this time interval, Eq. (3.101) can then be rewritten

$$Q_{B1} = (Y_{A0} + Y_{A1})/2 \; \Delta B_2(\Delta B_1) \tag{3.102}$$

Since under these conditions none of precursor $P$ has yet been converted to $A$ and/or $B$, $Y_{AO}$

I equal to zero. Consequently, Eq. (3.102) reduces to:

$$Q_{B1} = Y_{A1}/2(\Delta B_1) \tag{3.103}$$

where

$Q_{Bl}$ = amount of radiolabel incorporated into compound B by the end of time interval $t_0$–$t_l$.

$Ya_l$ = specific radioactivity of compound $A$ by the end of time interval $t_0$–$t_l$.

$\Delta B_l$ = amount of $B$ synthesized by the end of time interval $t_0$–$t_l$

### 3.9.3 Evaluation of the Contribution of "A" to the Formation of "B" in Pathway II

If the comparison of calculated and experimental results indicates that compound *B* is not formed from precursor *P* via compound *A* (Fig. 3.1a), then the question arises as to whether *B* is formed via pathway II or via pathway III. Furthermore, if *B* is found to be formed via pathway II, then the contribution of *A* to the formation of *B* needs to be assessed.

The determination of whether *B* is formed via pathway II or pathway III, can be achieved from in vitro investigations. In other words, if cell-free systems are available (see Chap. 4), for the particular pathway under consideration, then the presence or absence of precursor-product relationship between *A* and *B* can be readily demonstrated from conventional in vitro precursor-product conversions. For example, in investigating the precursor-product relationship between ALA, DV Pchlide and MV Pchlide in barley, (Tripathy and Rebeiz 1988) first determined that MV Pchlide was not formed from ALA via pathway I. Then they showed that MV Pchlide was formed from both ALA and DV Pchlide via pathway II and not via pathway III by demonstrating the conversion of DV Pchlide to MV Pchlide in vitro.

If pathway II is found to be operational as in Tripathy and Rebeiz (1988), then the contribution of *A* to the formation of *B* can be readily assessed from the difference between the theoretical and experimental $^{14}$C-incorporations of *P* into *B*, in vivo.

By assuming that the differences between the theoretical and experimental in vivo $^{14}$C-incorporation of *P* into *B* are due to the contribution of *A* to the formation of *B*, then the maximum possible percent conversion of *A* to *B* can be taken as:

$$\%\ \text{conversion} = 100 - (|\text{Exp} - Q_{Bx}|/\text{Exp})100 \qquad (3.104)$$

Where:

% conversion = maximum possible percent conversion of *A* to *B* during any time interval *x*.

Exp = actual $^{14}$C-incorporation into *B* by the end of time interval *x*, which is determined experimentally.

$Q_{BX}$ = theoretical $^{14}$C-incorporation into *B* by the end of time interval *x* which is calculated with Eq. (3.101).

$|\text{Exp} - Q_{Bx}|$ = absolute difference between the theoretical and experimental $^{14}$C-incorporation of $P$ into $B$ during time interval $x$

In this manner, when pathway I is operational, i.e. when the experimental (Exp) and theoretical ($Q_{Bx}$) $^{14}$C-incorporations into B are of equal magnitude, then, as expected, Eq. (3.104) returns a value of 100 % for the percent contribution of compound $A$ to the formation of compound $B$.

### 3.9.4 Sample Calculation

The following sample calculation was excerpted from Table II of (Tripathy and Rebeiz 1988). The purpose of the calculations was (a) to determine whether pathway II was operational in greening barley seedlings, and (b) if it was operational to determine the extent of the contribution of $A$ to the formation of $B$. In this particular case the parameters of Eq. (3.101) represented the following:

$P$ = $^{14}$C-δ-aminolevulinic acid

$A$ = Divinyl protochlorophyllide (DV Pchlide)

$B$ = Monovinyl protochlorophyllide (MV Pchlide)

The rise in specific radioactivity of DV Pchlide was a linear function of time (Tripathy and Rebeiz 1988).

Under these conditions, for the time interval $t_2$–$t_4$, Eq. (3.101) assumes the following form:

$$Q_{B4} = (Y_{A2} + Y_{A4})/2\ (\Delta B_4) \qquad (3.105)$$

where:

$Y_{A2}$, $Y_{A4}$ = the specific radioactivity of DV Pchlide by the end of time interval $t_1$–$t_2$and $t_2$–$t_4$, respectively, amounted to 163 and 418 dpm/pmol, respectively

($\Delta B_4$) = the increase in MV Pchlide by the end of time interval $t_2$–$t_4$ amounted to 1,490 pmol

Exp = the experimental $^{14}$C-incorporation into MV Pchlide by the end of time interval $t_2$–$t_4$ amounted to 1346.5 × $10^3$ dpm.

$Q_{B4}$ = the theoretical $^{14}$C-incorporation into MV Pchlide by the end of time interval $t_2$–$t_4$, as calculated from Eq. (3.105), amounted to 432.8 × $10^3$ dpm.

Since the theoretical ($Q_{BX}$) and experimental (Exp) $^{14}$C-ALA incorporations into MV Pchlide were drastically different, it was concluded that pathway I was Not operational in greening barley seedlings. On the other hand, since in vitro incubations exhibited a strong conversion of DV Pchlide to MV Pchlide, it was concluded that MV Pchlide was formed from δ-aminolevulinic acid in vivo via pathway II instead of pathway III.

Finally, the maximum possible percent conversion of DV Pchlide to MV Pchlide during time interval $t_2$–$t_4$ was calculated from Eq. (3.103):

$$\%\ \text{conversion} = 100–[(|1346.5-432.8|)/1346.5]100 = 32\ \%$$

# References


Bazzaz MB, Rebeiz CA (1978) Chloroplast culture: the chlorophyll repair potential of mature chloroplasts incubated in a simple medium. Biochim Biophys Acta 504:310–323

Bazzaz MB, Rebeiz CA (1979) Chloroplast culture V. Spectrofluorometric determination of chlorophyll(ide) *a* and *b* and pheophytin (or pheophorbide) *a* and *b* in unsegregated pigment mixtures. Photochem Photobiol 30:709–721

Belanger FC, Rebeiz CA (1982) Chloroplast biogenesis: detection of monovinyl magnesium protoporphyrin monoester and other monovinyl magnesium porphyrins in higher plants. J Biol Chem 257:1360–1371

Belanger FC, Rebeiz CA (1984) Chloroplast biogenesis 47: spectroscopic study of net spectral shifts induced by ligand coordination in metalated tetrapyrroles. Spectrochim Acta 40A:807–827

Belanger FC, Dugan JX, Rebeiz CA (1982) Chloroplast biogenesis: identification of chlorophyllide *a* (E458F674) as a divinyl chlorophyllide *a*. J Biol Chem 257:4849–4858

Daniell H, Rebeiz CA (1984) Bioengineering of photosynthetic membranes: requirement of magnesium for the conversion of chlorophyllide a to chlorophyll a during the greening of etiochloroplasts in vitro. Biotechnol Bioeng 26:481–487

Granick S (1948) Protoporphyrin 9 as a precursor of chlorophyll. J Biol Chem 172:717–727

Ioannides IM, Shedbalkar VP, Rebeiz CA (1997) Quantitative determination of 2-monovinyl protochlorophyll(ide) *b* by spectrofluorometry. Anal Biochem 249:241–244

Rebeiz CA, Belanger FC (1984) Chloroplast biogenesis 46: calculation of net spectral shifts induced by axial ligand coordination in metalated tetrapyrroles. Spectrochim Acta 40A:793–806

Rebeiz CA, Castelfranco P (1971a) Protochlorophyll biosynthesis in a cell-free system from higher plants. Plant Physiol 47:24–32

Rebeiz CA, Castelfranco P (1971b) Chlorophyll biosynthesis in a cell-free system from higher plants. Plant Physiol 47:33–37

Rebeiz CA, Lascelles J (1982) Biosynthesis of pigments in plants and bacteria. In: Govindgee (ed) Photosynthesis: energy conversion by plants and bacteria, vol 1. Academic, New York, pp 699–780

Rebeiz CA, Mattheis JR, Smith BB et al (1975a) Chloroplast biogenesis. Biosynthesis and accumulation of protochlorophyll by isolated etioplasts and developing chloroplasts. Arch Biochem Biophys 171:549–567

Rebeiz CA, Mattheis JR, Smith BB et al (1975b) Chloroplast biogenesis. Biosynthesis and accumulation of Mg-protoporphyrin IX monoester and longer wavelength metalloporphyrins by greening cotyledons. Arch Biochem Biophys 166:446–465

Rebeiz CA, Montazer-Zouhoor A, Daniell H (1984a) Chloroplast culture X: thylakoid assembly in vitro. Isr J Bot 33:225–235

Rebeiz CA, Montazer-Zouhoor A, Hopen HJ et al (1984b) Photodynamic herbicides: 1. Concept and phenomenology. Enzyme Microbiol Technol 6:390–401

Rebeiz CA, Tripathy BC, Mayasich JM (1988) Chloroplast biogenesis 61: kinetic analysis of precursor-product relationships in complex biosynthetic pathways. J Theor Biol 133:319–326

Shedbalkar VP, Ioannides IM, Rebeiz CA (1991) Chloroplast biogenesis. Detection of monovinyl protochlorophyll(ide) *b* in plants. J Biol Chem 266:17151–17157

Smith BB, Rebeiz CA (1977a) Spectrofluorometric determination of Mg-protoporphyrin monoester and longer wavelength metalloporphyrins in the presence of Zn-protoporphyrin. Photochem Photobiol 26:527–532

Smith BB, Rebeiz CA (1977b) Chloroplast biogenesis: detection of Mg-protoporphyrin chelatase in vitro. Arch Biochem Biophys 180:178–185

Thornber JP, Gregory RPF, Smith CA et al (1967) Studies on the nature of the chloroplast lamella. I. Preparation and some properties of two chlorophyll-protein complexes. Biochemistry 6:391–396

Tripathy BC, Rebeiz CA (1985) Chloroplast biogenesis. Quantitative determination of monovinyl and divinyl Mg-protoporphyrins and protochlorophyll(ides) by spectrofluorometry. Anal Biochem 149:43–61

Tripathy BC, Rebeiz CA (1988) Chloroplast biogenesis 60. Conversion of divinyl protochlorophyllide to monovinyl protochlorophyllide in green(ing) barley, a dark monovinyl/light divinyl plant species. Plant Physiol 87:89–94

Wu SM, Mayasich JM, Rebeiz CA (1989) Chloroplast biogenesis: quantitative determination of monovinyl and divinyl chlorophyll(ide) *a* and *b* by spectrofluorometry. Anal Biochem 178:294–300

# Chapter 4
# Development of Cell-Free Systems

*All truths are easy to understand once they are discovered. The point is to discover them (Galileo Galilei).*

## 4.1 Prologue

Originally, work on the biosynthesis of protochlorophyll(ide) [Pchl(ide)] and chlorophyll (Chl) in organello, started in 1967 in my laboratory at the National Research Institute in Tel-el-Amara, Lebanon (see Chap. 2) (Rebeiz 1967, 1968). At the time spectrophotometric instrumentation was used. Since I was aware that excised etiolated cucumber cotyledons greened very rapidly, within hours, in the light, I conjectured that if greening cotyledons were homogenized, I should be able to observe Pchl(ide) and Chl formation in the homogenate for a few minutes before the system fell apart. The first evidence of Chl biosynthesis in organello was observed in 1967 (Rebeiz 1967). However I soon realized that spectrophotometric techniques were not sensitive enough to observe consistent and reliable Pchl(ide) and Chl biosynthesis in organello. I therefore shifted to the use of $^{14}$C-δ-aminolevulinic acid ($^{14}$C-ALA) as a precursor of $^{14}$C-Chl. At the time ALA was known as a tetrapyrrole precursor (Granick 1961). We observed the first incorporation of $^{14}$C-ALA into $^{14}$C-Chl in my laboratory in 1969. The work was perfected in California at UC Davis in 1969–1970 when I joined Paul's Castelfranco Laboratory (Rebeiz and Castelfranco 1971a, b).

After 10 years of research the developed in organello and cell-free systems of Pchl(ide) and Chl biosynthesis were finally perfected in my laboratory at the University of Illinois, in Urbana-Champaign and the systematic investigations of the chlorophyll biosynthetic pathway became a possibility. An account of the gradual perfection of these systems will be described in this chapter.

C.A. Rebeiz, *Chlorophyll Biosynthesis and Technological Applications*,

DOI 10.1007/978-94-007-7134-5_4, © Springer Science+Business Media Dordrecht 2014

## 4.2 Total Protochlorophyll(ide) Biosynthesis in Organello

### 4.2.1 Radioactive Products of $^{14}$C-ALA Incubation

Greening etiolated cucumber cotyledons were rapidly homogenized with mortar and pestle at 4 °C and the homogenate was squeezed through cheese cloth. Incubation of $^{14}$C-ALA with the crude homogenate was performed for 16 h at 28 °C in the dark, with an array of cofactors listed below (Rebeiz and Castelfranco 1971a). The incubation produced a highly radioactive ether extract. Upon chromatography of the latter on thin layers of Silica Gel H, the radioactivity separated into four bands. The band at the origin consisted probably of free porphyrins and was not investigated any further. The remaining $^{14}$C-bands moved with the same chromatographic mobility of standard protochlorophyllide, Mg protoporphyrin monoester, and protochlorophyllide ester (Rebeiz and Castelfranco 1971a). Upon elution of these bands and rechromatography on Silica Gel H, they moved again with the same mobility as the standards. The $^{14}$C-Mg protoporphyrin monoester band was subsequently submitted to detailed chromatographic analysis. It coincided in every respect with standard Mg-protoporphyrin monoester.

In order to confirm the identity of the radioactive components in the two Protochl (ide) bands, the latter were submitted to further chromatographic analysis as described below.

### 4.2.2 Confirmation of the Nature of $^{14}$C-Protochlorophyllide

The $^{14}$C-protochlorophyllide band was eluted from Silica Gel H and rechromatographed as such on paper, in a variety of solvents, and after acidification on paper and on Silica Gel H. In toluene the $^{14}$C-protochlorophyllide band remained at the origin with standard protochlorophyllide while standard Mg-protoporphyrin monoester moved slightly from the origin, and standard Pchlide ester moved a little further (Rebeiz and Castelfranco 1971a). In this solvent the carotenoids move near the front. Upon acidification and chromatography in toluene, the $^{14}$C-Pchlide band cochromatographed with standard protopheophytin.

In 2,6-lutidine:0.05 N NH4OH (5:3.5 v/v) the $^{14}$C-Pchlide band cochromatographed with standard Pchlide. Upon acidification its chromatographic mobility decreased as expected for the Mg-free base in this solvent (Granick 1961), and it cochromatographed with standard protopheophytin.

In acetone: petroleum ether: acetic acid (3:7:0.01 v/v), spectroscopically pure, standard $^{14}$C-Pchlide and $^{14}$C-protopheophytin (Rebeiz 1967, 1968) gave rise to two major bands and one minor band (Rebeiz and Castelfranco 1971a). In this case too in vitro-biosynthesized $^{14}$C-Pchlide chromatographed in this solvent, before and after acidification, as standard $^{14}$C-Pchlide and[1]protopheophytin respectively. In this solvent Mg-protoporphyrin monoester (Mpe), protochlorophyllide ester, protoporphyrin monoester, and protopheophytin ester exhibited chromatographic mobilities strikingly different than $^{14}C$-Pchlide and $^{14}C$-protopheophytin. No efforts were made to determine whether the segregation of Pchlide and its Mg-free base into multiple bands was due to pigment degradation or to a separation of closely related, spectroscopically identical, compounds. A similar case was reported for radioisotopically and spectroscopically pure $^{14}C$-pheophorbide *a* and *b* chromatographed on icing sugar (Perkins and Roberts 1962). After acidification the $^{14}C$-Pchlide band cochromatographed on Silica Gel H in benzene:ethyl acetate: ethanol (8:2.5:5 v/v) with standard protopheophytin.

The foregoing results strongly suggested that the cell-free system did indeed synthesize $^{14}C$-Pchlide.

### *4.2.3 Confirmation of the Nature of $^{14}C$-Protoehlorophyllide Ester*

The $^{14}C$-Pchlide ester band was eluted in ether from Silica Gel H and rechromatographed as such and after acidification on paper. It was also chromatographed on Silica Gel H after partial acid hydrolysis.

In 2,6-lutidine:0.05N NH4OH (5:3.5 v/v) it moved differently than standard Pchlide and Mpe. It moved with the same mobility as standard Pchlide ester (Rebeiz and Castelfranco 1971a). In this solvent some standard Pchlide ester and $^{14}C$-Pchlide ester remained at the origin together with some carotene. It was conjectured that this may be due to interference by excess carotene in this solvent. After acidification, the $^{14}C$-Pchlide ester band moved with standard protopheophytin ester ahead of Mpe and protopheophytin. In acetone:petroleum ether:acetic acid (3:7:0.01 v/v) the $^{14}C$-Pchlide ester band moved with standard Pchlide ester, ahead of Mpe. After acidification it cochromatographed with standard protopheophytin ester ahead of Mpe and the multiple bands of protopheophytin. Upon partial hydrolysis of the $^{14}C$-Pchlide ester band in 12 N HC1 and chromatography on Silica Gel H in benzene:ethyl acetate:ethanol (8:2:5 v/v), the radioactivity exhibited the same mobility as standard protopheophytin ester and its hydrolysis product protopheophytin. In this case too, the results strongly suggested that the cell-free system was indeed synthesizing $^{14}C$-Pchlide ester.

### *4.2.4 Minimal Cofactor Requirement of the Tissue Homogenate Biosynthetic System*

The [illegible]al cofactor requirement for the incorporation of $^{14}C$-ALA into $^{14}C$-P[illegible] and $^{14}C$-Pchlide ester by the crude homogenate consisted of: CoA plus G[illegible]tassium phosphate, methyl alcohol, and $Mg^{2+}$. The absolute requirement for o[illegible]as also evident (Rebeiz and Castelfranco 1971a).

Other chemicals were tested for their effect on the biosynthetic activity of the crude homogenate namely: ATP, NAD, NADP, thiamine pyrophosphate, cytidine triphosphate, FAD, pyridoxal phosphate, NADPH, NADH, L-ascorbic acid, dehydroascrobic acid, D,L-methionine, cytochrome *c,* D-glucose + glucose oxidase, $Fe^{3+}$, $Fe^{2+}$, $Zn^{2+}$, $Co^{2+}$, vitamin $B_{12}$ and mannitol. None of these produced any stimulation in the cell-free crude homogenate system; some were slightly inhibitory.

#### 4.2.4.1 Effect of GSH and CoA on the Biosynthetic Activity of the Crude Homogenate

The individual effects of added GSH and CoA on the biosynthetic activity of the crude homogenate are described below. The omission of exogenous GSH from the reaction mixture depressed the $^{14}$C-incorporations into $^{14}$C-Mpe, $^{14}$C-Pchlide, and $^{14}$C-Pchlide ester regardless of the presence or absence of CoA. This finding suggested a general protective effect of GSH on sulfhydryl enzymes and porphyrinogen intermediates. However, the omission of exogenous CoA from the reaction mixture containing GSH resulted in decreased $^{14}$C-Pchlide biosynthesis without interference with $^{14}$C-Pchlide ester formation. It was conjectured that the CoA site of action along the biosynthetic pathway was probably located after Mpe formation and may be involved in the formation of $^{14}$C-Pchlide but not of $^{14}$C-Pchlide ester. These results supported the hypothesis that Pchlide and Pchlide ester were produced from a common precursor by two parallel and distinct biosynthetic routes (Rebeiz et al. 1970).

#### 4.2.4.2 Effect of $K^+$ and Pi on the Biosynthetic Activity of the Crude Homogenate

In order to determine which component of the potassium phosphate buffer had an effect on the biosynthetic activity of the system, the potassium phosphate buffer was replaced by a tris-Pi buffer and $K^+$ was added back to the reaction mixture as KCl.

Omission of exogenous phosphate had a depressing effect on the biosynthesis of both $^{14}$C-Mpe, $^{14}$C-Pchlide, and $^{14}$C-Pchlide ester (Rebeiz and Castelfranco 1971a). On the other hand, the omission of exogenous potassium from the complete mixture did not appear to affect the biosynthesis of $^{14}$C-Pchlide ester but depressed the biosynthesis of $^{14}$C-Mpe and $^{14}$C-Pchlide. The endogenous $K^+$ level in these crude homogenates was, unknown. Although these results did suggest an involvement of $K^+$ in the system, it was conjectured that additional experimental work with washed etioplasts preparations was needed before a specific cofactor role could be assigned to $K^+$ in Pchl(ide) biogenesis. The concentration of $Cl^-$ was without appreciable effect on the $^{14}$C-ALA incorporation into Mpe, Pchlide, or Pchlide ester (Rebeiz and Castelfranco 1971a).

#### 4.2.4.3 Effect of Methyl Alcohol and Other Aliphatic Alcohols on the Biosynthetic Activity of the Crude Homogenate

In order to study the alcohol specificity in the crude homogenate system, methanol was replaced in the reaction mixture by a short chain primary alcohol (ethanol) a secondary alcohol (isopropanol) or a tertiary alcohol (t-butanol). In all cases these alcohols were unable to substitute for methanol (Rebeiz and Castelfranco 1971a). The effect of higher concentrations of methanol on the biosynthetic activity of the system was subsequently investigated. It appeared that higher concentrations of methanol were inhibitory.

To determine whether methanol acted as a catalyst or a substrate, the system was incubated with $^{14}$C-methanol. Mpe as well as Pchlidee and Pchlide ester eluted from Silica Gel H were both labeled (Rebeiz and Castelfranco 1971a). Although these compounds were not purified to constant specific radioactivity, they remained radioactive after elution from Silica Gel H and rechromatography on paper in 2,6-lutidine:0.1 n NH4OH (5:3.5 v/v) or acetone:petroleum ether:acetic acid (3:7:0.01 v/v). These results suggested a substrate role for methanol in this system.

The differential incorporation of $^{14}$C-methanol into $^{14}$C-Pchlide and $^{14}$C-Pchlide ester (Rebeiz and Castelfranco 1971a) supported the hypothesis that these two compounds were produced via two separate pathways from a common precursor (Rebeiz et al. 1970). Alternatively, it was conjectured that a certain degree of transesterification of the alcohol of $^{14}$C-Pchlide ester with $^{14}$C-methanol may have taken place. The possibility that methanol might be preferentially incorporated into the phytol of Pchlide ester would also explain the observed incorporation data. However, we are not aware of any direct pathway leading from methanol to polyisoprenoids. The lower incorporation of $^{14}$C-methanol into $^{14}$C-Mpe and $^{14}$C-Pchlide as compared to the incorporation of $^{14}$C-ALA was expected under these experimental conditions. Indeed by assuming that methanol esterifies the propionic acid residue at the seventh position of the tetrapyrrole macrocycle, eight molecules of ALA would be incorporated into the macrocycle for every molecule of methanol utilized. Moreover, the specific radioactivity of the $^{14}$C-methanol was about 34 time lower than that of $^{14}$C-ALA.

#### 4.2.4.4 Effect of pH and Temperature on the Biosynthetic Activity of the Crude Homogenate

The multienzyme system in the crude homogenate appeared to have a pH optimum of about 7.7 for the biosynthesis of $^{14}$C-Mpe, and $^{14}$C-Pchl(ide) (Rebeiz and Castelfranco 1971a). Some of the initial experiments were carried out at pH 7.9 before it was recognized that the system was more active and more reproducible at pH 7.7.

The effect of two temperatures on the biosynthetic activity of the system was also investigated (Rebeiz and Castelfranco 1971a). At 20 °C more $^{14}$C-Mpe accumulated than at 28 °C whereas the incorporations into $^{14}$C-Pchlide and $^{14}$C-Pchlide ester were slightly depressed.

#### 4.2.4.5 Intracellular Localization of Mg Protoporphyrin and Pro-tochlorophyll Biosynthesis

In order to establish a connection between the porphyrin and phorbin biosynthetic activities and various subcellular fractions, the crude homogenate was fractionated by differential centrifugation into a crude etioplast preparation and a supernatant, enriched in soluble proteins, microsomes, mitochondria, and microbodies. The crude etioplast preparation was more active than either the crude homogenate or the supernatant (Rebeiz and Castelfranco 1971a).

Also a crude homogenate was prepared from etiolated cotyledons in a grinding buffer containing all the cofactors needed for maximal Pchl(ide) activity. From this homogenate a fortified etioplast pellet was prepared which was much more active than anything we had previously encountered and had a specific radioactivity much higher than the fortified crude homogenate (Rebeiz and Castelfranco 1971a). Upon washing these crude fortified etioplasts with the fortified grinding buffer their biosynthetic activity was remarkably well preserved (Rebeiz and Castelfranco 1971a). These results indicated that the in vitro biosynthesis of $^{14}$C-Mpe, $^{14}$C-Pchlide, and 14C-Pchlide ester was associated with the etioplasts.

#### 4.2.4.6 ATP and NAD Requirement for Maximal Biosynthetic Activity of Washed Fortified Etioplasts

The ability to prepare washed, active etioplasts presented a good opportunity for further studies of cofactor requirements in the presence of reduced levels of endogenous cofactors. Although a limited amount of experimentation was performed on this particular system, a requirement for ATP and NAD was established (Rebeiz and Castelfranco 1971a). It appeared that in the presence of ATP and NAD, the utilization of $^{14}$C-Mpe was increased as evidenced by reduced levels of the latter and increased levels of $^{14}$C-Pchlide. Both ATP and NAD appeared to be required for maximal accumulation of $^{14}$C-Pchlide by the washed fortified etioplasts (Rebeiz and Castelfranco 1971a). The presence of ATP did not appear to stimulate $^{14}$C-Pchlide ester accumulation. On the other hand, NAD alone, in the absence of ATP, resulted in a marked increase of $^{14}$C-Pchlide ester formation (Rebeiz and Castelfranco 1971a). These observations indicated that although both ATP and NAD were required for $^{14}$C-Pchlide biosynthesis, only NAD was needed for $^{14}$C-Pchlide ester biosynthesis.

## 4.3 Chlorophyll Biosynthesis in Organello

Once total $^{14}$CPchl(ide) biosynthesis in crude homogenates and in organello was achieved (Rebeiz and Castelfranco 1971a). We undertook the total biosynthesis of Chl in organello. Using the Pchl(ide) biosynthetic system it was possible to achieve total Chl biosynthesis in organello by carrying the incubations in the light instead of in darkness, since light is required for Chl biosynthesis.

### 4.3.1 *Radioactive Products of 14C-ALA Incubation with Homogenates Prepared from Etiolated and Greening Cotyledons*

Three types of cotyledons (Rebeiz 1967) were used for the preparation of crude homogenates:

(a) etiolated cotyledons still subject to a lag phase of chlorophyll *a* and *b* biosynthesis in the light,
(b) cotyledons irradiated for 2.5 h and capable of chlorophyll *a* biosynthesis in vivo and,
(c) cotyledons irradiated for 4.5 h and capable of both chlorophyll *a* and *b* biosynthesis in vivo (Rebeiz 1967).

When the homogenates prepared from 2.5-h irradiated cotyledons (green homogenate) or etiolated cotyledons (etiolated homogenate) were incubated with $^{14}$C-ALA for 16 h at 28 °C in the presence and absence of light, the $^{14}$C-Mpe $^{14}$C-Pchlide and $^{14}$C-Pchlide ester pools became radioactive (Rebeiz and Castelfranco 1971b). In both cases the ratios of $^{14}$C-Pchlide ester to $^{14}$C-Pchlide were two to three-fold higher when the incubations were carried out in the light instead of in the dark (Rebeiz and Castelfranco 1971b). Although chromatography on thin layers of Silica Gel H in benzene:ethyl acetate:ethanol (8:2:2, v/v) did not separate the $^{14}$C-Pchlide from $^{14}$C-chlorophyllide ($^{14}$C-Chlide) and the $^{14}$C-Pchlide ester from $^{14}$C-Chl *a*, these results suggested that in the light , $^{14}$C-Chl *a* might be the end product of the light-incubation rather than the usual mixture of $^{14}$C-Pchlide ester and $^{14}$C-Pchlide observed during dark-incubations (Rebeiz and Castelfranco 1971a). In order to prove that Point, the phytyl ester bands from the green reaction mixtures and the ones from the etiolated reaction mixtures were eluted from Silica Gel H and chromatographed on Whatman No. 1 paper in petroleum ether (60–90 °C): acetone: acetic acid (7:3:0.01, v/v). The radioactivity from the light incubation of the green homogenate moved with standard Chl *a*, while the radioactivity from the dark incubation of the etiolated homogenate moved with standard $^{14}$C-Pchlide ester (Rebeiz and Castelfranco 1971b). When the two phytyl ester bands derived from the light and dark incubations of the green homogenate were eluted, mixed and rechromatographed on paper, two bands appeared corresponding to standard Chl *a* and standard Pchlide ester (Rebeiz and Castelfranco 1971b). No efforts were made to separate $^{14}$C-Pchlide from $^{14}$C-Chlide by chromatography.

The $^{14}$C-Chl *a* from the light incubation of the green homogenate was further submitted to two-way paper chromatography in chloroform:petroleum ether (60–90 °C) (25:75, v/v) in the first direction followed by chromatography in petroleum ether (60–90 °C): acetone :n-propanol (90:10:0.45, v/v) in the second direction. $^{14}$C-Chl *a* cochromatographed with standard Chl *a* in both solvents. It moved with an $R_F$ of 0.78 in the first solvent and an $R_F$ of 0.63 in the second solvent. No $^{14}$C-Chl *b* could be detected on these chromatograms.

The purity of the $^{14}C$-Chl *a* fraction was tested further. A hexane extract from the light incubation of the green homogenate (2.5-h irradiated cotyledons) was mixed with carrier standard Chl *a*, and the specific radioactivity was determined at several stages of purification. It appeared, that the $^{14}C$-Chl *a* reached a constant specific radioactivity after the Silica Gel H purification (Rebeiz and Castelfranco 1971b).

These results strongly indicated that the green homogenate prepared from cotyledons exposed to light for 2.5 h was able to synthesize $^{14}C$-Chl *a* but not $^{14}C$-Chl *b* (Rebeiz 1967).

### 4.3.2 Biosynthesis of $^{14}C$-Chlorophyll a and b by Green Homogenates Prepared from Etiolated Cucumber Cotyledons Pre-irradiated for 4.5 h

It is well known that Chl *b* biosynthesis and accumulation becomes noticeable after etiolated tissues are partially greened. In etiolated cucumber cotyledons, that takes place after about 4 h of greening under white light (Rebeiz 1967). Thus when etiolated, excised cucumber cotyledons were irradiated with white fluorescent light for 4.5 h, they became partially green and capable of substantial Chl *b* biosynthesis in addition to Chl *a* (Rebeiz 1967). In order to find out whether homogenates prepared from such greening cotyledons were capable of Chl *b* biosynthesis, they were incubated in the light with $^{14}C$-ALA. The crude Chl *a* and Chl *b* fractions were both highly radioactive (Rebeiz and Castelfranco 1971b). The $^{14}C$-Chl *a* fraction was subsequently purified to constant specific radioactivity. Chromatography on Silica Gel H separated the $^{14}C$-Chl *a* from other $^{14}C$-porphyrins. This was accompanied by a strong decrease in specific radioactivity. Chromatography on cellulose MN 300 separated the $^{14}C$-Chl *a* efficiently from minor contamination by $^{14}C$-Chl *b* (Rebeiz and Castelfranco 1971b). Spectrophotometric measurements indicated a negligible Chl *b* contamination (about 2 %). Upon conversion into $^{14}C$-pheophytin *a* and rechromatography on cellulose MN 300, the specific radioactivity remained unchanged indicating that after the cellulose purification step, the $^{14}C$-Chl *a* was free of significant $^{14}C$-porphyrin, phorbin, or colorless radioactive contaminants (Jeffrey and Wright 1987; Perkins and Roberts 1962; Wickliff and Aronoff 1963). The $^{14}C$-pheophytin *a* fraction was subsequently degraded to pheophorbide *a* according to Perkins and Roberts (1960), and an aliquot was chromatographed on Silica Gel H in benzene: ethyl acetate:ethanol (8:2:5, v/v). As reported by Perkins and Roberts (1960), this procedure degraded $^{14}C$-pheophytin *a* extensively into $^{14}C$-pheophorbide *a* and two slow moving red fluorescent radioactive products, one of which was probably $^{14}C$-pyropheophorbide *a* (Perkins and Roberts 1960; Wickliff and Aronoff 1963).

The mixture of $^{14}C$-pheophorbide *a* and its $^{14}C$-tetrapyrrole derivatives was degraded further to derivatives of the individual pyrroles, that is to maleimides (Rebeiz and Castelfranco 1971b). The Maleimides quenched short wavelength ultraviolet light (254 nm) and appeared as blue spots on fluorescent thin layers viewed under ultraviolet light (Ellsworth and Aronoff 1968).

The crude $^{14}$C-Chl *b* fraction was also purified to constant specific radioactivity (Rebeiz and Castelfranco 1971b). After the first purification on thin layers of cellulose, the specific radioactivity dropped sharply. This was due to the separation of $^{14}$C-Chl *a* from $^{14}$C-Chl *b*. Upon rechromatography of the $^{14}$C-Chl *b* on cellulose the specific radioactivity remained unchanged indicating that after the first cellulose purification the $^{14}$C-Chl *b* fraction was free of significant amounts of $^{14}$C-porphyrins or $^{14}$C-Chl *a*. Spectrophotometric analysis indicated a negligible Chl *a* contamination (about 1.5 %) after the first cellulose purification.

After a second cellulose purification no contaminating Chl *a* could be detected by spectrophotometry. The radioactive shoulder running ahead of the bulk of the $^{14}$C-Chl *b* probably represented $^{14}$C-pheophytin *b* contamination.

## 4.4 Accumulation of Spectroscopically Detectable Amounts of Protochlorophyllide and Chlorophyll in Organello

After the achievement of $^{14}$C-Pchl(ide) and $^{14}$C-Chl *a* and *b* biosynthesis in organello, it was realized that progress in this field of research depended on the development of analytical techniques that allowed the detection of tetrapyrroles and the determination of their chemical structure by spectroscopic methods. Since tetrapyrrole were fluorescent, the research effort concentrated on the development of spectrofluorometric analytical techniques (Chap. 3). Thus by 1975 the first qualitative and quantitative spectrofluorometric techniques were developed (Rebeiz et al. 1975a). However the observed biosynthetic rates were still rather low. Then Paul Castelfranco and colleagues determined that concentrations of ATP higher than what was routinely used in previous cell-free systems (Rebeiz and Castelfranco 1971a, b) were needed for high rates of Mg-Proto biosynthesis in organello and demonstrated that ATP was a cofactor for Mg-Proto chelatase activity (Pardo et al. 1980). Building on this information the cell-free systems capable of Pchlide biosynthesis and accumulation were improved considerably as described below.

### *4.4.1 Development of in Organello Systems Capable of High Rates of Mg-Proto Monoester and Protochlorophyllide Biosynthesis in Organello*

By early 1982 the best in organello system capable of Mpe and Pchlide biosynthesis and accumulation is described in Table 4.1 displayed below.

The plastids were prepared from 4-day-oldetiolated cucumber cotyledons that had been pre-irradiated for 4 h with 320 μW/cm$^2$ of cool white fluorescent light.

**Table 4.1** Optimum concentrations of various cofactors needed for the Biosynthesis of Pchlide, Mpe and Proto by isolated plastids

| | Optimum cofactor concentration (mM) for: | | |
|---|---|---|---|
| Cofactors | Pchlide | MP(E) | Proto |
| Adenosine 5′-triphosphate | 20.0 | 20.0 | 0 |
| Nicotinamide adenine dinucleotide (oxidized) | 40.0 | 0 | 0 |
| Glutathione, reduced | 0 | 10.0 | 5.0 |
| Coenzyme A | 0 | 0 | 0 |
| Methyl alcohol | 1.25 | 2.5 | 25.0 |
| EDTA | 2.5 | 1.25 | 10 |
| $MgCl_2$ | 20.0 | 20.0 | 20.0 |
| $KH_2PO_4$ | 0 | 0 | 0 |
| Bovine serum albumin | 1 %[a] | 5 % | 0 |
| Sucrose | 330.0 | 330.0 | 330.0 |

[a]1 % (weight/volume)
Adapted from Rebeiz et al. (1982)

The tissue was hand-homogenized (ten strokes) at 4 °C and the plastids were incubated in the dark for 2 h in the absence or presence of different concentrations of various cofactors. Each incubation consisted of 2 ml of plastids (4–6 mg protein), 0.1 ml of 10 mM ALA, and 0.9 ml of $H_2O$. The Bovine serum albumin amounted to [a]1 % (weight/volume). Table 4.1, is displayed above.

### 4.4.2 *Effect of Kinetin in Enhancing the Synthesis and Accumulation of Protochlorophyllide in Organello*

We had previously proposed that etiochloroplasts differentiated less satisfactorily in organello than in vivo partly because Pchl(ide) and prothylakoid membranes accumulation appeared to be limited in vitro, by structural proteins synthesized in the cytoplasm and in the absence of which the massive formation of prothylakoids and grana was not possible (Rebeiz et al. 1973). It was therefore conjectured that should a method be found for obtaining etiochloroplasts containing excess Pchl (ide)-binding prothylakoids but lacking stochiometric amounts of membrane-bound Pchl(ide) then these plastids may be able to synthesize Pchlide, in organello, at very high rates in order to saturate the Pchl(ide) binding sites. We had indeed demonstrated earlier that in etiochloroplasts all of the Pchl(ide) was membrane bound (Smith and Rebeiz 1979). Several independent observations suggested that the forementioned goal may be experimentally feasible. First we noticed that when excised etiolated cucumber cotyledons were incubated overnight in the dark with an aqueous solution of kinetin, they underwent a 370 % increase in size. However, their Pchlide content increased only by about 128 % On the other hand, cytokinins are known to (a) promote the differentiation of plastids in vivo (Stetler and Laetsch 1965), and (b) as mentioned in (Daniell and Rebeiz 1982a) to increase the size and number of chloroplasts per cell and to increase the rate of RNA DNA and protein

**Table 4.2** Effect of kinetin-pretreatment of etiolated cotyledons upon the tetrapyrrole biosynthetic capacity of isolated plastids

| Experiment | Treatment | Δ Change after 2 h Pchlide MP(E) Proto in nmoles/100 mg plastid protein | | |
|---|---|---|---|---|
| | | Pchlide | MP(E) | Proto |
| A | Cotyledons were harvested in the dark and were pretreated either with pater a with kinetin; the plastids were isolatedin the dark, then were incubated with ALA | | | |
| | (a) Water pretreatment | 21.58 | 71.74 | 91.71 |
| | (b) Kinetin pretreatment | 56.40 | 8.86 | 451.40 |
| B | Cotyledons were harvested under 5 ftc (6 $\mu^2$ cm$^{-2}$) of white light and were pretreated either with water or with kinetin; the plastids were isolated under subdued laboratory light, and were incubated with ALA | | | |
| | (a) Water pretreatment | 18.24 | 73.86 | 267.36 |
| | (b) Kinetin pretreatment | 73.88 | 85.46 | 172.02 |

Cotyledons were harvested with hypocotyl hooks from 3-day old etiolated cucumber seedlings either in the dark or under subdued laboratory light (6 uw/cm$^2$). They were preincubated either with distilled water or with a 0.5 mM aqueous solution of kinetin for 20 h in the dark at 28 °C. The plastids were isolated either in the dark or under subdued laboratory light, were given a 30 s phototransforming light treatment (320 uw/cm$^2$ of white fluorescent light) then were incubated with ALA, in the dark. The Δ change refers to the pigment contents of the plastids at the end of the incubation minus the pigment content before incubation (Adapted from Daniell and Rebeiz 1982a)

biosynthesis in higher plants. Furthermore it was well known that in vivo, Pchl accumulation rapidly ceases in the dark due to feed back inhibition of ALA biosynthesis which may be relieved by addition of exogenous ALA (Beale and Castelfranco 1974; Sisler and Klein 1963). Altogether these observations raised the possibility that the forementioned kinetin treatment may have uncoupled the etioplast prothylakoid biosynthesis from Pchl(ide) biosynthesis which in turn resulted in the accumulation of excess kinetin-induced prothylakoid membranes devoid of stochiometric amounts of membrane-bound Pchl(ide).

In order to test the above hypothesis 3-day old etiolated cucumber Cotyledons were excised, with hypocotyl hooks, then were incubated either with distilled $H_2O$ or with a 0.5 mM aqueous kinetin solution, for 20 h in the dark at 28 °C.

The plastids were then isolated and their tetrapyrrole biosynthetic capability was determined by monitoring the conversion of exogenous ALA into Proto, Mpe and Pchlide (Daniell and Rebeiz 1982a). As shown in Table 4.2, A the Pchlide Net synthesis and accumulation capabilities of the plastids prepared from kinetin pretreated tissues was about 160 % higher than those of plastids prepared from the $H_2O$-pretreated control.

When the lag-phase of Chl biosynthesis was first eliminated (Rebeiz 1967) by exposing the cotyledons to laboratory light (4 h, 320 uw/cm$^2$) prior to the kinetin or $H_2O$ dark pretreatment, the biosynthetic capabilities of the plastids that were isolated from the kinetin pretreated tissues were about 400 % more active in Pchlide net synthesis and accumulation than the $H_20$ controls (Table 4.2, B). It is also apparent that the plastids prepared from kinetin-pretreated tissues were more potent in converting the nascent Proto into Mpe and Pchlide than the water controls as evidenced by the lower amounts of Proto accumulation (Table 4.2, Ba vs Bb). Altogether these results indicated that we may have succeeded with the fore-discussed treatment in uncoupling the simultaneous prothylakoid membrane biosynthesis from Pchl(ide) biosynthesis and in preparing etiochloroplasts containing excess Pchl(ide)-binding prothylakoid proteins.

It was conjectured that If the above hypothesis is correct and if the Pchlide biosynthesis-enhancing effect of the kinetin pretreatment is due to the pigment-uncoupled accumulation of prothylakoid proteins which are devoid of stochiometric amounts of bound Pchlide, then the addition of kinetin to incubated plastids should have no enhancing effect on the reactions of the Pchlide biosynthetic pathway per sec. This was found to be precisely the case (Daniell and Rebeiz 1982a).

### *4.4.3 Biosynthesis and Accumulation of Chlorophyll a at High Rates*

Once the biosynthesis and accumulation of Mpe and Pchlide was achieved as described above in Sect. 4.4.2 we directed our attention to the development of systems capable of high rates of Chl *a* biosynthesis and accumulation in organello. This effort is described below.

The high rates of Chl *a* net synthesis and accumulation were achieved by first preincubating 3-day-old etiolated cucumber cotyledons with an aqueous solution of 0.5 mM kinetin and 2 mM gibberellic acid for 20 h in darkness. The etiochloroplasts were then isolated as described in (Daniell and Rebeiz 1982b) and were resuspended in a medium modified from that reported in that publication. The medium and consisted of 0.5 M sucrose, 0.2 M Tris–HCl, pH 7.7, 20 mM $MgCl_2$, 2.5 mM EDTA, 1.25 mM methanol, 20 mM ATP, 40 mM NAD, 8 mM Methionine and 1 % BSA. Each incubation consisted of 1 ml of plastid suspension (12 mg plastid protein), one additional ml of the suspension medium, 0.1 ml of 10 mM ALA and 0.9 ml of $H_2O$. The plastids were irradiated with white light (320 $\mu w/cm^2$) for 30 s before incubation. Incubation was carried out at 28 °C for 2 h on a reciprocating water bath operated at 50 oscillations per min. Chlorophyll (ide) *a* [Chl(ide) *a*] net synthesis and accumulation was induced by exposing the plastids to an alternating light dark regime, which consisted of a 2.5 ms pulse of red actinic light, followed by 30 min of dark incubation. The red light pulse was generated by a Sunpack model Auto 611 photographic flash unit (Berkey Marketing Co., Woodside, NY) (Duggan and Rebeiz 1982b) shielded by a long wavelength cut-off red filter, Turner No. 25, that excluded light below 585 nm. In this manner, the Pchlide which was synthesized from the added ALA in the dark, was converted into Chl(ide) *a* by the brief red light treatment. During the subsequent dark incubation, Chlide *a* was converted into Chl *a* by esterification, and more Pchlide was regenerated during the following dark period.

**Table 4.3** Comparison of the rates of Chl(ide) *a* net synthesis by etiochloroplasts in vitro, with the highest rates observable during greening in vivo

| | | Δ change after 2 h incubation | | |
|---|---|---|---|---|
| | | nmoles/μmole chl(ide) present before incubation | | |
| Experiment | Treatments | Chlide *a* | Chl *a* | Chl(ide) *a* |
| A | Etiochloroplasts were isolated from kinetin and Gibberellic acid-pretreated cucumber cotyledons as described in Methods and incubated in the presence of 0.33 mM ALA under a repetitive light dark regime that consisted of red actinic light followed by 30 min of darkness | 2557.80 | 2172.96 | 4730.76 |
| B | Cotyledons pretreated with kinetin and Gibberellic acid were incubated in the presence of 0.33 mM ALA under a repetitive light dark regime that consisted of 2.5 ms of red actinic light followed by 30 min of darkness | 148.90 | −229.50 | −80.60 |
| C | Cotyledons pretreated with kinetin and Gibberellic acid were illuminated under 320 μW $cm^{-2}$ of cool white fluorescent light for 2 h | 36.01 | 2109.88 | 2145.89 |
| D | Cotyledons pretreated with water were illuminated under 320 μW $cm^{-2}$ of cool white fluorescent light for 2 h | −64.40 | 2327.37 | 2262.97 |
| E | Etiolated cucumber seedlings were exposed to 50 μW $cm^{-2}$ of cool white fluorescent light for 30 min followed by 3 h of dark incubation (15) to eliminate the lag phase in Chl(ide) a biosynthesis. The Cotyledons were excised with hypocotyl hooks and were illuminated under 320 μW $cm^2$ of white light for 2 h | 35.17 | 2466.87 | 2502.06 |

The Chls present in the acetone extracts, were extracted into hexane, while the Chlide remained in the hexane-extracted acetone fraction as described in (Bazzaz and Rebeiz 1979). The amounts of Chl *a* and Chlide *a* were determined by spectrofluorometry as described in (Bazzaz and Rebeiz 1979). The Δ change refers to the pigment content at the end of the incubation minus the pigment content before incubation. Chl(ide) refers to the total amount of Chl + Chlide *a*
Adapted from Daniell and Rebeiz (1982b)

The isolated etiochloroplasts approximately quadrupled their Chl(ide) *a* content during the 2 h incubation under the forementioned light–dark regime (Daniell and Rebeiz 1982b). For comparison purposes, the Chl(ide) *a* net synthesis and accumulation in excised cucumber cotyledons, which were greening at the highest rate observable in nature is compared quantitatively with the in organello rate in Table 4.3. For such a comparison to be meaningful, we have reported the rates on a unit Chl(ide) *a* present before incubation, which in effect normalized the biosynthetic rates to the same number of plastids in both systems (Rebeiz et al. 1982). It is apparent from Table 4.3, A, E that the isolated plastids accumulated Chl(ide) *a* at a rate about twice as high as the highest rate of greening achievable In vivo.

It should be noted that the isolated plastids and the cotyledonary tissue exhibited different requirements for achieving their highest greening rates. First, the isolated plastids required much lower light intensities than the cotyledons (Table 4.3, A, B, E). Second, while the plastids did very well in the presence of exogenous ALA [indeed no substantial tetrapyrrole biosynthesis occurs in the absence of added ALA in organello systems (Mattheis and Rebeiz 1977a, b)], the addition of ALA to the greening cotyledons was detrimental to Chl(ide) *a* accumulation, even under the low light intensities of exp. B. This is not surprising however as plant tissues are noteworthy for generating their own ALA during greening and for failing to accumulate substantial amounts of Chl(ide) *a* from exogenous ALA, under even moderate light intensities (Sisler and Klein 1963). Second, while pretreatment with hormones was required for achieving high Chl(ide) *a* biosynthetic rates in organello, this was not observed to be the case in vivo, as if the tissue generated its own hormonal requirements during greening in the light (Table 4.4, A, C, D).

The massive amounts of Chlide *a* detectable in vitro, was not generated by the hydrolysis of endogenous Chl *a* but was synthesized de novo from exogenous ALA as evidenced by the lack of Chlide *a* accumulation in dark controls, i.e., in etiochloroplasts incubated in complete darkness, with ALA, for 2 h. It could have arisen also from the newly formed Chl *a*. Finally the cell-free system described in exp. A (Table I) was not optimized for Chl(ide) *b* biosynthesis and accumulation. The in organello system capable of massive Chl(ide) *b* biosynthesis and accumulation will be discussed in Sect. 4.5.2 below.

## 4.5 Development of an in Organello System Capable of High Rates of Chlorophyll(ide) *b* Biosynthesis and Accumulation

### *4.5.1 Preparative Techniques*

After removing the hypocotyl hooks of light-pretreated cotyledons, 85–90 g batches of tissue were homogenized in a Waring blender (2 bursts, 5 s each) under subdued cool white fluorescent laboratory light (4.0 nmol/m$^2$/s) in 230 ml of a homogenization medium consisting of 0.5 M sucrose, 15 mM Hepes, 30 mM Tes, 1 mM MgC12, 1 mM EDTA, 5 mM cysteine, and 0.2 % bovine serum albumin (w/v) at room temperature, and pH 7.7 (Daniell and Rebeiz 1982b; Rebeiz et al. 1984). The homogenate was passed through four layers of cheese cloth and one layer of miracloth. The plastids were pelleted by centrifuging the homogenate at 200× g for 3 min followed by centrifuging the resulting supernatant for10 min at 1,500× g. The plastid pellet was gently resuspended in 10 ml of homogenization medium. The resuspended plastids were further purified by layering 6 ml of the suspension over 25 cm$^3$ of homogenization medium containing 35 % Percoll, in a 50 ml centrifuge tube and centrifugation at 6,000× g for 5 min in a Beckman JS-13 swinging bucket rotor at 1 °C. Intact plastids recovered as a pellet were gently resuspended in 5 $cm^3$ of medium consisting of 0.5 M sucrose, 0.2 M Tris–HCl, 20 mM MgCl2, 2.5 mM EDTA, 1.25 mM methanol, 20 mM ATP, 40 mM NAD, 8 mM methionine, and 5 μM phytol at a room temperature, pH 7.7 (Daniell and Rebeiz 1982b; Rebeiz et al. 1984).

Etioplast incubations consisted of 0.95 ml of plastid suspension (3–5 mg protein) and 0.05 ml of 10 mM ALA. Incubation was carried out at 28 °C for 15–60 min on a reciprocating water bath operated at 50 oscillation per min under 4 umol/$m^2$/s of cool white fluorescent light. Incubations were terminated by precipitation with 10 ml of acetone: 0.1 M $NH_4OH$ (9:1, v/v). The acetone extracts containing the tetrapyrrole pigments were cleared of insoluble lipoproteins by centrifugation at 39,000× g for 12 min. Chl *a,* a fully esterified tetrapyrrole, was removed from the aqueous acetone solution by extraction with 1 volume of hexane followed by a second extraction with 1/3 volume of hexane. The more polar monocarboxylic tetrapyrroles such as Pchlide *a* and Chlide *a* remained in the hexane-extracted aqueous acetone fraction. The amount of Pchlide *a* and Chlide *a* was determined spectrofluorometrically on aliquots of the hexane-extracted acetone fraction as described in (Rebeiz et al. 1975a). One ml aliquot of the hexane extract containing the Chl was dried under $N_2$ gas and the residue was redissolved in 4 ml of 80 % acetone. The amount of Chl *a* and *b* in the acetone solution was determined spectrofluorometrically as described in (Bazzaz and Rebeiz 1979) Fluorescence spectra were recorded as described in (Rebeiz et al. 1975a). The endogenous ALA content of the isolated plastids was determined as described by Mauzerall and Granick (1956).

### *4.5.2 Biosynthesis and Accumulation of Chlorophyll b*

Plastids were prepared from etiolated cucumber cotyledons at three different stages of greening. In the first set of experiments, etioplasts were prepared from etiolated cotyledons that were potentiated for Chl(ide) *b* biosynthesis by pretreatment with one 2.5 ms flash of "actinic white light" followed by 60 min of dark incubation. Such plastids were incapable, of Chl(ide) *b* net synthesis in vitro (Table 4.4, A).

In a second set of experiments, the cotyledons were greened for 24 h (120 nmol/$m^2$/s of metal halide radiation) prior to the preparation of etiochloroplasts. At this stage the cotyledons had accumulated large amounts of Chl *a* and *b*. The evaluation of the Chl *b* biosynthetic activity in organello was rather uncertain because of the high background of accumulated Chl *a* and *b* (Table 4.4, *B*). In a third set of experiments, etiolated cotyledons were greened for 4 h prior to etiochloroplast preparation. At this stage of greening, the lag-phase of Chl *b* biosynthesis had been removed and the tissue had just started active Chl *b* biosynthesis (Rebeiz 1967). As a consequence, although Chl *b* biosynthesis was fully potentiated, the amount of accumulated Chl(ide) *a* and *b* was not large enough to interfere with the

**Table 4.4** Conversion of exogenous 5-aminolevulinic acid (ALA) to Chl(ide) *a* and *b* (nmol per 100 mg plastid protein) by isolated etiochloroplasts

| | | | Incubation time [min] | | | |
|---|---|---|---|---|---|---|
| Exp. | Irradiation | Tetrapyrrole | 0 | 15 | 30 | 60 |
| *A* | 2.5 ms | MV Chlide *a* | 3.1 ± 0.4 | – | – | 3.1 ± 0.3 |
| | | MV Chl *a* | 21.8 ± 6.1 | – | – | 14.7 ± 2.1 |
| | | MV Chlide *b* | 0.3 ± 0.0 | – | – | 0.4 ± 0.2 |
| | | MV Chl *b* | 0.7 ± 0.2 | – | – | 0.8 ± 0.5 |
| *B* | 24 h | MV Chlide *a* | 45.1 ± 0.6 | – | – | 180.3 ± 13.7 |
| | | MV Chl *a* | 11849.5 ± 1137.7 | – | – | 12325.0 ± 205.1 |
| | | MV Chlide *b* | 12.5 ± 3.9 | – | – | 24.1 ± 0.8 |
| | | MV Chl *b* | 4380.4 ± 177.2 | – | – | 4577.4 ± 316.2 |
| *C* | 4 h | MV Chlide *a* | 14.3 ± 5.3 | – | 43.0 ± 8.3 | 64.7 ± 12.7 |
| | | MV Chl *a* | 97.3 ± 19.6 | – | 181.7 ± 11.7 | 108.6 ± 4.0 |
| | | MV Chlide *b* | 1.9 ± 0.3 | – | 4.3 ± 2.1 | 6.9 ± 1.3 |
| | | MV Chl *b* | 18.5 ± 4.4 | – | 33.2 ± 2.1 | 22.7 ± 0.1 |
| *D* | 4 h | MV Chlide *a* | 12.1 ± 0.7 | 25.3 ± 2.4 | 52.3 ± 1.2 | 55.0 ± 3.0 |
| | | MV Chl *a* | 138.0 ± 15.7 | 272.9 ± 8.8 | 244.7 ± 14.4 | 200.8 ± 4.5 |
| | | MV Chlide *b* | 1.8 ± 0.1 | 2.8 ± 0.1 | 5.9 ± 1.0 | 7.0 ± 0.2 |
| | | MV Chl *b* | 35.0 ± 1.9 | 59.4 ± 0.9 | 70.9 ± 2.0 | 62.1 ± 2.3 |

Etiochloroplasts were prepared from etiolated cucumber cotyledons pretreated with "white light" as follows: *A*: 2.5 ms actinic flash followed by 60 min of darkness; *B*: 120 nmol/m2/s for 24 h; C and *D*: 120 nmol/$m^2$/s for 4 h. Each incubation contained 0.05 ml of 10 mM ALA (about 10.4–13.8 nmol, depending on the protein content of the plastids). The endogenous ALA content amounted to an insignificant amount of about 0.003 nmol per 100 mg plastid protein
All values ± standard deviation are means of two replicates

determination of Chl(ide) *a* and *b* biosynthesis in organello. As shown in Table 4.4, C, D, net Chl(ide) *a* and *b* biosynthesis was observed after 15, 30, and 60 min of incubation.

## 4.6 Development of Cell-Free systems Capable of Supporting Partial Reactions of the Chlorophyll Biosynthetic Pathway

In what follows the development of in vitro systems capable of catalyzing partial reactions of the Chl biosynthetic pathway will be described.

### *4.6.1 Conversion of Protoporphyrin IX to Mg-Protoporphyrin IX*

Walker and Weinstein had described earlier a subplastidic system that overcame the apparently mandatory requirement of plastid intactness for Mg-Proto chelatase activity (Walker and Weinstein 1991b). The system was prepared from lysed pea (*Pisum sativum*) chloroplasts, and consisted of soluble and membrane-bound fractions. Attempts at preparing similar systems from cucumber chloroplasts were not successful (Walker and Weinstein 1991b). Below, the preparation of a stabilized subplastidic membrane fraction prepared from cucumber etiochloroplasts, capable of high rates of Mg insertion into exogenous Proto, without addition of a soluble stromal fraction, is described.

#### 4.6.1.1 Plastid Isolation

Four-day-old etiolated cucumber cotyledons were excised with hypocotyl hooks under subdued laboratory light (about 5 ft. candles). The excised cotyledons were incubated at 28 °C for 20 h in darkness in deep Petri dishes (80 × 100 mm), each containing 3 g of tissue and 9 ml of an aqueous solution composed of 2 mM potassium gibberellate and 0.5 mM kinetin, pH 4.3 (2, 11).

All further procedures were carried out under subdued laboratory light. After removal of the hypocotyl hooks, 20 g of pretreated cotyledons were hand-ground in a cold ceramic mortar containing 75 ml of homogenization medium. The latter consisted of 500 mM sucrose, 15 mM Hepes, 30 mM Tes, 1 mM MgCl2, 1 mM EDTA, 0.2 % (w/v) BSA, and 5 mM cysteine at a room-temperature pH of 7.7 (Lee et al. 1992). The homogenate was filtered through two layers of Miracloth (Calbiochem., La Jolla, CA.) and was centrifuged at 200 g for 5 min in a Beckman JA-20 angle rotor at 1 °C. The supernatant was decanted and centrifuged at 1,500 g for 20 min at 1 °C. The pelleted crude etiochloroplasts were gently resuspended in 5 ml of homogenization, suspension or lysing medium using a small paintbrush. The suspension medium was composed of 500 mM sucrose, 200 mM Tris, 20 mM MgCl2, 2.5 mM EDTA, 40 mM $NAD^+$, 20 mM ATP, 8 mM methionine, 1.25 mM methanol, and 0.1 % (w/v) BSA at a room-temperature pH of 7.7. Unless otherwise indicated, the lysing medium consisted of 25 mM Tris, 30 mM MgCl2, 7.5 mM EDTA, 40 mM NAD+, 20 mM ATP, 8 mM methionine, 37.5 mM methanol, and 4.5 mM glutathione at a room-temperature pH of 7.7 (Lee et al. 1992). For further plastid purification, the pelleted crude etiochloroplasts were resuspended in 5 ml of homogenization medium and were purified by Percoll density centrifugation (Lee et al. 1992). The pelleted, Percoll-purified etiochloroplasts were then resuspended either in the suspension or lysing medium.

#### 4.6.1.2 Preparation of Etiochloroplast Stroma and Membranes

To stabilize Mg-Proto chelatase activity, and unless otherwise indicated, 100 nmoles of Proto per 0.33 ml of membrane suspension were added to the lysed plastids immediately after lysis. The stroma and membranes fractions were then resolved following ultracentrifugation at 235,000 g for 1 h in a Beckman 80 Ti angle rotor at 1 °C (Lee et al. 1992).

#### 4.6.1.3 Mg-Proto Chelatase Assay

In a total volume of 1 ml containing 0.33 ml of crude etiochloroplasts, or purified etiochloroplasts, lysed plastids, stroma, or membrane fractions, the reaction mixture consisted of 100 μM Proto, 330 mM sucrose, 200 mM Tris, 20 mM MgCl2, 5 mM EDTA, 27 mM $NAD^+$, 15 mM ATP, 5 mM methionine, 25 mM methanol, 3 mM glutathione, and 0.1 % (w/v) BSA, at a room-temperature pH of 7.7. Incubation was in a flat-bottom glass tube. To each incubation tube was added 0.01 ml of 10 mM Proto (100 nmoles) except when the Proto had already been added to the lysed etiochloroplast suspension. The tubes were wrapped in aluminum foil and were incubated at 28 °C for 2 h in darkness in a shaking water bath operated at 50 oscillations/min.

Before and after incubation, pigments were extracted by the addition of 5 ml of cold acetone:0.1 N NH4OH (9:1 v/v) per ml reaction mixture. This was followed by centrifugation at 39,000 g for 10 min at 1 °C. The ammoniacal acetone extract was retained, and the pellet was discarded. Chlorophylls and other fully esterified tetrapyrroles were transferred from acetone to hexane by extraction with an equal volume of hexane, followed by a second extraction with one-third volume of hexane. The remaining hexane-extracted acetone residue containing Proto, Mg-Proto, and Pchlide, was used for quantitative determination of Mg-Proto by spectrofluorometry (Rebeiz et al. 1975b). The measured Mg-Proto pool consisted of Mg-Proto and smaller amounts of Mg-Proto monoester.

Fluorescence spectra were recorded at room temperature on a fully corrected photon-counting SLM spectrofluorometer Model 8000C, interfaced with an IBM model XT microcomputer. Determinations of Mg-Proto were performed on an aliquot of the hexane-extracted acetone fraction in a cylindrical microcell 3 mm in diameter. All spectra were recorded at emission and excitation bandwidth of 4 mm. The amount of Mg-proto was determined from its fluorescence amplitude at its emission maximum, upon excitation at 420 nm. Fluorescence amplitudes were converted to Mg-Proto concentrations by reference to a standard calibration curve. The digital spectral data were automatically converted by the computer into quantitative values.

Fluorescence polarization and anisotropy of Proto in different environments were measured by simultaneously observing the horizontal and vertical emission from the sample when exciting with horizontally and vertically polarized light as described in the SLM manual.

#### 4.6.1.4 Demonstration of Mg-Proto Chelatase Activity in Ruptured Etiochloroplasts

The purity of Percoll-purified cucumber etiochloroplasts and the efficacy of lysis by osmotic shock were evaluated in (Lee et al. 1991). Percoll-purified etiochloroplasts were more highly intact (87 %) than crude etiochloroplasts (68 %) and contamination by other subcellular organelles was reduced five to ninefold in comparison to the crude etiochloroplasts. Lysis of etiochloroplasts by osmotic shock was as efficient (98 %) as lysis by 0.1 % Triton X-100 (100 %) (Lee et al. 1991).

The activities of crude and Percoll-purified etiochloroplasts amounted to 325 and 540 nmoles respectively of Mg-Proto synthesized per 100 mg of plastid protein. These values were two to threefold larger than those reported by others for developing cucumber chloroplasts (Fuesler et al. 1981, 1984a; Walker and Weinstein 1991a). No significant differences in Mg-Proto chelatase activity between unlysed and lysed etiochloroplasts were observed, although the activity of purified plastids were significantly higher than the crude ones. It was therefore concluded that the Mg-Proto chelatase activities of either crude or Percoll-purified etiochloroplasts were not altered by plastid rupture. This in turn indicated that in cucumber etiochloroplasts, plastid intactness was not a mandatory requirement for the insertion of $Mg^{2+}$ into Proto by Mg-Proto chelatase. This was at variance with the results of others who found that any disruption of cucumber chloroplasts resulted in a drastic decrease in Mg-Proto chelatase activity (Fuesler et al. 1984b; Walker and Weinstein 1991a).

#### 4.6.1.5 Stabilization of Mg-Proto Chelatase Activity in a Subplastidic Membrane Fraction

Initial attempts aimed at recovering Mg-Proto chelatase activity in isolated subplastidic fractions met with limited success. Some activity was recovered in unstabilized plastid membranes and none was found in the plastid stroma

(Lee et al. 1992). In other words, although Mg-Proto chelatase activity survived etiochloroplast disruption, most of the activity was lost upon separating the plastid membranes from stroma by ultracentrifugation. Several attempts were made to stabilize Mg-Proto chelatase activity during ultracentrifugation. Success was achieved when Proto, the natural substrate for Mg-Proto chelatase, was added to lysed plastids immediately after lysis and prior to ultracentrifugation, at a concentration of 100 nmoles/0.33 ml of lysed plastid suspension (Lee et al. 1992). It is very likely that protection of Mg-Proto chelatase activity by adsorbed Proto involved stabilization of the enzyme by its substrate, a well documented phenomenon (Scopes 1982).

After ultracentrifugation all Mg-Proto chelatase activity was found in the membrane fraction. The stroma was inactive. The isolated plastid membranes contained the bulk of the added Proto. Although the observed membrane-bound Mg-Proto chelatase activity (85.20 nmol/2 h/100 mg protein) was only one-sixth that of purified etiochloroplasts, it was 145–450-fold higher than activities reported by others for cucumber subplastidic preparations (Smith and Rebeiz 1977a). It is worth noting that no improvement in Mg-Proto chelatase activity was observed upon recombining stroma and plastid membranes. On the contrary, the recombination resulted in a statistically significant drop in activity.

#### 4.6.1.6 Partition of the Exogenous Proto Substrate Between the Membrane and Stromal Fractions

As reported above, exogenous Proto had to be added to the lysed etiochloroplasts to stabilize the Mg-Proto chelatase activity during separation of plastid stroma from membranes. After ultracentrifugation, about 80 % of the added Proto was found to be associated with the membrane fraction, whereas the remaining 20 % was recovered with the stroma (Lee et al. 1992). To determine whether the adsorbed Proto was loosely or tightly bound to the membranes, the latter were resuspended in the lysing medium and were subjected to a second ultracentrifugation. Almost all the adsorbed Proto was recovered in the membrane fraction thus indicating that the Proto was tightly associated with the plastid membranes.

Further insight into the molecular environment of the membrane-bound Proto was derived from fluorescence polarization and anisotropy measurements. It is acknowledged that slow rotation of a fluorophore such as Proto, relative to the rapid emission of fluorescence, results in larger polarization and anisotropy values than if the fluorophore is rapidly undergoing rotation. On the other hand, the rate of rotation of a fluorophore depends on its molecular environment. For example a fluorophore in a viscous or rigid environment rotates much slower than in a more fluid environment. The polarization and anisotropy values of membrane-bound Proto were significantly higher than for the stromal Proto or for Proto dissolved in 80 % aqueous acetone or in the aqueous incubation medium (Lee et al. 1992). This suggested that the membrane-bound Proto was most probably solvated in a more rigid environment, such as the hydrophobic core of the membrane fraction.

Finally an attempt was made to determine whether the activity of Mg-Proto chelatase, with the Proto substrate already adsorbed to the plastid membranes, would increase upon addition of further exogenous Proto to the incubation medium. No significant differences in Mg-chelatase activities were observed with the addition of various amounts of Proto to the incubation medium. This indicated that the concentration of the Proto adsorbed to the membranes was high enough, to saturate the chelatase activity during 2 h of incubation.

#### 4.6.1.7 ATP Requirement for Subplastidic Membrane-Bound Mg-Proto Chelatase Activity

In intact cucumber etiochloroplasts, Mg-Proto chelatase activity, became saturated at about 10 mM ATP (Fuesler et al. 1984a). Two sets of experiments were designed to determine whether similar ATP concentrations would be required for optimal activity of the subplastidic membrane-bound chelatase.

In one set of experiments, ATP was omitted from the lysing medium, while in the other ATP was included in the lysing medium. When ATP was omitted from the lysing medium prior to separating plastid membranes from stroma, Mg-Proto chelatase activity was lost, irrespective of the amount of ATP subsequently added to the incubation medium (Lee et al. 1992). This indicated that ATP was required for enzyme stabilization during lysis and ultracentrifugation. When ATP was included in the lysing medium, membrane-bound Mg-Proto chelatase responded to addition of further ATP to the incubation medium in a manner similar to whole etiochloroplasts. It exhibited optimum activity at about 15 mM ATP.

#### 4.6.1.8 Mg requirement for Subplastidic Membrane-Bound Mg-Proto Chelatase Activity

As reported for intact etiochloroplasts (Fuesler et al. 1984a), added $Mg^{++}$ was also required for Mg-Proto chelatase activity. Washing with EDTA prior to demonstrating Mg requirement, as was reported by others for intact plastids (Fuesler et al. 1984a), was not necessary. After an initial lag phase that was overcome at concentrations of $MgCl_2$ larger than 5 mM, the activity of Mg-Proto reached a maximum at 10 mM $MgCl_2$. Higher $MgCl_2$ concentrations were inhibitory (Lee et al. 1992).

#### 4.6.1.9 EDTA Requirement for Subplastidic Membrane-Bound Mg-Proto Chelatase Activity

It has been our experience that optimal conversion of ALA to Proto, Mg-Porphyrins and protochlorophyllide requires the presence of EDTA in both the homogenization medium of greening tissues and the incubation medium of isolated etiochloroplasts

(Rebeiz et al. 1982). When EDTA was omitted from the homogenization, lysing, and incubation media, Mg-Proto chelatase activity was lost and membrane-bound Proto was converted to Zn-Proto (emission maximum at 590 nm, in hexane-extracted acetone, at room temperature), instead of being converted to Mg-Proto (emission maximum at 595 nm). The fluorescence properties of Mg-Proto and Zn-Proto have been described elsewhere (Smith and Rebeiz 1977b). Addition of 2.5–20 mM EDTA to the incubated plastids did not restore the Mg-Proto chelatase activity, but suppressed Zn-Proto formation. When EDTA was included in the homogenization, and lysing media, addition of EDTA (2.5–10 mM) to the incubation medium did not affect Mg-Proto chelatase activity, which appeared to proceed normally. However higher concentrations of added EDTA (15–20 mM) severely inhibited (74–83 %) Mg-Proto formation.

#### 4.6.1.10 Lack of Effect of Other Additives on Mg-Proto Chelatase Activity

The lysing medium also contained $NAD^+$, gluthathione, methanol and methionine, the effect of which upon Mg-Proto chelatase activity had not been determined. Percoll-purified plastids were therefore lysed in the normal lysing medium which contained $NAD^+$, glutathione, methanol and methionine, or in a lysing medium which lacked the above additives. The separated membranes were then incubated in the presence and absence of the aforementioned additives. These additives exhibited no measurable effect on Mg-Proto chelatase activity (Lee et al. 1992).

### *4.6.2 Development of a Cell-Free System Capable of the Conversion of Divinyl Mg-Protoporphyrin IX to Monovinyl Mg-Protoporphyrin IX*

Kim et al. (1997) had reported earlier that an interaction of plastid membranes, stroma, and NADPH was involved in the regulation of DV and MV Pchlide *a* biosynthesis. Although the combination of plastid stroma and plastid membranes resulted in higher total Pchlide *a* formation from exogenous DV Mg-Proto, the presence of stroma in the reaction mixture resulted in a 5 to16-fold decrease in the MV/DV Pchlide *a* ratio. This in turn suggested an inhibition of 4-vinyl reductase (4VR) reactions between DV Mg-Proto and Pchlide *a* by the plastid stroma. It was therefore conjectured that the study of 4VR activities in the absence of the plastid stroma, i.e. in isolated plastid membranes, may give a deeper insight into 4VR activities and may unmask some additional undetected 4VR activities.

It was therefore conjectured that a comparative study of various vinyl reductase (4VR) activities would be facilitated by solubilization of 4VR activities from

**Table 4.5** Detection of 4VMg-ProtoR and other 4VR activities in isolated plastid membranes and solubilized fractions prepared from etiolated barley leaves

| | Net change in | | | |
|---|---|---|---|---|
| | MV Mg-proto in 60 min | MV Mpe in 60 min | MV Pchlide *a* in 60 min | MV Chlide *a* in 5 min |
| Membrane fraction | (pmoles per mg protein) | | | |
| Membranes | 29.2 ± 14.1 | 48.2 ± 19.7 | 243.8 ± 16.8 | 47451 ± 345.1 |
| Solubilized fraction | 155.9 ± 30.6 | 235.9 ± 30.6 | 1375.1 ± 132.6 | 92211 ± 345.1 |
| Solubilized fraction as % of Membrane activity | 533.9 | 489.4 | 564.0 | 194.33 |
| 4VCR/4VR activities | 591 | 391 | 67 | 1 |

Percoll-purified etiochloroplasts were isolated from barley seedlings under laboratory light. The membranes and Chaps-solubilized fractions were prepared as described above

Values are means of 2–3 replicates ± standard deviation

The activity of 4VMg-ProtoR was 591 time less than that of 4VCR

Percoll-isolated plastids. In this effort barley etiochloroplast membranes were used. Barley etioplast membranes were prepared from crude and Percoll purified etioplasts as described elsewhere (Kolossov and Rebeiz 2001). Solubilization of 4VMg-ProtoR activity was performed as described below.

Four millimolar Chaps was used for solubilization. Such a concentration was successfully used in the past to solubilize 4-vinyl chlorophyllide reductase (4VCR) (Kolossov and Rebeiz 2001). Essentially, membranes from Percoll purified barley etioplasts were resuspended in solubilization buffer at a rate of 2 mg protein per ml. The solubilization buffer consisted of 20 mM Tris–HCl, 1 mM EDTA, 10 % of Glycerol and 4 mM 3-[(3-Cholamidopropyl)dimethylammonio]-1-Propanesulfonate (Chaps) adjusted to a pH of 7.7 at room temperature. The remaining steps were performed as described elsewhere (Kolossov and Rebeiz 2001).

Initial attempts at detecting 4-vinyl Mg-Proto reductase (4VMg-ProtoR or 4VMPR) in isolated etiochloroplast membranes under 4-vinyl Chlide reductase (4VCR) incubation conditions (Kolossov and Rebeiz 2001) were unsuccessful. After further experimentation, and adjustment of incubation conditions, it became possible to detect 4VMg-ProtoR activity in the isolated membrane preparation. However, the following adjustments in incubation conditions had to be made (a) the incubation time for 4VMg-ProtoR was raised from 5 min to 60 min, (b) the sample load was raised about tenfold; and, (c) the exogenous substrate concentration was lowered 4–8 times. For 4VMPR however, the buffer was diluted 1:1 (v/v) with distilled water in order to facilitate the extraction of Mg-Proto into diethyl ether. The other incubation conditions were kept unchanged as described in (Kolossov and Rebeiz 2001). The conversion of DV Mg-Proto to MV Mg-Proto by plastid membranes and the solubilized 4VMg-ProtoR activity is reported in column 2 of Table 4.5 which is displayed above.

### 4.6.3 *Development of a Cell-Free System Capable of the Conversion of Divinyl Mg-Proto Ester to Monovinyl Mg-Proto Ester*

This enzyme was first reported by Ellsworth and Hsing in a supernatant of etiolated wheat leaves homogenates (Ellsworth and Hsing 1973), but was never confirmed by others (Rebeiz et al. 2003).

In contrast to previous reports (Ellsworth and Hsing 1973), and for the first the time, 4VMpeR activity was unambiguously demonstrated as a membrane-bound enzyme. The activity was detected under the same preparative and incubation conditions described above for 4VMg-ProtoR. The 4V-MpeR activity is reported in column 3 of Table 4.5. The activity was 391 times weaker than that of 4VCR.

### 4.6.4 *Development of a Cell-Free System Capable of the Conversion of Divinyl Protochlorophyllide to Monovinyl Protochlorophyllide*

#### 4.6.4.1 Conversion of DV Pchlide *a* to MV Pchlide *a* in barley etiochloroplasts

The dark conversion of exogenous DV Pchlide to MV Pchlide in barley plastids poised in the DV monocarboxylic biosynthetic mode was investigated in barley etiochloroplasts as well as in barley chloroplasts. Isolated etiochloroplasts prepared from barley leaves preirradiated for 5 h were incubated in a medium that consisted of 0.5 M sucrose, 0.2 M Tris–HCl, 20 mM MgCl2, 2.5 mM $Na_2EDTA$, 20 mM ATP, 40 mM NAD, 8 mM methionine, 0.1 % bovine serum albumin (w/w), and 1.25 mM methanol at a room temperature PH of 7.7 (Tripathy and Rebeiz 1988). After 1 h of incubation. The etioplasts converted about 24–27 nmole per 100 mg plastid protein of DV Pchlide to MV Pchlide. Isolated barley chloroplasts converted 1.5 nmoles (Tripathy and Rebeiz 1988).

#### 4.6.4.2 Conversion of DV Pchlide *a* to MV Pchlide *a* in Isolated Barley Etiochloroplast Membranes

The rates of conversion of DV to MV Pchlide increased significantly in solubilized 4V-PideR prepared as described above in for 4V-Mg-Proto and its ester. The high rates of conversion are described in column 4 of Table 4.5 (Kolossov and Rebeiz 2010). The whole activity was confined to the inner barley etiochloroplast membranes, none was detected in the etiochloroplast envelope (Kolossov and Rebeiz 2010). Furthermore the activity was completely dependent on the presence of NADPH. The activity was also detected in barley chloroplasts (Kolossov and Rebeiz 2010).

### 4.6.5 *Development of a Cell-Free System Capable of the Conversion of Divinyl Chlorophyllide a to Monovinyl Chlorophyllide a*

#### 4.6.5.1 Conversion of DV Chlide a to MV Chlide a in Cucumber Etiochloroplast

The first in organello system capable of converting DV Chlide *a* to MV Chlide *a* was achieved in cucumber etiochloroplasts induced to accumulate DV Chlide *a* (Duggan and Rebeiz 1982b). The etiochloroplasts induced to accumulate DV Chlide *a* were isolated in a medium that consisted of 0.5 M sucrose, 0.2 M Tris–HCl, 1 mM $MgCl_2$, 2.5 mM EDTA, 1.25 mM methanol, 20 mM ATP, 1 mM NADP and 1 % BSA at a room temperature PH of 7.7. The plastids were then subjected to a 2.5-ms pulse of actinic white light followed by a few minutes of dark incubation during which the newly formed DV Chlide *a* was converted into MV Chlide *a* (Duggan and Rebeiz 1982b).

#### 4.6.5.2 Conversion of DV Chlide *a* to MV Chlide *a* in Isolated Cucumber Etiochloroplast Membranes

In 1995, the dependence of the reaction on NADPH was demonstrated in isolated cucumber cotyledon etiochloroplasts (Parham and Rebeiz 1992).

Plastid isolation was performed under a safe green light that transmitted light between 510 and 520 nm and which did not photoconvert DV Pchlide *a* to DV Chlide *a*. Five g batches of DV Pchlide *a*-enriched cotyledons (Duggan and Rebeiz 1982a) were hand homogenized in a cold mortar. The tissue was ground in 12.5 ml of homogenization buffer. The latter consisted of 500 mM sucrose, 15 mM Hepes, 1 mM MgCl2, 1 mM EDTA, 9 mM Tes, 5 mM cysteine, and 0.2 % BSA (w/v), adjusted with KOH to pH 8.0 at room temperature. The resulting homogenate was filtered through two layers of Miracloth (Calbiochem., La Jolla, CA) and centrifuged at 200 g for 5 min at 1 °C in a Beckman JA-20 angle rotor. The supernatant was decanted and centrifuged at 1,500 g for 10 min at 1 °C. The pelleted etiochloroplasts (about 5 mg protein) were gently resuspended with a paintbrush, in 5.0 ml of incubation buffer. Unless otherwise indicated the latter consisted of 500 mM Sucrose, 1.0 mM MgCl2, 2.5 mM EDTA, 20.0 mM ATP, 1.0 mM NAD+, 1.25 mM Methanol, 200.0 mM Tris and 0.2 % BSA (w/v) adjusted to pH 7.7 at room temperature.

Preparation of plastid stroma and membranes was also performed under the safe green light which did not photoconvert DV Pchlide *a* to DV Chlide *a*. Separation of plastid stroma from membranes was adapted from Lee et al. (1991). Etiochloroplasts (about 5 mg protein) were suspended in 5 ml of lysing buffer composed of 1.0 mM MgCl2, 2.5 mM EDTA, 20.0 mM ATP, 1.0 mM NAD+, 1.25 mM methanol, 0.2 % BSA (w/v), and unless otherwise indicated, 25 mM Tris–HCl. The pH was adjusted to 7.7 at room temperature with KOH and HCl. The lysed

**Table 4.6** NADPH requirement of membrane-bound 4VCR

| Etiochloroplast membranes[a] | Net change in DV Chlide *a* in 20 min (nmol/100 mg protein)[b] | Net change in MV Chlide *a* in 20 min |
|---|---|---|
| Without added reductants | −2.1 (−19.8 %)[c] | 2.9[d] |
| with added NADPH | −19.8 (−94.3 %) | 20.5 |
| With added NADH | −2.9 (−24.6 %) | 3.8 |
| With added GSH | −3.1(−25.8 %) | 3.0 |

[a]Etiochloroplasts were lysed in lysing buffer containing 25 mM Tris-HCl
[b]Mean of two replicates
[c]Values in parenthesis represent the net change in DV Chlide *a*, as a percent of total DVChlide *a* present before incubation
[d]In all cases, either no MV Chlide *a* or small amounts of MV Chlide *a* were detected at the beginning of dark incubation

plastid suspension was centrifuged at 235,000 g for 1 h in a Beckman 80 Ti fixed angle rotor at 1 °C. This centrifugation separated the suspension into a colorless soluble-protein supernatant (stroma) and a yellowish pellet (membranes) (Lee et al. 1991). The stromal fraction was decanted and the pelleted membranes were either resuspended in the lysing medium at a rate of 5 ml per 5 g of homogenized tissue or were washed once before use. To this effect, plastid membranes from 5 g of tissue (about 3 mg) were resuspended in 8 ml of 146 mM Tris–HCl adjusted to a room temperature pH of 7.7, and containing all the additives present in the lysing buffer. The suspension was centrifuged at 235,000 g for 0.5 h at 1 °C. The supernatant was decanted and the pelleted membranes were resuspended in 3 ml of lysing buffer containing or lacking cofactors.

DV Chlide *a* reduction was initiated by conversion of most of the DV Chlide *a* to MV Chlide *a* by a single, 2.5-ms flash of actinic white light followed by a dark incubation (Duggan and Rebeiz 1982a). The reduction of DV Chlide *a* to MV Chlide *a* was allowed to proceed in darkness for 20 min.

Conversion rates in the presence and absence of added NADPH are reported in Table 4.6, which is displayed above.

### 4.6.5.3 Conversion of exogenous DV Chlide *a* to MV Chlide *a* in Isolated Cucumber Etiochloroplast Membranes

Preparation of Plastid Membranes

Plastids were isolated essentially as previously described in (Parham and Rebeiz 1992). Essentially plastid isolation was performed under a safe green light that transmitted light between 510 and 520 nm, and which did not photoconvert Pchlide *a* to Chlide *a*. Five g batches of etiolated cucumber cotyledons, and etiolated barley, or corn leaves were hand homogenized in a cold mortar. The tissue was ground in 12.5 ml of homogenization buffer. The latter consisted of 500 mM sucrose, 15 mM

Hepes, 1 mM $MgCl_2$, 1 mM EDTA, 9 mM Tes, 5 mM cysteine, and 0.2 % BSA (w/v), adjusted with KOH to pH 8.0 at room temperature. The resulting homogenate was filtered through two layers of Miracloth (Calbiochem., La Jolla, CA) and centrifuged at 200 g for 5 min in a Beckman JA-20 fixed angle rotor at 1 °C. The supernatant was decanted and centrifuged at 1,500 g for 10 min at 1 °C. The pelleted etiochloroplasts (about 3–5 mg protein) were osmotically lysed in 5 ml of dilute (lysing) buffer consisting of 25 mM Tris–HCl, and 0.2 % BSA. The pH was adjusted to 7.7 at room temperature. The resulting suspension was centrifuged at 235,000 *g* for 1.0 h in a Beckman 80 Ti fixed-angle rotor at 1 °C. The pelleted membrane fraction was separated from the stroma and resuspended with a paint-brush in incubation buffer at a ratio of 5 ml/5 g of tissue. Unless otherwise indicated the incubation buffer consisted of 40 mM citrate monohydrate:80 mM $K_2HPO_4$, 0.2 % BSA, and 0.55–0.80 mM NADPH, at pH 6.3. For determination of the optimum pH of the enzyme, other incubation buffers were used as indicated below.

### Assay of [4-vinyl] Chlorophyllide *a* Reductase Using Exogenous DV Chlide *a*

All steps were carried out under a green safelight that transmitted light between 510 and 520 nm, and which did not photoconvert Pchlide *a* to Chlide *a*. Etioplast membranes devoid of any endogenous Chlide *a* or *b* were isolated and resuspended in 1 ml of an NADPH-fortified incubation buffer as described above. To achieve temperature equilibration, the reaction mixture was pre-incubated for 5 min at 30 °C before initiating the reaction. The reaction was triggered by addition of 25–30 μl of DV Chlide *a* substrate dissolved in 80 % acetone, to a final concentration of 1 μM, and was monitored spectrofluorometrically (see below) by the appearance of MV Chlide *a*, the reaction product. Depending on the particular experiment, the reaction was allowed to proceed for 45 s to several min in darkness, and was terminated by addition of 7 ml of cold ammoniacal acetone. The resulting mixture was centrifuged at 39,000 g for 12 min at 1 °C. Chlorophylls and other fully esterified tetrapyrroles were transferred from acetone to hexane by extraction with an equal volume of hexane, followed by a second extraction with one-third volume of hexane. The remaining hexane-extracted acetone residue, which contained monocarboxylic and dicarboxylic tetrapyrroles, was used for pigment determination by spectrofluorometry at room temperature and at 77 K. Spectrofluorometric determinations at 77 K were performed after transfer of the pigments to ether. The amount of DV and MV Chlide *a* was determined from the total amount of Chlide *a* and from the proportion of DV and MV Chlide *a*, as described below (see Chap. 3). The total amount of Chlide *a* was determined on an aliquot of the hexane-extracted acetone fraction from the room temperature Soret excitation maximum at 433 nm. The room temperature fluorescence excitation spectrum was recorded at an emission wavelength of 674 nm on a fully corrected photon counting, high-resolution SLM spectrofluorometer Model 8000C, interfaced with an IBM microcomputer Model 60. The hexane-extracted acetone aliquot was placed in a cylindrical microcell 3 mm in diameter. Emission and excitation bandwidths of 4 nm were used. The photon count was integrated for 0.5 s at each 1 nm increment. The fluorescence excitation amplitude was converted to concentration by reference to a calibration curve of known amounts of DV Chlide *a* versus fluorescence excitation amplitudes. In constructing the calibration curve, the amount of DV Chlide *a* was determined by room temperature absorption spectroscopy in 80 % acetone, using a molar extinction coefficient of $69.29 \times 10^3$ at 663 nm (Shedbalkar and Rebeiz 1992). The proportion of DV and MV Chlide *a* was determined by 77 K fluorescence excitation spectroscopy after transfer to diethyl ether as described by Wu et al. (1989) and in Chap. 3. 77 K fluorescence excitation spectra in ether were recorded at emission bandwidths that varied from 0.5 to 4 nm depending on signal intensity.

### Conversion of Exogenous DV Chlide *a* to MV Chlide *a* by Membrane-Bound [4-vinyl] Chlorophyllide *a* Reductase

Membrane-bound 4VCR was very active towards exogenous DV Chlide *a*. In eight experiments, reaction rates ranged from 18.15 to 195.21 nmoles of MV Chlide *a* formed per 100 mg membrane protein per min. This range reflects biological variations between various plastid membrane preparations. In each experiment the samples were run in duplicate, and the difference between values for the two samples averaged 5.70 ± 5.32 nmoles per 100 mg membrane protein per min. For the eight reported experiments, the mean values of substrate (DV Chlide *a*) disappearance and product (MV Chlide *a*) formation were 74.11 ± 9.14 and 62.80 ± 4.04 nmoles per 100 mg membrane protein per min respectively. Product formation from exogenous substrate was accompanied by about 15 % substrate destruction. These rates are about 50–300-fold higher than the reported rates of [4-vinyl] Pchlide *a* reductase towards exogenous Pchlide *a* in barley etioplasts (Parham and Rebeiz 1995; Tripathy and Rebeiz 1988).

### Temperature-Dependence of [4-Vinyl] Chlorophyllide *a* Reductase Using Exogenous DV Chlide *a*

Temperature-dependence of the enzyme was determined by incubating cucumber etioplast membranes at 0, 20, 30, 40, and 60 °C in 146 mM Tris–HCl buffer (pH 7.7) containing 0.2 % BSA, and 0.55 mM NADPH. After temperature equilibration for 5 min, 4VCR activity was initiated by addition of DV Chlide *a*, to a final concentration of 1 μM. Enzyme activity was monitored over a period of 2 min. Under these conditions, maximal activity was observed at 30 °C. Complete inhibition was observed at 60 °C. After 2 min incubation at 60 °C, no substrate degradation was observed.

pH-Dependence of [4-Vinyl] Chlorophyllide *a* Reductase
Using Exogenous a DV Chlide a Substrate

Cucumber etioplast membranes were incubated at 30 °C in various buffers covering a pH range of 5.3–11.0. Mc Ilvaine buffer (40 mM citrate monohydrate/80 mM $K_2HPO_4$) was used for pH values ranging from 5.3 to 6.9. Tris–HCl (146 mM) was used in the pH range of 7.0–9.0, and 0.25 mM K2HPO4/NaOH was used at pH 11.0. All incubation buffers contained 0.2 % BSA and 0.55 mM NADPH. 4VCR exhibited high activity between pH 6.0 and 7.0. No activity was observed at pH 11.0. After 2 min incubation at pH 11.0, no substrate degradation was observed at a pH of 6.3 and a temperature of 30 °C, enzymatic activity in cucumber etioplast membranes was quasilinear for the first 60 s of incubation (Parham and Rebeiz 1995).

[4-vinyl] Chlorophyllide a Reductase Activity is Expressed
in Other Plant Species

The occurrence of 4VCR in other plant species such as corn and barley was also investigated. Unlike cucumber, which is a dark DV/light DV plant species (see Chap. 14), corn and barley are two dark MV/light DV plant species (Ioannides et al. 1994). In darkness, these monocotyledonous species accumulate MV Pchlide *a,* and in the light they form MV Chl *a*. 4VCR activity was strongly expressed in etiolated corn and barley. Contrary to the immediate onset of activity observed in cucumber, an induction period of about 15 s was observed in barley and corn etioplasts. The final levels of activity in corn and barley were higher than in cucumber (Parham and Rebeiz 1995).

#### 4.6.5.4 Solubilization and Partial Purification of 4-vinyl Chlorophyllide *a* Reductase

Preparation of Etioplast Membranes

Percoll-purified etioplasts were suspended in lysing buffer at a rate of 30 ml per pellet from 60 g of tissue. The lysing buffer consisted of 20 mM Tris–HCl, 1 mM EDTA, 20 mM sucrose and 4 mM dithiothreitol, at a room temperature pH of 7.7. Etioplasts membranes were pelleted by centrifugation at 39,000 g for 12 min at 2EC. One hundred g of etiolated barley leaves yielded 3.5–4.5 mg of membrane protein.

Solubilization of 4VCR

Etioplast membranes were re-suspended in solubilization buffer at a rate of 2–3 mg protein per ml. The solubilization buffer consisted of 20 mM Tris–HCl, 1 mM

EDTA, and various amounts of glycerol and CHAPS as indicated in specific experiments. The pH was adjusted to 7.7 at room temperature. Solubilization was carried out on ice with continuous stirring for 30 min. After 30 min, the suspension was centrifuged at 100,000 g for 1 h in a Beckman Ti80 fixed angle rotor at 2 °C. The resulting supernatant containing 1–2 mg solubilized membrane protein per ml was stored at −80 °C until use. No detectable loss of 4VCR activity was observed after several months of storage at −80 °C (Kolossov and Rebeiz 2001).

#### Chromatography of Solubilized Etioplast Membrane Proteins on DEAE-Sephacel

Partial purification of 4VCR was achieved by chromatography on DEAE-Sephacel. About 7 ml of the solubilized membrane protein fraction were applied to a 5 ml column of DEAE-Sephacel which was pre-equilibrated with a solution of 5 % glycerol, 20 mM Tris–HCl, 0.1 mM DTA, and 4 mM CHAPS, adjusted to a room temperature pH of 7.7. Elution of 4VCR activity was achieved at a speed of 0.4 ml per min, with a linear 0.02 M ammonium sulfate gradient dissolved in the pre-equilibration buffer. Two-ml fractions were collected, and 4VCR activity was monitored. The active fractions were pooled and concentrated in disposable Millipore Centriplus-e0 Centrifuge cartridges, until a retentate volume of 1.5–2.0 ml was collected. The retentate contained about 1–1.5 mg solubilized protein per ml (Kolossov and Rebeiz 2001).

#### Further Purification of 4VCR Activity on Cibacron Blue 3GA-1000 Agarose

Further purification of 4VCR activity was achieved by chromatography on Cibacron Blue 3GA-1000 agarose. This resin acts as a non-specific affinity medium. The column (1.5 ml bed volume) was made up of a 5 ml disposable pipette tip. It was pre-equilibrated with column buffer that consisted of 20 mM Tris–HCl, 0.1 mM EDTA, 5 % glycerol, 200 mM ammonium sulfate and 2.5 mM CHAPS, adjusted to a room temperature pH of 7.7. After applying the concentrated retentate (see above) to the Cibacron column, the column was washed with 3 bed volumes of column buffer. 4VCR activity was eluted with column butter adjusted to 5 mM CHAPS concentration (Kolossov and Rebeiz 2001).

#### Purification of Solubilized 4VCR

Solubilization of 4VCR resulted in 1.5–2.0-fold purification (Kolossov and Rebeiz 2001). Further purification was achieved by column chromatography on DEAE-Sephacel. DEAE is a weak base that acquires a net positive charge when ionized. It therefore binds and exchanges anions. An additional two to threefolds purification of 4VCR was achieved in the ammonium sulfate eluate of the DEAE-Sephacel column.

Further purification of the DEAE-Sephacel 4VCR eluate was achieved by chromatography on Cibacron Blue 3GA-1000 Agarose.

Cibacron Blue 3GA-1000 agarose is a non-specific affinity chromatography medium. Upon chromatography of the concentrated DEAE-Sephacel eluate on Cibacron Blue 3GA-100 agarose, about 65 % of the proteins were not retained by Cibacron and passed through the column. Elution of adsorbed 4VCR in column buffer containing 5 mM CHAPS resulted in an overall purification of about 20–21-folds (Kolossov and Rebeiz 2001). 4VCR yields ranged from 11 to 17 % of the total original activity. 4VCR activity of the Cibacron eluate was stable for several months at −81 °C.

The electrophoretic profiles of the membrane and solubilized fractions as well as that of the Cibacron eluate are depicted in Kolossov and Rebeiz (2001).

Demonstration of 4VCR Activity in Barley Chloroplast Membranes

It was previously assumed that 4VCR activity disappeared or decreased to undetectable levels in photoperiodically grown plants (Abd-El-Mageed et al. 1997). That conclusion was based on experimentation involving isolated chloroplasts having a full complement of Chl. It has now become apparent that in addition to 4VR inhibition by the plastid stroma, the high concentration of Chl interfered with the 4VR spectrofluorometric assays. Upon solubilization of the 4VR activities from chloroplast membranes, the stromal inhibition was relieved, most of the Chl was left behind in the membranes, and the 4VCR activities became unmasked. It amounted to 7.6 nmoles of MV Chlide *a* formed per mg of purified protein after 5 min of incubation (Kolossov and Rebeiz 2010).

### *4.6.6 Development of a Cell-Free System Capable of the Conversion of Chlorophyllide a to Chlorophyll a*

#### 4.6.6.1 Preparatory Techniques

Etiolated cucumber cotyledons were excised without hypocotyl hooks, and were placed in a large Petri dish, 15 cm in diameter prior to light treatment. One 100 g of etiolated tissues were harvested in darkness under a safe green light. The tissue was homogenized using a blender (Waring blender 7010, model 31BL91), in 250 ml of homogenization buffer. Homogenization consisted of two bursts of 3 s each. The homogenization buffer consisted of 500 mM HEPES, 1 mM $MgCl_2$, 1 mM EDTA, 30 mM TES, 5 mM cysteine, and 0.2 % BSA (w/w), adjusted with KOH to pH 7.7 at room temperature. The resulting homogenate was filtered through four layers of cheesecloth and centrifuged at 200 g for 5 min at 1 °C in a Beckman JA-20 angle rotor. The supernatant containing etioplasts was decanted and centrifuged at 1,500 g for 10 min at 1 °C. The pelleted etioplasts were gently resuspended with a paintbrush in 2-ml of incubation buffer. The latter consisted of 500 mM sucrose, 20 mM $MgCl_2$, 2.5 mM EDTA, 20 mM ATP, 40 mM NAD, 1.25 mM methanol, 200 mM Tris. HCl, 8 mM methionine, and 0.1 % BSA (w/v) adjusted with HCl to a pH of 7.7 at room temperature. One ml of etioplast suspension was placed in a 25-ml beaker prior to light treatment.

The Petri dish containing excised, etiolated cucumber cotyledons, or the beaker containing isolated etioplasts, were placed 10 cm below two 135 W second Rokunar AC Studio Flash-Model 150 that generate a synchronized 2.5 ms actinic white light flash. Reflecting mirrors were placed below and on the sides of the sample in order to increase the amount of incident light reflected back onto the sample area. Immediately after the flash, *i.e.* within less than 1 s, the treated tissues were frozen in liquid $N_2$ (control), or left in darkness for different periods of time. At each dark-time interval the reaction was stopped rapidly by pouring liquid nitrogen over the tissue. For isolated etioplasts, the reaction was stopped by addition of 10 ml of acetone:0.1 N $NH_4$ OH. Controls consisted of etiolated tissues or etioplasts that were not subjected to the light treatment.

### Effect of Exogenous Mg2$^{+}$ on the Conversion of Chlorophyllide *a* to Chlorophyll *a*

$Mg^{2+}$ is a cofactor which is involved in the conversion of: ALA to Pchlide *a* (Rebeiz and Castelfranco 1971a) of Proto to Mg-Proto, (Fuesler et al. 1981) and is usually an adjunct cofactor in reactions involving ATP. It was therefore deemed desirable to probe the effect of various concentrations of added $MgCl_2$ on the tetrapyrrole biosynthetic capabilities of kinetin and GA-pretreated etiochloroplasts and on the conversion of Chlide *a* to Chl *a*.

Pretreated etiochloroplasts responded very favorably to the addition of exogenous $MgCl_2$ to the incubation medium. The rates of Pchlide and Proto net synthesis and accumulation reached a maximum and a minimum, respectively, at an added $Mg^{2+}$ concentration of 20 mM while the highest rate of Mpe accumulation was observed at an exogenous concentration of 40 mM. The requirement of different $Mg^{2+}$ concentrations for achieving optimal rates of Pchlide and Mpe accumulation indicated that in addition to insertion into Proto, Mg may also involved as a cofactor in additional biosynthetic reactions between Mpe and Pchlide, most probably in the formation of the cyclopentanone ring of Pchlide.

To our surprise, the conversion of Chlide *a* to Chl *a* was also found to dependent on the addition of exogenous $Mg^{2+}$ (Daniell and Rebeiz 1984). This novel observation in turn indicated that in addition to ATP, $Mg^{2+}$ was also involved in the conversion of Chlide *a* to Chl *a*. However, even at the highest rates of Chlide *a* esterification, i.e. at a 20 mM concentration of added $Mg^{2+}$, the conversion of

Chlide *a* to Chl *a* was only partial (38 %) while higher amounts of added $Mg^{2+}$ were inhibitory (Daniell and Rebeiz 1984). This in turn suggested that something was still amiss from the used incubation medium.

Effect of Exogenous Geranylgeraniol Pyrophosphate and Phytol on the Conversion of Chlorophyllide *a* to Chlorophyll *a*

Etiochloroplasts were isolated from kinetin and GA-pretreated cotyledons and were incubated in the fortified incubation medium in the absence (control) or presence of granylgranyl pyropohosphate (GGDP) or phytol. The addition of the isoprenoid alcohols did not seriously depress the Pchlide accumulation capabilities of the plastids. On the contrary, a 52-fold excess of added phytol ($50 \times 10^{-3}$ mM) with respect to the endogenous Chlide *a* pool ($0.96 \times 10^{-3}$ mM) significantly enhanced the rates of Chlide *a* mobilization by the added isoprenoid alcohols. In the presence of a 344-fold excess of GGPP (0.33 mM) or 5.2-fold excess of phytol (0.005 mM) to endogenous Chlide *a* ($0.96 \times 10^{-3}$ mM), the Chlide *a* pool was depleted by 92.4 and 89.3 %, respectively, and its level at the end of incubation was similar to the levels of endogenous Chlide *a* encountered in plastids freshly isolated from greening tissues. The bulk of the Chlide *a* that disappeared was converted to Chl *a* (Daniell and Rebeiz 1984). Some of it also appeared to be converted to Chlide *b* (Daniell and Rebeiz 1984).

The joint requirement of ATP, $Mg^{2+}$ and phytol for achieving high rates of conversion of Chlide *a* to Chl *a* in organello is described in Daniell and Rebeiz (1984)

Etiochloroplasts were isolated from kinetin and GA-pretreated cotyledons and were incubated in the fortified incubation medium in the absence (control) or presence of GGPP or phytol. the observed enhancement of Chlide a mobilization by the added isoprenoid alcohols. In the presence of a 344-fold excess of GGPP (0.33 mM) or 5.2-fold excess of phytol (0.005 mM) to endogenous Chlide a ($0.96 \times 10^{-3}$ mM), the Chlide a pool was depleted by 92.4 and 89.3 %, respectively, and its level at the end of incubation was similar to the levels of endogenous Chlide a encountered in plastids freshly isolated from greening tissues. The bulk of the Chlide a that disappeared was converted to Chl a. Some of it also appeared to be converted either to Chlide b (Daniell and Rebeiz 1984).

## References

Abd-El-Mageed HA, El Sahhar KF, Robertson KR et al (1997) Chloroplast biogenesis 77. Two novel monovinyl and divinyl light–dark greening groups of plants and their relationship to the chlorophyll *a* biosynthetic heterogeneity of green plants. Photochem Photobiol 66:89–96

Bazzaz MB, Rebeiz CA (1979) Chloroplast culture V. Spectrofluorometric determination of chlorophyll(ide) *a* and *b* and pheophytin (or pheophorbide) *a* and *b* in unsegregated pigment mixtures. Photochem Photobiol 30:709–721

Beale SI, Castelfranco PA (1974) The biosynthesis of δ-aminolevulinic acid in higher plants. II. Formation of $^{14}C$-δ-aminolevulinic acid from labelled precursors in greening plant tissues. Plant Physiol 53:291–296

Daniell H, Rebeiz CA (1982a) Chloroplast culture VIII. A new effect of kinetin in enhancing the synthesis and accumulation of protochlorophyllide in vitro. Biochem Biophys Res Commun 104:837–843

Daniell H, Rebeiz CA (1982b) Chloroplast culture IX. Chlorophyll(ide) *a* biosynthesis in vitro at rates higher than in vivo. Biochem Biophys Res Commun 106:466–470

Daniell H, Rebeiz CA (1984) Bioengineering of photosynthetic membranes: requirement of magnesium for the conversion of chlorophyllide a to chlorophyll a during the greening of etiochloroplasts in vitro. Biotechnol Bioeng 26:481–487

Duggan JX, Rebeiz CA (1982a) Chloroplast biogenesis 37: induction of chlorophyllide *a* (E459F675) accumulation in higher plants. Plant Sci Lett 24:27–37

Duggan JX, Rebeiz CA (1982b) Chloroplast biogenesis 42. Conversion of DV chlorophyllide a to monovinyl chlorophyllide *a* in vivo and in vitro. Plant Sci Lett 27:137–145

Ellsworth RK, Aronoff S (1968) Investigations on the biogenesis of chlorophyll *a*. III. Biosynthesis of Mg-vinylporphine $a_5$ methylester from Mg-protoporphine IX monomethyl ester as observed in *Chlorella* mutants. Arch Biochem Biophys 125:269–277

Ellsworth RK, Hsing AS (1973) The reduction of vinyl-side chains of Mg-protoporphyrin IX monomethyl ester in vitro. Photosynthetica 313:119–129

Fuesler TP, Wright LA, Castelfranco PA (1981) Properties of magnesium chelatase in greening etioplasts. Metal ion specificity and effect of substrate concentration. Plant Physiol 67:246–249

Fuesler TP, Castelfranco PA, Wong YS (1984a) Formation of Mg-containing chlorophyll precursors from protoporphyrin IX, d-aminolevulinic acid, and glutamate in isolated, photo synthetically competent, developing chloroplasts. Plant Physiol 74:928–933

Fuesler TP, Wong YS, Castelfranco PA (1984b) Localization of Mg-chelatase and Mg-protoporphyrin IX monomethyl ester (oxidative) cyclase activities within isolated, developing chloroplasts. Plant Physiol 75:662–664

Granick S (1961) Magnesium protoporphyrin monoester and protoporphyrin monomethyl ester in chlorophyll biosynthesis. J Biol Chem 236:1168–1172

Ioannides IM, Fasoula DM, Robertson KR et al (1994) An evolutionary study of chlorophyll biosynthetic heterogeneity in green plants. Biochem Syst Ecol 22:211–220

Jeffrey SW, Wright SW (1987) A new spectrally distinct component in preparations of chlorophyll *c* from micro-alga (Prymnesiophyceae). Biochim Biophys Acta 894:180–188

Kim JS, Kolossov V, Rebeiz CA (1997) Chloroplast biogenesis 76: regulation of 4-vinyl reduction during conversion of divinyl Mg-protoporphyrin IX to monovinyl protochlorophyllide *a* is controlled by plastid membrane and stromal factors. Photosynthetica 34:569–581

Kolossov VL, Rebeiz CA (2001) Chloroplast biogenesis 84. Solubilization and partial purification of membrane-bound [4-vinyl] chlorophyllide *a* reductase from etiolated barley leaves. Anal Biochem 295:214–219

Kolossov VL, Rebeiz CA (2010) Evidence for Various 4-vinyl reductase activities in higher plants. In: Rebeiz CA, Benning C, Bohnert HJ et al (eds) The chloroplast: basics and applications. Springer, Netherlands, pp 25–38

Lee HJ, Ball M, Rebeiz CA (1991) Intraplastidic localization of the enzymes that convert delta-aminolevulinic acid to protoporphyrin IX in etiolated cucumber cotyledons. Plant Physiol 96:910–915

Lee HJ, Ball MD, Parham R et al (1992) Chloroplast biogenesis 65. Enzymic conversion of protoporphyrin IX to Mg-protoporphyrin IX in a subplastidic membrane fraction of cucumber etiochloroplasts. Plant Physiol 99:1134–1140

Mattheis JR, Rebeiz CA (1977a) Chloroplast biogenesis. Net synthesis of protochlorophyllide from protoporphyrin IX by developing chloroplasts. J Biol Chem 252:8347–8349

Mattheis JR, Rebeiz CA (1977b) Chloroplast biogenesis. Net synthesis of protochlorophyllide from magnesium protoporphyrin monoester by developing chloroplasts. J Biol Chem 252:4022–4024

Mauzerall D, Granick S (1956) The occurrence and determination of delta-aminolevulinic acid and porphobilinogen in urine. J Biol Chem 219:435–446

Pardo AD, Chereskin BM, Castelfranco PA et al (1980) ATP requirement for Mg chelatase in developing chloroplasts. Plant Physiol 65:956–960

Parham R, Rebeiz CA (1992) Chloroplast biogenesis: [4-vinyl] chlorophyllide *a* reductase is a divinyl chlorophyllide *a*-specific NADPH-dependent enzyme. Biochemistry 31:8460–8464

Parham R, Rebeiz CA (1995) Chloroplast biogenesis 72: a [4-vinyl] chlorophyllide *a* reductase assay using divinyl chlorophyllide *a* as an exogenous substrate. Anal Biochem 231:164–169

Perkins HJ, Roberts DWA (1960) Chlorophyll biosynthesis in wheat leaves. Biochem Biophys Acta 45:613–620

Perkins EJ, Roberts DWA (1962) Purification of chlorophylls, pheophytins, an pheophorbides for specific activity determinations. Biochim Biophys Acta 58:486–498

Rebeiz CA (1967) Studies on chlorophyll biosynthesis in etiolated excised cotyledons of germinating cucumber at different stages of seedling development. Magon Serie Scientifique 13:1–21

Rebeiz CA (1968) The chloroplast pigments of etiolated and greening cucumber cotyledons. Magon Serie Scientifique 21:1–25

Rebeiz CA, Castelfranco P (1971a) Protochlorophyll biosynthesis in a cell-free system from higher plants. Plant Physiol 47:24–32

Rebeiz CA, Castelfranco P (1971b) Chlorophyll biosynthesis in a cell-free system from higher plants. Plant Physiol 47:33–37

Rebeiz CA, Yaghi M, Abou Haidar M et al (1970) Protochlorophyll biosynthesis in cucumber (Cucumis sativus, L.) cotyledons. Plant Physiol 46:57–63

Rebeiz CA, Larson S, Weier TE et al (1973) Chloroplast maintenance and partial differentiation in vitro. Plant Physiol 51:651–659

Rebeiz CA, Mattheis JR, Smith BB et al (1975a) Chloroplast biogenesis. Biosynthesis and accumulation of protochlorophyll by isolated etioplasts and developing chloroplasts. Arch Biochem Biophys 171:549–567

Rebeiz CA, Mattheis JR, Smith BB et al (1975b) Chloroplast biogenesis. Biosynthesis and accumulation of Mg-protoporphyrin IX monoester and longer wavelength metalloporphyrins by greening cotyledons. Arch Biochem Biophys, 166:446–465

Rebeiz CA, Daniell H, Mattheis JR (1982) Chloroplast bioengineering: the greening of chloroplasts in vitro. In: Scott CD (ed) Biotechnology bioengineering symposium, vol 12, John Wiley, New York, pp 414–439

Rebeiz CA, Montazer-Zouhoor A, Daniell H (1984) Chloroplast culture X: thylakoid assembly in vitro. Isr J Bot 33:225–235

Rebeiz CA, Kolossov VL, Briskin D et al (2003) Chloroplast biogenesis: chlorophyll biosynthetic heterogeneity, multiple biosynthetic routes and biological spin-offs. In: Nalwa HS (ed) Handbook of photochemistry and photobiology, vol 4. American Scientific Publishers, Los Angeles, pp 183–248

Scopes RK (1982) Protein purification; principles and practice. Springer, New York, pp 185–200

Shedbalkar VP, Rebeiz CA (1992) Chloroplast biogenesis: determination of the molar extinction coefficients of divinyl chlorophyll *a* and *b* and their pheophytins. Anal Biochem 207:261–266

Sisler EC, Klein W (1963) The effect of age and various chemicals on the lag phase of chlorophyll synthesis in dark grown bean seedlings. Physiol Plant 16:315–322

Smith BB, Rebeiz CA (1977a) Chloroplast biogenesis: detection of Mg-protoporphyrin chelatase in vitro. Arch Biochem Biophys 180:178–185

Smith BB, Rebeiz CA (1977b) Spectrofluorometric determination of Mg-protoporphyrin monoester and longer wavelength metalloporphyrins in the presence of Zn-protoporphyrin. Photochem Photobiol 26:527–532

Smith BB, Rebeiz CA (1979) Chloroplst biogenesis XXIV. Intrachloroplastic localization of the biosynthesis and accumulation of protoporphyrin IX, magnesium protoporphyrin IX,

Magnesium-protoporphyrin monoester and longer wavelength metalloporphyrins during greening. Plant Physiol 63:227–231
Stetler DA, Laetsch WM (1965) Kinetin-induced chloroplast maturation in cultures of tobacco tissue. Science 149:1387–1388
Tripathy BC, Rebeiz CA (1988) Chloroplast biogenesis 60. Conversion of divinyl protochlorophyllide to monovinyl protochlorophyllide in green(ing) barley, a dark monovinyl/light divinyl plant species. Plant Physiol 87:89–94
Walker CJ, Weinstein JD (1991a) In vitro assay of the chlorophyll biosynthetic enzyme Mg-chelatase: resolution of the activity into soluble and membrane-bound fractions. Proc Natl Acad Sci U S A 88:5789–5793
Walker CJ, Weinstein JD (1991b) Further characterization of the magnesium chelatasein isolated developing cucumber chloroplasts. Plant Physiol 95:1189–1196
Wickliff JL, Aronoff S (1963) Turnover of chlorophyll a in mature soybean leaves. Plant and Cell Physiol:441–449
Wu SM, Mayasich JM, Rebeiz CA (1989) Chloroplast biogenesis: quantitative determination of monovinyl and divinyl chlorophyll(ide) *a* and *b* by spectrofluorometry. Anal Biochem 178:294–300

# Chapter 5
# Reactions Between δ-Aminolevulinic Acid and Protoporphyrin IX

*A solid edifice has to be built on solid foundations*
*(Constantin A. Rebeiz).*

The reactions between ALA and Proto are shared between heme and Chl biosynthesis. Since most of the Chl biosynthetic heterogeneity is rooted in reactions further down the Chl biosynthetic pathway, the reactions between ALA and Proto will be briefly discussed.

All tetrapyrroles that will be discussed in this review are derivatives either of Porphin or Phorbin (Fig. 5.1). Because of its simplicity, the Fischer nomenclature and numbering systems will be used throughout this overview. The sequence of reactions between δ-aminolevulinic acid (ALA) and protoporphyrin IX (Proto) are depicted in Fig. 5.2.

In higher plant thylakoids, the reactions between ALA and Proto are considered to take place in five different environments and may involve both spatial and chemical biosynthetic heterogeneities. Chlorophyll biosynthetic heterogeneity (see Synopsis and Chap. 14) refers (a) either to spatial biosynthetic heterogeneity, (b) to chemical biosynthetic heterogeneity, or (c) to a combination of spatial and chemical biosynthetic heterogeneities (Rebeiz et al. 2003). Spatial biosynthetic heterogeneity refers to the biosynthesis of an anabolic tetrapyrrole or end product by identical sets of enzymes, at several different locations of the thylakoid membranes. On the other hand, chemical biosynthetic heterogeneity refers to the biosynthesis of an anabolic tetrapyrrole or end product at several different locations of the thylakoid membranes, via different biosynthetic routes, each involving at least one different enzyme. This hypothesis is based on the detection of resonance excitation energy transfer from Proto to various Chl–protein complexes in multiple thylakoid environments (Table 5.1). It is also based on the observed resonance excitation energy transfer from Proto to Pchlide *a*. In addition, Averina and coworkers have proposed the existence of at least two types of Chl biosynthesis centers which differ in their ability of form ALA (Averina et al. 1993; Averina 1998).

C.A. Rebeiz, *Chlorophyll Biosynthesis and Technological Applications*,
DOI 10.1007/978-94-007-7134-5_5, 

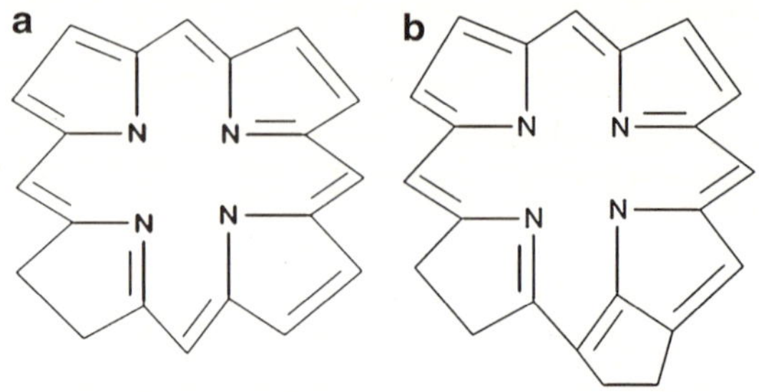


**Fig. 5.1** (**a**) Porphin and (**b**) phorbin nuclei

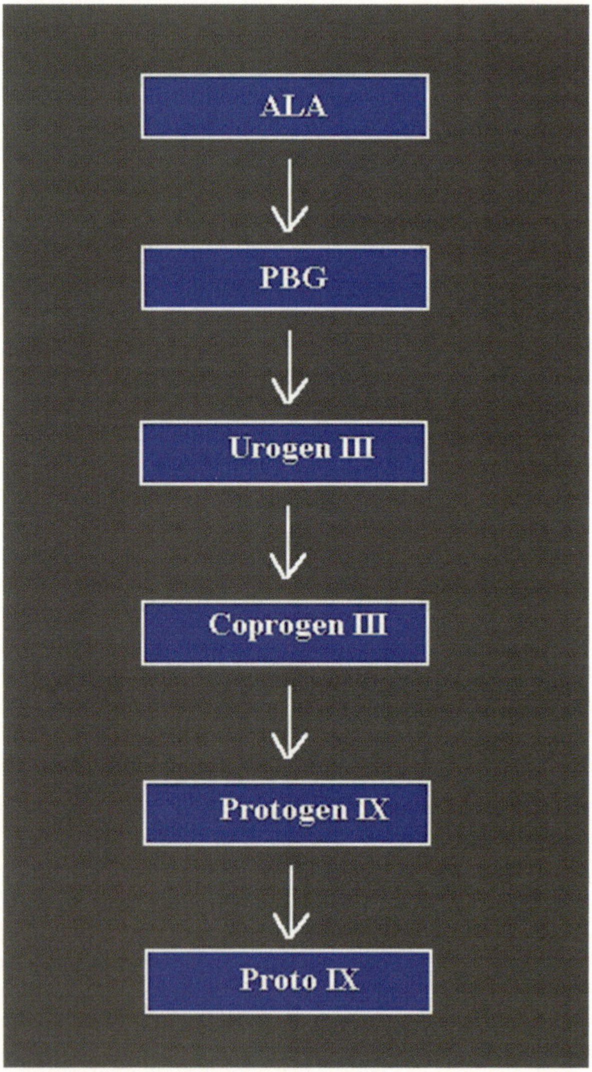


**Fig. 5.2** Sequence of biosynthetic reactions between ALA and Proto. *ALA* = δ-aminolevulinic acid, *PBG* = porphobilinogen, *Urogen III* = uroporphyrinogen III, *Coprogen III* = coproporphyrinogen III, *DV Protogen* = divinyl protoporphyrinogen IX, *DV proto* = divinyl Proto

**Table 5.1** Mapping of excitation resonance energy transfer maxima to Chl *a* F686, Chl *a* F694 and Chl *a*F738-742 *in situ*

| | | Undil donor | Dil donor | Excitation resonance energy maxima to: | | | Conc | | |
|---|---|---|---|---|---|---|---|---|---|
| | | Conc | Conc | Chl *a* F686 | Chl *a* F694 | Chl *a* F738 | ALA | Dpy | |
| Plant species | Major donor | (pmoles/ml suspension) | | (nm) | | | (nM) | | Incub (h) |
| Cucumber | Proto | 1,620 | 54 | 397p, 402p, 410p, 415p | 390s, 400p, 409p | 390s, 395s, 408p, 417p | 4.5 | 3.7 | 6 |
| Cucumber | Proto | 1,242 | 83 | 387p, 402p, 412p | 392p, 406p | 388p, 399p, 403p, 410p, 415p | 20 | 4 | 6 |
| Cucumber | Proto | 1,374 | 92 | 390p, 399p, 405p, 412p | 399p, 409p, 412s | 399p, 400p, 416p | 20 | 0 | 6 |
| Cucumber | Proto | 5,640 | 376 | 395p, 404s, 411p, 416p | 395p,406p, 414p | 393p, 400s, 407p | 20 | 16 | 6 |
| Cucumber | Proto | 3,138 | 1,046 | 402s, 411p | 404p, 410s, 416p, | 399s, 405s, 411p | 20 | 0 | 12 |
| Barley | Proto | 390 | 13 | 391, 398s, 404s, 411p | 389s, 395p, 406p, 414p | 390s, 393p, 400s, 406p, 412p, 416s, | 4.5 | 3.7 | 6 |
| Barley | Proto | 1,492 | 61 | 389p, 396s, 404s, 410p, 412p | 396p, 406p, 412p | 389s, 395p, 406s, 410p, | 20 | 16 | 6 |
| Barley | Proto | 966 | 64 | 395s, 400p, 405s, 413p | 389p, 397s, 403p, 412p | 388s, 393p, 400s, 406p, 412p | 20 | 0 | 6 |
| Barley | Proto | 1,015 | 68 | 389p, 396p, 412p, 413s | 389p, 398p, 409p | 396s, 400p, 412p, 414s | 20 | 4 | 6 |

Peak (*p*) and shoulders (*s*) of excitation resonance energy transfer from Proto to various Chl-protein complexes are interpreted as transfer from different environments to the Chl-protein complexes. *Undil* donor concentration before dilution, *Dil* donor concentration after dilution, *s* shoulder, *p* peak
Adapted from Kolossov et al. (2003)

## 5.1 Biosynthetic Heterogeneity of Delta-Aminolevulinic Acid (ALA)

δ-Aminolevulinic acid (5-aminolevulinic acid) is the building block of all tetrapyrroles in nature. It is synthesized via a different pathway in animal cells and lower plants than in green plants (Fig. 5.3).

### 5.1.1 Biosynthesis of ALA in Animal Cells

In animal cells, ALA is formed by condensation of glycine and succinyl-CoA (Gibson et al. 1958). The reaction is catalyzed by ALA synthetase and takes place in the mitochondria. The biosynthesis of succinyl-CoA from succinic acid is catalyzed by succinyl-thiokinase in the presence of $Mg^{++}$ and ATP, and takes place as well in the mitochondria. ALA is exported to the cytoplasm for further metabolism (Granick 1963). In animal mitochondria ALA is mainly destined for the biosynthesis of Proto and heme (Rebeiz et al. 1996) (Fig. 5.4).

### 5.1.2 Biosynthesis of ALA in Lower Plants

In *Rhodopseudomonas spheroides*, a bacterium (Chen et al. 1981), *Scenedesmus obliquus* (Klein and Senger 1978) unicellular green alga, and in *Euglena gracilus* (Beale et al. 1981) a unicellular green flagellate alga, two pathways using, either glycine and succinyl-CoA (see above), or incorporating the whole C-5-skeleton of glutamate into ALA (see below) are functional in the biosynthesis of ALA. In *Scenedesmus* both pathways appear to contribute to Chl formation in the light. In *Euglena*, it was proposed that ALA synthesis via glycine and succinyl-CoA is responsible for non-plastid tetrapyrrole biosynthesis (Beale et al. 1981). These results stress the need to investigate in more depth the molecular basis and biological significance of the ALA biosynthetic heterogeneity in a wider range of lower and higher plants.

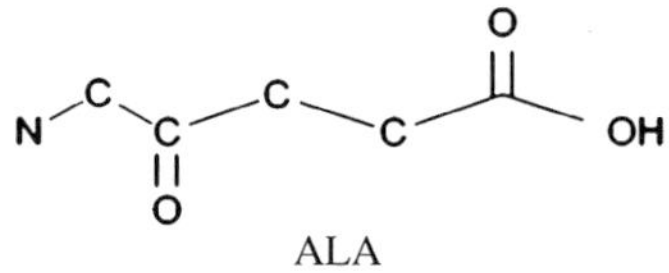


**Fig. 5.3** δ-Aminolevulinic acid (ALA)

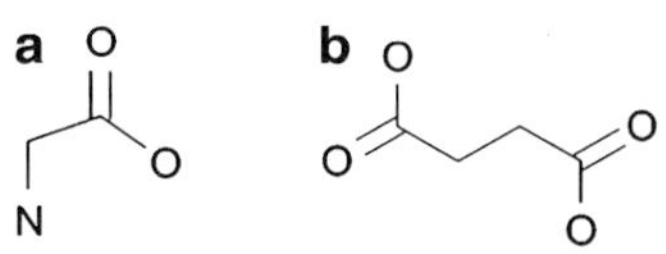


**Fig. 5.4** (**a**) Glycine and (**b**) succinate molecules

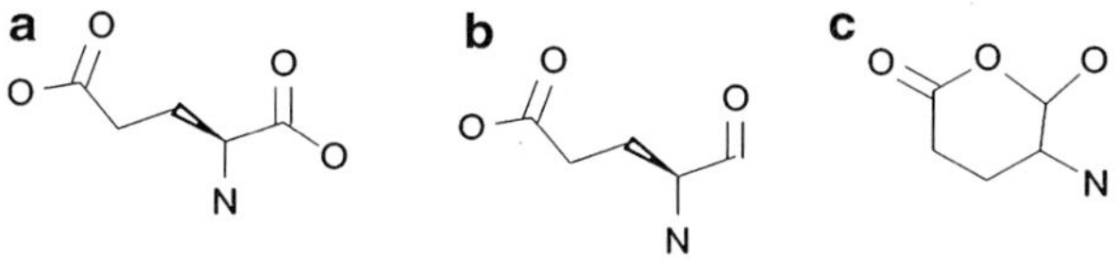


**Fig. 5.5** (**a**) Glutamate, (**b**) glutamate semialdehyde and (**c**) Hydroxyaminotetrahydropyranone (HAT) molecules

#### 5.1.2.1 ALA Synthetase

A query for ALA synthetase in Animal cells and lower Plants addressed to the SwissProt and PIR protein databases via the Biology Workbench, yielded 16 distinct sequences which are depicted on the LPBP website at http://www.vlpbp.org/greening/xv/, Sequenced Enzymes/ALA Synthetase, as well as in Appendix I.

### *5.1.3 Biosynthetic Heterogeneity of ALA in Higher Plants*

In higher plants ALA is formed from glutamic acid (Beale and Castelfranco 1974) via three reactions (Kannangara et al. 1984). In a first reaction, glutamate- $tRNA^{Glu}$ ligase catalyzes the ligation of glutamate to tRNA in the presence of ATP and $Mg^{++}$. In a second reaction, the glutamyl-tRNA complex is converted into a linear glutamate semialdehyde (GSA) by NADPH:Glu-tRNA(oxido)reductase (also called glutamyl-tRNA dehydrogenase) (Kannangara et al. 1984) or into a cyclic GSA (hydoroxyaminotetrahydropyranone, HAT for short) (Jordan et al. 1989). Finally, GSA amino-transferase converts GSA to ALA in the presence of vitamin $B_6$ and pyridoxal phosphate. These reactions take place in the stroma of the plastid (Fig. 5.5).

The understanding of ALA biosynthetic heterogeneity in higher plants is still at a primitive stage although reports are surfacing in support of that notion. For example a reported differential inhibition of ALA formation by gabaculine in black pine (*Pinus nigra* L.) during seed germination (Drazic and Bogdanovic 2000) strongly suggests that in black pine, ALA is formed at least via two different routes, one of which is inhibited by gabaculine. Also Averina and coworkers have proposed the existence of at least two types of Chl biosynthesis centers which differ in their ability to form ALA (Averina et al. 1993; Averina 1998). Recently, multiple resonance excitation energy transfer sites from Proto to various Chl-protein complexes have been detected in higher green plants (Table 5.1), which led to the extension of the Proto biosynthetic heterogeneity all the way to ALA formation. Therefore, ALA biosynthesis is proposed to take place in five different thylakoid environments, via various routes. It is uncertain at this stage whether or not similar reactions in the biosynthetic routes between ALA and Proto are catalyzed by identical enzymes or not. In other words it is still uncertain whether the spatial heterogeneity of ALA formation is accompanied by chemical biosynthetic heterogeneity or not.

#### 5.1.3.1 Glutamate tRNA Ligase

A query for Glutamate t-RNA Ligase addressed to the SwissProt and PIR protein databases via the Biology Workbench, yielded 38 distinct sequences which are depicted on the LPBP website at http://www.vlpbp.org/greening/xv/, Sequenced Enzymes/Glu-tRNA Ligase, as well as in Appendix I.

#### 5.1.3.2 Glutamate tRNA Oxido Reductase

A query for Glutamate t-RNA (Oxido) Reductases addressed to the PIR, PRODOM, SwissProt, and TREMBL protein databases via the Biology Workbench, yielded many distinct sequences which are depicted on the LPBP website at http://www.vlpbp.org/greening/xv/, Sequenced Enzymes/Glu-tRNA Reductase, as well as in Appendix I.

#### 5.1.3.3 Glutamate Semialdehyde Aminotransferase

A query for GSA-Aminotransferase addressed to the PIR, PRODOM, SwissProt, and TREMBL protein databases via the Biology Workbench, yielded 28 unique sequences which are depicted on the LPBP website at "http://www.vlpbp.org/greening/xv/, Sequenced Enzymes/GSA-Aminotransferase", as well as in Appendix I.

## 5.2 Biosynthesis of Porphobilinogen (PBG)

PBG is the precursor of uroporphyrinogen III (Urogen III) that is the precursor of all intermediates of the heme and Chl biosynthetic pathways (Fig. 5.6). It is formed from two molecules of ALA; in the process two molecules of water are liberated (Fig. 5.7). The dimerization reaction is catalyzed by ALA dehydratase also known as PBG synthase (Gibson et al. 1955; Schmid and Shemin 1955). The enzyme binding sites of the two ALA substrates have been designated the **A** and **P** sites. The **A** site gives rise to the acetic side chain, while the **P** site gives rise to the propionic side chain of PBG. The first substrate binds to the **P** site where it forms an Schiff base with the enzyme. The second ALA molecule interacts with the **A** site (Jordan and Seehra 1980) to form an enzyme-two ALA substrate complex. The precise mechanism by which the 5-membered PBG ring is formed from the enzyme-two substrate complex is still uncertain.

Fig. 5.6 (a) Schiff base-enzyme intermediate and (b) PBG

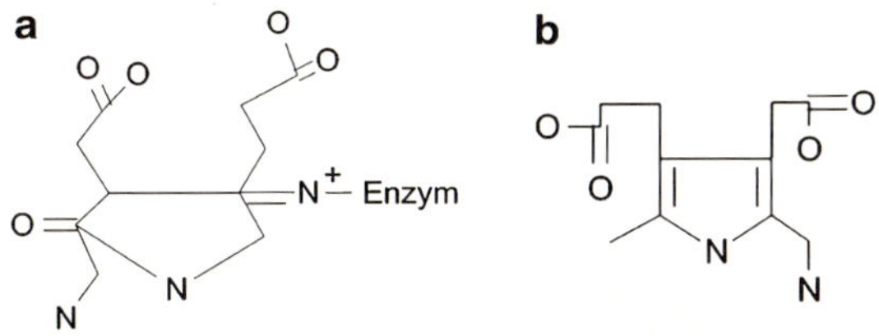


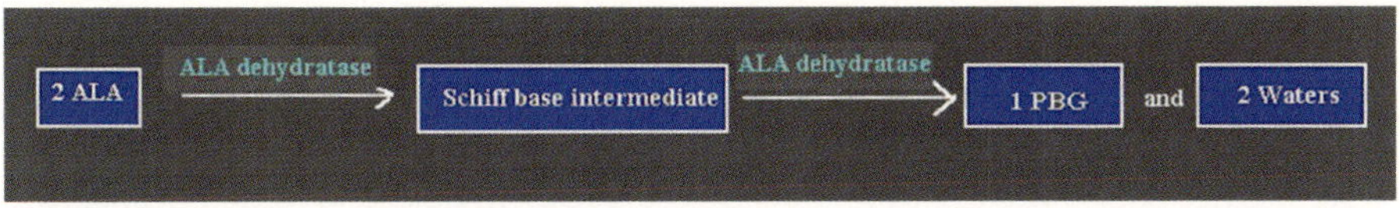


**Fig. 5.7** Conversion of ALA to PBG

In *E. coli* ALA Dehydratase contains two metal binding sites that have been designated α and β (Spencer and Jordan 1994). The α-site binds preferentially a $Zn^{2+}$ ion that is essential for catalytic activity. The β-site is exclusively a transition-metal-ion-binding site thought to be involved in protein conformation.

In animal cells PBG is formed from ALA in the cytoplasm (Rebeiz et al. 1996). In plants, PBG synthase is loosely bound to the plastid membranes (Lee et al. 1991). Beyond the possibility that in higher plants PBG may contribute to the formation of Proto in five different environments (Table 5.1), no specific efforts have been made to document the nature and extent of PBG biosynthetic heterogeneity in plants.

### *5.2.1 ALA Dehydratase*

A query for ALA dehydratase addressed to various protein databases listed in the Biology Workbench, yielded 30 unique records which are depicted under which are depicted on the LPBP website at http://www.vlpbp.org/greening/xv/, Sequenced Enzymes/ ALA Dehydratase, as well as in Appendix I.

## 5.3 Biosynthesis of Uroporphyrinogen III (Urogen III)

Urogen III is the universal precursor of all metabolic tetrapyrroles (Neve and Labbe 1956). It is the branching point where the biosynthesis of vitamin $B_{12}$ diverges from that of heme and Chl. Its biosynthesis from PBG requires the cooperation of two enzymes, PBG deaminase (Bogorad 1958a) and Urogen III synthase also known as cosynthetase (Bogorad 1958b). In *E. coli* PBG deaminase is coded for by the gene *hemC* (Thomas and Jordan 1986). The apoprotein consists of 353 amino acids with a molecular weight of 34245. The active site contains two constitutive PBG molecules (dipyrromethane cofactor) attached to the apoprotein by a cysteine

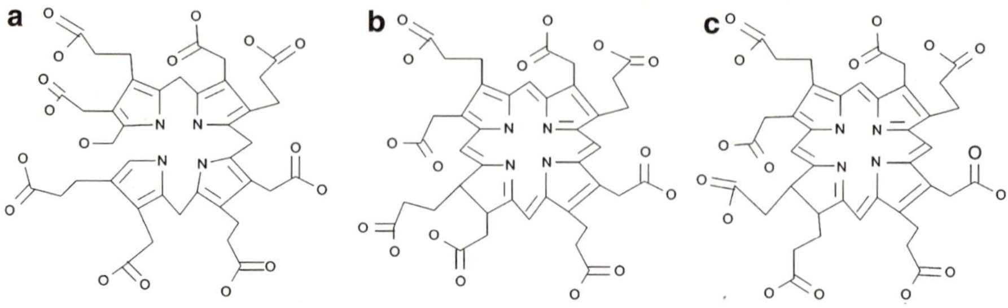

**Fig. 5.8** (**a**) Hydroxymethylbilane, (**b**) uroporphyrinogen I, (**c**) uroporphyrinogen III

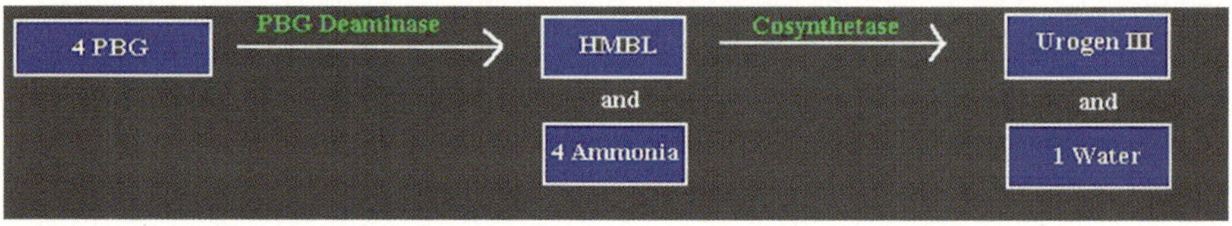


**Fig. 5.9** Conversion of PBG to uroporphyrinogen III

residue (Cys-242) (Hart et al. 1987; Jordan and Warren 1987). In a first step, one PBG molecule binds to the deaminase. A covalent bond is formed between the second constitutive PBG molecule and the first PBG substrate, and one molecule of ammonia is released. This first condensation leads to the formation of ring A of Urogen III. This step is repeated three more times and results in the formation of an open chain tetrapyrrole which is displaced from the enzyme by water to yield 1-hydroxymethylbilane (HMBL) (also called preuroporphyrinogen) (Fig. 5.8) (Battersby et al. 1979, 1982a, b, 1983; Jordan and Seehra 1979). Hydroxymethylbilane is unstable and in the absence of the cosynthetase cyclises at neutral pH to yield Urogen I (Fig. 5.8). In the presence of the cosynthetase, hydroxymethylbilane is rapidly converted into Urogen III (Battersby et al. 1982b). This reaction involves inversion of ring D of HMBL and cyclization with the release of one water molecule. In *E. coli* the cosynthetase is coded for by the gene *hemD* (Jordan et al. 1988). The apoprotein consists of 246 amino acids with a molecular weight of 27766. *HemC* and *hemD* occur in tandem and overlap by one base pair. In animal cells, Urogen III is formed in the cytoplasm (Rebeiz et al. 1996). In plant cells, PGB deaminase and the cosynthetase are loosely bound to the plastid membranes (Lee et al. 1991) (Fig. 5.9).

Beyond the possibility that in higher plants Urogen III may contribute to the formation of Proto in five different environments (Table 5.1), no specific efforts have been made to document the nature and extent of Urogen III biosynthetic heterogeneity in plants.

## 5.4 Biosynthesis of Coproporphyrinogen III (Coprogen III)

Coprogen III is the precursor of protoporphyrinogen IX. It is formed from Urogen III by decarboxylation, a reaction catalyzed by Urogen decarboxylase which converts Urogen III to Coprogen III (Granick and Mauzerall 1958; Mauzerall and Granick 1958). Stepwise decarboxylation of the 4 acetate side chains and the resulting structures of the intermediates led to the proposal that the acetate side chains on rings D, A, B. and C are decarboxylated in a clockwise fashion starting with ring D (Jackson et al. 1976, 1980). Although this appears to be the case in patients suffering from porphyria cutanea tarda, a random rather than an ordered decarboxylation appears to prevail in normal individuals (Luo and Lim 1993). These observations led to the proposal that the substrate binding site has such a flexible architecture that at low Urogen concentrations, decarboxylation may be ordered, while at high substrate concentrations it may be random (Akhtar 1994). The DNA coding for Urogen III decarboxylase in humans (Romeo Romeo et al. 1986) and rats (Romana et al. 1987) has been cloned and sequenced. The human enzyme consists of 367 amino acids with a molecular weight of 40,831. In animal cells Coprogen III is formed in the cytoplasm (Rebeiz et al. 1996)). In plants Urogen III decarboxylase appears to be loosely bound to the plastid membranes (Lee et al. 1991). Beyond the possibility that in higher plants Coprogen III may contribute to the formation of Proto in five different environments (Table 5.1), no specific efforts have been made to document the nature and extent of Coprogen III biosynthetic heterogeneity in plants (Figs. 5.10 and 5.11).

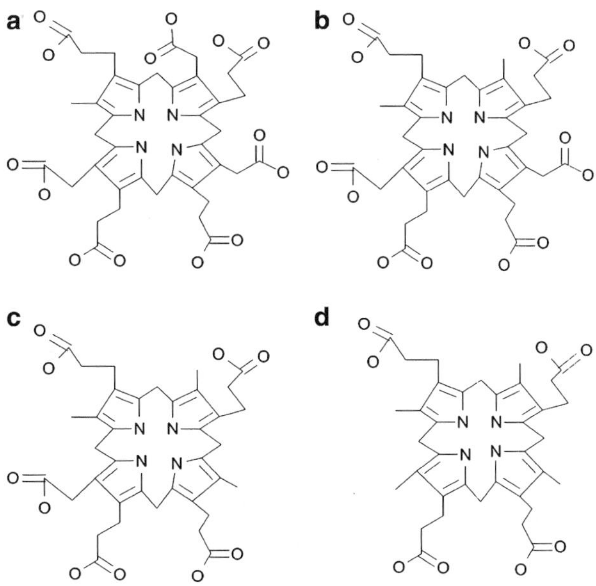


**Fig. 5.10** (**a**) Heptaporphyrinogen III, (**b**) hexaporphyrinogen III, (**c**) pentaporphyrinogen III and (**d**) coproporphyrinogen III

## 5.5 Biosynthesis of Protoporphyrinogen IX (Protogen IX)

Protogen IX is the precursor of protoporphyrin IX (Proto). Conversion of Coprogen III to Protogen IX involves oxidative decarboxylation of the two propionate side chains on rings A and B and their conversion to vinyl groups (Sano and Granick 1961). The mammalian enzyme has an absolute requirement for oxygen, but requires no reducing agent. Recent studies indicate that the mammalian enzyme is a dimer of two 37,000 subunits (Kohno et al. 1983). The observation of harderopoprphyrinogen accumulation (one vinyl at position 2 and one propionate at position 4) (Sano and Granick 1961) and its subsequent isolation (Kennedy et al. 1970) led to the proposal that the decarboxylation of ring A occurs before that of ring B. The precise mechanism of oxidative decarboxylation is still uncertain. In animal cells, cytoplasmic Coprogen III is transported to the mitochondria in an ATP-dependent process where it is converted to Protogen IX (Rebeiz et al. 1996). In plant cells, Coproporphyrinogen oxidase appears to be loosely bound to the plastid membranes (Lee et al. 1991).

Beyond the possibility that in higher plants Protogen IX may contribute to the formation of Proto in five different environments (Table 5.1), no specific efforts have been made to document the nature and extent of Protogen IX biosynthetic heterogeneity in plants (Fig. 5.12).

**Fig. 5.11** Formation of coproporphyrinogen III

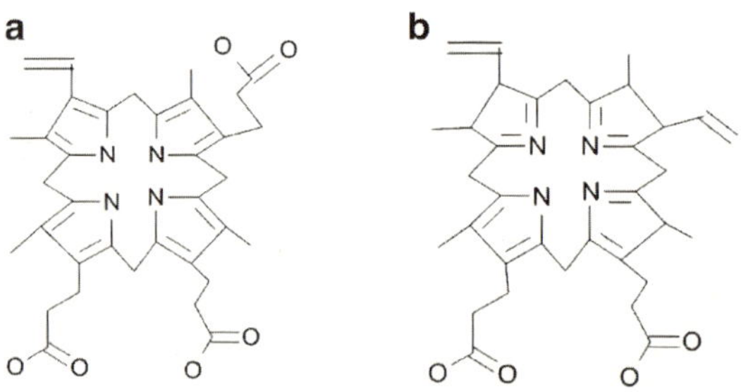


Harderoporphyrinogen Protoporphyrinogen IX

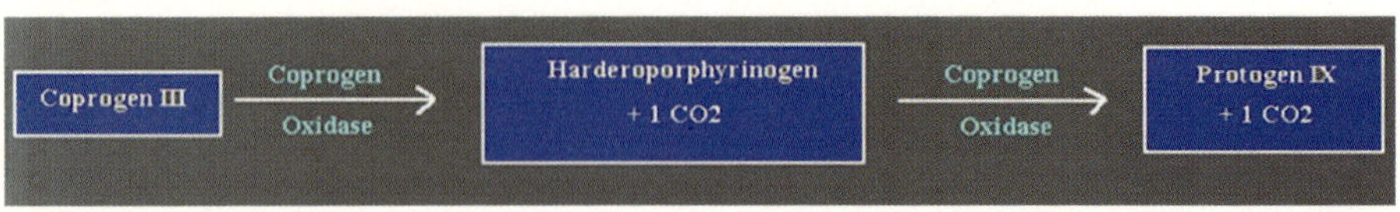


**Fig. 5.12** Conversion of coproporphyrinogen III to protoporphyrinogen IX

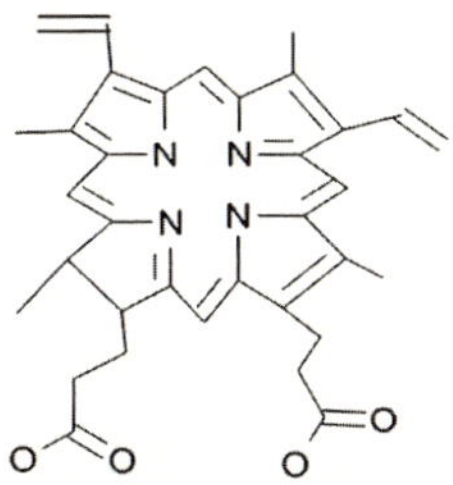

Protoporphyrin IX

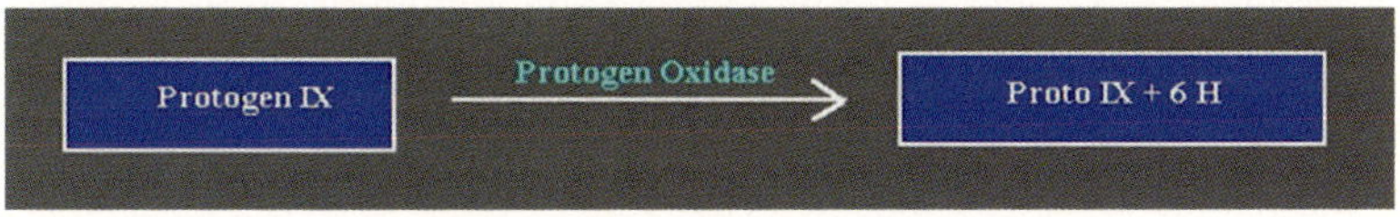


**Fig. 5.13** Conversion of protoporphyrinogen IX to protoporphyrin IX

## 5.6 Biosynthesis of Protoporphyrin IX (Proto)

Proto is the immediate precursor of Mg-Proto, which is the first committed intermediate of the Chl biosynthetic pathway. The role of Proto as an intermediate in the Chl biosynthetic pathways was based on the detection of Proto in X-ray Chlorella mutants inhibited in their capacity to form Chl (Granick 1948). It was conjectured that since the mutants had lost the ability to form Chl but accumulated Proto, the latter was a logical precursor of Chl. The unambiguous role of Proto as a precursor of all Mg-porphyrins and phorbins including Chl was established (a) by conversion of exogenous Proto to Mg-Proto monomethyl ester (Mpe) by Rhodopseudomonas spheroides in the presence of ATP and Mg (Gorchein 1972) and (b) by conversion of exogenous $^{14}$C- and unlabeled-Proto to Pchlide *a* [the immediate precursor of Chlide *a*] in organello (Mattheis and Rebeiz 1977), using a cell-free system capable of the conversion of $^{14}$C-ALA to $^{14}$C- Pchlide *a*, $^{14}$C-Pchlide ester *a* and $^{14}$C-Chl *a* and *b* (Rebeiz and Castelfranco 1971a, b), and capable of the net conversion of exogenous ALA to Mg-Protoporphyrins and Pchlide *a* (Rebeiz et al. 1975) (Fig. 5.13).

### *5.6.1 Biosynthesis of Protoporphyrin IX (Proto) via Oxidation of Protogen IX*

Protoporphyrinogen IX oxidase (Protox for short) catalyzes the conversion of Protogen IX to Proto (Jacobs and Jacobs 1987; Poulson and Polglase 1975). The oxidation involves the removal of 4 peripheral (meso) hydrogens and two inner hydrogens from the pyrrole nitrogens. In aerobic organisms, oxygen is the oxidant. Removal of the hydrogens appears to be stereospecific (Battersby et al. 1976). The enzyme has been purified to apparent homogeneity from bovine liver (Siepker et al. 1987). It appears to be a monomer with a molecular weight of approximately

65,000. The bovine enzyme has a tightly bound FAD prosthetic group. The plant enzyme has been partially characterized (Jacobs and Jacobs 1987). Protox was recently purified form spinach chloroplasts (Watanabe et al. 2000). Its molecular weight was estimated at 60 kDa by SDS-PAGE.

Diphenyl ether herbicides inhibit Protox (Matringe and Scalla 1988) and result in the accumulation of Protogen IX which translocates to various part of the cell (Lehnen et al. 1990). When Protogen is converted to Proto at various cellular sites, cell death ensues in the light in a typical porphyrin-dependent photodynamic herbicidal phenomenon (Rebeiz et al. 1984). In animal cells, during heme biosynthesis, Protogen IX is converted to Proto in the mitochondria. In plant cells, Protox appears to be loosely bound to the plastid membranes (Lee et al. 1991).

### *5.6.2 Biosynthetic Heterogeneity of Protoporphyrin IX*

In cucumber cotyledon and barley leaves chloroplasts, multiple Proto sites appear to be involved in resonance excitation energy transfer to the Chl *a* of LHCII, the major thylakoid LHC antenna, to LHCI-680, one of the inner LHC antennae of PSI, to Chl *a* of CP47 and CP29, two PSII antennae and to LHCI-73 an inner PSI antenna. This has led to the conclusion that the various Proto pools exist in different environments in the thylakoid membranes (Table 5.1) and (Kopetz et al. 2004). This observation together with an apparent ALA biosynthetic heterogeneity (Averina et al. 1993; Averina 1998), has led to the proposal of five different Proto biosynthesis sub-locations in thylakoid membranes. It is uncertain at this stage whether or not the 5 sub-locations contain identical enzymes that catalyze the conversion of ALA to Proto or not. In other words it is still uncertain whether the spatial heterogeneity of Proto formation is accompanied by biosynthetic heterogeneity.

## References

Akhtar M (1994) The modification of acetate and propionate side chains during the biosynthesis of haem and chlorophylls: mechanistic and stereochemical studies. In: The biosynthesis of tetrapyrrole pigments. Ciba Foundation symposium, vol 180. Wiley, Chichester, pp 131–155

Averina NG (1998) Mechanisms of regulation and interplastid location of chlorophyll biosynthesis. Biol Membr 15(5):504–516

Averina NG, Rudoi AB, Fradkin LI (1993) Centers of chlorophyll biosynthesis – current notions. Biofizika 38(6):1082–1086

Battersby AR, Donald MC, Redfern JR et al (1976) Biosynthesis of porphyrins and related macrocycles. V. Structural integrity of the type III porphyrinogen macrocycle in an active biological system; studies on the aromitazation of protoprphyrin-IX. J Chem Soc Perkins Trans 1:266–273

Battersby AR, Fookes CJR, Matcham GWJ et al (1979) Order of assembly of the four pyrrole rings during the biosynthesis of the natural porphyrins. J Chem Soc Chem Commun:539–541

Battersby AR, Fookes CJR, Gustafson-Potter KE et al (1982a) Biosynthesis of porphyrins and related macrocycles. Part 18. Proof by spectroscopy and synthesis that unarranged hydroxybilane is the product from deaminase and the substrate for cosynthetase in the biosynthesis of uroporphyrinogen-III. J Chem Soc Perkins Trans 1:2427

Battersby AR, Fookes CJR, Gustafson-Potter KE et al (1982b) Biosynthesis of porphyrins and related macrocycles. XVII. Chemical and enzymic transformation of isomeric aminoethylbilanes into uroporphyrinogens: proof that unrearranged bilnae is the preferred enzymic substrate and detection of a transient intermediate. J Chem Perkin Trans 1:2413

Battersby AR, Fookes CJR, Matcham GWJ et al (1983) Biosynthesis of porphyrins and related macrocycles. Part 20. Purification of deaminase and studies on its mode of action. J Chem Soc Perkins Trans 1:3031

Beale SI, Castelfranco PA (1974) The biosynthesis of δ-aminolevulinic acid in higher plants. II. Formation of $^{14}$C-δ-aminolevulinic acid from labelled precursors in greening plant tissues. Plant Physiol 53:291–296

Beale SI, Foley T, Dzelzkalns V (1981) δ–Aminolevulinic acid synthetase from *Euglena gracilis*. Proc Natl Acad Sci U S A 78(3):1666–1669

Bogorad L (1958a) The enzymic synthesis of porphyrins from porphobilinogen. I. Uroporphyrinogen I. J Biol Chem 233:501–509

Bogorad L (1958b) The enzymic synthesis of porphyrins from porphobilinogen. II. Uroporphyrin III. J Biol Chem 233:510–515

Chen J, Miller GW, Takemoto JY (1981) Biosynthesis of δ–aminolevulinic acid in *Rhodopseudomonas spaeroides*. Arch Biochem Biiophys 208(1):221–228

Drazic G, Bogdanovic M (2000) Gabaculine does not inhibit cytokinin-stimulated biosynthesis of delta-aminolevulinic acid. Plant Sci 154(1):23–29

Gibson KD, Neuberger A, Scott JJ (1955) The purification and properties of 5-aminolevulinic acid dehydratase. Biochem J 70:618–629

Gibson HD, Laver WG, Neuberger A (1958) Initial stages in the biosynthesis of porphyrins. II. The formation of 5-aminolevulinic acid from glycine and succinyl-CoA by particles from chicken erythrocytes. Biochem J 70:71–81

Gorchein A (1972) Magnesium protoporphyrin chelatase activity in Rhodopseudomonas spheroides. Studies with whole cells. Biochem J 127:97–106

Granick S (1948) Protoporphyrin 9 as a precursor of chlorophyll. J Biol Chem 172:717–727

Granick S (1963) The pigments of the biosynthetic chain of chlorophyll and their interactions with light. In: Proceedings of the fifth international congress of biochemistry, vol VI. Pergmon Press, New York, pp 176–186

Granick S, Mauzerall D (1958) Porphyrin biosynthesis in erythrocytes. II. Enzymes converting delta-aminolevulinic acid to coproporphyrinogen. J Biol Chem 232:1119–1140

Hart GJ, Miller AD, Leeper FJ et al (1987) Biosynthesis of the natural porphyrins: proof that hydroxymethylbilane synthase (porphobilinogen deaminase) uses a novel binding group in its catalytic action. J Am Chem Soc Chem Commun:1762–1765

Jackson AH, Sancovich HA, Ferramola AM et al (1976) Macrocyclic intermediates in the biosynthesis porphyrins. Philos Trans R Soc Lond B Biol Sci 273:191–206

Jackson AH, Sancovich HA, Ferramola AM (1980) Synthetic and biosynthetic studies of porphyrins. III. Structures of intermediates between uroporphyrinogen III and coproporphyrinogen III: synthesis of fourteen heptacarboxylic, hexacarboxylic and pentacarboxylic porphyrins related to uroporphyrin III. Bioorg Chem 9:71–120

Jacobs JM, Jacobs NJ (1987) Oxidation of protoporphyrinogen to protoporhyrin a step in chlorophyll and haem biosynthesis. Biochem J 244:219–224

Jordan PM, Seehra JS (1979) The biosynthesis of uroporphyrinogen III: order of assembly of the four porphobilinogen molecules in the formation of the tetrapyrrole ring. FEBS Lett 104:364–366

Jordan PM, Seehra JS. (1980) $^{13}$C NMR as a probe for the study of enzyme catalyzed reactions. Mechanism of action of 5-aminolevulinic acid dehydratase. FEBS Lett 114:283–286

Jordan PM, Warren MJ (1987) Evidence for a dipyrromethane cofactor at the catalytic site of E. coli porphobilinogen deaminase. FEBS Lett 225:87–92

Jordan PM, Mgbeje BIA, Thomas SD et al (1988) Nucleotide sequence of the hemD gene of Escherichia coli encoding uroporphyrinogen III synthetase and initial evidence for a hem operon. Biochem J 249:613–616

Jordan PM, Cheung RP, Sharma RP et al (1989) A cyclic intermediate, 2-hydroxy-3-aminotetrahydropyran-1-one (HAT) as a precursor for 5-aminolevulinic acid in greening barley. Tet Lett 34:1177

Kannangara CG, Gough SP, Oliver RP et al (1984) Biosynthesis of δ-aminolevulinic acid in greening barley leaves. VI. Activation of glutamate by ligation to RNA. Carlsberg Res Commun 49:417–437

Kennedy GY, Jackson AH, Kenner GW et al (1970) Isolation, structure and synthesis of a tricarboxylic porphyrin from harderian glands of rat. FEBS Lett 7:205–206

Klein O, Senger H (1978) Two biosynthetic pathways to δ−aminolevulinic acid in a pigment mutant of the green alga *Scenedesmus obliquus*. Plant Physiol 62:10–13

Kohno H, Furukawa T, Yoshihaga T et al (1983) Coproporphyrinogen oxidase. J Biol Chem 268:21359–21363

Kolossov VL, Kopetz KJ, Rebeiz CA (2003) Chloroplast biogenesis 87: evidence of resonance excitation energy transfer between tetrapyrrole intermediates of the chlorophyll biosynthetic pathway and chlorophyll *a*. Photochem Photobiol 78:184–196

Kopetz KJ, Kolossov VL, Rebeiz CA (2004) Chloroplast biogenesis 89: development of analytical tools for probing the biosynthetic topography of photosynthetic membranes by determination of resonance excitation energy transfer distances separating metabolic tetrapyrrole donors from chlorophyll *a* acceptors. Anal Biochem 329:207–219

Lee HJ, Ball M, Rebeiz CA (1991) Intraplastidic localization of the enzymes that convert delta-aminolevulinic acid to protoporphyrin IX in etiolated cucumber cotyledons. Plant Physiol 96:910–915

Lehnen LPJ, Sherman TD, Beceril JM et al (1990) Tissue and cellular localization of acifluorfen-induced porphyrins in cucumber cotyledons. Pest Biochem Physiol 37:239–248

Luo J, Lim CK (1993) Order of uroporphyrinogen III decarboxylation on incubation of porphobilinogen and uroporphyrinogen III with erythrocyte uroporphyrinogen decarboxylase. Biochem J 289:529–532

Matringe M, Scalla R (1988) Studies on the mode of action of acifluorfen-methyl in nonchlorophyllous soybean cells. Effects of acifluorfen-methyl on cucumber cotyledons: porphyrin accumulation. Plant Physiol 86:619–622

Mattheis JR, Rebeiz CA (1977) Chloroplast biogenesis. Net synthesis of protochlorophyllide from protoporphyrin IX by developing chloroplasts. J Biol Chem 252:8347–8349

Mauzerall D, Granick S (1958) Porphyrin biosynthesis in erythrocytes. III. Uroporphyrinogen and its decarboxylation. J Biol Chem 232:1141–1162

Neve RA, Labbe RF (1956) Reduced uroporphyrinogen III in the biosynthesis of heme. J Am Chem Soc 78:691–692

Poulson R, Polglase WJ (1975) The enzymic conversion of protoporphyrinogen IX to protoporphyrin IX. Protoporphyrinogen oxidase activity in mitochondrial extracts of Saccharomyces cerevisiae. J Biol Chem 250:1269–1274

Rebeiz CA, Castelfranco P (1971a) Protochlorophyll biosynthesis in a cell-free system from higher plants. Plant Physiol 47:24–32

Rebeiz CA, Castelfranco P (1971b) Chlorophyll biosynthesis in a cell-free system from higher plants. Plant Physiol 47:33–37

Rebeiz CA, Mattheis JR, Smith BB et al (1975) Chloroplast biogenesis. Biosynthesis and accumulation of protochlorophyll by isolated etioplasts and developing chloroplasts. Arch Biochem Biophys 171:549–567

Rebeiz CA, Montazer-Zouhoor A, Hopen HJ et al (1984) Photodynamic herbicides: 1. Concept and phenomenology. Enzyme Microbiol Technol 6:390–401

Rebeiz N, Arkins S, Kelley KW et al (1996) Enhancement of coproporphyrinogen III transport into isolated leucocyte mitochondria by ATP. Arch Biochem Biophys 333:475–481

Rebeiz CA, Kolossov VL, Briskin D et al (2003) Chloroplast biogenesis: chlorophyll biosynthetic heterogeneity, multiple biosynthetic routes and biological spin-offs. In: Nalwa HS (ed) Handbook of photochemistry and photobiology, vol 4. American Scientific Publishers, Los Angeles, pp 183–248

Romana M, LeBoulch P, Romeo PH (1987) Rat uroporphyrinogen decarboxylase cDNA: nucleotide sequence and comparison to human uroporphyrinogen decarboxylase. Nucleic Acids Res 15:7211

Romeo PH, Raich N, Duhart A et al (1986) Molecular cloning and nucleotide sequence of a complete human uroporphyrinogen decarboxylase cDNA. J Biol Chem 261:9825–9831

Sano S, Granick S (1961) Mitochondrial coproporphyrinogen oxidase and protoporphyrin formation. J Biol Chem 236:1173–1180

Schmid R, Shemin D (1955) The enzymic formation of porphobilinogen from 5-aminolevulinic acid and its conversion to protoporphyrin. J Am Chem Soc 77:506–508

Siepker LJ, Ford M, de Kock R et al (1987) Purification of bovine protoporphyrinogen oxidase: immunological cross-reactivity and structural relationship to ferrochelatase. Biochim Biophys Acta 913:349–358

Spencer P, Jordan PM (1994) Investigation of the nature of the two metal-binding sites in 5-amiolevulinic acid dehydratase from *Escherichia coli*. Biochem J 300:373–381

Thomas SD, Jordan PM (1986) Nucleotide sequence of the hemC locus encoding porphobilinogen deaminase of Escherichia coli K12. Nucleic Acids Res 14:6215–6226

Watanabe N, Che F-S, Terashima K et al (2000) Purification and properties of protoporphyrinogen oxidase from spinach. Plant Cell Physiol 41(7):880–892

# Chapter 6
# The Iron and Magnesium Branches of the Porphyrin Biosynthetic Pathway

*In 1844 Verdiel, suggested a relationship between chlorophyll and heme upon chemical conversion of chlorophyll to a red pigment.*

## 6.1 The Iron Branch of the Porphyrin Biosynthetic Pathway: Biosynthesis of Heme

Protoporphyrin IX is a branching point for heme and Chl biosynthesis. Insertion of ferrous iron into Proto leads to the formation of protoheme (Fig. 6.1), while insertion of Mg into the Proto macrocycle, leads to the formation of Mg-proto which is the precursor of all Mg-porphyrins and Chls in nature (see below).

The terminal step in heme biosynthesis involves insertion of ferrous iron into Proto by ferrochelatase to yield protoheme (Goldberg et al. 1956). In animal cells, conversion of Proto to protoheme takes place in the mitochondria. In *Euglena*, It has been reported that protoheme is formed in the mitochondria from Proto formed from ALA which is formed via the glycine-succinate pathway, and in the plastid from Proto formed from ALA which is formed via the C5-pathway (Weinstein and Beale 1983). In higher plants ferrochelatase is found in the mitochondria and the plastids, which strongly suggest that protoheme biosynthesis takes place in both organelles (Little and Jones 1976).

Ferrochelatase was first purified from rat liver (Taketani and Tokunaga 1981). Insertion of $Fe^{++}$ into Proto is accompanied by the release of two protons from the pyrrole nitrogens (Fig. 6.2). Mammalian ferrochelatase has a reported molecular weight of about 40,000. Specificity of the enzyme for Proto is not absolute, as the enzyme is able to handle a variety of porphyrin IX isomers, with substituents at the 2 and 4 positions of rings A and B that are smaller than hydroxyethyl in size and are uncharged. In is not clear whether the same situation prevails in higher plants.

C.A. Rebeiz, *Chlorophyll Biosynthesis and Technological Applications*, DOI 10.1007/978-94-007-7134-5_6, 

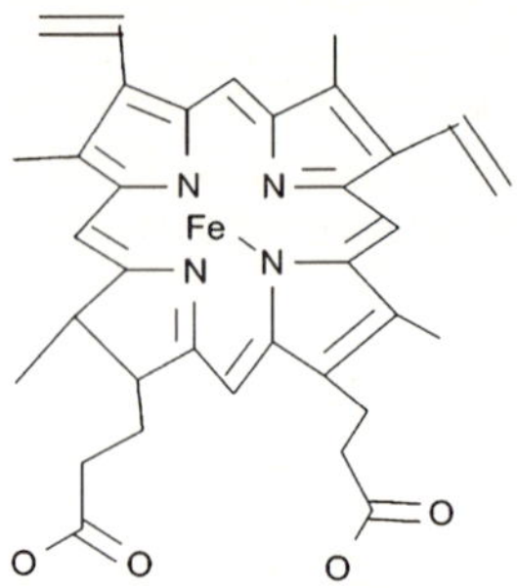


**Fig. 6.1** Protoheme

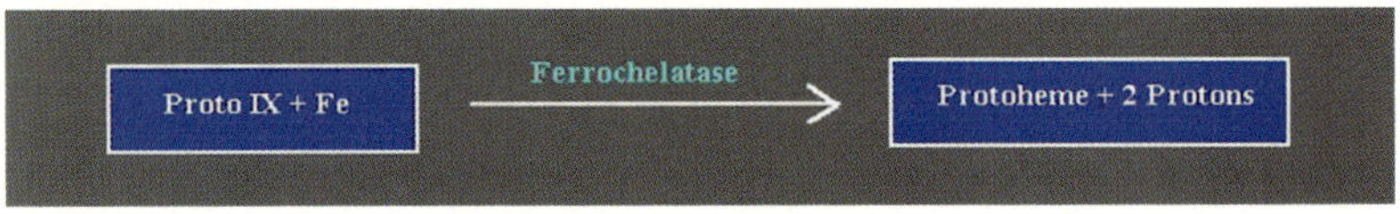


**Fig. 6.2** Conversion of Proto to protoheme

## 6.2 The Mg-Branch of the Porphyrin Biosynthetic Pathway

Most of the Chl in nature is formed from divinyl (DV) Proto via divinyl (DV) and monovinyl (MV) Mg-carboxylic biosynthetic routes. These routes are referred to as carboxylic routes because the Mg-tetrapyrrole intermediates all have one or two free carboxylic groups. Furthermore, most of these routes are heterogeneous. That is the biosynthesis of most of the intermediates can proceed via more than one path. This phenomenon is a manifestation of the overall Chl biosynthetic heterogeneity (Rebeiz et al. 2003) that permeates the whole Chl biosynthetic pathway. Chlorophyll biosynthetic heterogeneity was discovered when it was realized that most of the carboxylic and fully esterified tetrapyrrole pools of plants consist of DV and MV components. The biological significance of this phenomenon is becoming clearer as the Chl biosynthetic pathway is increasingly viewed in the context of the structural and functional heterogeneity of photosynthetic membranes.

Since some of the early biochemical work was done before discovery of the DV and MV Chl biosynthetic heterogeneity, and before development of appropriate analytical methodologies (Rebeiz 2002), it is not certain whether the investigated reactions of this early biochemical work involved only DV or both DV and MV tetrapyrroles. To emphasize this ambiguity, the MV and DV connotation will be omitted from the discussion of the early work. In other words, in this context, Mg-Proto would refer either to DV Mg-Proto, to MV Mg-Proto or to a mixture of both. On the other hands, DV and MV Mg-Proto would refer specifically to the DV and MV tetrapyrrole species respectively. The Mg-tetrapyrrole intermediates of the various Chl biosynthetic routes will be discussed in various section of this monograph.

Since the Chl biosynthetic heterogeneity will be discussed at length throughout this monograph, a brief overview of this heterogeneity will be presented below. Three figures and one table that will be referred to throughout this monograph will be presented below.

### *6.2.1 Biosynthetic Heterogeneity of the Chlorophyll Biosynthetic Pathway: An Overview*

Since the 1963 seminal review of Smith and French (1963), our understanding of the Chl biosynthetic pathway has changed dramatically. Several factors have contributed to this phenomenon, among which: (a) development of systems capable of Chl and thylakoid membrane biosynthesis in organello and in vitro, (Daniell and Rebeiz 1982; Kolossov et al. 1999; Rebeiz and Castelfranco 1971a, b; Rebeiz et al. 1984), (b) powerful analytical techniques that allowed the qualitative and quantitative determination of various intermediates of the Chl biosynthetic pathway (Rebeiz 2002), (c) recognition that the greening process proceeds differently in etiolated and green tissues, in darkness and in the light and in plants belonging to different greening groups (Abd-El-Mageed et al. 1997; Carey and Rebeiz 1985; Ioannides et al. 1994) and (d) recognition of the probability that the structural and functional complexity of thylakoid membranes is rooted in a multibranched, heterogeneous Chl biosynthetic pathway (Rebeiz et al. 1999).

Chlorophyll biosynthetic heterogeneity (Rebeiz et al. 1981, 1983, 1994, 1995, 2003) refers either (a) to spatial biosynthetic heterogeneity, (b) to chemical biosynthetic heterogeneity, or (c) to a combination of spatial and chemical biosynthetic heterogeneities.

Spatial biosynthetic heterogeneity refers to the biosynthesis of an anabolic tetrapyrrole or end product by identical sets of enzymes, at several different locations of the thylakoid membranes. On the other hand, chemical biosynthetic heterogeneity refers to the biosynthesis of an anabolic tetrapyrrole or end product at several different locations of the thylakoid membranes, via different biosynthetic routes, each involving at least one different enzyme.

#### 6.2.1.1 Chemical Heterogeneity

As mentioned above, spatial biosynthetic heterogeneity refers to the biosynthesis of an anabolic tetrapyrrole or end product by identical sets of enzymes, at several different locations of the thylakoid membranes

Figure 6.3 summarizes the Chl biosynthetic routes that take place in DDV-LDV-LDDV plant species such as cucumber (Kolossov and Rebeiz 2010). Figure 6.4 summarizes the Chl biosynthetic routes that take place in DMV-LDV-LDMV plant species such as corn, wheat and barley (Kolossov and Rebeiz 2010). Figure 6.5, summarizes the fully esterified Chl biosynthetic routes that take place in all greening groups (Rebeiz et al. 2003).

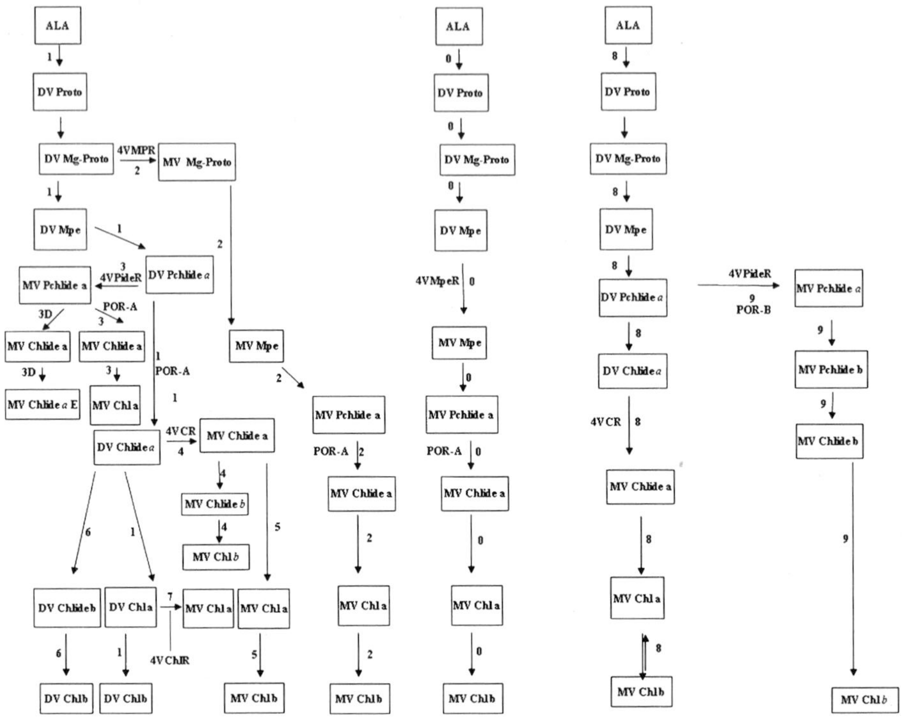

**Fig. 6.3** Monocarboxylic divinyl Chl *a* and *b* biosynthetic route of DDV-LDV-DLDV plants. (Adapted from Kolossov and Rebeiz 2010). *DV* divinyl, *MV* monovinyl, *ALA* δ-aminolevulinic acid, *Proto* protoporphyrin IX, *Mpe* Mg-Proto monomethyl ester, *Pchlide* protochlorophyllide; *Chlide* chlorophyllide, *Chl* chlorophyll, *4VMPR* [4-vinyl] Mg-Proto reductase, *4VMpeR* [4-vinyl] Mg-Proto monoester reductase, *4VPideR* [4-vinyl] protochlorophyllide *a* reductase, *4VCR* [4-vinyl] chlorophyllide *a* reductase, *4VChlR* [4-vinyl] Chl reductase, *POR* Pchlide *a* oxidoreductase. *Arrows* joining DV and MV routes refer to reactions catalyzed by [4-vinyl] reductases. The new biosynthetic route labeled 0 was called for by the discovery of 4VMpeR. All other routes are designated by Arabic numerals as described in Rebeiz et al. (2003)

### 6.2.1.2 Spatial Chl Biosynthetic Heterogeneity

As just mentioned, chemical biosynthetic heterogeneity refers to the biosynthesis of an anabolic tetrapyrrole or end product at several different locations of the thylakoid membranes, via different biosynthetic routes, each involving at least one different enzyme. We recently tested for spatial Chl biosynthetic heterogeneity by monitoring resonance excitation energy transfer between anabolic tetrapyrrole intermediates of the Chl biosynthetic pathway and various thylakoid Chl-protein complexes (Kolossov et al. 2003).

Fluorescence resonance energy transfer involves the transfer of excited state energy from an excited donor "D*" to an unexcited acceptor "A" (Calvert and Pitts 1967; Lakowicz 1999; Turro 1965). The transfer is the result of dipole-dipole

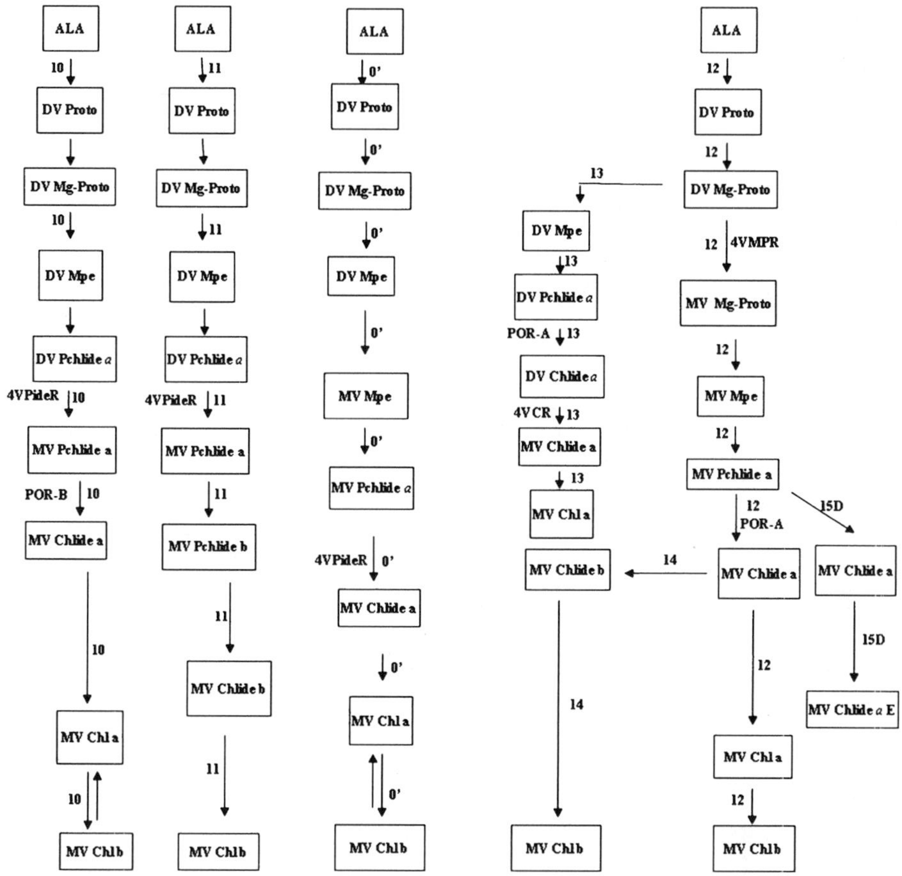


**Fig. 6.4** Monocarboxylic monovinyl Chl *a* and *b* biosynthetic routes of DMV-LDV-DLMV plants (Adapted from Kolossov and Rebeiz 2010). The new biosynthetic routes labeled 0′ was called for by the discovery of 4VMpeR and 4VCR activity in barley chloroplasts. All other routes are numbered as described in Rebeiz et al. (2003). *D* reaction occurring in darkness. All abbreviations are as in Rebeiz et al. (2003)

interaction between donor and acceptor and does not involve the exchange of a photon. The rate of energy transfer depends upon (a) the extent of overlap of the emission spectrum of the donor and the absorption spectrum of the acceptor, (b) the relative orientation of the donor and acceptor transition dipoles, and (c) the distance between donor and acceptor molecules. As soon as the excited donor "D*" and unexcited acceptor "A" states are coupled by an appropriate interaction, they become degenerate if there is an excited state of the acceptor "A", which requires exactly the same excitation energy available in "D*". When such a condition exists, excitation of one of the degenerate states leads to a finite probability that the same excitation will appear in the other degenerate state (Turro 1965). This probability increases with time but is inversely proportional to the sixth power of the fixed

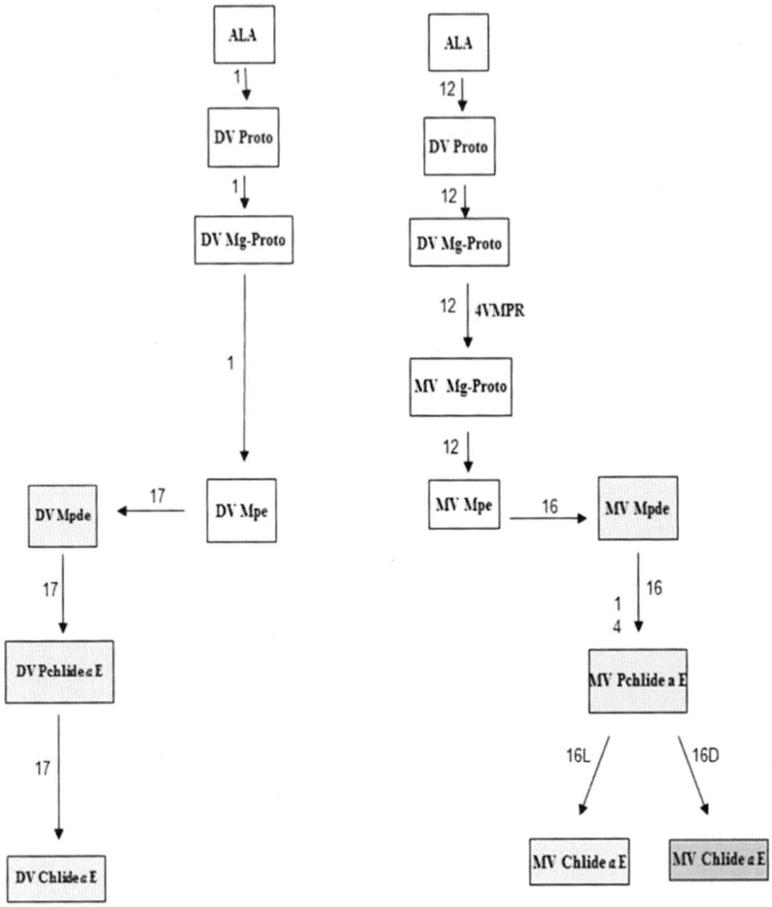


**Fig. 6.5** Biosynthetic pathway (Adapted from Rebeiz et al. 2003) and depicting the fully esterified Chl biosynthetic routes. To facilitate understanding of the text, various biosynthetic routes are designated by the numbers 16–17. All abbreviation are as in Rebeiz et al. (2003)

distance separating the centers of the donor and acceptor molecules. It has been estimated that dipole-dipole energy transfer between donor and acceptor molecules may occur up to a separation distance of 50–100 Å (Calvert and Pitts 1967).

Resonance excitation energy transfer from three tetrapyrrole donors to the Chl *a* of Chl-protein complexes were monitored, namely: from Proto, DV Mg-Proto and its methyl ester and MV and DV Pchlide *a*. DV Proto is a common precursor of heme and Chl. It is the immediate precursor of DV Mg-Proto. As such, it is an early intermediate along the Chl biosynthetic chain. Biosynthetically, it is several steps removed from the Chl end product. Mg-Proto is a mixed MV-DV, dicarboxylic tetrapyrrole pool, consisting of DV and MV Mg-Proto. It is the precursor of DV and MV Pchlide a. The protochlorophyll(ide) [(Pchl(ide)] of higher plants consists of about 95 % Pchlide *a* and about 5 % Pchlide *a* ester (Pchlide a E). The latter is esterified with long chain fatty alcohols (LCFAs) at position 7 of the macrocycle. While Pchlide a E consists mainly of MV Pchlide *a* ester, Pchlide *a* consists of DV and MV Pchlide *a*. The latter are the immediate precursors of DV and MV chlorophyllide (Chlide) *a*. Accumulation of the various tetrapyrrole donors was induced by incubation of green tissues with delta-aminolevulinic acid (ALA) and/or 2,2′-dipyridyl (Rebeiz et al. 1988).

The task of selecting appropriate Chl a-protein acceptors was facilitated by the fluorescence properties of green plastids. At 77 °K, emission spectra of isolated chloroplasts exhibit maxima at 683–686 nm (~F685), 693–696 nm (~F695), and 735–740 nm (~F735). It is believed that the fluorescence emitted at ~ F685 nm arises from the Chl *a* of LHCII, the major thylakoid LHC antenna, and LHCI-680, one of the LHC antennae of PSI (Bassi et al. 1990). That emitted at ~F695 nm is believed to originate mainly from the Chl *a* of CP47 and CP29, two PSII antennae (Bassi et al. 1990). That emitted at ~F735 nm is believed to originate primarily from the Chl *a* of LHCI-730, a PSI antenna (Bassi et al. 1990). Since these emission maxima are readily observed in the fluorescence emission spectra of green tissues and are associated with definite thylakoid Chl *a*-protein complexes, it was conjectured that they would constitute a meaningful resource for monitoring excitation resonance energy transfer between anabolic tetrapyrroles and representative Chl *a*-protein complexes. To monitor the possible occurrence of resonance energy transfer between the accumulated anabolic tetrapyrroles and Chl *a*-protein complexes, excitation spectra were recorded at 77 °K at the respective emission maxima of the selected Chl *a* acceptors, namely at ~685, ~695, and ~735 nm. It was conjectured that if resonance energy transfers were to be observed between the tetrapyrrole donors and the selected Chl a acceptors, definite excitation maxima would be observed. These excitation maxima would correspond to absorbance maxima of the various tetrapyrrole donors, and would represent the peaks of the excitation resonance energy transfer bands (Kolossov et al. 2003).

Pronounced excitation resonance energy transfer bands from Proto, Mp(e), and Pchl(ide) *a* to Chl *a* ~F685, ~F695, and ~F735 were detected as shown in Table 6.1, which is depicted below.

It was proposed that the various intra-membrane environment of Proto, Mp (e) and Pchl(ide) which were manifested by a differential donation of excitation resonance energy transfer to different Chl *a*-apoprotein complexes represent evidence of Chl spatial biosynthetic heterogeneity (Kolossov et al. 2003)

### 6.2.2 Why Is Tetrapyrrole Metabolism Important

Heme and chlorophyll (Chl) are porphyrins. Porphyrins (also referred to as tetrapyrroles) are essential for life in the biosphere). Chlorophyll catalyzes the conversion of solar energy to chemical energy via the process of photosynthesis.

**Table 6.1** Mapping of excitation resonance energy transfer maxima to Chl *a* F686, Chl *a* F694 and Chl *a*F738-742 *in situ*. A dash represents missing data. *s* shoulder, *p* peak. Only the barley spectra depicted in Fig. 8 were recorded at the observed peak of Chl *a* emission at F642 nm (Adapted from Kolossov et al. 2003)

| Plant species | Major donor | Donor conc (pmoles/ml suspension) | Excitation resonance energy maxima to: Chl a F686 | Chl a F694 | Chl a F738 | Conc ALA (mM) | Conc Dpy (mM) | Incub (h) |
|---|---|---|---|---|---|---|---|---|
| Cucumber | Proto | 54 | 397p, 402p, 410p, 415p | 390s, 400p, 409p | 390s, 395s, 408p, 417p | 4.5 | 3.7 | 6 |
| Cucumber | Proto | 83 | 387p, 402p, 412p | 392p, 406p | 388p, 399p, 403p, 410p, 415p | 20 | 4 | 6 |
| Cucumber | Proto | 92 | 390p, 399p, 405p, 412p | 399p, 409p, 412s | 399p, 400p, 416p | 20 | 16 | 6 |
| Cucumber | Proto | 376 | 395p, 404s, 411p, 416p | 395p,406p, 414p | 393p, 400s, 407p | 20 | 16 | 6 |
| Cucumber | Proto | 1,046 | 402s, 411p | 404p, 410s, 416p, | 399s, 405s, 411p | 20 | 0 | 12 |
| Barley | Proto | 13 | 391, 398s, 404s, 411p | 389s, 395p, 406p, 414p | 390s, 393p, 400s, 406p, 412p, 416s, | 4.5 | 3.7 | 6 |
| Barley | Proto | 61 | 389p, 396s, 404s, 410p, 412p | 396p, 406p, 412p | 389s, 395p, 406s, 410p, | 20 | 4 | 6 |
| Barley | Proto | 64 | 395s, 400p, 405s, 413p | 389p, 3974, 403p, 412p | 388s, 393p, 400s, 406p, 412p | 20 | 0 | 6 |
| Barley | Proto | 68 | 389p, 396p, 412p, 413s | 389p, 398p, 409p, | 396s, 400p, 412p | 20 | 0 | 6 |
| Cucumber | Mp(e) | 26 | 419p, 431p | 422p, 429p, 434p | – | 20 | 0 | 6 |
| Cucumber | Mp(e) | 83 | 422p, 432p | 420p, 425p | 417p, 424s, 427s, 429p | 4.5 | 3.7 | 6 |
| Cucumber | Mp(e) | 91 | 418s, 424p, 433p | 419p, 426p | 414p, 423p | 20 | 4 | 6 |

(continued)

**Table 6.1** (continued)

| Plant species | Major donor | Donor conc (pmoles/ml suspension) | Excitation resonance energy maxima to: Chl *a* F686 | Chl *a* F694 | Chl *a* F738 | Conc ALA (mM) | Dpy | Incub (h) |
|---|---|---|---|---|---|---|---|---|
| Cucumber | Mp(e) | 185 | 421p, 427s, 430s | 421p, 428s | 421p, 430p | 20 | 0 | 12 |
| Cucumber | Mp(e) | 618 | 421p, 427s, 430s | 421p, 427s, 430s | 421p, 430p | 20 | 0 | 12 |
| Barley | Mp(e) | 10 | 420p, 428s | 424p, 430s | 426s, 432s | 4.5 | 3.7 | 6 |
| Barley | Mp(e) | 11 | 423p | 418p, 422s, 427p | 422s, 426p, 431s | 20 | 0 | 6 |
| Barley | Mp(e) | 25 | 423p, 428s | 418p, 430p | 426s, 432p | 20 | 4 | 6 |
| Cucumber | DV Pch-lide *a* | 133 | 438p, 446p, 453s, 460s, 467p | 440s, 448p, 454s, 460p | 448p, 453p, 461p | 20 | 4 | 6 |
| Cucumber | DV Pch-lide *a* | 153 | 443p, 449p, 457p | 436s, 442p, 453p, 463p | 439p, 453p, 457p, 460p | 4.5 | 3.7 | 6 |
| Cucumber | DV Pch-lide *a* | 412 | 437p, 444s, 452p, 458p | 435p, 441p, 451p, 462p | 437p, 447s, 454s, 457p, 463s | 20 | 0 | 6 |
| Cucumber | DV Pch-lide *a* | 1,030 | 438s, 447p, 452p, 456s, 462s | 441s, 447p, 452p, 459p | 436p, 448s, 454s, 458p, 462p | 20 | 0 | 6 |
| Cucumber | DV Pch-lide *a* | 1,435 | 435p, 447s, 453p, 460s | 438s, 445s, 452p, 456s, 460s, 462s | 436s, 444s, 452s, 458p, 462s | 20 | 0 | 12 |
| Cucumber | DV Pch-lide *a* | 4,784 | 440s, 449p, 455s, 460s | 434p, 440s, 447s, 452p, 459s | 434s, 440p, 447s, 462p | 20 | 0 | 12 |
| Barley | MV Pch-lide *a* | 26 | 434s, 441p, 452p, 460p | 438s, 445p, 449p, 463p | 440p, 449p, 458s, 468p, | 4.5 | 3.7 | 6 |

(continued)

**Table 6.1** (continued)

| Plant species | Major donor | Donor conc (pmoles/ml suspension) | Excitation resonance energy maxima to: Chl *a* F686 | Chl *a* F694 | Chl *a* F738 | Conc ALA (mM) | Dpy | Incub (h) |
|---|---|---|---|---|---|---|---|---|
| Barley | MV Pch-lide *a* | 104 | 439p, 445s, 450p, 458p, 463s | 436s, 447p, 455p, 463s | 440p, 450p, 458p | 20 | 4 | 6 |
| Barley | MV Pch-lide *a* | 193 | 439s, 444p, 451p, 462p, 467p | 435s, 440s, 446p, 453p, 460p | 438s, 453p, 457p, 464s | 20 | 0 | 6 |

Organic life in the biosphere is made possible by consumption of the chemical energy generated by photosynthesis.

Hemes are the prosthetic groups of cytochromes which are involved in electron transport during oxidative phosphorylation and photosynthetic phosphorylation which generate ATP and NADPH. The latters are essential for many cellular functions.

The importance of heme and Chl to life in the biosphere will be stressed in several chapters of this monograph.

## *6.2.3 Mg-Protoporphyrin IX Chelatase*

### 6.2.3.1 Mg Insertion into Protoporphyrin IX

The enzymatic insertion of Mg into Proto by Mg-Proto chelatase, to yield Mg-Proto was achieved in organello (Smith and Rebeiz 1977a, b). At the low ATP concentrations used in this system, the biosynthesis of Mg-Proto was accompanied by the formation of Zn-Proto. Simultaneous equations were used in order to deconvolute the fluorescence spectra and be able to determine the amounts of Mg-Proto in the presence of Zn-Proto contamination (Smith and Rebeiz 1977a, b). Later on, interference from Zn-Proto was eliminated when it was realized that ATP was a mandatory cofactor for Mg-insertion into Proto and that higher concentration of added ATP eliminated the Zn-Proto problem formation (Pardo et al. 1980) (Fig. 6.6)

In cucumber etiochloroplasts, Mg-Proto chelatase is bound to the plastid membranes (Lee et al. 1991, 1992; Smith and Rebeiz 1979). The activity of the

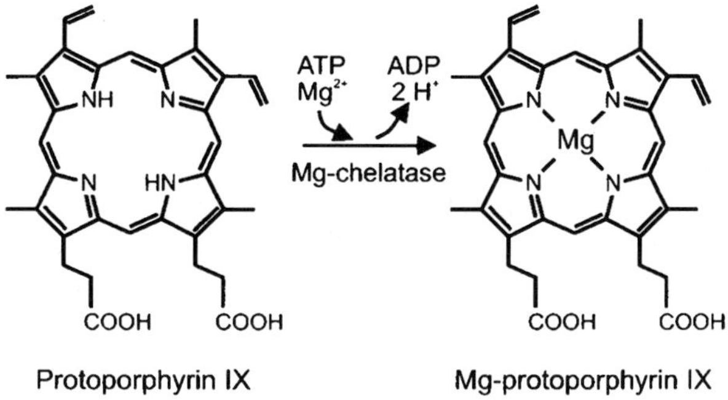


**Fig. 6.6** Incorporation of $Mg^{2+}$ into protoporphyrin IX by Mg-chelatase (Adapted from Kannangara and von Wettstein 2010)

membrane-bound enzyme increased upon addition of exogenous Mg (Lee et al. 1992). In pea chloroplasts, contrary to what was observed in cucumber plastids, both stroma and membranes were needed to reconstitute Mg-Proto chelatase activity (Walker and Weinstein 1991). It is not known whether the discrepancy between the cucumber and pea results is due to differences in preparatory methodologies or not. Indeed it has been reported that the separation of plastid stroma from plastid membranes may result in the solubilization of membrane components if appropriate precautions are overlooked (Lee et al. 1991). In cucumber but not in pea, Mg-Proto chelatase was stabilized by its substrate (i. e. exogenous Proto) before separation of the stroma from the plastid membranes (Lee et al. 1992).

#### 6.2.3.2 Molecular Biology of Mg-Proto Chelatase

Mutational analysis of the *Rb. Capsulatus* photosynthesis gene cluster suggested that three sequenced genes, namely *bchH*, *bchD* and *bchI* were involved in Mg-chelation (Suzuki et al. 1997). The *bchH*, *bchI* and *bchD* genes from *R. spheroides* were expressed in *E. coli*. When cell-free extracts from strains containing the gene products *BchH*, *BchI*, and *BchD* were combined, the mixture was able to catalyze the insertion of Mg into Proto in an ATP-dependent manner (Gibson et al. 1995). The authors suggested that *BchH* binds Proto prior to the insertion of the Mg atom. Also genes from *Synechocystis* PCC6803 a cyanobacterium, with homology to the *bchH*, *bchD* and *bchI* genes, namely *chlH*, *chlD* and *chlI*, were cloned and overexpressed in *E. coli* (Jensen et al. 1995). In this case too, the combined cell-free extracts containing the *ChlH*, *ChlI* and *ChlD* gene products were able to catalyze the insertion of $Mg^{2+}$ into Proto in an ATP-dependent manner. The N-terminal half of the ChlD protein exhibited a 40–41 % homology to *Rhodobacter BchI* and *Synechocystis ChlI*, whereas the C-terminal half displayed a 33 % homology to *Rhodobacter BchD*. The authors suggested the existence of an evolutionary relationship between the *I* and *D* genes.

It is now acknowledged that insertion of $Mg^{2+}$ into Proto appears to be a two–step reaction, consisting of activation followed by $Mg^{2+}$ insertion (Jensen et al. 1999). The activation step requires ATP and the ChlI and ChlD subunits and results in the formation of a ChlI-ChlD-ATP complex. Insertion of $Mg^{2+}$ into Proto also requires ATP and the ChlH subunit. It was observed however that during formation of the ChlI-ChlD-ATP complex, ATP may be replaced by a slowly hydrolysable analog such as 5′-[γ-thio]triphosphate, by a non-hydrolysable ATP analog such as adenosine 5;-[β,γ -imido] triphosphate, or to a lesser extent by ADP. There was an absolute requirement, however for ATP hydrolysis during $Mg^{2+}$ insertion by the ChlH protein.

In Arabidopsis thaliana, the *ChlH* gene product was observed to undergo a dramatic diurnal variation, rising almost to its maximum level by the end of the dark period, increasing slightly at the onset of the light period and declining steadily to a minimum by the end of the light period (Gibson et al. 1996). It was proposed that the ChlH protein plays a role in regulating the levels of chlorophyll during the daily dark–light cycle. Furthermore immunoblotting showed that the distribution of the ChlH protein between the stroma and chloroplast membranes varied depending on the concentration of $Mg^{2+}$. For example in soybean, the ChlH protein was found either in the stroma at low $Mg^{2+}$ concentration in the lysing buffer, or on the chloroplast envelope at high lysing buffer $Mg^{2+}$ concentration (Nakayama et al. 1998).

The three subunits of Mg-Proto chelatase namely the 40 kDa, the 70 kDA and the 140 kDa subunits are described in great details by Kannangara and von Wettstein (2010)

## References

Abd-El-Mageed HA, El Sahhar KF, Robertson KR et al (1997) Chloroplast biogenesis 77. Two novel monovinyl and divinyl light–dark greening groups of plants and their relationship to the chlorophyll *a* biosynthetic heterogeneity of green plants. Photochem Photobiol 66:89–96

Bassi R, Rigoni F, Giacometti GM (1990) Chlorophyll binding proteins with antenna function in higher plants and green algae. Photochem Photobiol 52:1187–1206

Calvert JG, Pitts JN (1967) Photochemistry. Wiley, New York

Carey EE, Rebeiz CA (1985) Chloroplast biogenesis 49. Difference among angiosperms in the biosynthesis and accumulation of monovinyl and divinyl protochlorophyllide during photoperiodic greening. Plant Physiol 79:1–6

Daniell H, Rebeiz CA (1982) Chloroplast culture IX. Chlorophyll(ide) *a* biosynthesis in vitro at rates higher than in vivo. Biochem Biophys Res Commun 106:466–470

Gibson LCD, Willows RD, Kannangara CG et al (1995) Magnesium-protoporphyrin chelatase of *Rhodobacter sphaeroides*: reconstitution of activity by combining the products of the *bchH, -I,* and *-D* genes expressed in *Escherichia coli*. Proc Natl Acad Sci U S A 92:1941–1944

Gibson LCD, Marrison JL, Leech RM et al (1996) A putative Mg chelatase subunit from *Arabidopsis thaliana* cv C24. Plant Physiol 111:61–71

Goldberg A, Ashenbrucker M, Cartwright GE et al (1956) Studies on the biosynthesis of heme in vitro by avian erythrocytes. Blood 11:821–833

Ioannides IM, Fasoula DM, Robertson KR et al (1994) An evolutionary study of chlorophyll biosynthetic heterogeneity in green plants. Biochem Syst Ecol 22:211–220

Jensen PE, Gibson LCD, Henningsen KW et al (1995) Expression of the *chlI, chlD* and *chlH* genes from the cyanobacterium *Synechocystis* PCC6803 in *Escherichia coli* and demonstration that the three cognate proteins are required for magnesium-protoporphyrin chelatase activity. J Biol Chem 271(28):1662–1667

Jensen PE, Gibson LCD, Hunter CN (1999) ATPase activity associated with the magnesium-protoporphyrin IX chelatase enzyme of Synechocystis PCC6803: evidence for ATP hydrolysis during $Mg^{2+}$ insertion, and the MgATP-dependant interaction of the ChlI and ChlD subunits. Biochem J 339(1):127–134

Kannangara CG, Von Wettstein D (2010) Magnesium chelatase. In: Rebeiz CA, Benning C, Bohnert HJ (eds) The chloroplast: basics and applications, vol 31. Springer, Dordrecht, pp 79–88

Kolossov VL, Rebeiz CA (2010) Evidence for various 4-vinyl reductase activities in higher plants. In: Rebeiz CA, Benning C, Bohnert HJ (eds) The chloroplast: basics and applications. Springer, Dordrecht, pp 25–38

Kolossov V, Ioannides IM, Kulur S et al (1999) Chloroplast biogenesis 82: development of a cell-free system capable of the net synthesis of chlorophyll(ide) *b*. Photosynthetica 36:253–258

Kolossov VL, Kopetz KJ, Rebeiz CA (2003) Chloroplast biogenesis 87: evidence of resonance excitation energy transfer between tetrapyrrole intermediates of the chlorophyll biosynthetic pathway and chlorophyll *a*. Photochem Photobiol 78:184–196

Lakowicz JR (1999) Principles of fluorescence spectroscopy. Kluwer Academic/Plenum Press, New York

Lee HJ, Ball M, Rebeiz CA (1991) Intraplastidic localization of the enzymes that convert delta-aminolevulinic acid to protoporphyrin IX in etiolated cucumber cotyledons. Plant Physiol 96:910–915

Lee HJ, Ball MD, Parham R et al (1992) Chloroplast biogenesis 65. Enzymic conversion of protoporphyrin IX to Mg-protoporphyrin IX in a subplastidic membrane fraction of cucumber etiochloroplasts. Plant Physiol 99:1134–1140

Little HN, Jones OTG (1976) The subcellular localization and properties of the ferrochelatase of etiolated barley. Biochem J 156:309–314

Nakayama M, Masuda T, Bando T et al (1998) Cloning and expression of the soybean *chlH* gene encoding a subunit of Mg-chelatase and localization of $Mg^{2+}$ concentration-dependent ChlH protein within the chloroplast. Plant Cell Physiol 39(3):275–284

Pardo AD, Chereskin BM, Castelfranco PA et al (1980) ATP requirement for Mg chelatase in developing chloroplasts. Plant Physiol 65:956–960

Rebeiz CA (2002) Analysis of intermediates and end products of the chlorophyll biosynthetic pathway. In: Smith A, Witty M (eds) Heme chlorophyll and bilins, methods and protocols. Humana Press, Totowa, pp 111–155

Rebeiz CA, Castelfranco P (1971a) Chlorophyll biosynthesis in a cell-free system from higher plants. Plant Physiol 47:33–37

Rebeiz CA, Castelfranco P (1971b) Protochlorophyll biosynthesis in a cell-free system from higher plants. Plant Physiol 47:24–32

Rebeiz CA, Belanger FC, McCarty SA et al (1981) Biosynthesis and accumulation of novel chlorophyll *a* and *b* chromophoric species in green plants. In: Akoyunoglou G (ed) Chloroplast development, Photosynthesis. Balaban International Services, Philadelphia, pp 197–212

Rebeiz CA, Wu SM, Kuhadje M et al (1983) Chlorophyll *a* biosynthetic routes and chlorophyll *a* chemical heterogeneity. Mol Cell Biochem 58:97–125

Rebeiz CA, Montazer-Zouhoor A, Daniell H (1984) Chloroplast culture X: thylakoid assembly in vitro. Isr J Bot 33:225–235

Rebeiz CA, Montazer-Zouhoor A, Mayasich JM et al (1988) Photodynamic herbicides. Recent developments and molecular basis of selectivity. Crit Rev Plant Sci 6:385–434

Rebeiz CA, Parham R, Fasoula DA et al (1994) Chlorophyll biosynthetic heterogeneity. In: Chadwick DJ, Ackrill K (eds) The biosynthesis of the terapyrrole pigments. Wiley, New York, pp 177–193

Rebeiz CA, Gut LJ, Keywan L et al (1995) Photodynamics of porphyric insecticides. Crit Rev Plant Sci 14:329–366

Rebeiz CA, Ioannides IM, Kolossov V et al (1999) Chloroplast biogenesis 80. Proposal of a unified multibranched chlorophyll a/b biosynthetic pathway. Photosynthetica 36:117–128

Rebeiz CA, Kolossov VL, Briskin D et al (2003) Chloroplast biogenesis: chlorophyll biosynthetic heterogeneity, multiple biosynthetic routes and biological spin-offs. In: Nalwa HS (ed) Handbook of photochemistry and photobiology, vol 4. American Scientific Publishers, Los Angeles, pp 183–248

Smith JHC, French CS (1963) The major accessory pigment in photosynthesis. Annu Rev Plant Physiol 14:181–224

Smith BB, Rebeiz CA (1977a) Chloroplast biogenesis: detection of Mg-protoporphyrin chelatase in vitro. Arch Biochem Biophys 180:178–185

Smith BB, Rebeiz CA (1977b) Spectrofluorometric determination of Mg-protoporphyrin monoester and longer wavelength metalloporphyrins in the presence of Zn-protoporphyrin. Photochem Photobiol 26:527–532

Smith BB, Rebeiz CA (1979) Chloroplst biogenesis XXIV. Intrachloroplastic localization of the biosynthesis and accumulation of protoporphyrin IX, magnesium protoporphyrin IX, magnesium-protoporphyrin monoester and longer wavelength metalloporphyrins during greening. Plant Physiol 63:227–231

Suzuki JY, Bollivar DW, Bauer CE (1997) Genetic analysis of chlorophyll biosynthesis. Annu Rev Genet 31:61–89

Taketani S, Tokunaga R (1981) Rat liver ferrochelatase. Purification, properties and stimulation by fatty acids. J Biol Chem 256:12748–12753

Turro NJ (1965) Molecular photochemistry. Benjamin, London

Walker CJ, Weinstein JD (1991) In vitro assay of the chlorophyll biosynthetic enzyme Mg-chelatase: resolution of the activity into soluble and membrane-bound fractions. Proc Natl Acad Sci U S A 88:5789–5793

Weinstein JD, Beale SI (1983) Separate physiological roles and subcellular compartments for two tetrapyrrole biosynthetic pathways in *euglena gracilis*. J Biol Chem 258:6799–6807

# Chapter 7
# The Chl *a* Carboxylic Biosynthetic Routes: Reactions Between Mg-Protoporphyrin IX and Protochlorophyllide *a*

*A man has to resolve either to put out nothing new or to become a slave to defend it.*

Faraday

## 7.1 The Mg-Protoporphyrin IX (Mg-Proto) Pool

Mg-protoporphyrin (Mg-Proto) (Fig. 7.1) is the immediate precursor of Mg-proto monomethyl ester (Mpe). The proposed role of Mg-Proto as an intermediate in the Chl biosynthetic pathway was based on the detection of Mg-Proto in X-ray Chlorella mutants inhibited in their capacity to form Chl (Granick 1948). It was conjectured that since the mutants had lost the ability to form Chl but accumulated Mg-Proto, the latter was a logical precursor of Chl. On the basis of absorbance spectroscopic determinations the accumulated Mg-Proto was assigned by Granick a divinyl (DV) Chemical structure (Fig. 7.1, I), with vinyl groups at positions 2 and 4 of the tetrapyrrole macrocycle.

Mg-Proto as a precursor of other Mg-porphyrins and of Pchlide *a* was demonstrated by conversion of $^{3}$H-Mg-Proto to $^{3}$H-Pchlide *a*, the immediate precursor of chlorophyllide (Chlide) *a*, by crude homogenates of etiolated wheat (Ellsworth and Hervish 1975).

### *7.1.1 Heterogeneity of the Mg-Proto Pools*

When more powerful fluorescence spectroscopic techniques were used to reinvestigate the chemical structure of the Mg-Proto pool of plants it was discovered that it was chemically heterogeneous and consisted of DV and monovinyl (MV) components (Fig. 7.1) (Belanger and Rebeiz 1982). The chemical structure of

C.A. Rebeiz, *Chlorophyll Biosynthesis and Technological Applications*, 
DOI 10.1007/978-94-007-7134-5_7, 

**Fig. 7.1** The heterogeneous Mg-Proto Pool

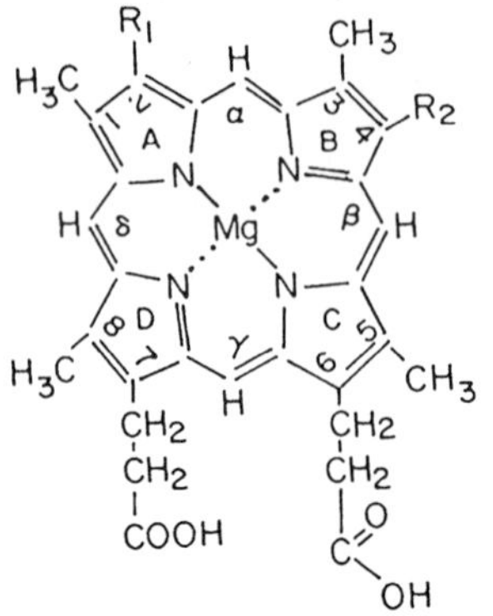


**Mg-Protoporphyrin IX (Mg-Proto)**

| | R1 | R2 | Compound |
|---|---|---|---|
| I | Vinyl | Vinyl | DV Mg-Proto |
| II | Vinyl | Ethyl | MV Mg-Proto |

the structure of Mg-Proto was ascertained by chemical derivatization coupled to fluorescence spectroscopy. The conversion of DV and MV Mg-proto to DV and MV Pchlide *a* respectively was demonstrated by Tripathy and Rebeiz (1986).

Molecular biological studies of Mg-Proto chelatase have not yet addressed the problem of the spatial and chemical heterogeneities of Mg-Proto formation. The proportion of DV to MV Mg-Proto biosynthesis depends upon the greening group affiliation of plants, the plants species, and pretreatment of plant tissues. For example cucumber cotyledons a dark divinyl- light divinyl-light dark divinyl (DDV-LDV-LDDV) plant tissue(Abd-El-Mageed et al. 1997), pretreated with 2,2-′-dipyridyl (Dpy) accumulate more DV than MV Mg-Proto in darkness. On the other hands, more MV Mg-Proto is formed in dark monovinyl-light divinyl-light dark-monovinyl (DMV-LDV-LDMV) plants such as etiolated corn or barley (Tripathy and Rebeiz 1986).

It has recently become apparent that the DV-MV biosynthetic heterogeneity of the carboxylic Chl *a* biosynthetic routes originates in the various Mg-Proto pools instead of the Proto pool (Kim and Rebeiz 1996). Indeed with the development of improved techniques for the extraction and determination of DV and MV Proto (Rebeiz 2002), it was shown that under no circumstances was it possible to induce the formation of MV Proto from ALA or from DV Proto in higher plants tissues. However the conversion of exogenous DV Mg-Proto to MV Mg-Proto in organello (Kim and Rebeiz 1996) was readily achieved.

#### 7.1.1.1 The Divinyl (DV) Mg-Proto Pools

Divinyl Mg-Proto has two vinyl groups at positions 2 and 4 of the tetrapyrrole macrocycle (Fig. 7.2). The DV nature of the DV Mg-Proto component of the

**Fig. 7.2** DV Mg-Proto

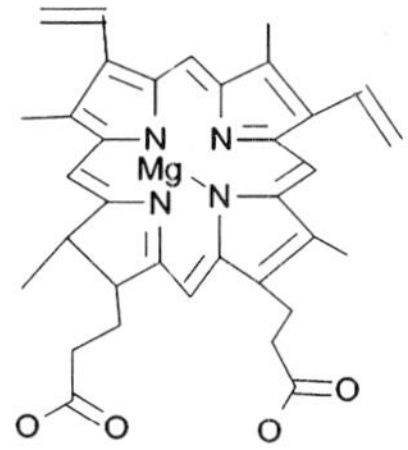


Mg-Proto pool of higher plants was determined by chemical derivatization coupled to analytical fluorescence spectroscopy at 77 K (Belanger and Rebeiz 1982).

### Biosynthetic Heterogeneity of Divinyl (DV) Mg-Proto in DDV-LDV-LDDV Plants

The biosynthesis of DV Mg-Proto from DV Proto was first reported in isolated cucumber etiochloroplasts in the presence of added ATP and Mg (Tripathy and Rebeiz 1986). In Fig. 7.3, the biosynthesis of DV Mg-proto from DV Proto is depicted to occur in three different thylakoid environments as suggested by multiple resonance energy transfer from Mp(e) to various Chl *a*-Protein complexes (Table 6.1, Chap. 6) and (Kolossov et al. 2003), as well as further conversions to Pchlides and Chls as will be discussed later. It is unclear at this stage whether the spatial biosynthetic heterogeneity of DV Mg-Proto is accompanied by chemical biosynthetic heterogeneity or not. In other words, it is unclear whether the proposed biosynthesis of DV Mg-Proto from DV Proto via routes 1, 0, and 8 is catalyzed by identical Mg-Proto chelatases or by different Mg-Proto chelatase isozymes. These biosynthetic routes will be discussed further later on.

### Biosynthetic Heterogeneity of DV Mg-Proto in DMV-LDV-LDMV Plant Species Like Barley

The accumulation of DV Mg-Proto in LDV-DDV-LDMV plant species such as Corn and other monocots treated with ALA and ALA +Dpy has been reported earlier (Rebeiz 1991). To our knowledge the direct conversion of DV Proto to DV Mg-proto in organello or in vitro, in DMV-LDV-LDMV plants has not been reported, and in my opinion is an oversight.

In Fig. 7.4, the biosynthesis of DV Mg-proto from DV Proto in DMV-LDV-LDMV plants is visualized to occur in four different thylakoid environments from DV Proto via routes 10, 11, 0′ and 12. This was suggested by multiple resonance energy transfer from Mp(e) to various Chl *a*-Protein complexes (Table 6.1, Chap. 6) (Kolossov et al., 2003), and by further conversions of DV Mg-Proto to Pchlides and Chls, as will be discussed later.

It is unclear at this stage whether the spatial biosynthetic heterogeneity of DV Mg-Proto is accompanied by chemical biosynthetic heterogeneity or not. In other

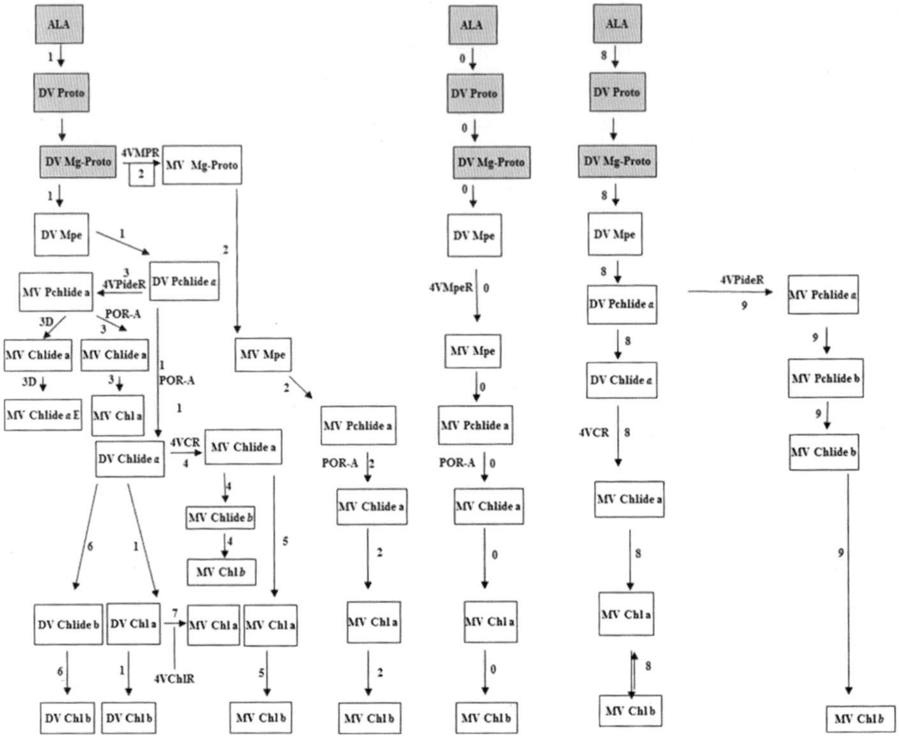


**Fig. 7.3** Biosynthetic routes 1, 0 and 8 which are responsible for the formation of DV Mg-Proto from DV Proto in LDV-DDV-LDDV plant species. Routes 1, 0 and 8 re highlighted in *light grey* (Adapted from Fig. 6.3 of Chap. 6, and from Kolossov and Rebeiz 2010)

words, it is unclear whether the proposed biosynthesis of DV Mg-Proto from DV Proto via routes 10, 11, 0′and 12 is catalyzed by identical Mg-Proto chelatases or by different Mg-Proto chelatase isozymes. These biosynthetic routes will be discussed further later on.

## Metabolism of DV Mg-Proto in DDV-LDV-LDDV Plant Species

The specific role of DV Mg-Proto as a precursor of DV Pchlide *a* in DDV-LDV-LDV plant species was demonstrated by conversion of exogenous DV Mg-Proto to DV Pchlide *a* in isolated etiochloroplasts of cucumber, a DDV-LDV-LDDV plant species. In cucumber etiochloroplasts the MV and DV proportions of synthesized Pchlide amounted to 89 % DV Pchlide *a* and 11 % MV Pchlide *a* (Tripathy and Rebeiz 1986).

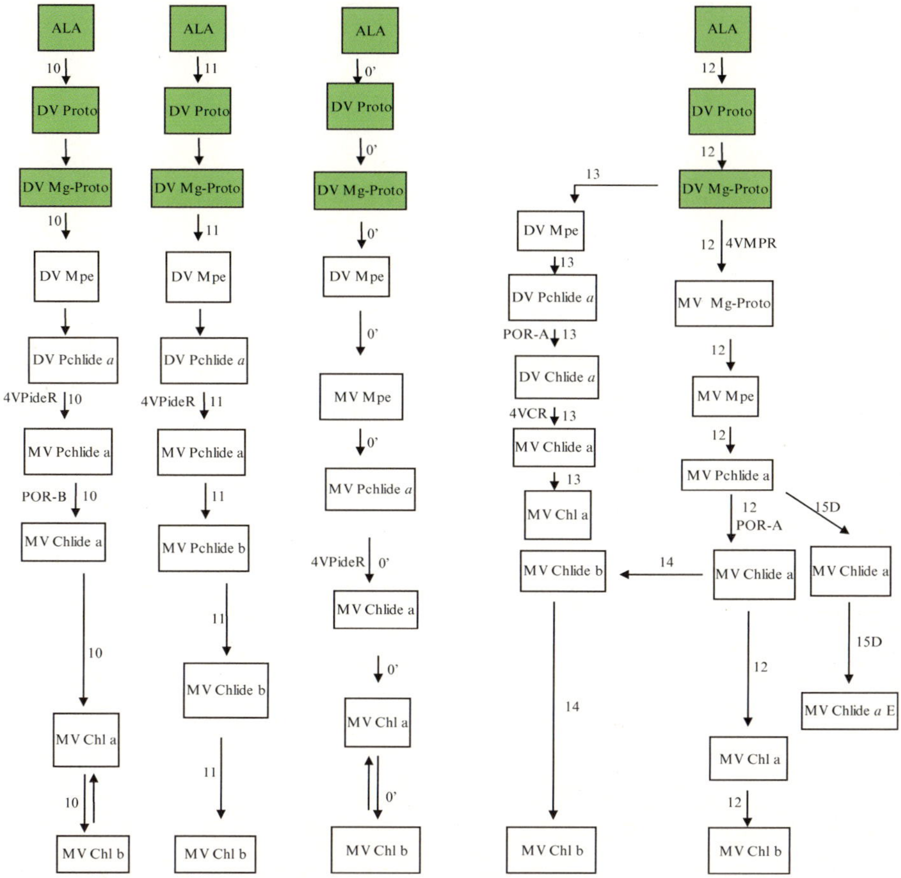


**Fig. 7.4** Biosynthetic routes 10, 11, 0′ and 12 which are responsible for the formation of DV Mg-Proto from DV Proto in LMV-DDV-LDMV plant species. These routes are highlighted in *green* (Adapted from Fig. 6.4 of Chap. 6, and from Kolossov and Rebeiz 2010)

Metabolism of DV Mg-Proto in DMV-LDV-LDMV Plant Species

The specific role of DV Mg-Proto as a precursor of DV Pchlide *a* in DMV-LDV-LDMV plant species was demonstrated by conversion of exogenous DV Mg-Proto to MV Pchlide *a* in isolated etioplasts of barley, a DMV-LDV-LDMV. In barley etioplasts the MV and DV proportions amounted to 16 % DV Pchlide *a* and 84 % MV Pchlide *a* (Tripathy and Rebeiz 1986).

#### 7.1.1.2 The Monovinyl (MV) Mg-Proto Pool

MV Mg-Proto has one vinyl group at position 2 and one ethyl group at position 4 of the tetrapyrrole macrocycle (Fig. 7.5). The MV nature of the MV Mg-Proto

**Fig. 7.5** MV Mg-Proto

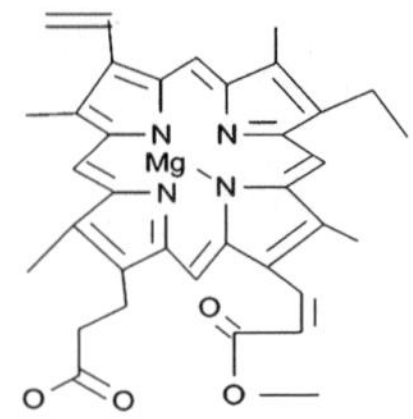


component of the Mg-Proto pool of higher plants was determined by chemical derivatization coupled to analytical fluorescence spectroscopy at 77 K (Belanger and Rebeiz 1982).

## Biosynthesis of MV Mg-Proto in DDV-LDV-LDDV Plant Species

In DDV-LDV-LDDV plant species, MV Mg-Proto is formed from DV Mg-Proto by reduction of the vinyl group to ethyl at position 4 (ring B) of the macrocycle in one thylakoid environment (Fig. 7.6) (Kim and Rebeiz 1996; Kolossov and Rebeiz 2010). By similarity with DDV-LDV-LDMV plants such as barley (see below), The reaction is probably catalyzed by a [4-vinyl] Mg-Proto reductase (4VMPR) (Kim and Rebeiz 1996).

The presence of this enzyme in DDV-LDV-LDDV plant species such as cucumber is strongly suggested by the biosynthesis and accumulation of MV Mg-proto during incubation of etiolated cucumber cotyledons with DV Proto in isolated cucumber etiochloroplasts (Tripathy and Rebeiz 1986) and by the metabolism of MV Mg-Proto in organello in DDV-LDV-LDDV Plant species (see below).

## Biosynthesis of MV Mg-Proto in DMV-LDV-LDMV Plant Species

In DMV-LDV-LDMV plant species, MV Mg-Proto is formed from DV Mg-Proto by reduction of the vinyl group to ethyl at position 4 (ring B) of the macrocycle in one thylakoid locations (Fig. 7.7), (Kim and Rebeiz 1996; Kolossov and Rebeiz 2010). The reaction is catalyzed by a [4-vinyl] Mg-Proto reductase (4VMPR). This enzyme was detected in isolated barley etiochloroplasts, a DMV-LDV-LDMV, and appears to be bound to the plastid membranes. A positive response of 4VMPR to added NADPH has been observed (Kim and Rebeiz 1996). It is very probable that 4VMPR is distinct from [4-vinyl] Pchlide *a* reductase (4PideR), which converts DV Pchlide *a* to MV Pchlide *a* (Tripathy and Rebeiz 1988); from [4-vinyl] Chlide *a* reductase (4VCR), which converts DV Chlide *a* to MV Chlide *a* (Kolossov and Rebeiz 2001; Pardo et al. 1980; Parham and Rebeiz 1992, 1995), and from [4-vinyl] Chl *a* reductase (4VChlR) (Adra and Rebeiz 1998; Wang et al. 2010). For example, *Rhodobacter capsulatus* in which the *bchJ* gene which codes for DV Pchlide *a* reductase, has been deleted, accumulates massive amounts of MV Mg-Proto and its monoester (precursors of Pchlide *a*) in addition to the accumulation of DV

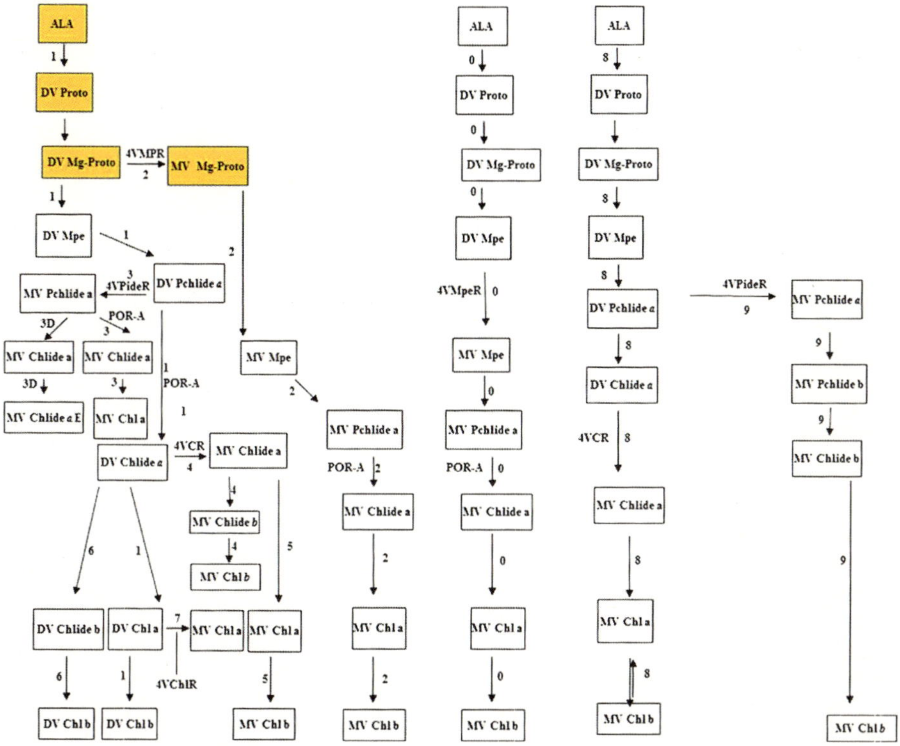


**Fig. 7.6** Biosynthetic route 2 which is responsible for the putative formation of MV Mg-Proto from DV Proto in DDV-LDV-LDDV plant Species (Adapted from Fig. 6.3 of Chap. 6 and from (Kolossov and Rebeiz 2010)

Pchlide *a* (Suzuki and Bauer 1995). This in turn indicates that separate [4-vinyl] reductases are active before DV Pchlide *a*, DV Chlide *a* and DV Chl *a* vinyl reduction at position 4 of the macrocycle. It should be pointed that contrary to other MV intermediates which can be formed via multiple biosynthetic routes, MV Mg-Proto can only be formed from DV Mg-Proto via two routes namely routes 2 and 12 (Figs. 7.6 and 7.7).

### Metabolism of MV Mg-Proto in DDV-LDV-LDDV Plant Species

The specific role of MV Mg-Proto as a precursor of MV Pchlide *a* was demonstrated by conversion of exogenous MV Mg-Proto to MV Pchlide *a* in isolated cucumber etiochloroplasts (Tripathy and Rebeiz 1986). Conversion of MV Mg-Proto to MV Pchlide *a* was not accompanied by formation of DV Pchlide *a*. This in turn strongly suggested that in DMV-LDV-LDDV plant species, at the level of MV Mg-Proto, further metabolism can only proceed exclusively via MV biosynthetic routes (Fig. 7.6).

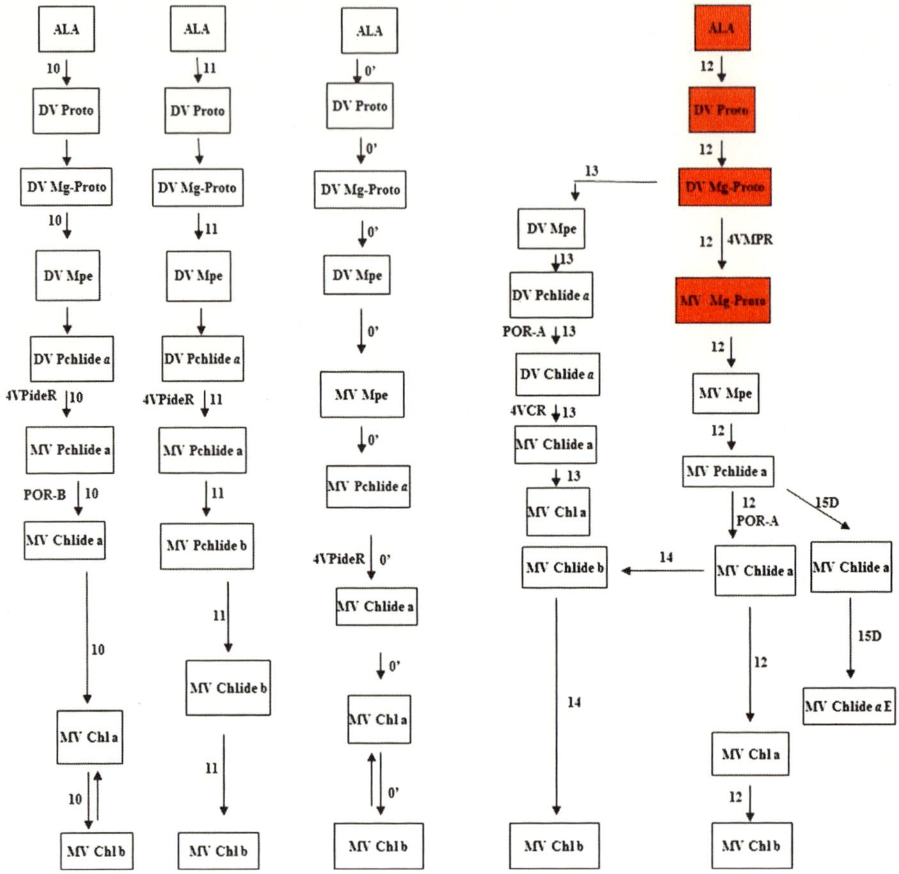


**Fig. 7.7** Biosynthetic route 12 which is responsible for the formation of MV Mg-Proto from DV Proto in DMV-LDV-LDMV plant Species such as Barley (Adapted from Fig. 6.4 of Chap. 6 and from Kolossov and Rebeiz 2010)

## Metabolism of MV Mg-Proto in DMV-LDV-LDMV Plant Species

The specific role of MV Mg-Proto as a precursor of MV Pchlide *a* in DMV-LDV-LDMV plant species was demonstrated by conversion of exogenous MV Mg-Proto to MV Pchlide *a* in barley etiochloroplasts (Tripathy and Rebeiz 1986). Conversion of MV Mg-Proto to MV Pchlide *a* was not accompanied by formation of DV Pchlide *a*. This in turn indicates that in DMV-LDV-LDMV plant species, at the level of MV Mg-Proto, further metabolism can only proceed via exclusive MV biosynthetic routes Fig. 7.7).

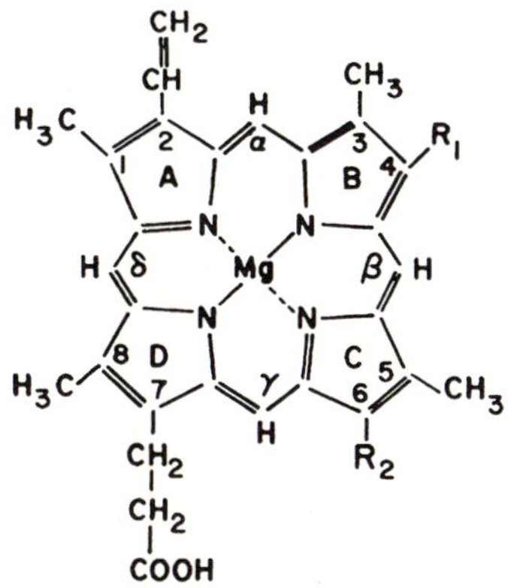


**Mg-Protoporphyrin IX monomethyl ester (Mpe)**

| | R1 | R2 | Compound |
|---|---|---|---|
| I | Vinyl | Methyl Propionate | DV Mpe |
| II | Ethyl | Methyl Propionate | MV Mpe |

**Fig. 7.8** The Mg-Proto monomethyl ester (Mpe) pool

**Fig. 7.9** Conversion of Mg-Proto to Mpe by SAMMT

## 7.2 The Mg-Proto Monomethyl Ester (Mpe) Pool

Protoporphyrin IX monomethyl ester (Mpe) pool is the precursor of the Pchlide *a* pool. The role of Mpe as an intermediate in the Chl biosynthetic pathway was based on the detection of Mpe in X-ray *Chlorella* mutants inhibited in their capacity to form Chl (Granick 1961). It was conjectured that since the mutants had lost the ability to form Chl but accumulated Mpe, the latter was a logical precursor of Chl. On the basis of absorbance spectroscopic analysis, Mpe was assigned by Granick a DV chemical structure. Mpe was also detected in barley leaves incubated with ALA and 2,2′-dipyridyl (Dpy) (Granick 1961). In this case too, the accumulated Mpe was assigned a divinyl (DV) chemical structure (Fig. 7.8).

Mg-Proto is converted to Mpe by transfer of a methyl group from (−) S-adenosyl-L- methionine (SAM) to Mg-Proto. The reaction results in the methyl esterification of the propionic acid residue at position 6 (ring C) of the macrocycle. The reaction is catalyzed by (−) S-adenosyl-L-methionine-magnesium protoporphyrin methyl transferase (SAMMT) (Fig. 7.9).

The occurrence of SAMMT was first reported in *Rhodopseudomonas spheroides* (Gibson et al. 1963). The enzyme was confined to the chromatophores to which it was firmly bound. Substrate specificity was lax since in addition to Mg-Proto, zinc proto, calcium Proto, Mg-mesoporphyrin and Mg-deuteroporphyrin also acted as substrates. S-adenosyl homocysteine and S-adenosylethionine inhibited the reaction competitively. The enzyme has also been detected in corn (*Zea mays*) chloroplasts (Radmer and Bogorad 1967). A 1600-fold purification of the *R. spheroides* enzyme was achieved by affinity chromatography (Hinchigeri et al. 1984). The purified enzyme exhibited an equilibrium-ordered sequential Bi Bi mechanism with Mg-Proto as the obligatory first substrate, and SAM as the second substrate. The nucleotide sequence of the *R. capsulatus* enzyme has been reported (Bollivar and Bauer 1992).

Originally, in *R. capsulatus*, SAMMT was believed to be coded for by the *bchH* gene, while the *bchM* gene was believed to code for a polypeptide involved in the formation of the cyclopentanone ring (ring E) of Pchlide *a* (Bauer et al. 1993). Later on, the *bchM* gene of *R. capsulatus* was expressed in *E. coli* and the gene product was subsequently demonstrated by enzymatic analysis to catalyze methylation of Mg-proto to form Mpe (Bollivar et al. 1994). Activity required the substrates Mg-proto and S-adenosyl-L-methionine. To our knowledge, no higher plant SAMMT gene has been isolated. A query for SAMMT addressed to the various protein databases listed in the Biology Workbench, yielded 3 unique records which are depicted on the VLPBP website at "http://www.vlpbp.org/greening/XVI. Sequenced Enzymes/ SAM-Mg-proto MT". These sequences can be viewed and used for sequence similarity searches or other manipulations using the Biology Workbench.

The metabolic function of Mpe as a precursor of Pchlide *a* was demonstrated by conversion of exogenous [$^{14}$C]-Mpe and unlabeled Mpe to [$^{14}$C]-Pchlide *a*, and Pchlide *a* respectively, in organello (Mattheis and Rebeiz 1977). Pchlide *a* is the immediate precursor of Chlide *a*. In this undertaking, an in organello system capable of the converting $^{14}$C-ALA to $^{14}$C- Pchlide *a*, $^{14}$C-Pchlide ester *a* and $^{14}$C-Chl *a* and *b* (Rebeiz and Castelfranco, 1971a, b), and capable of the net conversion of exogenous ALA to Mg-Protoporphyrins and Pchlide *a* (Rebeiz et al. 1975) was used.

### *7.2.1 Biosynthetic Heterogeneity of the Mpe Pool*

When more powerful fluorescence spectroscopic techniques were used to reinvestigate the chemical nature of the Mpe pool of plants it was found to be chemically heterogeneous and to consist of DV and monovinyl (MV) components (Belanger and Rebeiz 1982). Substrate amounts of MV Mpe are now routinely prepared by incubation of etiolated barley leaves with ALA and Dpy (Rebeiz 2002).

The proportion of DV to MV Mpe biosynthesis depended upon the greening group affiliation of plants, the plants species, and pretreatment of plant tissues. For example cucumber cotyledons a DDV-LDV-LDDV plant tissue(Abd-El-Mageed et al. 1997), pretreated with 2,2′-dipyridyl (Dpy) accumulate more DV than MV Mpe in darkness. On the other hands, more MV Mpe was formed in DMV-LDV-LDMV plants such as etiolated corn or barley.

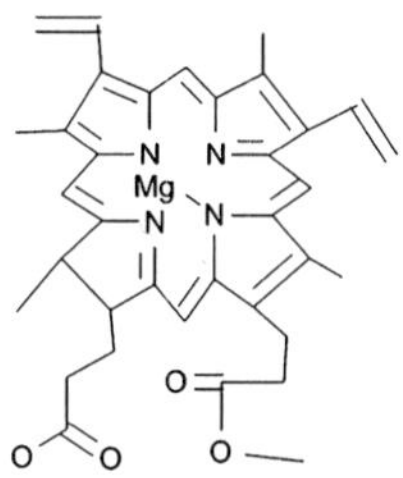


**Fig. 7.10** DV Mg-Proto monomethyl ester (DV Mpe)

#### 7.2.1.1 The Divinyl Mpe Pool

Divinyl Mg-Proto monomethyl ester has two vinyl groups at positions 2 and 4 of the tetrapyrrole macrocycle (Fig. 7.10). The DV nature of the DV Mg-Proto monomethyl ester component of the Mg-Proto monomethyl ester pool of higher plants was determined by chemical derivatization coupled to analytical fluorescence spectroscopy at 77 K (Belanger and Rebeiz 1982).

Biosynthetic Heterogeneity of DV Mpe in LDV-DDV-LDDV Plants Species

In Fig. 7.11, Three DV Mpe pools are depicted to be formed in three different thylakoid environments. The assignment of three DV Mpe biosynthetic pools to three different thylakoid locations is based on the detection of multiple resonance excitation transfer bands between Mp(e) and various Chl-protein complexes (Table 6.1, Chap. 6) and considerations related to the biosynthesis of DV and MV Pchlide *a* which will be discussed later.

It is unclear at this stage whether the spatial biosynthetic heterogeneity of DV Mpe is accompanied by chemical biosynthetic heterogeneity or not. In other words, it is unclear whether the proposed biosynthesis of DV Mpe from DV Mg-Proto via routes 1, 0, and 8 is catalyzed by identical SAMMT or by different SAMMT isozymes.

Biosynthetic Heterogeneity of DV Mpe in LMV-DDV-LDMV Plants Species

In Fig. 7.12, four DV Mpe pools are depicted to be formed in four different thylakoid environments. The assignment of four DV Mpe biosynthetic pools to four different thylakoid locations is based on the detection of multiple resonance excitation transfer bands between Mp(e) and various Chl-protein complexes (Table 6.1, Chap. 6) and considerations related to the biosynthesis of DV Pchlide *a* which will be discussed later.

It is unclear at this stage whether the spatial biosynthetic heterogeneity of DV Mpe in DMV-LDV-LDMV plant species is accompanied by chemical biosynthetic heterogeneity or not. In other words, it is unclear whether the proposed biosynthesis of DV Mpe from DV Mg-Proto via routes 10, 11, 0′, and 13 is catalyzed by identical SAMMT or by different SAMMT isozymes.

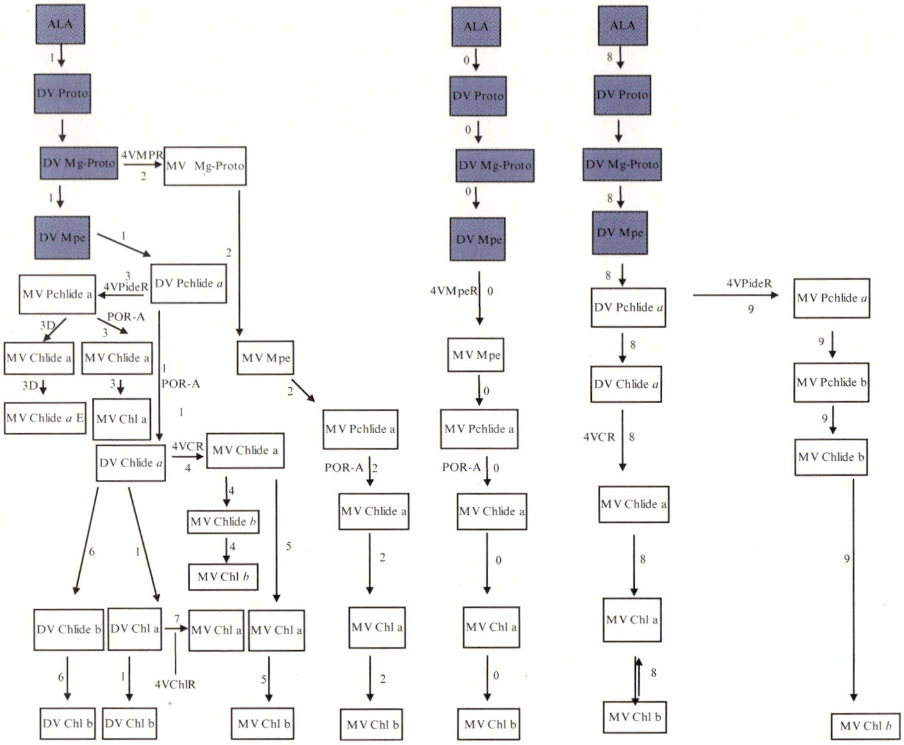


**Fig. 7.11** Biosynthetic routes 1, 0 and 8 which are responsible for the formation of DV Mpe in LDV-DDV-LDDV plant species. Routes 1, 0 and 8 re highlighted in *blue* (Adapted from Fig. 6.3 of Chap. 6, and from Kolossov and Rebeiz 2010)

## Metabolism of DV Mpe in LDV-DDV-LDDV Plants Species

As was observed for Mg-Proto, the proportion of DV to MV Mg-Proto biosynthesis depended on the greening group affiliation, plant species and pretreatment of plant tissues. For example cucumber cotyledons a DDV-LDV-LDDV plant tissue (Abd-El-Mageed et al. 1997), pretreated with Dpy accumulate more DV than MV Mpe in darkness (Belanger et al. 1982).

The specific role of DV Mpe as a precursor of DV Pchlide *a* was demonstrated by conversion of exogenous DV Mpe to DV Pchlide *a* in isolated etiochloroplasts of cucumber (Tripathy and Rebeiz 1986), a DDV-LDV-LDDV plant species. In cucumber etioplasts, DV Mpe was converted into 83 % DV Pchlide *a*, and 17 % MV Pchlide *a*.

To our knowledge, no kinetic studies have been performed on SAMMT purified to homogeneity, with pure DV Mpe. Since the mechanism of action of SAMMT has been reported to vary *i.e.* ping pong (Ellsworth and Pierre 1974), random Bi Bi (Ebbon and Tait 1969), or ordered Bi Bi (Hinchigeri et al. 1984) depending on the

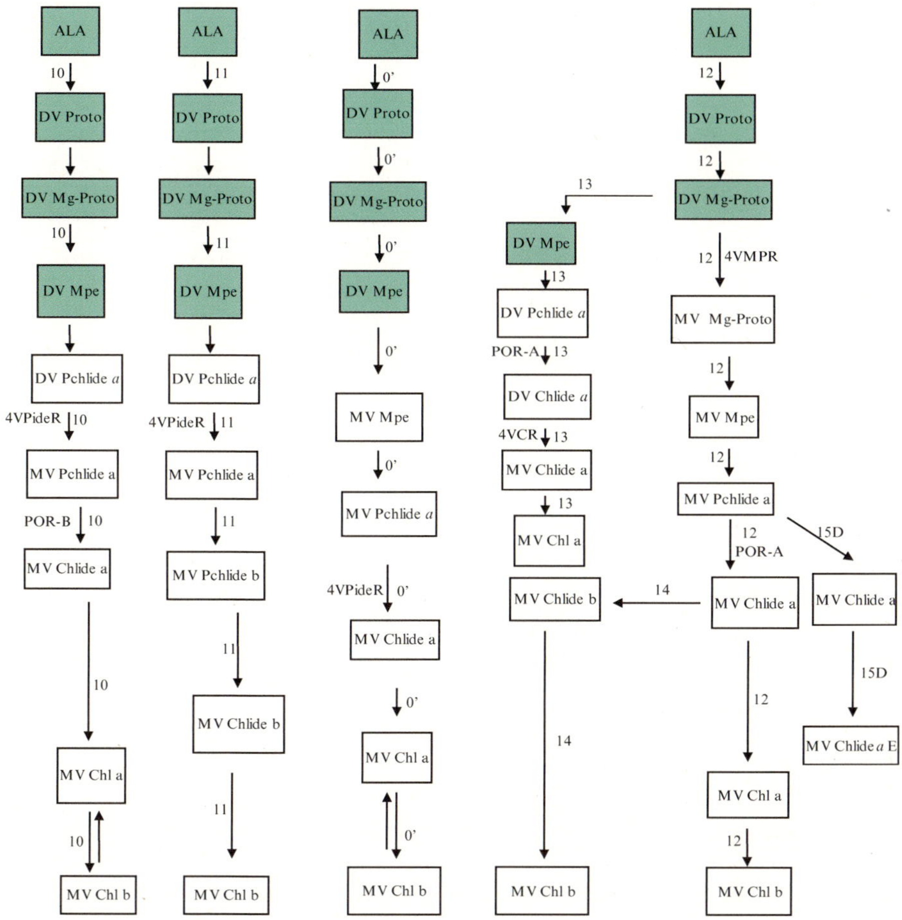


**Fig. 7.12** Biosynthetic routes 10, 11, 0′ and 13 which are responsible for the formation of DV Mpe in LMV-DDV-LDMV plant species. Routes 10, 11, 0′ and 13 are highlighted in *green* (Adapted from Fig. 6.4 of Chap. 6, and from Kolossov and Rebeiz 2010)

source of enzyme, it is not possible to assign with certainty a precise mechanism for its action without precise knowledge of the DV or MV nature of the Mpe substrate.

In Fig. 7.11, three DV Mpe pools are depicted as being formed from DV Mg-Proto via routes 1, 0 and 8. At this stage it is unclear whether the spatial biosynthetic heterogeneity indicated by multiple resonance excitation energy transfer bands (Table 6.1, Chap. 6) is accompanied by chemical biosynthetic heterogeneity or not. In other words, it is unclear whether the biosynthesis of DV Mpe from DV Mg-Proto via routes 1, 0 and 8 is catalyzed by identical SAMMTs or by SAMMT isozymes.

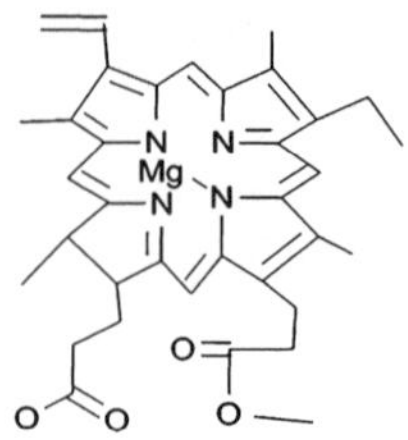

**Fig. 7.13** MV Mg-Proto monomethyl ester (MV Mpe)

Metabolism of DV Mpe in LMV-DDV-LDMV Plants Species

The specific role of DV Mpe as a precursor of DV Pchlide *a* in DDMV-LDV-LDMV plant species was demonstrated by conversion of exogenous DV Mpe to DV Pchlide *a* in isolated etiochloroplasts of barley (Tripathy and Rebeiz 1986), a DMV-LDV-LDMV plant species (Abd-El-Mageed et al. 1997). In barley etioplasts, DV Mpe was converted into 44 % DV Pchlide *a*, and 56 % MV Pchlide *a*.

To our knowledge, no kinetic studies have been performed on SAMMT purified to homogeneity, with pure MV Mpe. Since the mechanism of action of SAMMT has been reported to vary *i. e.* ping pong (Ellsworth et al. 1974), random Bi Bi (Ebbon and Tait 1969), or ordered Bi Bi (Hinchigeri et al. 1984) depending on the source of enzyme, it is not possible to assign with certainty a precise mechanism for its action without precise knowledge of the DV or MV nature of the Mpe substrate.

#### 7.2.1.2 The MV Mpe Pool

Monovinyl Mg-Proto monomethyl ester has one vinyl groups at positions 2 and one ethyl group at position 4 of the tetrapyrrole macrocycle (Fig. 7.13). The MV nature of the MV Mpe pool was ascertained by chemical derivatization coupled to analytical fluorescence spectroscopy at 77 K (Belanger and Rebeiz 1982).

Biosynthesis of MV Mpe in DDV-LDV-LDDV Plant Species

In DDV-LDV-LDDV plant species, MV Mg-Proto is formed from DV Mg-Proto by reduction of the vinyl group to ethyl at position 4 (ring B) of the macrocycle in one thylakoid environment (Fig. 7.14) (Kim and Rebeiz 1996; Kolossov and Rebeiz 2010). By similarity with DDV-LDV-LDMV plants such as barley (see below), The reaction is probably catalyzed by a [4-vinyl] Mpe reductase (4VMPR) (Kolossov and Rebeiz 2010).

The presence of this enzyme in DDV-LDV-LDDV plant species such as cucumber is strongly suggested by the biosynthesis and accumulation of MV Pchlide *a* during incubation of MV Proto in isolated cucumber etiochloroplasts (Tripathy and Rebeiz 1986).

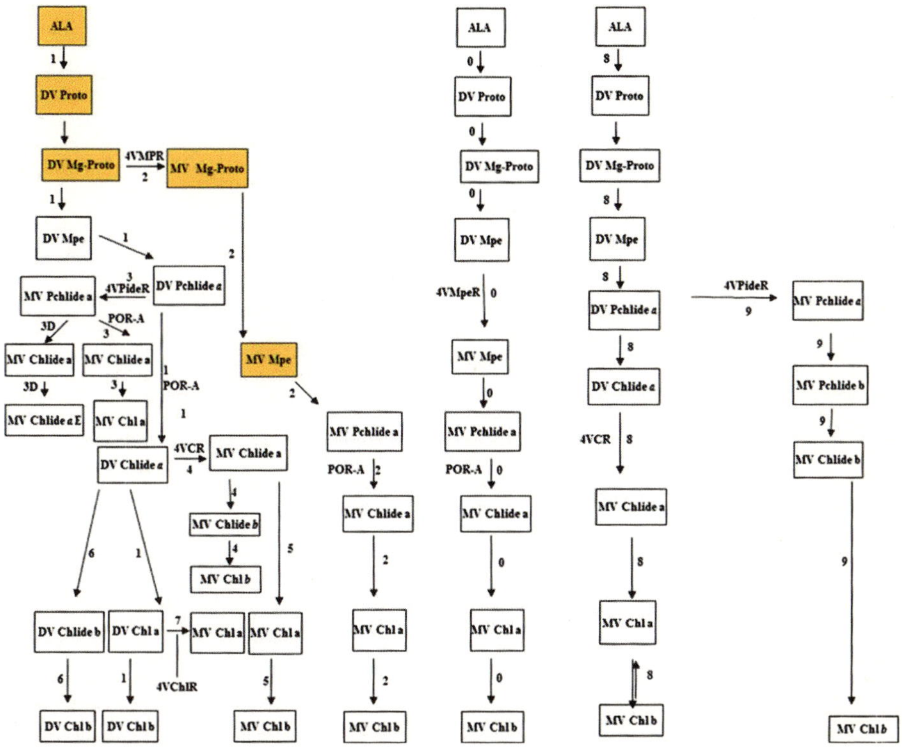


**Fig. 7.14** Biosynthetic route 2 which is responsible for the formation of MV Mpe from MV Mg-Proto in DDV-LDV-LDDV plant Species. Route 2 is in *dark tan* (Adapted from Fig. 6.3 of Chap. 6 and from Kolossov and Rebeiz 2010)

### Biosynthesis of MV Mpe in DMV-LDV-LDMV Plant Species

In DMV-LDV-LDMV plant species, MV MPE is formed from DV Mg-Proto by reduction of the vinyl group to ethyl at position 4 (ring B) of the macrocycle in one thylakoid environment (Fig. 7.15) (Kim and Rebeiz 1996; Kolossov and Rebeiz 2010). The reaction is catalyzed by a [4-vinyl] Mg-Proto reductase (4VMPR) (Kim and Rebeiz 1996).

The presence of this enzyme in DMV-LDV-LDMV plant species such as barley was ascertained by detection of the enzyme in isolated barley etiochloroplasts (Kolossov and Rebeiz 2010). The enzyme activity was solubilized from the barley etiochloroplasts membranes by Chaps (Kolossov and Rebeiz 2010).

### Metabolism of MV Mpe in DDV-LDV-LDDV Plant Species

The specific role of MV Mpe as a precursor of MV Pchlide *a* in DDV-LDV-LDDV plant species was demonstrated by the strong conversion of exogenous MV Mpe to

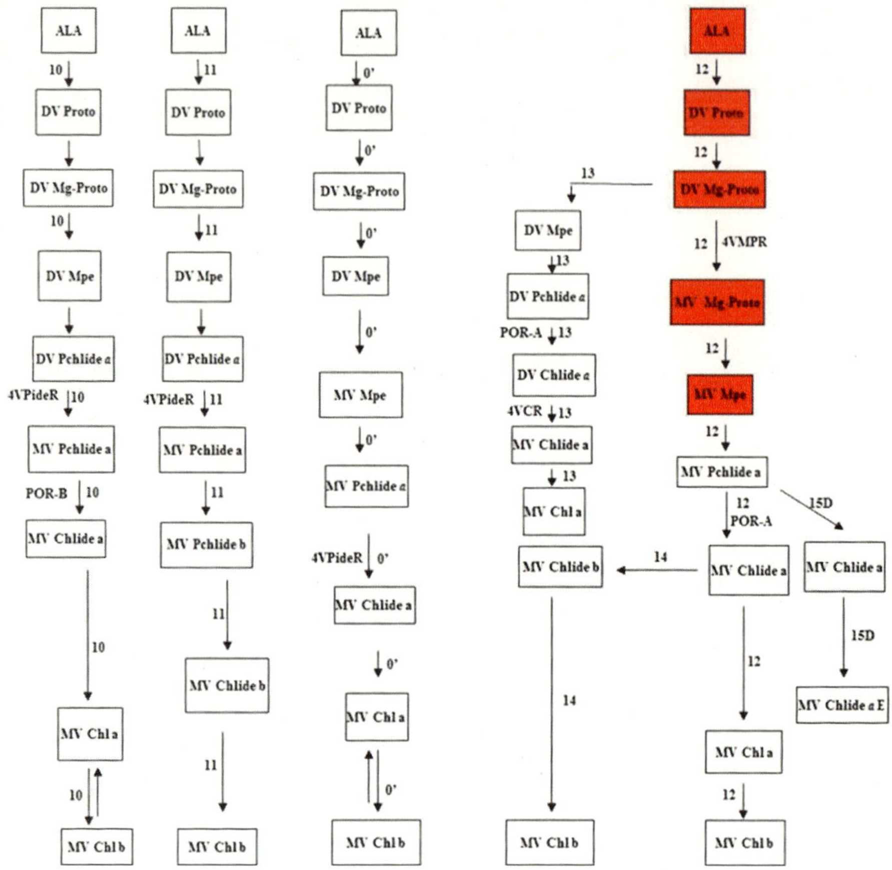


**Fig. 7.15** Biosynthetic route 12 which is responsible for the formation of MV Mpe from MV Mg-Proto in DMV-LDV-LDMV plant Species. Route 12 is highlighted in *red* (Adapted from Fig. 6.4 of Chap. 6 and from Kolossov and Rebeiz 2010)

MV Pchlide *a* in isolated cucumber etiochloroplasts (Tripathy and Rebeiz 1986). Conversion of MV Mg-Proto to MV Pchlide *a*, was accompanied by formation of trace amounts of DV Pchlide *a* within the margin of error of the assay (Rebeiz 2002).

### Metabolism of MV Mpe in DMV-LDV-LDMV Plant Species

The specific role of MV Mg-Proto as a precursor of MV Pchlide *a* in DMV-LDV-LDMV plant species was demonstrated by the strong conversion of exogenous MV Mpe to MV Pchlide *a* in isolated barley etiochloroplasts (Tripathy and Rebeiz 1986). Conversion of MV Mpe to MV Pchlide *a*, was accompanied by formation of trace amounts DV Pchlide *a* within the range of experimental error of the equations used to determine the amount of formed MV Pchlide (Rebeiz 2002, Rebeiz et al. 2003).

# References

Abd-El-Mageed HA, El Sahhar KF, Robertson KR et al (1997) Chloroplast biogenesis 77. Two novel monovinyl and divinyl light–dark greening groups of plants and their relationship to the chlorophyll *a* biosynthetic heterogeneity of green plants. Photochem Photobiol 66:89–96

Adra AN, Rebeiz CA (1998) Chloroplast biogenesis 81. Transient formation of divinyl chlorophyll *a* following a 2.5 ms light flash treatment of etiolated cucumber cotyledons. Photochem Photobiol 68:852–856

Bauer EC, Bollivar DW, Suzuki TY (1993) Genetic analyses of photopigment biosynthesis in eubacteria: a guiding light for algae and plants. J Bacteriol 175:3919–3925, July

Belanger FC, Rebeiz CA (1982) Chloroplast biogenesis: detection of monovinyl magnesium protoporphyrin monoester and other monovinyl magnesium porphyrins in higher plants. J Biol Chem 257:1360–1371

Belanger FC, Dugan JX, Rebeiz CA (1982) Chloroplast biogenesis: identification of chlorophyllide *a* (E458F674) as a divinyl chlorophyllide *a*. J Biol Chem 257:4849–4858

Bollivar DW, Bauer CE (1992) Nucleotide sequence of S-adenosyl-L-methionine:magnesium-protoporphyrin methyltransferase from Rhodobacter capsulatus. Plant Physiol 98:408–410

Bollivar DW, Jiang Z-Y, Bauer CE et al (1994) Heterologous expression of the *bchM* gene product from *Rhodobacter capsulatus* and demonstration that it encodes S-adenosyl-L-methionine:Mg-protoporphyrin IX methyltransferase. J Bacteriol 176:5290–5296, Sept

Ebbon JG, Tait GH (1969) Biochem J 111:573–582

Ellsworth RK, Hervish PV (1975) Biosynthesis of protochlorophyllide a from Mg-protoporphyrin IX in vitro. Photosynthetica 9:15–139

Ellsworth RK, Dullaghan JP, St. Pierre ME (1974) The reaction mechanism of S-adenosyl-L-methionine:magnesium protoporphyrin IX methyltransferase of wheat. Photosynthetica 8:376–383

Gibson KD, Neuberger A, Tait GH (1963) Studies on the biosynthesis of porphyrins and bacteriochlorophyll by Rhodopseudomonas spheroides. S-adenosylmethionine-magnesium protoporphyrin methyltransferase. Biochem J 88:325–334

Granick S (1948) Magnesium protoporphyrin as a precursor of chlorophyll in *Chlorella*. J Biol Chem 175:333–342

Granick S (1961) Magnesium protoporphyrin monoester and protoporphyrin monomethyl ester in chlorophyll biosynthesis. J Biol Chem 236:1168–1172

Hinchigeri SB, Nelson DW, Richards WR (1984) The purification and reaction mechanism of S-adenosyl-L-methionine:magnesium protoporphyrin methyltransferase from hodopseudomonas spheroides. Photosynthetica 18:168–178

Kim JS, Rebeiz CA (1996) Origin of the chlorophyll *a* biosynthetic heterogeneity in higher plants. J Biochem Mol Biol 29:327–334

Kolossov VL, Rebeiz CA (2001) Chloroplast biogenesis 84. Solubilization and partial purification of membrane-bound [4-vinyl] chlorophyllide *a* reductase from etiolated barley leaves. Anal Biochem 295:214–219

Kolossov VL, Rebeiz CA (2010) Evidence for various 4-vinyl reductase activities in higher plants. In: Rebeiz CA, Benning C, Bohnert HJ et al (eds) The chloroplast: basics and applications. Springer, Dordrecht, pp 25–38

Kolossov VL, Kopetz KJ, Rebeiz CA (2003) Chloroplast biogenesis 87: evidence of resonance excitation energy transfer between tetrapyrrole intermediates of the chlorophyll biosynthetic pathway and chlorophyll *a*. Photochem Photobiol 78:184–196

Mattheis JR, Rebeiz CA (1977) Chloroplast biogenesis. Net synthesis of protochlorophyllide from magnesium protoporphyrin monoester by developing chloroplasts. J Biol Chem 252:4022–4024

Pardo AD, Chereskin BM, Castelfranco PA et al (1980) ATP requirement for Mg chelatase in developing chloroplasts. Plant Physiol 65:956–960

Parham R, Rebeiz CA (1992) Chloroplast biogenesis: [4-vinyl] chlorophyllide *a* reductase is a divinyl chlorophyllide *a*-specific NADPH-dependent enzyme. Biochemistry 31:8460–8464

Parham R, Rebeiz CA (1995) Chloroplast biogenesis 72: a [4-vinyl] chlorophyllide *a* reductase assay using divinyl chlorophyllide *a* as an exogenous substrate. Anal Biochem 231:164–169

Radmer R, Bogorad L (1967) (−) S-adenosyl-L-methionine-magnesium protoporphyrin methyltransferase, an enzyme of the biosynthetic pathway of chlorophyll in *Zea mays*. Plant Physiol 42:463–465

Rebeiz CA (1991) Tetrapyrrole-dependent photodynamic herbicides and the chlorophyll biosynthetic pathway. In: Pell E, Steffen K (eds) Active oxygen/oxidative stress and plant metabolism. American Society of Plant Physiology, Rockville, pp 193–203

Rebeiz CA (2002) Analysis of intermediates and end products of the chlorophyll biosynthetic pathway. In: Smith A, Witty M (eds) Heme chlorophyll and bilins, methods and protocols. Humana Press, Totowa, pp 111–155

Rebeiz CA, Castelfranco P (1971a) Protochlorophyll biosynthesis in a cell-free system from higher plants. Plant Physiol 47:24–32

Rebeiz CA, Castelfranco P (1971b) Chlorophyll biosynthesis in a cell-free system from higher plants. Plant Physiol 47:33–37

Rebeiz CA, Mattheis JR, Smith BB et al (1975) Chloroplast biogenesis. Biosynthesis and accumulation of protochlorophyll by isolated etioplasts and developing chloroplasts. Arch Biochem Biophys 171:549–567

Rebeiz CA, Kolossov VL, Briskin D et al (2003) Chloroplast biogenesis: chlorophyll biosynthetic heterogeneity, multiple biosynthetic routes and biological spin-offs. In: Nalwa HS (ed) Handbook of photochemistry and photobiology, vol 4. American Scientific Publishers, Los Angeles, pp 183–248

Suzuki JY, Bauer CE (1995) Altered monovinyl and divinyl protochlorophyllide pools in bchJ mutants of rhodobacter capsulatus. Possible monovinyl substrate discrimination of light-independent protochlorophyllide reductase. J Biol Chem 270:3732–3740

Tripathy BC, Rebeiz CA (1986) Chloroplast biogenesis. Demonstration of the monovinyl and divinyl monocarboxylic routes of chlorophyll biosynthesis in higher plants. J Biol Chem 261:13556–13564

Tripathy BC, Rebeiz CA (1988) Chloroplast biogenesis 60. Conversion of divinyl protochlorophyllide to monovinyl protochlorophyllide in green(ing) barley, a dark monovinyl/light divinyl plant species. Plant Physiol 87:89–94

Wang P, Gao J, Chunmei w et al (2010) Divinyl chlorophyll(ide) *a* can be converted to monovinyl chlorophyl(lide) *a* by a divinyl reductase in rice. Plant Physiol 153:994–1003

# Chapter 8
# The Chl *a* Carboxylic Biosynthetic Routes: Protochlorophyllide *a*

*Great spirits have always encountered violent opposition from mediocre minds (Albert Einstein).*

## 8.1 Protochlorophyllide *a* (Pchlide *a*) Pool

Protochlorophyllide *a* (Pchlide *a*) (Fig. 8.1) is the immediate precursor chlorophyllide *a* (Chlide *a*). The proposed role of Pchlide *a* as an intermediate in the Chl biosynthetic pathway was based on the detection of Pchlide *a* in X-ray Chlorella mutants inhibited in their capacity to form Chl (Granick 1950a). It was conjectured that since the mutants had lost the ability to form Chl but accumulated Pchlide *a*, the latter was a logical precursor of Chl*a*. On the basis of absorbance spectroscopic determinations the accumulated Mg-Proto was assigned by Granick a monovinyl (MV) chemical structure (Fig. 8.1, II), Pchlide with an ethyl group at positions 2 and a vinyl group at position 4 of the tetrapyrrole macrocycle.

Granick proposed that Pchlide *a* was the immediate precursor of Pchlide *a* phytyl ester, a fully esterified Pchlide, which was wrongly believed at that time to be the only precursor of Chl *a* in nature (Granick 1948, 1950a; Koski 1950). The biosynthetic function of Pchlide *a* as the precursor of chlorophyllide (Chlide) *a*, one of the immediate precursors of Chl *a*, was not recognized till 7 years later (Wolff and Price 1957) when Pchlide *a* as a precursor of Chlide *a* was demonstrated by conversion of Pchlide *a* to Chlide *a*.

C.A. Rebeiz, *Chlorophyll Biosynthesis and Technological Applications*, 
DOI 10.1007/978-94-007-7134-5_8, 

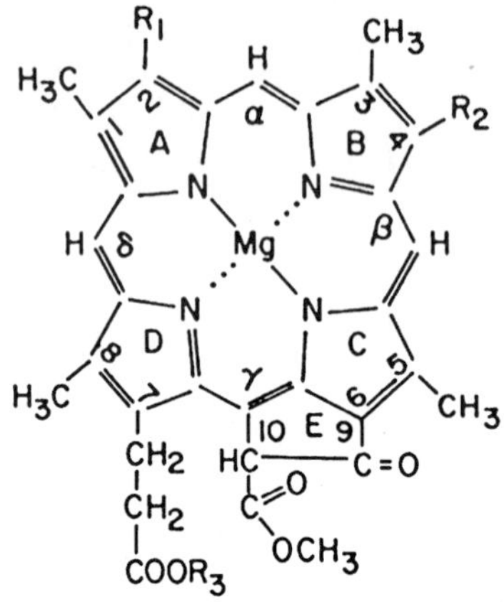


**Protochlorophyllide *a* (Pchlide *a*)**

| | R1 | R2 | R3 | Compound |
|---|---|---|---|---|
| I | Vinyl | Vinyl | H | DV Pchlide *a* |
| II | Vinyl | Ethyl | H | MV Pchlide *a* |

**Fig. 8.1** The Divinyl (*DV*)-Monovinyl (*MV*) Pchlide *a* pool

### 8.1.1 Chemical Heterogeneity of the Pchlide a Pool

Biosynthetic Heterogeneity of the Pchlide *a* Chromophore.

When more powerful fluorescence spectroscopic techniques were used to reinvestigate the chemical structure of the Pchlide *a* of green plants, it was discovered that it was chemically heterogeneous and consisted of DV and monovinyl (MV) components (Fig. 8.1) (Belanger and Rebeiz 1980a). The chemical structure of the Pchlide *a* pool was ascertained by chemical derivatization coupled to fluorescence spectroscopy, as well as by NMR and field desorption mass spectroscopy (Belanger and Rebeiz 1980b; Wu and Rebeiz 1984). The conversion of DV and MV Pchlide a to DV and MV Chlide*a* respectively was demonstrated by Dugan and Rebeiz (1982).

In discussing Pchlide *a* biosynthetic heterogeneity distinctions will be made between (a) etiolated and green plants, (b) between biosynthesis during the dark and light phases of the photoperiod in green plants, and (c) between DDV-LDV-LDDV and DMV-LDV-LDMV plant species. As will be discussed in Chap. 14, DDV-LDV-LDDV plant species such as cucumber accumulate mainly DV Pchlide *a* in darkness and in the light. In the light, Chl biosynthesis proceeds mainly via regenerated DV Pchlide *a*. On the other hand, DMV-LDV-LDMV plant species such as corn wheat and barley accumulate mainly MV Pchlide *a* in darkness. In the light, some DV Pchlide *a* is formed, but Chl biosynthesis proceeds mainly via regenerated MV Pchlide *a* (Abd-El-Mageed et al. 1997).

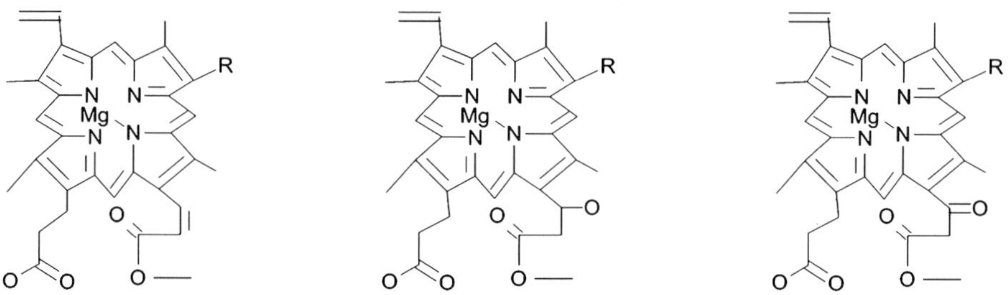

**R = Vinyl: DV– Acrylic**
**R= Ethyl: MV– Acrylic**

**R = Vinyl: DV–OH**
**R= Ethyl: MV–OH**

**R = Vinyl: DV–Keto**
**R= Ethyl: MV–Keto**

**Fig. 8.2** Intermediate Acrylic, OH and Keto derivatives of Mpe involved in the formation of the cyclopentanone ring of Pchlide *a*. The Residue *R* represents a vinyl group for the DV intermediates and an ethyl group for the MV intermediates

#### 8.1.1.1 Formation of the Cyclopentanone Ring

Formation of the cyclopentanone ring (ring E) during the proposed conversion of Mpe to Pchlide *a* was suggested in 1950 to involve a beta-oxidation of a putative methyl Propionate side chain to a 3-keto derivative (Granick 1950b). Later on, detection of putative DV and MV *metal-free* acrylic, hydroxy and keto derivatives in ultraviolet *Chlorella* mutants, led to the proposal that in lower plants, the formation of DV and MV Pchlide *a* involves a beta-oxidation sequence of the methyl propionate of DV and MV Mpe, at position 6 of the macrocycle and the formation of acrylic, hydroxy, and keto propionate Mpe intermediates (Ellsworth and Aronoff 1969). The authors suggested that the DV and MV keto methyl propionate species cyclized automatically to yield DV and MV Pchlide *a* respectively (Ellsworth and Aronoff 1969). This work was met with skepticism and the putative intermediates were considered to be artifacts. This feeling was reinforced by the inability of the techniques, used by Ellsworth and Aronoff as well as of other analytical techniques available at that time, to detect the proposed MV Mpe substrate (MV Mpe), the proposed DV and MV acrylic, OH, and keto intermediates, and the DV Pchlide *a* end product in normal, green, lower and higher plants (Fig. 8.2).

A new phase in the study of the cyclopentanone ring formation was ushered by the introduction of powerful in organello systems capable of the massive net synthesis of Pchlide *a* from exogenous ALA and tetrapyrrole substrates (Daniell and Rebeiz 1982a, b; Mattheis and Rebeiz 1977a, b; Rebeiz et al. 1975; Tripathy and Rebeiz 1986), and the development of sensitive analytical fluorescence methodologies that allowed the demonstration of the DV and MV heterogeneity of the metabolic pools between Mg-Proto and Chl *a* (Rebeiz et al. 2003). With the use of similar techniques, the reactions between Mpe and Pchlide *a* have been reinvestigated by Castelfranco and collaborators (Walker et al. 1988; Wong and Castelfranco 1985). In a series of experiments involving the conversion of the added putative tetrapyrrole intermediates to Pchlide *a*, in organello, it was shown that (a) both the OH and keto methyl propionate derivatives of Mpe could be converted to Pchlide *a*, (b) that substrates with unesterified and esterified propionic acid residues at position 7 of the macrocycle were active, (c) that both, the DV and MV OH and keto methyl propionate derivatives also served as substrates, (d) that 2-ethyl,4-vinyl analogs were inactive, (e) that the 6-methyl acrylate derivative was also inactive, (e) that only one of the two 6-hydroxy enantiomers was active, (f) that only one of the two MV 6-keto derivative was active, and (g) and that the MV 6-keto derivative was 4 times more active that the DV analog, whereas DV and MV Mg-Proto were equally active. It is unfortunate that in this work, no efforts were made to distinguish between the conversion of the various substrates into DV and MV Pchlide *a* end products.

On the basis of the above results, it has been suggested that in plants, the formation of the cyclopentanone ring involves conversion of Mpe to OH and keto methyl propionate derivatives, with stereospecificity at the level of the keto derivative. The enzymatic activity has been referred to as Pchlide cyclase.

It also appears that the reactions between Mpe and Pchlide *a* require molecular oxygen and iron (Spiller et al. 1982), and are inhibited by N-ethylmaleimide, dithiothreitol, and beta-mercaptoethanol (Wong and Castelfranco 1985). It has also been reported that the conversion of Mpe to Pchlide *a* requires both the membrane and stromal fractions of the plastids (Walker and Weinstein 1991). In our hands however, excellent cyclopentanone ring synthetase activity is observed with isolated plastid membranes without the need of a stromal factor. Along with NADPH, the stromal factor appears to be involved in the regulation of the proportions of DV and MV Pchlide *a* formation (Kim et al. 1997).

#### 8.1.1.2 The Divinyl Pchlide *a* Pools

In 1963, Jones reported the detection of DV Pchlide *a* in *R. spheroides* cultures, in which the biosynthesis of bacteriochlorophyll was inhibited by incubation with 8-hydroxyquinoline (Jones 1963a, b). Jones proposed that DV Pchlide *a* was a transient immediate precursor of MV Pchlide *a* in all plants. However DV Pchlide *a* could not be detected in higher plants till 1979 (Belanger and Rebeiz 1979).

Divinyl Pchlide *a*, has two vinyl groups at positions 2 and 4 of the tetrapyrrole macrocycle (Fig. 8.3). It differs from Mpe by having a cyclopentanone ring at positions 5–6 of the macrocycle instead of a methyl propionate residue. The mechanism of the reactions involved in cyclopentanone ring formation during conversion of Mpe to Pchlide *a* was discussed above. The DV nature of the DV Pchlide *a* component of the heterogeneous Pchlide *a* pool of higher plants was originally ascertained by chemical derivatization coupled to 77 K analytical fluorescence spectroscopy (Belanger and Rebeiz 1980b). It was also confirmed by $^1$H nuclear magnetic resonance (NMR) and fast atom bombardment (FAB) mass spectroscopy (Wu and Rebeiz 1984).

**Fig. 8.3** DV Pchlide *a*

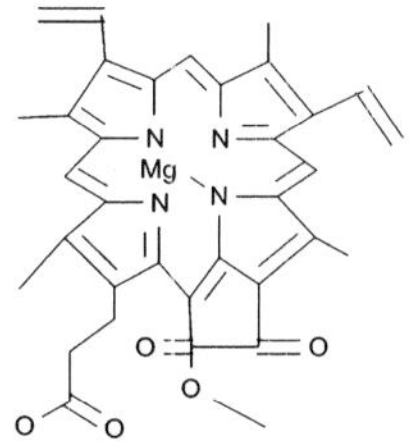


## Biosynthetic Heterogeneity of Divinyl Pchlide *a* in DDV-LDV-LDDV Plants

In Fig. 8.4, the biosynthesis of DV Pchlide *a* from DV Mpe is depicted to occur in two different thylakoid environments as suggested by multiple resonance energy transfer from DV Pchlide a to various Chl *a*-Protein complexes (Table 6.1, Chap. 6) and (Kolossov et al. 2003), as well as further conversions to Chlides and Chls as will be discussed later. It is unclear at this stage whether the spatial biosynthetic heterogeneity of DV Pchlide *a*, is accompanied by chemical biosynthetic heterogeneity or not. In other words, it is unclear whether the proposed biosynthesis of DV Pchlide *a*, from DV Mpe via routes 1, and 8 is catalyzed by identical Pchlide cyclases or by different cyclase isozymes. These biosynthetic routes will be discussed further later on.

### *Biosynthesis of DV Pchlide a via Biosynthetic Route 1 in Etiolated DDV- LDV-LDDV Plants in Darkness, and in Greening DDV-LDV-LDDV Plants During the Light Phases of the Photoperiod*

Formation of the cyclopentanone ring via biosynthetic route 1 (Fig. 8.4) in etiolated DDV-LDV-LDDV plants was demonstrated by conversion of exogenous DV Mpe to DV Pchlide *a* in isolated cucumber etiochloroplasts (Tripathy and Rebeiz 1986). DV Mpe was converted into 83 % DV Pchlide *a* and 17 % MV Pchlide *a*. The formation of much smaller amounts of MV Pchlide *a* can be accounted for by biosynthetic route 2 (Fig. 8.4) and suggests that biosynthetic route 1 (Fig. 8.4) is the predominant biosynthetic route in etiolated DDV-LDV-LDDV plant species in darkness. biosynthetic route 1 is also active in greening DDV-LDV-LDDV plants during the first few dark phases of the photoperiod when Pchlide *a* accumulation is substantial (Cohen et al. 1977; Cohen and Rebeiz 1978).

### *Biosynthesis of DV Pchlide a via Biosynthetic Route 8 in DDV-LDV-LDDV Plants During the Light and Dark Phases of the Photoperiod*

In green DDV-LDV-LDDV plant species such as cucumber, DV Pchlide *a* is continuously present during the light cycles of the photoperiod and only trace amounts of MV Pchlide *a* are formed (Carey et al. 1985; Ioannides et al. 1994). Interruption of the light cycle by a brief dark period (LD) indicated that such plant species form most of their Chl via regenerated DV Pchlide *a* (Abd-El-Mageed et al. 1997). These

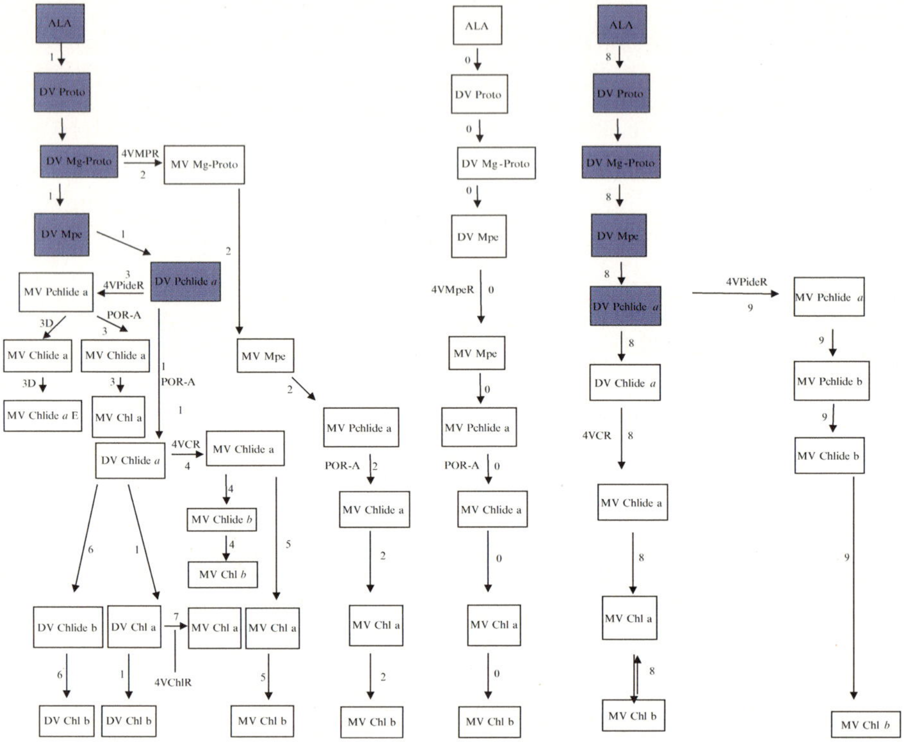


**Fig. 8.4** Biosynthetic routes 1, and 8 which are responsible for the formation of DV Pchlide *a* from DV Mpe in LDV-DDV-LDDV plant species. Routes 1, and 8 are highlighted in *blue* (Adapted from Fig. 6.3 of Chap. 6, and from Kolossov and Rebeiz 2010)

observation suggest very strongly that during the light cycles of the photoperiod, green DDV-LDV-LDDV plant species form most of their MV Chl *a* via DV Pchlide *a*, DV Chlide *a* and MV Chlide *a* as depicted in route 8.

## Biosynthetic Heterogeneity of DV Pchlide *a* in DMV-LDV-LDMV Plant Species Like Barley and Corn

The accumulation of DV Pchlide *a* in LDV-DDV-LDMV plant species such as Corn and other monocots treated with ALA and ALA +Dpy has been reported earlier (Rebeiz et al. 1991).

In Fig. 8.5, the biosynthesis of DV Pchlide *a* from DV Mpe in DMV-LDV-LDMV plants is visualized to occur in three different thylakoid environments via routes 10, 11, and 13. This was suggested by multiple resonance energy transfer from Mp (e) to various Chl *a*-Protein complexes (Table 6.1, Chap. 6) (Kolossov et al. 2003), and by further conversions of DV Pchlide *a* to Pchlides and Chls, in DMV-LD-LDMV Plant species as will be discussed below.

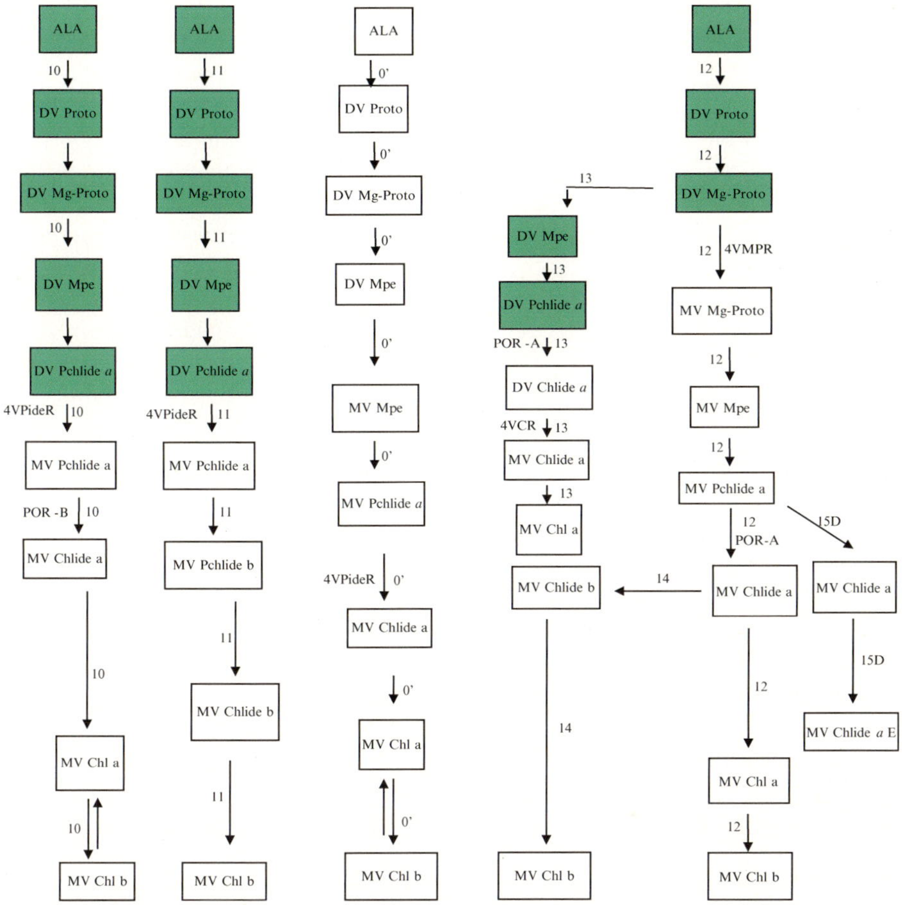


**Fig. 8.5** Biosynthetic routes 10, 11, and 13 which are responsible for the formation of DV Pchlide *a* from DV Mpe in LMV-DDV-LDMV plant species. These routes are highlighted in *green* (Adapted from Fig. 6.4 of Chap. 6, and from Kolossov and Rebeiz 2010)

It is unclear at this stage whether the spatial biosynthetic heterogeneity of DV Pchlide *a* is accompanied by chemical biosynthetic heterogeneity or not. In other words, it is unclear whether the proposed biosynthesis of DV Pchlide *a* from DV Mpe via routes 10, 11, and 13 is catalyzed by identical Pchlide *a* cyclases or by different Pchlide *a* isozymes. These biosynthetic routes will be discussed further later on.

*Biosynthesis of DV Pchlide a via Biosynthetic Routes 10, and 11 in Etiolated DMV-LDV-LDMV Plants in Darkness and in Green DMV-LDV-LDMV Plants During the Dark Phases of the Photoperiod*

DV Pchlide *a* is formed via routes 10 and 11 (Fig. 8.5) in dark-grown DMV-LDV-LDMV plant species and during the dark phases of the photoperiod.

The direct conversion of DV Mpe to DV Pchlide *a* in organello, in DMV-LDV-LDMV plants has been reported (Tripathy and Rebeiz 1986). DV Mpe was converted to 17 % DV and 83 % MV Pchlide *a*. These results suggested that the fate of DV Mpe in DMV-LDV-LDMV in etiolated plants and during the dark phases of the photoperiod is conversion to MV Pchlide *a* rather than DV Pchlide *a*, via routes 10 and 11 as shown in Fig. 8.5. This is achieved by reduction of DV Pchlide *a* to MV pchlide *a* (Tripathy and Rebeiz 1988). This in turn suggests that biosynthetic routes 10 and 11 are highly active in etiolated tissues of this greening group of plants. These results also caution against assuming that conversion of exogenous DV substrates to Pchlide *a* in organello is likely to yield DV Pchlide *a* exclusively, without specific analysis of DV and MV components as has been often assumed (Walker and Weinstein 1991; Wong and Castelfranco 1985). The operation of routes 10 and 11 in DMV-LD-LDMV plant species is also compatible with the detection of strong 4-vinyl Pchlide *a* reductase (4VPideR) activity that converts DV Pchlide *a* to MV Pchlide *a* in barley etiochloroplasts (Tripathy and Rebeiz 1988).

*Biosynthesis of DV Pchlide a via Biosynthetic Route 13 in Greening DMV-LDV-LDMV Plants During the Light Phases of the Photoperiod*

DV Pchlide *a* is formed via route 13 in light-grown DMV-LDV-LDMV plant species during the light phases of the photoperiod. Biosynthetic route 13 is active in greening DMV-LDV-LDMV plants during the first few light phases of the photoperiod when Pchlide a accumulation is substantial (Cohen and Rebeiz 1978).

### Metabolism of DV Pchlide *a* in DDV-LDV-LDDV Plant Species via Routes 1 and 8

Precursor-product relationships between DV Pchlide *a* and DV Chlide *a* (Fig. 8.4, routes 1, and 8) were established by demonstrating the photoreduction of DV Pchlide *a* to DV Chlide *a* in etiolated cucumber cotyledons induced to accumulate DV Pchlide *a* exclusively (Duggan and Rebeiz 1982). DV Pchlide *a* is probably converted to DV Chlide *a* by PORA in etiolated DDV-LDV-LDDV plants (Armstrong et al. 1995). During photoperiodic greening, in the light, DV Pchlide *a* is converted to DV Chlide *a* (Abd-El-Mageed et al. 1997) probably by PORB which is active in the light in light grown DDV-LDV-LDDV plants (1981; Runge et al. 1996).

### Metabolism of MVPchlide *a* in DMV-LDV-LDMV Plant Species via Biosynthetic Route 13

During photoperiodic greening of LMV-DDV-LDMV plant species small amounts of DV Pchlide *a* are always detectable (Abd-El-Mageed et al. 1997) and are probably converted to MV Chlide *a* via DV Chlide *a* (Fig. 8.5).

**Fig. 8.6** MV Pchlide *a*

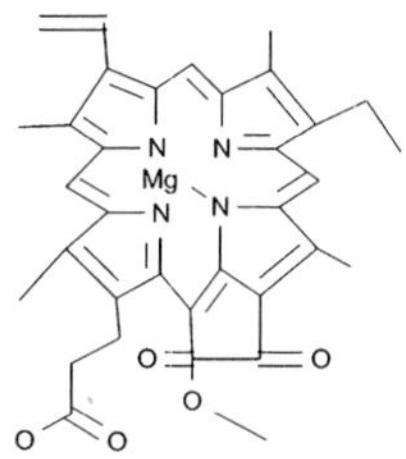


#### 8.1.1.3 The Monovinyl (MV) Pchlide *a* Pool

MV Pchlide *a* has one vinyl group at position 2 and one ethyl group at position 4 of the tetrapyrrole macrocycle (Fig. 8.6). The MV nature of the MV Pchlide *a* component of the Pchlide *a* pool of higher plants was determined by chemical derivatization coupled to analytical fluorescence spectroscopy at 77 K (Belanger and Rebeiz 1980a). It was confirmed by NMR spectroscopy and field desorption mass spectroscopy (Wu and Rebeiz 1984). The mechanism of the reactions involved in cyclopentanone ring formation during conversion of Mpe to Pchlide *a* was discussed above in a previous section.

Biosynthetic Heterogeneity of MV Pchlide *a* in DDV-LDV-LDDV Plants

In Fig. 8.7, the biosynthesis of MV Pchlide *a* is depicted to occur in four different thylakoid environments as suggested by multiple resonance energy transfer from MV Pchlide *a* to various Chl *a*-Protein complexes (Table 6.1, Chap. 6) and (Kolossov et al. 2003), as well as further conversions to Chlides and Chls as will be discussed later. It is unclear at this stage whether the spatial biosynthetic heterogeneity of MV Pchlide *a*, is accompanied by chemical biosynthetic heterogeneity or not. In other words, it is unclear whether the proposed biosynthesis of MV Pchlide *a* via routes 2, 3, 0 and 9 is catalyzed by identical Pchlide cyclases or by different cyclase isozymes. These biosynthetic routes will be discussed further below.

In addition to DV Pchlide *a*, etiolated and green DDV-LDV-LDDV plant species such as cucumber, form smaller amounts of MV Pchlide *a* during prolonged dark incubation (Carey and Rebeiz 1985; Ioannides et al. 1994). This Pchlide *a* formation can be accounted for by routes 2 and 3. In route 3, MV Pchlide *a* formation can be accounted for by a slow conversion of DV Pchlide *a* to MV Pchlide *a*, a reaction catalyzed by 4VPideR during prolonged dark-incubations (Tripathy and Rebeiz 1988). Probably, biosynthetic routes 2 and 3 are also active in greening DDV-LDV-LDDV plants during the first few dark phases of the photoperiod when Pchlide *a* accumulation is substantial (Cohen et al. 1977; Cohen and Rebeiz 1978).

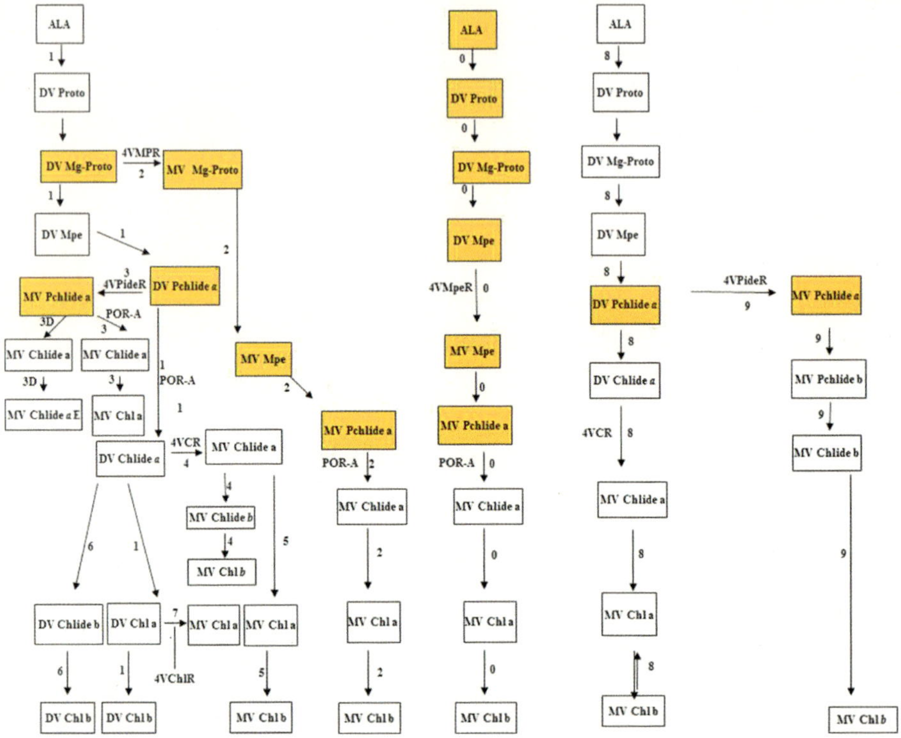


**Fig. 8.7** Biosynthetic routes 2, 3, 0, and 9 which are responsible for the formation of MV Pchlide *a* in LDV-DDV-LDDV plant species. The routes are highlighted in *yellow* (Adapted from Fig. 6.3 of Chap. 6, and from Kolossov and Rebeiz 2010)

### *Biosynthesis of MV Pchlide a via Biosynthetic Route 2 in Etiolated and Greening DDV-LDV-LDDV Plants During the Dark and Light Phases of the Photoperiod*

Biosynthetic route 2 is initiated by reduction of the vinyl group at position 4 of the DV Mg-proto macrocycle to ethyl, and conversion of DV Mg-proto to MV Mg-proto. The reaction is catalyzed by 4VMPR (Kim et al. 1997; Kolossov and Rebeiz 2010). The nascent MV Mg-Proto is then converted into MV Mpe and MV Pchlide *a* as depicted in route 2 (Tripathy and Rebeiz 1986).

### *Biosynthesis of MV Pchlide a via Biosynthetic Route 3 in Etiolated DDV-LDV-LDDV Plants in Darkness and in Green DDV-LDV-LDDV Plants During the Dark and Light Phases of the Photoperiod*

As depicted in Fig. 8.7, MV Pchlide *a* can also be formed by reduction of the vinyl group of DV Pchlide *a* to ethyl at position 4 of the macrocycle, via route 3. MV Pchlide *a* formation can be accounted for by a slow conversion of DV Pchlide *a* to MV Pchlide *a*, a reaction catalyzed by 4VCPideR during prolonged dark incubations (Tripathy and Rebeiz 1988; Walker et al. 1988).

*Biosynthesis of MV Pchlide* a *via Biosynthetic Route 0 in Etiolated DDV-LDV-LDDV Plants in Darkness, and in Green DDV-LDV-LDDV Plants During the Dark Phases of the Photoperiod*

4VMpeR was detected in barley not in cucumber. Therefore this route should be deleted if further research fails to detect 4VMpeR in DDV-LDV-LDDV plant species.

*Biosynthesis of MV Pchlide* a *via Biosynthetic Route 9 in Etiolated DDV-LDV-LDDV Plants in Darkness, and in Green DDV-LDV-LDDV Plants During the Dark Phases of the Photoperiod*

Biosynthetic route 9 branches from route 8 and is proposed to account for the formation of MV Pchlide *b* in green DDV-LDV-LDDV plant species in the light. In this route MV Pchlide *a* is considered to be formed from DV Pchlide *a* by the action of 4VPideR (Tripathy and Rebeiz 1988). It should be pointed out that even during the light phases of the photoperiod DDV-LDV-LDDV plants form very small amounts of MV Pchlide *a* in addition to the formation of massive amounts of DV Pchlide *a* (Carey et al. 1985). It should also be stressed that in DDV-LDV-LDDV plants, formation of MV Pchlide *a* via route 9 is destined exclusively for the biosynthesis of MV Pchlide *b* in the light (Ioannides et al. 1997).

Biosynthetic Heterogeneity of MV Pchlide *a* via Routes 10, 11, 0′, 12 in DMV-LDV-LDMV Plant Species

In Fig. 8.8, the biosynthesis of MV Pchlide *a* in DMV-LDV-LDMV plants is depicted to occur in four different thylakoid environments as suggested by multiple resonance energy transfer from MV Pchlide a to various Chl *a*-Protein complexes (Table 6.1, Chap. 6) and (Kolossov et al. 2003), as well as further conversions to Chlides and Chls as will be discussed later. It is unclear at this stage whether the spatial biosynthetic heterogeneity of MV Pchlide *a*, is accompanied by chemical biosynthetic heterogeneity or not. In other words, it is unclear whether the proposed biosynthesis of MV Pchlide *a* via routes 10, 11, 0′ and 12 is catalyzed by identical Pchlide cyclases or by different cyclase isozymes. These biosynthetic routes will be discussed further below.

*Biosynthesis of MV Pchlide a in DMV-LDV-LDMV Plant Species via Routes 10*

As depicted in Fig. 8.8, MV Pchlide *a* in DMV-LDV-LDMV etiochloroplasts in route 10 can be formed by reduction of the vinyl group of DV Pchlide *a* to ethyl at position 4 of the macrocycle. The operation of route 10 in etiolated DMV-LDV-LDMV plants is based on the fact that in isolated barley etiochloroplasts DV Pchlide *a* is actively converted to MV Pchlide a by 4VPideR (Tripathy and Rebeiz 1988).

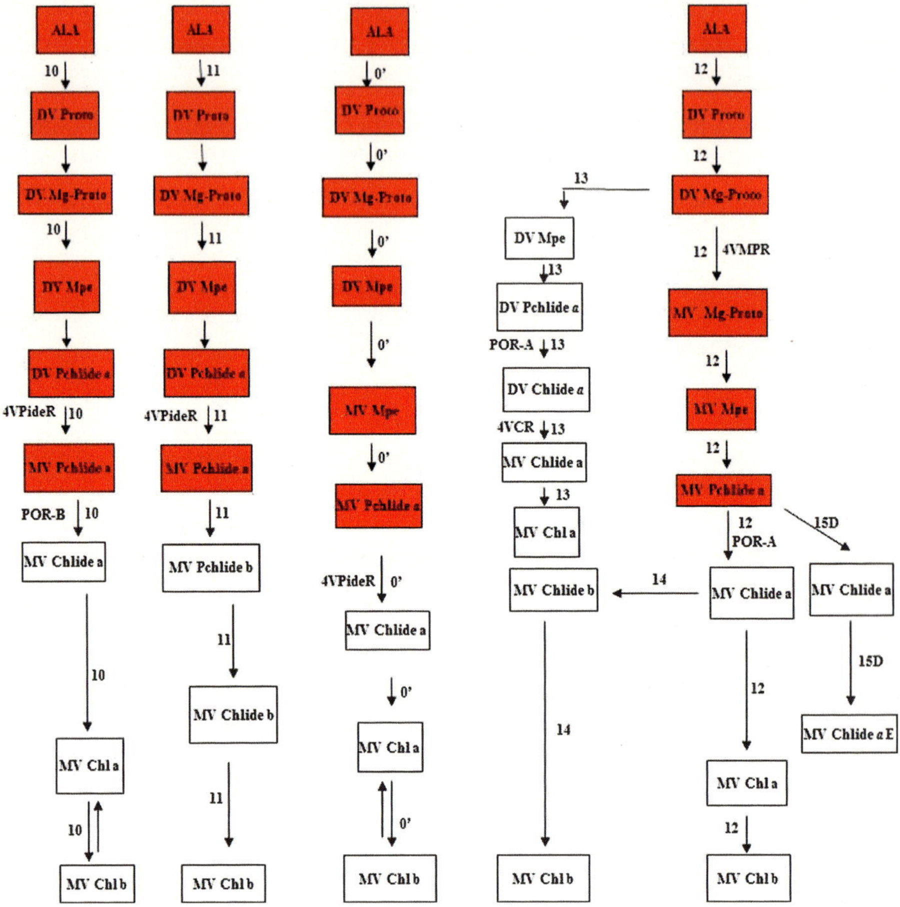


**Fig. 8.8** Biosynthetic routes 10, 11, 0′, and 12 which are responsible for the formation of MVPchlide *a* in LMV-DDV-LDMV plant species. The routes are highlighted in *red* (Adapted from Fig. 6.4 of Chap. 6, and from Kolossov and Rebeiz 2010)

*Biosynthesis of MV Pchlide a in DMV-LDV-LDMV Plant Species via Routes 11*

As depicted in Fig. 8.8, MV Pchlide *a* can also be formed by reduction of the vinyl group of DV Pchlide *a* to ethyl at position 4 of the macrocycle, via route11, a reaction catalyzed by 4VPideR (Tripathy and Rebeiz 1988). This in turn is conjectured to lead to the formation of MV Pchlide *b* in greening DMV-LDV-LDMV plants (Kolossov and Rebeiz 2003).

*Biosynthesis of MV Pchlide a in DMV-LDV-LDMV Plant Species via Routes 0′*

In this biosynthetic route, MV Pchlide *a* is formed from MV Mpe via VPideR. The Operation of biosynthetic route 0′ in DMV-LDV-LDMV plants during photoperiodic greening is justified by the detection and solubilization of 4-Vinyl

Mpe reductase (4VMpeR) in greening barley etiochloroplasts (Kolossov and Rebeiz 2010). Such etiochloroplasts can actively convert MV Mpe to MV Pchlide *a* (Tripathy and Rebeiz 1986).

*Biosynthesis of MV Pchlide a in DMV-LDV-LDMV Plant Species via Routes 12*

MV Pchlide *a* is formed via biosynthetic route 12 from MV Mg-Proto and MV Mpe (Fig. 8.8). Indeed in green DMV-LDV-LDMV plants MV Pchlide *a* formation is very active (Abd-El-Mageed et al. 1997) and barley etiochloroplasts are capable of strongly converting MV Mpe to MV Pchlide *a* (Tripathy and Rebeiz 1986).

## 8.2 Pchlide-Protein Complexes

Further metabolism of DV and MV Pchlide *a* takes place only if the Pchlide is complexed to an apoprotein. The Pchlide-apoprotein complex is referred to as a Pchl-holochrome (Pchl-Hs). Various Pch-Hs holochromes will be discussed in detail below.

### *8.2.1 Heterogeneity of Pchlide a-Protein Complexes*

As we have mentioned previously, Pchlides *a* are chemically heterogeneous. That heterogeneity is expressed chemically as DV, and MV substitutions at the level of the tetrapyrrole chromophore at position 4 of the macrocycle. It is also expressed at the level of the esterifying group at position 7 of the macrocycle as will be discussed in Chap. 9, and the chromophore-protein complexes referred to as Pchl-Hs. Since in most cases the chromophore of Pchl-Hs consist mostly of Pchlide *a* and much smaller amounts of Pchlide *a* ester (Chap. 9), it is appropriate to refer to these holochromes as Pchl(ide)-Hs. The overall heterogeneity is also expressed by the detection of multiple resonance excitation energy transfer bands between Pchl(ide)-Hs and various Chl-protein complexes (Table 6.1, Chap. 6).

Discussion of the heterogeneity of Pchl(ide)-Hs will focus (a) on the possible nature of the chromophore-apoprotein association, (b) on the spectroscopic properties of various Pchl(ide) *a*-Hs *in situ*, and (c) on the properties of purified Pchl(ide)-Hs.

#### 8.2.1.1 Nature of the Chromophore-Protein Association of Pchl(ide)-Hs

In etiolated tissues, Pchlide *a* is the most abundant tetrapyrrole (90–95 %) followed by less abundant Pchlide *a* ester (Pchlide *a* E) (5–10 %) (Rebeiz et al. 1970). Pchlide *a* and its ester will be referred to collectively as Pchl(ide) *a*. Upon binding to apoproteins, the spectroscopic properties of Pchl(ide) *a* chromophores change drastically (see below). It should be kept in mind that the bulk of Pchl(ide) *a*-Hs is made up of Pchlide *a*-Hs.

In Pchl(ide) *a*-Hs, various Pchl(ide) *a* chromophores are bound to different apoproteins by non-covalent forces. This is evidenced by the ready extraction of the Pchl(ide) *a* chromophores by organic solvents such as acetone. Association of the chromophores with apoproteins, probably involve (a) axial coordination of the Pchl(ide) *a* central Mg-atom to nucleophyllic amino acid side chains (Kolossov et al. 2003; Rebeiz and Belanger 1984), and (b) hydrogen bonding between the keto group of the cyclopentanone ring of the Pchl(ide) *a* chromophore and appropriate amino acid side chains (Kolossov et al. 2003; Rebeiz and Belanger 1984). Pigment-pigment interaction may involve axial coordination of the keto group of the cyclopentanone ring of one Pchl(ide) *a* chromophore to the central Mg-atom of another Pchl(ide) *a* chromophore as suggested by Katz et al. (1966) for Chl-Chl association in hydrophobic environments, as well as Π−Π interactions of Pchl(ide) *a* chromophores (Boucher and Katz 1967). Axial coordination of the histidine nitrogen of apoproteins to the central Mg-atom (Deisenhofer and Michel 1991) of Pchl(ide) *a* has not been established for various Pchl(ide) a-Hs.

#### 8.2.1.2 Spectroscopic Properties of Various Pchlide *a*-Hs

The existence of at least two spectroscopically different Pchl(ide) *a*-Hs was first reported by Hill and coworkers (Hill et al. 1953). Using a Zeiss microspectroscope they observed that in etiolated barley leaves, a band absorbing at 650 nm disappeared (was phototransformed, i.e. was photoconverted to a Chl-like compound) as the light was turned on and was replaced by the appearance of two new absorbance bands: one near 670 nm which corresponded to newly formed Chl *a*-like compound, and one at 635 nm, which did not appear to be convertible to Chl. These results gave rise to the notion that etiolated tissues contained two spectroscopically different Pchl(ide) *a*-H complexes. A longer wavelength (LW), phototransformable (t) complex absorbing at 650 nm, and a shorter wavelength (SW), non-phototransformable (nt) complex, absorbing at 635 nm.

To explain the difference between the LW and SW Pchl(ide) *a*-Hs, Butler and Briggs proposed, on the basis of freezing and thawing treatments of plant tissues, that aggregation of pigment molecules in etioplasts shifts the absorption maximum to longer wavelengths, while disaggregation of pigment molecules shifts the absorption maximum to shorter wavelengths (Butler and Briggs 1966). Using freezing and thawing as well as extraction, heat and acid treatments, Dujardin and Sironval (1970) suggested the presence of three universal Pchl(ide) *a*-Hs in plants, namely: an aggregated, phototransformable species absorbing at 647–648 nm that involves pigment-protein and pigment-pigment interactions, a second phototransformable species absorbing at 639–640 nm which involves only pigment-protein interactions, and a non-phototransformable species absorbing at 627–628 nm, which is loosely bound to proteins. They also proposed that pigment-pigment interaction is not required for phototransformation while binding to a specific protein is required. Using absorption, fluorescence emission and excitation spectroscopy at 77 °K, Kahn et al. (1970), further characterized the three Pchl(ide) *a*-Hs as consisting of (a) a nt-fluorescent species with a red excitation maximum at 628 nm and a red fluorescence emission maximum at 630 nm [nt-Pchl(ide) *a* (E628 F630)], (b) a t, nonfluorescent species with a red excitation maximum at 639 nm [(t-Pchl(ide) *a* E639 F00)] which transfers its excitation energy to a Pchl(ide) *a*-H with a red excitation maximum at 650 nm and a red fluorescence emission maximum at 655 nm [t-(Pchl (ide) *a* E650 F655)]. The latter is the predominant Pchl(ide) *a*-H in etiolated tissues. It is now known that this Pchl(ide) *a*-H species is a ternary complex of Pchlide *a* with NADPH and Pchlide a oxidoreductase.

Using high resolution 77 °K spectrofluorometry coupled to matrix analysis, Cohen and Rebeiz (1981) carried out a detailed studies of the Pchl(ide) *a*-Hs that accumulate in etiolated cucumber a DDV-LDV-LDDV species, and bean a DMV-LDV-LDMV plant species. The various Pchl(ide) *a*-H species were assigned Soret excitation maxima (E), and fluorescence emission maxima (F). The following Pchl(ide) *a*-H species were detected in etiolated cucumber cotyledons: nt-SW Pchl(ide) *a*-H (E440 F630), t-SW Pchl(ide) *a*-H (E443 F633), -(E444 F636) and -(E445 F640), and t-LW Pchl(ide) *a* (E450 F657), which was the predominant species in etiolated cucumber. In red-kidney bean, the following Pchl(ide) *a*-H species were detected nt-SW Pchl (ide) *a* H (E440 F630), t-SW Pchl(ide) *a*-H (E441 F633), -(E442 F636) and -(E443 F640), and t-LW Pchl(ide) *a* (E447 F657), which was the predominant species in etiolated bean. It is now known that this Pchl(ide) *a*-H species is a ternary complex of Pchlide *a* with NADPH and Pchlide *a* oxidoreductase.

The contribution of SW and LW Pchl(ide) *a*-Hs to the natural greening process was assessed during photoperiodic greening, i.e. during greening under alternating light/dark photoperiods (Cohen and Rebeiz 1978). The following observations were made (a) SW Pchl(ide) *a*- H species appeared within the first 24 h of germination of cucumber seedlings, (b) subsequently, LW Pchl(ide) *a* H species appeared then disappeared, (c) The ratio of LW/SW Pchl(ide) *a* H species reached a maximum of 3:1 by the end of the second dark cycle and reached a value of zero by the end of the 6th dark cycle, (d) SW Pchl(ide) *a* H species were continuously present during the dark and light cycles and appeared to contribute actively to the greening process, and finally (e) primary corn and bean leaves exhibited a similar pattern of Pchl(ide) *a* H formation.

#### 8.2.1.3 Purification of Pchlide *a*-Hs

Early work dealing with the purification of Pchl(ide) *a*-Hs was described by Boardman (1966). The partially purified Pchl(ide) *a*-H (MW = 600,000) exhibited a red absorption maximum at 637.5 nm. Upon illumination, part of the Pchl(ide) *a* was converted into Chlide *a* with a red absorption maximum at 681 nm which after 2 min in darkness shifted to 675 nm. This preparation, however, did not preserve the heterogeneous spectral properties observed in vivo. A purer preparation from etiolated bean leaves

(MW = 300,000) was described by Schopfer and Siegelman (1968). The purified Pchl(ide) *a*-H exhibited a red absorption maximum at 639 nm, which also did not reflect the spectral heterogeneity observed in vivo. In the light the red absorption maximum shifted to a Chlide *a* red absorption maximum at 678 nm, which drifted to 672 nm in darkness. More purified Pchl(ide) a-Hs were prepared from etiolated barley (MW 63,000) and bean (MW 100,000) by Henningsen and Kahn (1971). Photoconversion yielded a Chl(ide) *a* complex with a red absorption maximum at 678 nm. In this case too, the spectral properties of the purified Pchl(ide) *a*-H did not reflect the spectral heterogeneity observed in vivo.

## References


Abd-El-Mageed HA, El Sahhar KF, Robertson KR et al (1997) Chloroplast biogenesis 77. Two novel monovinyl and divinyl light-dark greening groups of plants and their relationship to the chlorophyll a biosynthetic heterogeneity of green plants. Photochem Photobiol 66:89–96

Armstrong GA, Runge S, Frick G et al (1995) Identification of NADPH: protochlorophyllide oxidoreductases A and B: a branched pathway for light-dependent chlorophyll biosynthesis in *Arabidopsis thaliana*. Plant Physiol 108:1505–1517

Belanger FC, Rebeiz CA (1979) Chloroplast biogenesis XXVII. Detection of novel chlorophyll and chlorophyll precursors in higher plants. Biochem Biophys Res Commun 88:365–472

Belanger FC, Rebeiz CA (1980a) Chloroplast biogenesis. Detection of divinyl protochlorophyllide in higher plants. J Biol Chem 255:1266–1272

Belanger FC, Rebeiz CA (1980b) Chloroplast biogenesis. Detection of divinyl protochlorophyllide ester in higher plants. Biochemistry 19:4875–4883

Boardman NK (1966) Protochlorophyll. In: Vernon LP, Seeley GR (eds) The chlorophylls. Academic, New York, pp 437–479

Boucher LJ, Katz JJ (1967) Aggregation of metalloporphyrins. J Am Chem Soc 89:4703–4708

Butler R, Briggs WR (1966) The relation between structure and pigments during the first stages of proplastid greening. Biochim Biophys Acta 112:45–53

Carey EE, Rebeiz CA (1985) Chloroplast biogenesis 49. Difference among angiosperms in the biosynthesis and accumulation of monovinyl and divinyl protochlorophyllide during photoperiodic greening. Plant Physiol 79:1–6

Carey EE, Tripathy BC, Rebeiz CA (1985) Chloroplast biogenesis 51. Modulation of monovinyl and divinyl protochlorophyllide biosynthesis by light and darkness in vitro. Plant Physiol 79:1059–1063

Cohen CE, Rebeiz CA (1978) Chloroplast biogenesis 22. Contribution of short wavelength and long wavelength protochlorophyll species to the greening of higher plants. Plant Physiol 61:824–829

Cohen CE, Rebeiz CA (1981) Chloroplast biogenesis 34. Spectrofluorometric characterization in situ of the protochlorophyll species in etiolated tissues of higher plants. Plant Physiol 67:98–103

Cohen CE, Bazzaz MB, Fullet SE et al (1977) Chloroplast biogenesis XX. Accumulation of porphyrin and phorbin pigments in cucumber cotyledons during photoperiodic greening. Plant Physiol 60:743–746

Daniell H, Rebeiz CA (1982a) Chloroplast culture IX. Chlorophyll(ide) *a* biosynthesis in vitro at rates higher than in vivo. Biochem Biophys Res Commun 106:466–470

Daniell H, Rebeiz CA (1982b) Chloroplast culture VIII. A new effect of kinetin in enhancing the synthesis and accumulation of protochlorophyllide in vitro. Biochem Biophys Res Commun 104:837–843

Deisenhofer J, Michel H (1991) Crystallography of chlorophyll proteins. In: Scheer H (ed) Chlorophylls. CRC press, Boca Raton, pp 613–625

Duggan JX, Rebeiz CA (1982) Chloroplast biogenesis 37: induction of chlorophyllide *a* (E459F675) accumulation in higher plants. Plant Sci Lett 24:27–37

Dujardin E, Sironval C (1970) The reduction of protochlorphyllide into chlorophyllide III. The phototransformation of the forms of the protochlorophyllide-lipoprotein complex found in darkness. Photosynthetica 4:129–138

Ellsworth RK, Aronoff S (1969) Investigations of the biogenesis of chlorophyll *a*. IV. Isolation and partial characterization of some biosynthetic intermediates between Mg-protoporphine IX monomethyl ester and Mg-vinylpheoporphine $a_5$, obtained from *Chlorella* mutants. Arch Biochem Biophys 130:374–383

Granick S (1948) Magnesium protoporphyrin as a precursor of chlorophyll in *Chlorella*. J Biol Chem 175:333–342

Granick S (1950a) Magnesium vinyl pheoporphyrin $a_{5,}$ another intermediate in the biological synthesis of chlorophyll. J Biol Chem 183:713–730

Granick S (1950b) The structural and functional relationships between heme and chlorophyll. Harvey Lect 44:220–245

Henningsen KW, Kahn A (1971) Photoactive subunits of protochlorophyll(ide) holochrome. Plant Physiol 47:685–690

Hill R, Smith JHC, French CS (1953) The absorption and fluorescence properties of natural protochlorophylls. Yearb Carneg Inst 52:153–155

Ioannides IM, Fasoula DM, Robertson KR (1994) An evolutionary study of chlorophyll biosynthetic heterogeneity in green plants. Biochem Syst Ecol 22:211–220

Ioannides IM, Shedbalkar VP, Rebeiz CA (1997) Quantitative determination of 2-monovinyl protochlorophyll(ide) *b* by spectrofluorometry. Anal Biochem 249:241–244

Jones OTG (1963a) Magnesium 2,4-divinyl phaeoporphyrin $a_5$ monomethyl ester, a protochlorophyll-like pigment produced by *Rhodopseudomonas spheroides*. Biochem J 89:182–189

Jones OTG (1963b) The inhibition of bacteriochlorphyll biosynthesis in *Rhodopseudomonas spheroides* by 8-hydroxyquinoline. Biochem J 88:335–343

Kahn A, Boardman NK, Thorne SW (1970) Energy transfer between protochlorophyllide molecules: evidence for multiple chromophores in the photoactive protochlorophyllide-protein complex in vivo and in vitro. J Mol Biol 48:85–101

Katz JJ, Dougherty RC, Boucher LC (1966) Infrared and nuclear magnetic resonance spectroscopy of chlorophyll. In: Vernon LP, Seely GR (eds) The chlorophylls. Academic, New York, pp 185–251

Kim JS, Kolossov V, Rebeiz CA (1997) Chloroplast biogenesis 76: regulation of 4-vinyl reduction during conversion of divinyl Mg-protoporphyrin IX to monovinyl protochlorophyllide *a* is controlled by plastid membrane and stromal factors. Photosynthetica 34:569–581

Kolossov VL, Rebeiz CA (2003) Chloroplast biogenesis 88. Protochlorophyllide *b* occurs in green but not in etiolated plants. J Biol Chem 278(50):49675–49678

Kolossov VL, Rebeiz CA (2010) Evidence for various 4-vinyl reductase activities in higher plants. In: Rebeiz CA, Benning C, Bohnert HJ et al (eds) The chloroplast: basics and applications. Springer, Dordrecht, pp 25–38

Kolossov VL, Kopetz KJ, Rebeiz CA (2003) Chloroplast biogenesis 87: evidence of resonance excitation energy transfer between tetrapyrrole intermediates of the chlorophyll biosynthetic pathway and chlorophyll *a*. Photochem Photobiol 78:184–196

Koski VM (1950) Chlorophyll formation in seedlings of *Zea mays* L. Arch Biochem 29:339–343

Mattheis JR, Rebeiz CA (1977a) Chloroplast biogenesis. Net synthesis of protochlorophyllide from protoporphyrin IX by developing chloroplasts. J Biol Chem 252:8347–8349

Mattheis JR, Rebeiz CA (1977b) Chloroplast biogenesis. Net synthesis of protochlorophyllide from magnesium protoporphyrin monoester by developing chloroplasts. J Biol Chem 252:4022–4024

Rebeiz CA, Belanger FC (1984) Chloroplast biogenesis 46: calculation of net spectral shifts induced by axial ligand coordination in metalated tetrapyrroles. Spectrochim Acta 40A:793–806

Rebeiz CA, Yaghi M, Abou Haidar M et al (1970) Protochlorophyll biosynthesis in cucumber (Cucumis sativus, L.) cotyledons. Plant Physiol 46:57–63

Rebeiz CA, Mattheis JR, Smith BB et al (1975) Chloroplast biogenesis. Biosynthesis and accumulation of protochlorophyll by isolated etioplasts and developing chloroplasts. Arch Biochem Biophys 171:549–567

Rebeiz CA, Nandihalli UB, Reddy K (1991) Photodynamic herbicides and chlorophyll biosynthesis modulators. In: Baker NR, Percival M (eds) Herbicides. Elsevier, Amsterdam, pp 173–208

Rebeiz CA, Kolossov VI, Briskin D et al (2003) Chloroplast biogenesis 86: chlorophyll biosynthetic heterogeneity, multiple biosynthetic routes and biotechnological spin-offs. In: Nalwa N (ed) Handbook of photochemistry and photobiology. American Scientific Publisher, Los Angeles, pp 183–248

Runge S, Ulrich S, Frick J et al (1996) Distinct roles for light-dependent NADP: ptotochlorophyllide oxidoreductase (POR) A and B during greening in higher plants. Plant J 9(4):513–523

Schopfer P, Siegelman HW (1968) Purification of protochlorophyllide holochrome. Plant Physiol 43:990–996

Spiller SC, Castelfranco AM, Castelfranco PA (1982) Effect of iron and oxygen on chlorophyll biosynthesis. Plant Physiol 69:107–111

Tripathy BC, Rebeiz CA (1986) Chloroplast biogenesis. Demonstration of the monovinyl and divinyl monocarboxylic routes of chlorophyll biosynthesis in higher plants. J Biol Chem 261:13556–13564

Tripathy BC, Rebeiz CA (1988) Chloroplast biogenesis 60. Conversion of divinyl protochlorophyllide to monovinyl protochlorophyllide in green(ing) barley, a dark monovinyl/light divinyl plant species. Plant Physiol 87:89–94

Walker CJ, Weinstein JD (1991) In vitro assay of the chlorophyll biosynthetic enzyme Mg-chelatase: resolution of the activity into soluble and membrane-bound fractions. Proc Natl Acad Sci U S A 88:5789–5793

Walker CJ, Mansfield KE, Rezzano IN et al (1988) The magnesium-protoporphyrin IX (oxidative) cyclase system. Studies of the mechanism and specificity of the reaction sequence. Biochem J 255:685–692

Wolff JB, Price L (1957) Terminal steps of chlorophyll *a* biosynthesis in higher plants. Arch Biochem Biophys 72:293–301

Wong Y-S, Castelfranco PA (1985) Properties of the Mg-protoporphyrin IX monomethyl ester (oxidative) cyclase system. Plant Physiol 79:730–733

Wu SM, Rebeiz CA (1984) Chloroplast biogenesis 45: molecular structure of protochlorophyllide (E443 F625) and of chlorophyllide *a* (E458 F674). Tetrahydron 40(4):659–664

# Chapter 9
# The Chl *a* Carboxylic Biosynthetic Routes: (Photo) Conversion of Protochlorophyllides (Pchlides) *a* to Chlorophyllide (Chlide) *a*

*One of the true measures of greatness resides in the willingness and ability of truly great men to recognize excellence when they encounter it*

DV and MV Chlides *a* are the main immediate precursors of Chl *a* (Fig. 9.1). They are formed via multiple light-dependent and light-independent biosynthetic routes from Pchlide *a*. In all cases, the reaction involves reduction of the double bond at position 7–8 of the macrocycle by addition of two trans-hydrogens. Most of the investigations of the photoreduction of Pchlide *a* have dealt with transformable long wavelength Pchlide *a*-HochromeE650 F657 [(t-LW-Pchlide *a*-H (E650 F657))]. The latter is a ternary complex of Pchlide *a*, NADPH and Pchlide *a* oxidoreductase, a shuttling photoenzyme.

The notion that the t-LW-Pchl(ide) *a* H apoprotein acts as a shuttling photoenzyme that catalyzes the conversion of Pchlide *a* to Chlide *a* was first proposed by Sironval et al. (1967). In this work the authors reported that Pchl (ide) *a* (E647 F 657), with a red excitation maximum at 647 nm and a red emission maximum at 657 nm, is photoconverted to Chlide *a* (E676 F690). The latter shifts in darkness to a Chlide *a* (E682 F697) species. At this stage, the authors suggested that the apoprotein discharges the newly formed Chlide *a* and picks up another Pchlide *a* which may be photoconverted to Chlide *a* via a similar cycle. The spectral shifts described by Sironval et al. were confirmed by Gassman et al. (1968), and Bonner (1969). The concept of shuttling photoenzyme was also compatible with the photoconversion of photo-inactive Pchlide *a* 633 to phototransformable Pchlide *a* 650 reported by Gassman (1973).

Also discussed will be the light-independent conversion of Pchlide *a* to Chlide *a*. Although light-independent Pchlide *a* reduction has been known to occur in gymnosperms its occurrence in angiosperms is novel (Adamson and Packer 1984; Adamson et al. 1997).

C.A. Rebeiz, *Chlorophyll Biosynthesis and Technological Applications*, DOI 10.1007/978-94-007-7134-5_9, © Springer Science+Business Media Dordrecht 2014

**Fig. 9.1** DV and MV chlorophyllide (Chlide) *a*

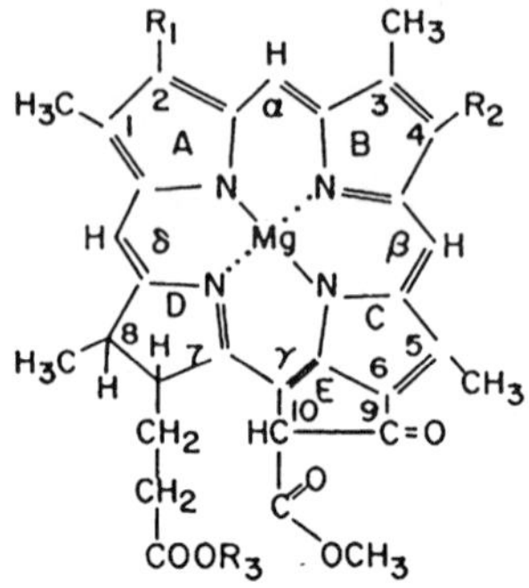


**Chlorophyllide *a* (Chlide *a*)**

| | R1 | R2 | R3 | Compound |
|---|---|---|---|---|
| I | Vinyl | Vinyl | H | DV Chlide *a* |
| II | Vinyl | Ethyl | H | MV Chlide *a* |

## 9.1 Formation of Chlide *a* via Light-Independent Pchlide *a* Reductase(s)

Algae, ferns, mosses, and the cotyledons of most gymnosperms, all with a DDV-LDV-LDDV greening group affiliation (Ioannides et al. 1994) are capable of converting Pchlide *a* to Chlide *a* in the absence of light (Kirk and Tilney-Basset 1967; Rudiger and Schoch 1991; Shulz and Senger 1993), via a reaction catalyzed by a light-independent Pchlide *a* reductase. Most probably light-independent Chlide *a* formation is via a nt-SW-Pchlide a species (Schoefs and Franck 1998).

Although in angiosperms light is required for the formation of photosynthetic pigment-protein complexes and the accumulation of massive amounts of Chl, Adamson and coworkers pioneered the notion that in this phylum, a certain amount of Chl *a* biosynthesis can also take place in darkness via a light-independent Pchlide *a* reductase (Adamson et al. 1997). Dark-incorporation of $^{14}$C-glutamate and $^{14}$C-ALA into $^{14}$C-Chl *a* in barley leaves and barley etiochloroplasts appear to confirm Adamson's contention (Tripathy and Rebeiz 1987). It is therefore proposed that biosynthetic routes 3D and 15D (Figs. 3 and 4 in Chap. 4) are also functional in angiosperms such as cucumber and barley. It should be emphasized however that the amount of Chl *a* formed via Chlide *a* in darkness is very small, and its biological significance is unknown.

Genetic and sequence analysis have indicated that in *R. capsulatus*, three genes, *bchL*, *bchN*, and *bchB* appear to be involved in Pchlide *a* reduction in darkness (Suzuki et al. 1997). The three open frames exhibited significant sequence similarity to the three subunits of nitrogenase, which led to the proposal that light-independent Pchlide *a* reductase and nitrogenase share a common evolutionary ancestor. Expression of the *bchL*, *bchN*, and *bchB* genes has been however unsuccessful. Very recently, Yuichi and Bauer reported demonstration of dark-Pchlide *a* reductase activity in reconstituted systems from *R. capsulatus*, a purple nonsulfur photosynthetic bacterium (Yuichi and Bauer 2000). Two of the putative three subunits, BchL and BchN were expressed in *R. capsulatus* as S tag fusion proteins. The third subunit, BchB, copurified with the *BchN* protein, thus indicating that the BchN and BchB proteins form a tight complex. Dark Pchlide *a* reductase activity was shown to be dependent on the presence of all three subunits, on ATP, and on the reductant dithionite. In angiosperms, the corresponding gene products ChlL, ChlN, and ChlB, also appear to be evolutionarily related to the subunits of the eubacterial nitrogenase enzyme complex (Armstrong 1998).

## 9.2 Kinetics of the Photoconversion of Pchlide *a*-H (E650 F657) to Chlide *a*

### *9.2.1 Action Spectrum of the Photoconversion*

Pchl(ide) *a* H (E650 F657) is the photoreceptor for its own photoconversion to Chlide *a* (Koski et al. 1951). In an albino corn mutant lacking carotenoids, the action spectrum exhibited two prominent peaks, one at 650 nm and one at 445 nm that corresponded to the absorption spectrum of LW t-Pchlide *a* H of the mutant.

### *9.2.2 Effect of Temperature on the Photoconversion*

The phototransformation of LW t-Pchlide *a* H to Chlide *a* was completely inhibited at −195 °C (Smith and Benitez 1954). Partial photoconversion took place at −70 °C. At temperatures beyond 50 °C, photoconversion was progressively inhibited due to apoprotein denaturation. Dependency of the photoconversion upon temperature indicates that the phototransformation is not a purely photochemical reaction but also involves a thermochemical component.

### *9.2.3 Quantum Yield of the Photoconversion*

The average quantum yield of the photoconversion at 642 nm amounts to about 0.6 (Smith and Benitez 1954). Therefore it is not clear from this work whether one or two quanta of light are required for the photoconversion. On the other hands, Thorne (1971) proposed a two quantum process for the photoconversion.

### 9.2.4 *Effect of Environment on the Photoconversion*

The rate of photoconversion expressed as a percentage of the photoconvertible protochlorophyllide was found to be independent of the initial concentration of the holochrome and was not influenced by the viscosity of the medium (Boardman 1962). This led to the proposal that the photoconversion did not involve a collision process between independent protein molecules or between a protein molecule and a hydrogen donor molecule. Instead, Boardman (1962) proposed a restricted collision process between the photo-activated Pchl molecule and the hydrogen donor. However since the rate of phototransformation was temperature-dependent, it seemed likely that the hydrogenation involved some vibrational or rotational movement of that part of the protein molecule in close proximity to the Pchlide *a* chromophore.

### 9.2.5 *Photoconversion Kinetics*

While Smith and Benitez (1954), opted for a bimolecular reaction with respect to Pchlide *a*, Thorne and Boardman (1972) suggested that by allowing for energy transfer within molecular groups, the true kinetics of the photoconversion was first order, which is still compatible with the restricted collision hypothesis (Boardman 1962).

## 9.3 The Multiple Light-Dependent Pchlide *a* Oxidoreductases (PORs)

It has been proposed that at least four different POR isozymes may be present in plants (Dehesh et al. 1986; Ikeuchi and Murakami 1982). In *Arabidopsis thaliana* and Barley, two different genes with about 75 % homology, PorA and PorB, have been shown to code for two different Pchlide *a* oxidoreductases, namely PORA and PORB (Armstrong et al. 1995; Holtorf et al. 1995). PORA is synthesized in the dark and constitutes the bulk of the crystalline prolamellar body of etioplasts. However the transcription of its gene is turned off in the light and the enzyme is rapidly degraded by a light-induced protease (Reinbothe et al. 1995, 1999). On the other hand, the PorB gene is transcribed in darkness and in the light, and the transcripts are translated continuously into the enzyme which is responsible for the bulk of Chl *a* biosynthesis and accumulation in daylight. More recently, a gene that encodes a third POR in *Arabidopsis thaliana* has been reported (Oosawa et al. 2000). The Protein has been named PORC. This enzyme is expressed only during the light phase of the photoperiod. Since PORA, B and C respond to light differently (see below), it has been suggested that the function of the three PORs of *Arabidopsis* are not redundant, but may allow the plant to adapt its needs for Chl biosynthesis according to the prevailing light regime (Su et al. 2001). In our opinion, adaptation of Chl biosynthesis to different light conditions, proceeds via multiple and different Chl biosynthetic routes.

### *9.3.1 NADPH-Protochlorophyllide a (Photo) Oxidoreductase A (PORA, or PCR)*

As pointed out above, most of the early investigations of the photoreduction of Pchlide *a* dealt with t-LW-Pchlide H (E650 F657). The notion that the Pchlide *a* H apoprotein of t-LW-Pchlide *a* H (E650 F657) (i.e. PORA) acts as a shuttling photoenzyme that catalyzes the conversion of Pchlide *a* to Chlide *a* was first proposed by Sironval et al. (1967). In this work the authors reported that t-LW-Pchlide *a* (E647 F 657), with a red excitation maximum at 647 nm and a red emission maximum at 657 nm, is photoconverted by light to Chlide *a* (E676 F690). The latter is converted in darkness to Chlide *a* (E682 F697). At this stage of the reaction, the authors suggested that the apoprotein discharges the newly formed Chlide *a* (E682 F697) and picks up another Pchlide *a* chromophore that may be photoconverted to Chlide *a* via a similar cycle. The spectral shifts described by Sironval et al. were confirmed by Gassman et al. (1968), and Bonner (1969). The concept of a shuttling Pchlide *a* reductase photoenzyme was also compatible with the reported conversion of nt-SW-Pchlide *a* (F633) to t-LW-Pchlide *a* (F650) during the photoreduction process (Gassman 1973).

A significant step in the understanding of Pchlide *a* photoreduction was achieved with the realization that NADPH is the hydrogen donor for the reaction (Griffiths 1974). This was followed by the proposal that the shuttling photoenzyme (POR, now called PORA), NADPH, and Pchlide *a* formed a photoactive ternary Pchlide *a*-NADPH-enzyme complex with a red absorption maximum at 652 nm (Apel et al. 1980). Equally important was the purification of PORA from etiolated barley (Apel 1981). The purified enzyme consisted of one polypeptide (Mr 36000) with two to three bound Pchlide *a* chromophores. It is synthesized in the cytoplasm as a precursor protein of about 44 kDa. The transit sequence of about 8 kDa is hydrolyzed when the enzyme is transported into the plastid (Apel 1981). The size of PORA reported by various authors depends on the plant species and varies from 33 to 38 kDa (Shulz and Senger 1993). More recently, pigment-free PORA was purified from barley etioplasts by solubilization with n-octyl-B-D-glucoside and chromatography on DEAE-cellulose (Klement et al. 1999). Using pigment and protein analysis it was shown that barley etioplasts contained a one-to-one PORA and Pchlide *a*. The enzyme was twice as active towards MV than toward DV Pchlide *a* (Klement et al. 1999).

It has also been demonstrated that during the greening of etiolated tissues a rapid decline of PORA is observed. For example, after 6 h of continuous illumination, when the rate of Chl *a* accumulation is at its peak, only traces of the PORA protein are detected (Santel and Apel 1981). The disappearance of PORA from etiolated tissues during greening was confirmed by Kay and Griffiths (1983). These observation and further experimentation have led to the proposal that in etiolated tissues, although PORA functions only for a short period of time after the onset illumination, it is required for normal greening (Runge et al. 1996).

### 9.3.2 Protochlorophyllide a Oxidoreductase B (PORB)

It has been proposed that PORB provides the means to sustain light-dependent Chl biosynthesis in fully greened mature plants, in the absence of PORA and t-LW-Pchlide a H (E450 F655) (Runge et al. 1996). In other words, it was suggested that in some t-SW-Pchlide *a* H species, the apoprotein consists of PORB.

The photoreduction of Pchlide *a* by purified PORB overexpressed heterologously in *E. coli* has recently been described (Lebedev and Timko 1999). The PORB reaction is described as consisting of two steps. In a first photochemical step, a single quantum mechanism leads to the formation of an unstable tetrapyrrole intermediate with a putative free electron. In a second step, the free radical intermediate is spontaneously converted to Chlide *a*. Both steps appear to proceed at subzero temperatures. At room temperature, the rate of the reaction depends linearly on enzyme and substrate concentrations, and the reduction kinetics are consistent with one mole of substrate bound per active PORB monomer.

### 9.3.3 Protochlorophyllide a Photooxidoreductase C (PORC)

Like PORA and PORB, PORC is light-and-NADPH-dependent. In contrast to the *PORA* and *PORB* mRNAs, the *PORC* mRNA accumulates only after the beginning of illumination (Oster et al. 2000). In light-adapted mature plants only *PORB* and *PORC* mRNAs were detectable, and the amounts of both mRNAs exhibited pronounced diurnal rhythmic fluctuations (Su et al. 2001). However, differences were observed between *PORB* and *PORC*. The differences can be summarized as follows: (a) While the oscillations of *PORB* mRNA are under the control of the circadian clock, that of *PORC* is not, (b) Upon transferring to darkness seedlings grown under continuous white light, the concentration of *PORC* mRNA rapidly declined and became undetectable, while *PORB* mRNA did not, (c) When seedlings were exposed to different light intensities, the amounts of PORB mRNA remained the same, while the mRNAs of *PORA* and *PORC* were modulated in an inverse way by light intensity changes.

### *9.3.4 Contribution of t-LW-Pchlide a (PORA) and t-SW-Pchlide a (PORB) to Photoperiodic Greening*

Under natural photoperiodic greening conditions, Pchlide *a* accumulates during the dark cycles of the photoperiod and contributes to Chl *a* biosynthesis and accumulation at the onset of light (i.e. at dawn) (Cohen et al. 1977). Furthermore, Pchlide *a* is always present in green tissues during the light phases of the photoperiod (Abd-El-Mageed et al. 1977; Carey et al. 1985; Cohen et al. 1977). During photoperiodic greening Pchlide *a* (E650 F655), also known as t-LW-Pchlide *a* H [(E450 F657)], and its apoprotein, PORA, are transient species that peak during the 7th dark cycle and become undetectable by the 11th dark cycle (Cohen and Rebeiz 1978). On the other hand SW Pchl(ide) *a* H species and their apoproteins, PORB and/or PORC are present throughout the photoperiodic greening process (Cohen and Rebeiz 1978). In other words although t-LW-Pchlide *a* H (E450 F657) contributes to prolamellar body reformation and Chl *a* formation during the first few dark cycles, it is SW t-Pchlides *a* Hs and PORB/PORC that prevail during the light cycles of photoperiodic greening, and contribute the bulk of Chl *a* accumulation under normal field conditions (Cohen and Rebeiz 1978).

The significance of the accumulation patterns of t-SW Pchl(ide) *a* Hs, and PORB/C during photoperiodic greening, to the Chl *a* biosynthetic process, rests upon the direct photoconvertibility of t-SW Pchlide *a* Hs to Chlide *a* without prior conversion to t-LW-Pchl(ide) *a* (E450 F657) and subsequent photoreduction by PORA. This was reported to be the case by Cohen and Rebeiz (1978). However, the photoconversion of t-SW-Pchlide *a* H, was slower than that of t-LW-Pchlide *a* H (E450 F657) which is catalyzed by PORA.

In conclusion, since etiolation and prolamellar body formation are not abnormal phenomena, but are part of the natural succession of the dark (night) and light (daylight) cycles during photoperiodic greening (Cohen and Rebeiz 1978; Rebeiz and Rebeiz 1986), it is very plausible for PORA and PORB/C, to play definite but distinct roles during Chl *a* biosynthesis in darkness and in the light (see below).

## 9.4 Heterogeneity of the Photoconversion of the Pchlide *a* Chromophore to Chlide *a*

### *9.4.1 Photoconversion of DV Pchlide a to DV Chlorophyllide a in DDV-LDV-LDDV Plants via Routes 1 and 8*

The biosynthesis of DV Chlide *a* from DV Pchlide *a* was first reported in isolated cucumber cotyledons induced to accumulate massive amounts of DV Pchlide *a* (Duggan and Rebeiz 1982a). In Fig. 9.2, the biosynthesis of DV Chlide *a* from DV Pchlide *a* in DDV-LDV-LDDV plants is depicted to occur in two different thylakoid

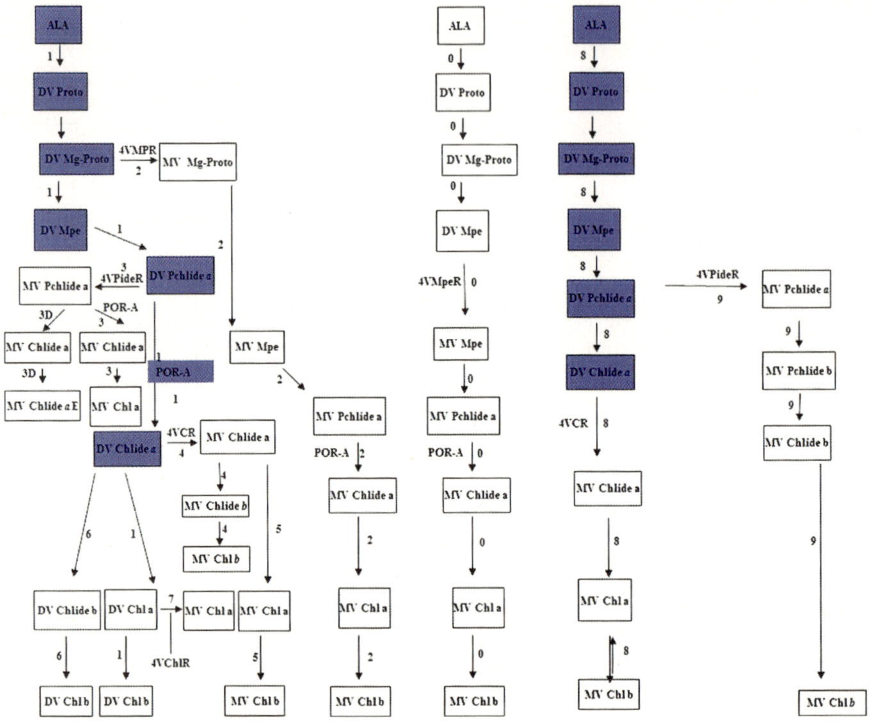


**Fig. 9.2** Photoconversion of DV Pchlide *a* to DV Chlide *a* in DDDV-LDV-LDDV plants via routes 1 and 8 (Adapted from Fig. 6.3 of Chap. 6)

environments via routes 1 and 8 as suggested by multiple resonance energy transfer from DV Pchlide a to various Chl *a*-Protein complexes (Table 6.1, Chap. 6) and (Kolossov et al. 2003), as well as further conversions to Chlide *a* and Chls as will be discussed later. It is unclear at this stage whether the spatial biosynthetic heterogeneity of DV Chlide *a* is accompanied by chemical biosynthetic heterogeneity or not. In other words, it is unclear whether the proposed biosynthesis of DV Chlide *a* from DV Pchlide *a* via routes 1, and 8 is catalyzed by identical POR A photoenzymes or by different POR A isozymes. These biosynthetic routes will be discussed further below.

#### 9.4.1.1 Photoconversion of DV Pchlide *a* to DV Chlorophyllide *a* in DDDV-LDV-LDDV Plants via Route 1

Photoconversion of DV Pchlide *a* to DV Chlide *a* via biosynthetic route 1 (Fig. 9.2) in etiolated DDV-LDV-LDDV plants was demonstrated in vivo and in organello by conversion of DV Pchlide *a* to DV Chlide *a* (Duggan and Rebeiz 1982a, b). Biosynthetic route 1 is also probably active in greening DDV-LDV-LDDV plants during the first few dark phases of the photoperiod when DV Pchlide a accumulation is substantial (Cohen et al. 1977; Cohen and Rebeiz 1978).

#### 9.4.1.2 Photoconversion of DV Pchlide *a* to DV Chlorophyllide *a* in DDDV-LDV-LDDV Plants via Route 8 During Photoperiodic Greening

In green DDV-LDV-LDDV plant species such as cucumber, DV Pchlide *a* is continuously present during the light cycles of the photoperiod and only trace amounts of MV Pchlide *a* are formed (Carey et al. 1985; Ioannides et al. 1994). Interruption of the light cycle by a brief dark period (LD) indicated that such plant species form most of their Chl via regenerated DV Pchlide *a* (Abd-El-Mageed et al. 1977). These observation suggested very strongly that during the light cycles of the photoperiod, green DDV-LDV-LDDV plant species formed most of their MV Chl via DV Pchlide *a*, and DV Chlide *a* as depicted in route 8.

### *9.4.2 Photoconversion of MV Pchlide a to MV Chlorophyllide a in DDV-LDV-LDDV Plants via Routes 2, 3 and 0*

In DDV-LDV-LDDV etiolated tissues, or in greening DDV-LDV-LDDV tissues during the first few dark-cycles of the photoperiod small amounts of MV Pchlide *a* are formed from MV Mpe via route 2 (Tripathy and Rebeiz 1986), or via route 3 by 4-vinyl reduction of DV Pchlide *a* during prolonged dark-incubations (Tripathy and Rebeiz 1988). It can also take place via route 0 as discussed below (Fig. 9.3).

#### 9.4.2.1 Photoconversion of MV Pchlide *a* to MV Chlorophyllide *a* in DDDV-LDV-LDDV Plants via Route 2

In DDV-LDV-LDDV etiolated tissues, or in greening DDV-LDV-LDDV tissues during the first few dark-cycles of the photoperiod small amounts of MV Pchlide *a* are formed from MV Mpe via route 2 (Tripathy and Rebeiz 1986). Such t-LW-MV Pchlides *a* are readily photoconvertible to MV Chlide *a* by PORA (Belanger and Rebeiz 1980). Biosynthesis of MV Chlide *a* via biosynthetic routes 2, also takes place in greening DDV-LDV-LDDV plants during the first few dark phases of the photoperiod when t-LW-Pchlide *a* accumulation is substantial (Cohen et al. 1977; Cohen and Rebeiz 1978).

#### 9.4.2.2 Photoconversion of MV Pchlide *a* to MV Chlorophyllide *a* in DDDV-LDV-LDDV Plants via Route 3

In route 3, MV Pchlide *a* is formed from DV Pchlide *a* by reduction of the vinyl side chain at position 4 to ethyl then photoreduction of the MV Pchlide *a* thus formed to MV Chlide *a*.

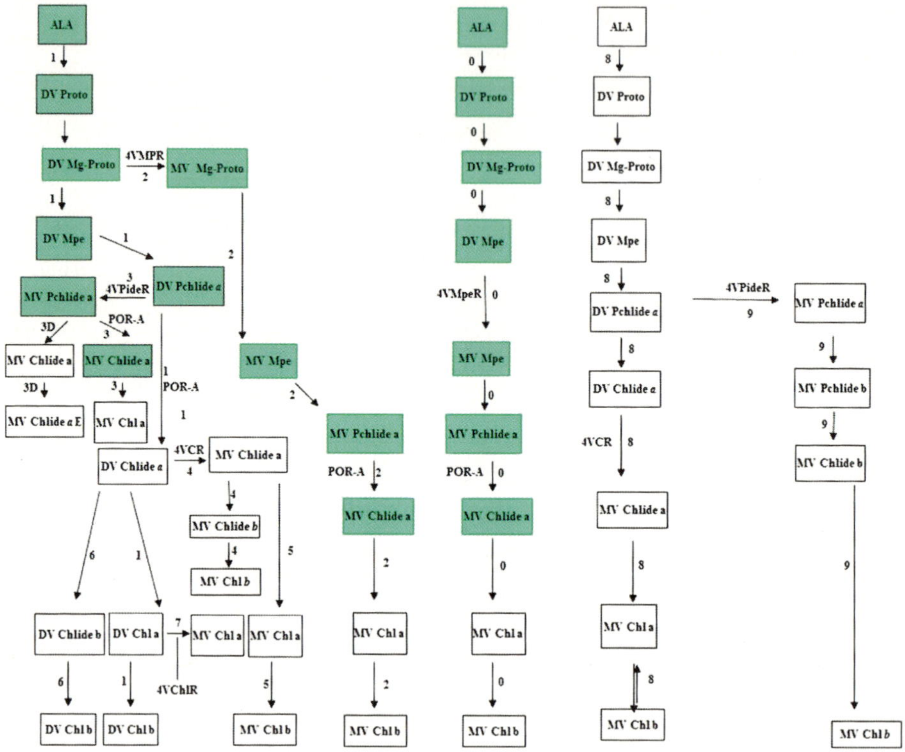


**Fig. 9.3** Photoconversion of MV Pchlide *a* to MV Chlide *a* in DDDV-LDV-LDDV plants via routes 3, 2 and 0 (Adapted from Fig. 6.3 of Chap. 6)

Biosynthesis of MV Chlide *a* via biosynthetic route 3 also takes place in greening DDV-LDV-LDDV plants during the first few dark phases of the photoperiod when t-LW-Pchlide *a* accumulation is substantial (Cohen et al. 1977; Cohen and Rebeiz 1978).

#### 9.4.2.3 Photoconversion of MV Pchlide *a* to MV Chlorophyllide *a* in DDDV-LDV-LDDV Plants via Route 0

In biosynthetic route 3, DV Mpe may be converted to MV Mpe by 4-vinyl Mpe Reductase (4VMpeR) if this reductase is found in DDDV-LDV-LDDV Plants as is the case in DMV-LDV-LDMV plants (Muir and Neuberger 1949).

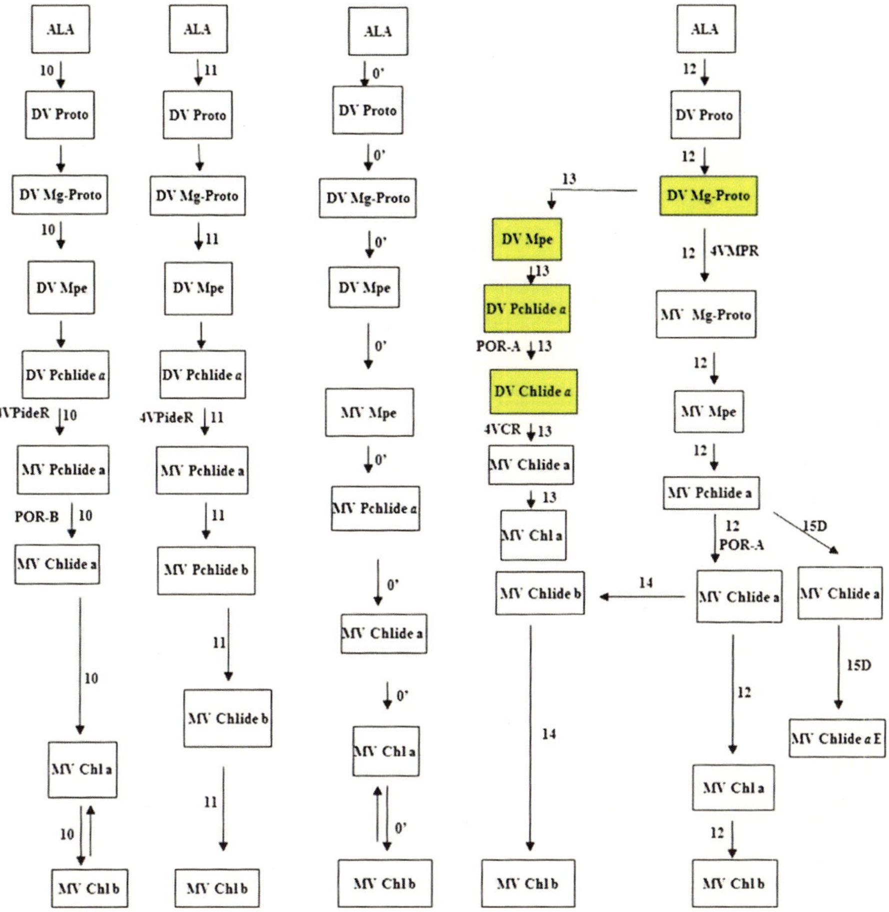


**Fig. 9.4** Photoconversion of DV Pchlide *a* to DV Chlide *a* in DMV-LDV-LDMV plants via route 13 (Adapted from Fig. 6.4 of Chap. 6)

### *9.4.3 Photoconversion of DV Pchlide a to DV Chlorophyllide a in DMV-LDV-LDMV Plants via Route 13, During the Light Phase of the Photoperiod*

DV Chlide *a* is formed via route 13 by photoconversion of DV Pchlide *a* in light-grown DMV-LDV-LDMV plant species during the light phases of the photoperiod. The reaction is probably catalyzed by POR-A. Biosynthetic route 13 is active in greening DDV-LDV-LDDV plants during the first few light phases of the photoperiod when Pchlide a accumulation is substantial (Cohen and Rebeiz 1978) (Fig. 9.4).

## 9.4.4 Photoconversion of MV Pchlide a to MV Chlorophyllide a in DMV-LDV-LDMV Plants via Routes 10, 0′ and 12

### 9.4.4.1 Photoconversion of MV Pchlide *a* to MV Chlorophyllide *a* via Route 10 in Greening DMDV-LDV-LDMV Plants During the Light Phase of the Photoperiod

During the light phases of the photoperiod, in DMV-LDV-LDMV plant species such as corn wheat and barley, MV Pchlide *a* can be formed from DV Pchlide *a* via route 10, which involves reduction of the vinyl group of DV Pchlide *a* at position 4 of the macrocycle to ethyl, a reaction catalyzed by 4VPideR (Tripathy and Rebeiz 1986). The nascent MV Pchlide *a* can then be rapidly photoconverted to MV Chlide *a* most probably by PORB which predominates in the light. The latter is active in green tissues, and is assigned to route 10 on the basis of the continuous detection of DV Pchlide *a* in the light, which in DMV-LDV-LDMV plant species is rapidly converted to MV Pchlide *a* and MV Chlide *a* (Abd-El-Mageed et al. 1977) (Fig. 9.5).

### 9.4.4.2 Photoconversion of MV Pchlide *a* to MV Chlorophyllide *a* in DMV-LDV-LDMV Plants via Route 0′

In this biosynthetic route, MV Chlide *a* is formed by photoconversion of MV Pchlide. The Operation of biosynthetic route 0′ in DMV-LDV-LDMV plants during photoperiodic greening is justified by the detection and solubilization of 4-Vinyl Mpe reductase (4VMpeR) in greening barley etiochloroplasts (Kolossov and Rebeiz 2010). Such etiochloroplasts can actively convert MV Mpe to MV Pchlide *a* (Tripathy and Rebeiz 1986).

### 9.4.4.3 Photoconversion of MV Pchlide *a* to MV Chlorophyllide *a* in DMV-LDV-LDMV Plants via Route 12 in Etiolated DMV-LDV-LDMV Plants in Darkness, and in Greening DMV-LDV-LDMV Plants During the Initial Dark Phases of the Photoperiod

Etiolated DMV-LDV-LDMV tissues accumulate massive amounts of MV Pchlide *a* in darkness and form most of their Chl via regenerated MV Pchlide *a* during the dark and light cycles of the photoperiod (Abd-El-Mageed et al. 1977; Carey and Rebeiz 1985; Tripathy and Rebeiz 1986). In etiolated barley for example, the resulting MV Pchlide *a* pool can be rapidly photoconverted to MV Chlide *a* (Belanger et al. 1982). In etiolated tissues and during the initial dark phases of the photoperiod when t-LW-Pchlide *a* accumulation is substantial (Cohen et al. 1977; Cohen and Rebeiz 1978), the photoreduction of MV Pchlide *a* is most probably catalyzed by PORA which predominates in etiolated tissues.

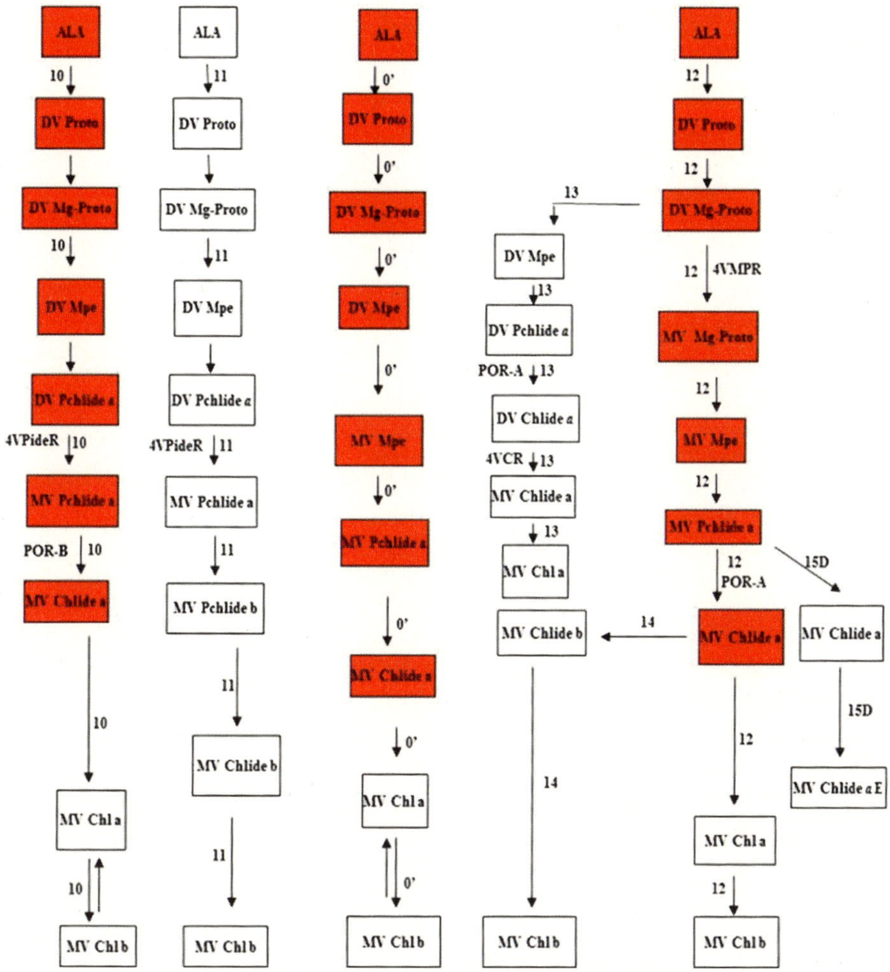


**Fig. 9.5** Photoconversion of MV Pchlide *a* to MV Chlide *a* in DMV-LDV-LDMV plants via routes 10, 0′ and 12 (Adapted from Fig. 6.4 of Chap. 6)

## 9.4.5 Formation of MV Chlide a by Vinyl Reduction of DV Chlide a via Routes 4 and 8 in DDV-LDV-LDDV Plants

In DDV-LDV-LDDV etiolated tissues, or in green DDV-LDV-LDDV tissues during the initial dark-cycles of the photoperiod, MV Chlide *a* is formed by reduction of the vinyl group of DV Chlide *a* at position 4 of the macrocycle to ethyl via routes 4 and 8 (Fig. 9.6). The reduction of the vinyl group of DV Chlide *a* to ethyl is catalyzed in darkness and in the light by a very potent enzyme, 4VCR (Duggan and Rebeiz 1982b; Parham and Rebeiz 1992). 4VCR is a fast acting

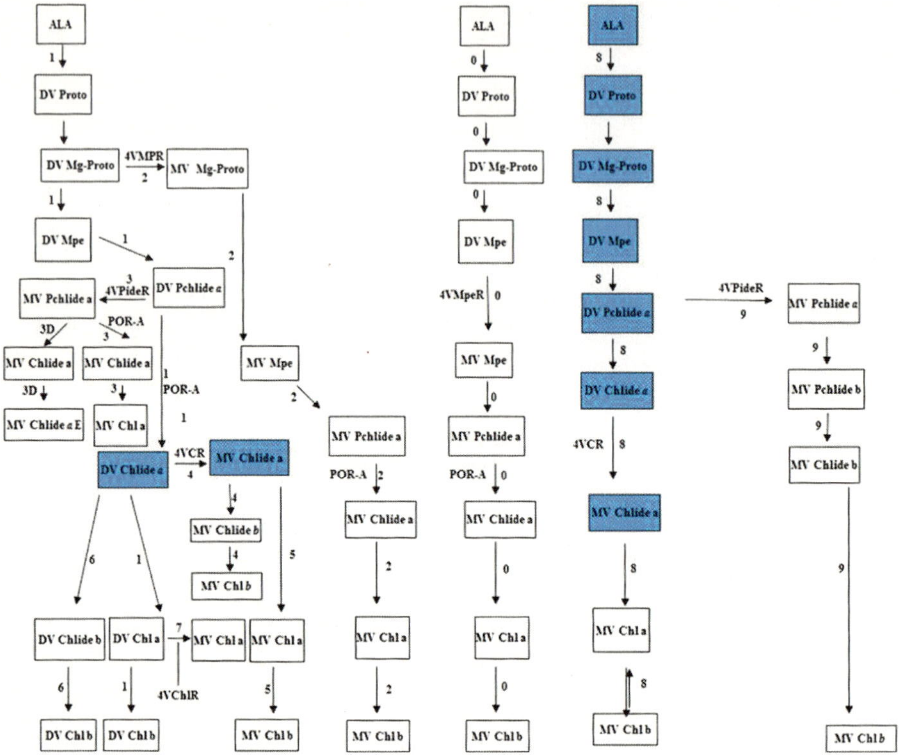

**Fig. 9.6** Formation of MV Chlide *a* by vinyl reduction of DV Chlide *a* via routes 4 and 8 in DDV-LDV-LDDV plants (Adapted from Fig. 6.3 of Chap. 6)

membrane-bound, NADPH-dependent enzyme (Parham and Rebeiz 1992; Parham and Rebeiz 1995). It is a stable enzyme that has been solubilized and purified about 20 fold (Kolossov and Rebeiz 2001).

Discussion of Biosynthetic routes 4 and 8 in DDV-LDV-LDDV plants is given below.

#### 9.4.5.1 Formation of MV Chlide *a* by Vinyl Reduction of DV Chlide *a* via Route 4 in Etiolated DDV-LDV-LDDV Plants in Darkness, and in Greening DDV-LDV-LDDV Plants During the Initial Dark Phases of the Photoperiod

The most spectacular manifestation of MV Chlide *a* formation is via route 4 is in DDV-LDV-LDDV etiolated tissues induced to accumulate DV Pchlide a exclusively, by successive dark–light treatments (Duggan and Rebeiz 1982a). The accumulated DV Pchlide *a* is photoconverted by a short light flash to DV Chlide *a* and the latter is very rapidly converted to MV Chlide *a* in vivo and in organello (Duggan and Rebeiz 1982b).

#### 9.4.5.2 Formation of MV Chlide *a* by Vinyl Reduction of DV Chlide *a* via Route 8 in Greening DDV-LDV-LDDV Plants During the Light Phases of the Photoperiod

Formation of MV Chlide *a* by rapid vinyl reduction of transient DV Chlide *a* via route 8, is justified by the detection of 4VCR activity in photoperiodically grown green tissues, which has been recently documented by Abd El-Mageed et al. (1977), and more recently by Kolossov and Rebeiz (2001).

### *9.4.6 Formation of MV Chlide a by Vinyl Reduction of DV Chlide a via Route 13 in Greening DMV-LDV-LDMV Plants During the Light Phases of the Photoperiod*

Formation of MV Chlide *a* by rapid vinyl reduction of transient DV Chlide *a* via route 13, has been documented in barley etioplasts after a light flash of 2.5 ms (Adra and Rebeiz 1998) (Fig. 9.7).

## 9.5 Photoreduction Intermediates and Spectral Shifts During Photoreduction of Protochlorophyll(ide) *a* H (E550 F655)

In etiolated tissues, the photoreduction of t-LW-Pchlide *a* H (E550 F655) by PORA is accompanied by complex spectral shifts (Fig. 9.8) of intermediates and end products that eventually result in the conversion of the crystalline prolamellar body and prothylakoids to thylakoid membranes. It appears that some of the spectral shifts may be regulated by protein phosphorylation (Wiktorsson et al. 1996). At the present stage, the significance of the postillumination spectral shifts to Chl biosynthetic heterogeneity is unclear (Fig. 9.8).

### *9.5.1 Spectral Shift I*

Spectral shift I is light-dependent. It was reported by Thorne (1971) in etiolated bean leaves. It occurs as a result of fractional photoconversion of LW t-Pchlide *a* H (E650 F655), to a dark-stable pigment-apoprotein complex (E668 F674), with 77 K red excitation and maxima at 668 and 674 nm, respectively. This intermediate yields a mixture of Pchlide *a* and Chlide *a* after dark-ethanol extraction. The photoconversion rate for Chlide *a* H (E668 F674) was twice the rate for the photoconversion of the next intermediate, thus suggesting that, in vivo, photoconversion of Pchlide *a* to Chlide *a* is a two step two photon process.

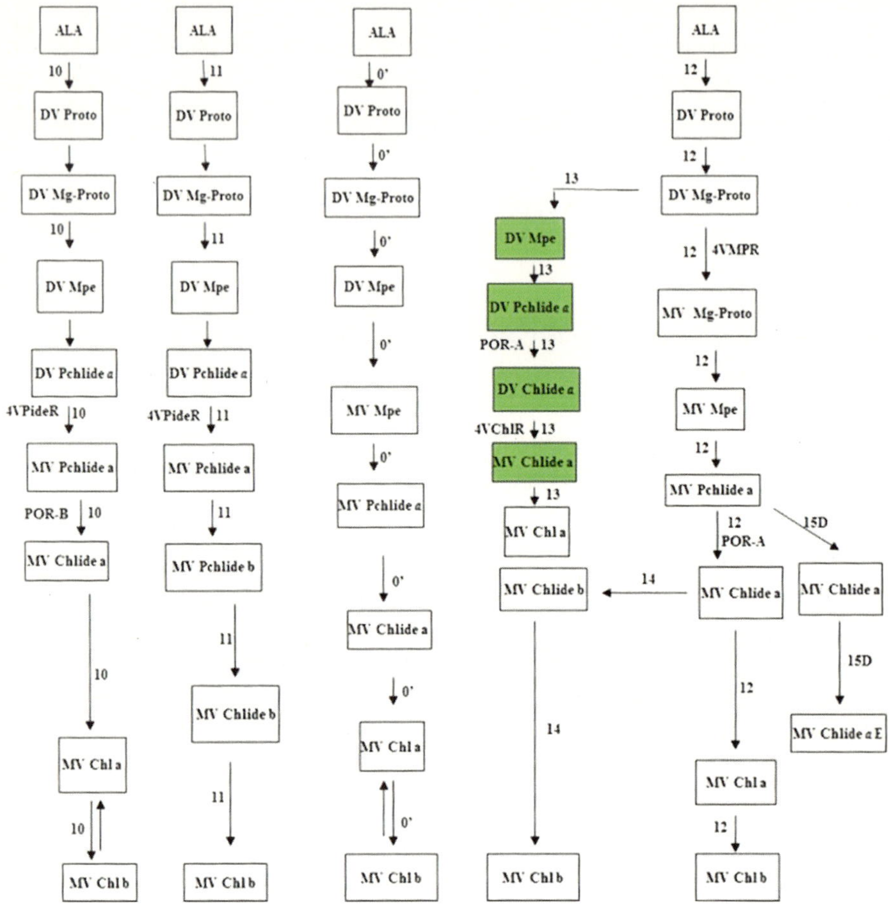


**Fig. 9.7** Formation of MV Chlide *a* by vinyl reduction of DV Chlide *a* via route 13 in DMV-LDV-LDM plants (Adapted from Fig. 6.4 of Chap. 6)

### 9.5.2 *Spectral Shift II*

Light-dependent spectral shift II was first described by Sironval et al. (1967) as a photoconversion of t-LW-Pchlide *a* (E650 F655) to a Chlide *a* (E676 F690)-apoprotein complex. Later on, that Pchlide *a*-protein complex was referred to as Chlide *a* (E676 F688) by Sironval and Kuyper (1972). Then in 1971, Thorne reported that the photoprecursor of Chlide *a* (E676 F690) was Chlide *a* (E668 F674) instead of t-LW-Pchlide *a* (E650 F655). The chemical nature of the chromophore in Chlide *a* (E668 F674) is not clear however. Sironval et al. (1967), and Sironval and Kuyper (1972) initially proposed that it was some kind of Pchlide *a*-Chlide *a* intermediate of an ill defined nature. Thorne (1971), proposed however, that the chromophore consisted exclusively of Chlide *a*. Shift II was also confirmed by Gassman et al. (1968), and by Bonner (1969).

**Fig. 9.8 Spectral Shifts of the Ternary Pchlide *a*-NADPH-Apoprotein Complex During the Photoconversion of Pchlide *a* to Chl(ide) *a*.** "E" and "F" refer to the red fluorescence excitation and emission maxima respectively at 77 K

### *9.5.3 Spectral Shift III*

Shift III, which converts Chlide *a* (E676 F690) to Chlide *a* (E682 F697) takes place very rapidly in darkness. It was considered by Sironval et al. (1967) to lead to the formation of a mature Chlide *a*-apoprotein complex that releases Chlide *a* from the Pchlide a oxidoreductase complex. This in turn allows PORA to pick up another Pchlide *a* chromophore to yield t-LW-Pchlide *a*-H (E650 F655).

### *9.5.4 Spectral Shift IV*

The conversion of Chlide *a* (E682 F697) to Chl(ide) *a* (E672 F680) species was the first spectral shift to be described during the conversion of Pchlide *a* to Chl(ide) *a*. It was reported by Shibata in 1957, as a spectral shift that took place in darkness or in the light in about 10–20 min after the onset of illumination, depending on the age of the etiolated tissue, and the plant species. During this shift Chlide *a* is esterified with geranylgeraniol, which is reduced stepwise to phytol (see section dealing with the reactions between Chlide *a* and Chl *a*).

### *9.5.5 Spectral Shift V*

The fifth shift was also described by Shibata (1957). It takes place either in the light or in darkness, and corresponds to the final integration of Chl *a* into various pigment proteins of the thylakoid membranes. On the basis of energy transfer from Pchlide *a* (E650 F655) to Chlide *a* (E682 F697) at fractional or partial photoconversions, Thorne (1971) concluded that diffusion of the chromophore from the apoprotein occurs at the level of Chl(ide) *a* (E674 F683) (shift V) instead of Chlide *a* (E682 F697) as proposed by Sironval et al. (1967). From the maximal appearance of the (E668 F6740) photointermediate and the Pchlide *a* and Chlide *a* content Thorne (1971) proposed that the Pchlide *a* aggregate comprised about 20 molecules of Pchlide *a*.

# References

Abd-El-Mageed HA, El Sahhar KF, Robertson KR, Parham R, Rebeiz CA (1977) Chloroplast biogenesis 77. Two novel monovinyl and divinyl light-dark greening groups of plants and their relationship to the chlorophyll a biosynthetic heterogeneity of green plants. Photochem Photobiol 66:89–96

Adamson H, Packer N (1984) Dark-synthesis of chlorophyll in vivo and dark reduction of protochlorophyll ide in vitro by pea chloroplasts. In: Sironval C, Brouers M (eds) Protochlorophyllide reduction and greening. Martinus Nijhoff, Boston, pp 353–363

Adamson HY, Hiller HJ, Walmsley J (1997) Protochlorophyllide reduction and greening in angiosperms: an evolutionary perspective. J Photochem Photophys 41:201–221

Adra AN, Rebeiz CA (1998) Chloroplast biogenesis 81. Transient formation of divinyl chlorophyll *a* following a 2.5 ms light flash treatment of etiolated cucumber cotyledons. Photochem Photobiol 68:852–856

Apel K (1981) The protochlorophyllide holochrome of barley (Hordeum vulgare L.). Phytochrome-induced decrease of the translatable mRNA coding for the NADPH:protochlorophyllide:oxidoreductase. Eur J Biochem 120:89–93

Apel K, Santel HJ, Redlinger TE, Falk H (1980) The protochlorophyll ide holochrome of barley (Hordeum vulgare L.). Isolation and characterization of the NADPH:protochlorophyll ide oxidoreductase. Eur J Biochem 111:251–258

Armstrong GA (1998) Greening in the dark: light-independent chlorophyll biosynthesis from anoxygenic photosynthetic bacteria to gymnosperms. J Photochem Photobiol B Biol 43:87–100

Armstrong GA, Runge S, Frick G, Sperlink U, Apel K (1995) Identification of NADPH: protochlorophyll ide oxidoreductases A and B: a branched pathway for light-dependent chlorophyll biosynthesis in *Arabidopsis thaliana*. Plant Physiol 108:1505–1517

Belanger FC, Rebeiz CA (1980) Chloroplast biogenesis 30. Chlorophyll(ide) (E459 F675) and Chlorophyll(ide) (E449 F675) the first detectable products of divinyl and monovinyl protochlorophyll photoreduction. Plant Sci Lett 18:343–350

Belanger FC, Dugan JX, Rebeiz CA (1982) Chloroplast biogenesis: identification of chlorophyllide *a* (E458F674) as a divinyl chlorophyllide *a*. J Biol Chem 257:4849–4858

Boardman NK (1962) Biochim Biophys Acta 64:279–288

Bonner B (1969) A short-lived intermediate form in the in vivo conversion of protochlorphyll ide 650 to chlorophyllide 684. Plant Physiol 44:739–747

Carey EE, Rebeiz CA (1985) Chloroplast biogenesis 49. Difference among angiosperms in the biosynthesis and accumulation of monovinyl and divinyl protochlorophyllide during photoperiodic greening. Plant Physiol 79:1–6

Carey EE, Tripathy BC, Rebeiz CA (1985) Chloroplast biogenesis 51. Modulation of monovinyl and divinyl protochlorophyllide biosynthesis by light and darkness in vitro. Plant Physiol 79:1059–1063

Cohen CE, Rebeiz CA (1978) Chloroplast biogenesis XXII. Contribution of short wavelength and long wavelength protochlorophyll species to the greening of higher plants. Plant Physiol 61:824–829

Cohen CE, Bazzaz MB, Fullet SE, Rebeiz CA (1977) Chloroplast biogenesis XX. Accumulation of porphyrin and phorbin pigments in cucumber cotyledons during photoperiodic greening. Plant Physiol 60:743–746

Dehesh KM, Hauser M, Apel K (1986) Light-induced changes in the distribution of the 36,000 Mr polypeptide of NADPH:protochlorophyllide oxidoreductase within different cellular compartments of barley (*Hordeum vulgare*, L.). I. Localization by immunoblotting in isolated plastids and total extracts. Planta 169:162–171

Duggan JX, Rebeiz CA (1982a) Chloroplast biogenesis 37. Induction of chlorophyllide a (E459 F675) accumulation in higher plants. Plant Sci Lett 24:27–37

Duggan JX, Rebeiz CA (1982b) Chloroplast biogenesis 42. Conversion of DV chlorophyllide a to monovinyl chlorophyllide a in vivo and in vitro. Plant Sci Lett 27:137–145

Gassman M (1973) The conversion of photoinactive protochlorophyllide $_{633}$ to phototransformable protochlorophyll ide $_{650}$ in etiolated bean leaves treated with delta-aminolevulinic acid. Plant Physiol 52:590–594

Gassman M, Granick S, Mauzerall D (1968) A rapid spectral shift change in etiolated red kidney leaves following phototransformation of protochlorophyllide. Biochem Biophys Res Commun 32:295–300

Griffiths WT (1974) Source of reducing equivalents for the in vitro synthesis of chlorophyll from protochlorophyll. FEBS Lett 46:301–304

Holtorf R, Reinbothe S, Reinbothe C, Bereza B, Apel K (1995) Two routes of chlorophyllide synthesis that are differentially regulated by light in barley (Hordeum vulgare L.). Proc Natl Acad Sci U S A 92:3254–3258

Ikeuchi M, Murakami S (1982) Measurement and identification of NADPH:protochlorophyll ide oxidoreductase solubilized with Triton-X-100 from etioplast membranes of squash cotyledons. Plant Cell Physiol 23:1089–1099

Ioannides I, Fasoula DA, Robertson KR, Rebeiz CA (1994) An evolutionary study of chlorophyll biosynthetic heterogeneity in green plants. Biochem Syst Ecol 22:211–220

Kay SA, Griffiths WT (1983) Light-induced breakdown of NADPH-protochlorophyllide oxidoreductase in vitro. Plant Physiol 72:229–236

Kirk JTO, Tilney-Basset RAE (1967) The plastids: their chemistry, structure, growth, and inheritance. Freeman, London, pp 504–506

Klement H, Helfrich M, Oster U, Schoch S, Rudiger W (1999) Pigment-free NADPH:protochlorophyllide oxidoreductase from *Avena sativa* L: purification and substrate specificity. Eur J Biochem 265(3):862–874

Kolossov VL, Rebeiz CA (2001) Chloroplast biogenesis 84. Solubilization and partial purification of membrane-bound [4-vinyl] chlorophyllide *a* reductase from etiolated barley leaves. Anal Biochem 295:214–219

Kolossov VL, Rebeiz CA (2010) Evidence for various 4-vinyl reductase activities in higher plants. In: Rebeiz CA, Benning C, Bohnert HJ et al (eds) The chloroplast: basics and applications. Springer, Dordrecht/London, pp 25–38

Kolossov VL, Kopetz KJ, Rebeiz CA (2003) Chloroplast biogenesis 87: evidence of resonance excitation energy transfer between tetrapyrrole intermediates of the chlorophyll biosynthetic pathway and chlorophyll *a*. Photochem Photobiol 78:184–196

Koski VM, French CS, Smith JHC (1951) The action spectrum for the transformation of protochlorophyll to chlorophyll in normal and albino corn seedlings. Arch Biochem Biophys 31:1–17

Lebedev N, Timko MP (1999) Protochlorophyllide oxidoreductase B-catalyzed protochlorophyllide photoreduction in vitro. Proc Natl Acad Sci U S A 96(17):9954–9959

Muir HM, Neuberger M (1949) The biogenesis of porphyrins. The distribution of 15N in the ring system. Biochem J 45:163

Oosawa N, Masuda T, Awai K, Fusada N, Shimada H, Ohta H, Takmiya K (2000) Identification and light-induced expression of a novel gene of NADPH-protochlorophyllide oxidoreductase isoform in *Arabidopsis thaliana*. FEBS Lett 474:133–136

Oster U, Tanaka R, Rudiger W (2000) Cloning and functional expression of the gene encoding the key enzyme for chlorophyll *b* biosynthesis (CAO) from *Arabidopsis thaliana*. Plant J 21:305–310

Parham R, Rebeiz CA (1992) Chloroplast biogenesis: [4-vinyl] chlorophyllide *a* reductase is a divinyl chlorophyllide *a*-specific NADPH-dependent enzyme. Biochemistry 31:8460–8464

Parham R, Rebeiz CA (1995) Chloroplast biogenesis 72: a [4-vinyl] chlorophyllide *a* reductase assay using divinyl chlorophyllide *a* as an ecogenous substrate. Anal Biochem 231:164–169

Rebeiz CC, Rebeiz CA (1986) Chloroplast biogenesis 53: ultrastructural study of chloroplast development during photoperiodic greening. In: Akoyunoglou G, Senger H (eds) Regulation of chloroplast differentiation. Alan Liss, New York, pp 389–396

Reinbothe S, Reinbothe C, Holtorf H, Apel K (1995) Two NADPH:protochlorophyllide oxidoreductases in barley: evidence for the selective disappearance of POR A during the light-induced greening of etiolated seedlings. Plant Cell 7:1933–1940

Reinbothe C, Lebedev N, Reinbothe S (1999) A protochlorophyllide light-harvesting complex involved in de-etiolation of higher plants. Nature 397:80–84

Rudiger W, Schoch S (1991) The last steps of chlorophyll biosynthesis. In: Scheer H (ed) Chlorophylls. Academic Press, New York, pp 451–464

Runge S, Ulrich S, Frick J, Apel K, Armstrong GA (1996) Distinct roles for light-dependent NADP:ptotochlorophyllide oxidorectase (POR) A and B during greening in higher plants. Plant J 9:513–523

Santel HJ, Apel K (1981) The protochlorophyll ide Holochrome of Barley (Hordeum vulgare L.). The effect of light on the NADPH:protochlorophyllide oxidoreductase. Eur J Biochem 120:95–103

Schoefs B, Franck F (1998) Chlorophyll synthesis in dark-grown pine primary needles. Plant Physiol 118:1159–1168

Shibata K (1957) Spectroscopic studies on chlorophyll formation in intact leaves. J Biochem 44:147–172

Shulz R, Senger H (1993) Protochlorophyllide reductase: a key enzyme in the greening process. In: Sundqvist C, Ryberg M (eds) Pigment-protein complexes in plastids: synthesis and assembly. Academic Press, New York, pp 179–218

Sironval C, Kuyper Y (1972) The reduction of protochlorophyllide into chlorophyllide. Photosynthetica 6:254–275

Sironval C, Kuyper Y, Michel JM, Brouers M (1967) The primary photoact in the conversion of protochlorphyll ide into chlorophyllide. Stud Biophys 5:43–50

Smith JHC, Benitez A (1954) The effect of temperature on the conversion of protochlorophyll to chlorophyll a in etiolated barley leaves. Plant Physiol 29:135–143

Su Q, Frick G, Armstrong G et al (2001) POR C of *Arabidopsis thaliana:* a third light-and NADPH-dependent protochlorophyllide oxidoreductase that is differently regulated by light. Plant Mol Biol 47:805–813

Suzuki JY, Bollivar DW, Bauer CE (1997) Genetic analysis of chlorophyll biosynthesis. Ann Rev Genet 31:61–89

Thorne SW (1971) The greening of etiolated bean leaves I. The initial photoconversion process. Biochim Biophys Acta 226:113–127

Thorne SW, Boardman NK (1972) Biochim Biophys Acta 267:104–110

Tripathy BC, Rebeiz CA (1986) Chloroplast biogenesis. Demonstration of the monovinyl and divinyl monocarboxylic routes of chlorophyll biosynthesis in higher plants. J Biol Chem 261:13556–13564

Tripathy BC, Rebeiz CA (1987) Non-equivalence of glutamic acid and delta-aminolevulinic acid as substrates for protochlorophyllide and chlorophyll biosynthesis in darkness. In: Biggins J (ed) Progress in photosynthesis research, vol IV. Martinus Nijhoff, Amsterdam, pp 439–443

Tripathy BC, Rebeiz CA (1988) Chloroplast biogenesis 60. Conversion of divinyl protochlorophyllide to monovinyl protochlorophyllide in green(ing) barley, a dark monovinyl/light divinyl plant species. Plant Physiol 87:89–94

Wiktorsson B, Ryberg M, Sundqvist C (1996) Aggregation of NADPH-protochlorophyllide oxidoreductase-pigment complexes is favoured by protein phosphorylation. Plant Physiol Biochem 34:23–34

Yuichi F, Bauer CE (2000) Reconstitution of light-independent protochlorophyllide reductase from purified *BchL* and *BchB* subunits. In vitro confirmation of nitogenase-like features of a bacteriochlorophyll biosynthesis enzyme. J Biol Chem 275:23583–23588

# Chapter 10
# The Chl *a* Carboxylic Biosynthetic Routes: Conversion of Chlide *a* to Chl *a*

*The first step in accomplishing anything is the belief that it can be done.*

Most of the chlorophyll *a* (Chl *a*) in higher and lower plants is formed by esterification of chlorophyllide *a* (Chlide *a*) (Fig. 10.1). A minor Chl *a* fraction esterified with long chain fatty acids (LCFA) other than phytol is also formed from MV protochlorophyllide *a* E (Pchlide *a* E) as described in Chap. 11. In this section emphasis will be placed on the biosynthetic heterogeneity of Chl *a* formed by esterification of Chlide *a* with phytol.

## 10.1 Chlorophyll *a* Biosynthetic Heterogeneity

The biosynthetic heterogeneity of the Chl *a* of green plants is extremely complex. In addition to the DV and MV chemical heterogeneity of the Chl *a* chromophore, and the chemical and spatial biosynthetic heterogeneity of its immediate precursor, Chlide *a* (see Chap. 9), another layer of biosynthetic heterogeneity is imposed by the esterification process. Indeed, although in green plants, most of the Chl *a* is esterified with phytol ($C_{20}H_{39}OH$), conversion of Chlide *a* to Chlide *a*-phytol appears to follow different routes in etiolated and green tissues (*vide infra*).

C.A. Rebeiz, *Chlorophyll Biosynthesis and Technological Applications*, DOI 10.1007/978-94-007-7134-5_10, 

Fig. 10.1 The Chl *a* pool

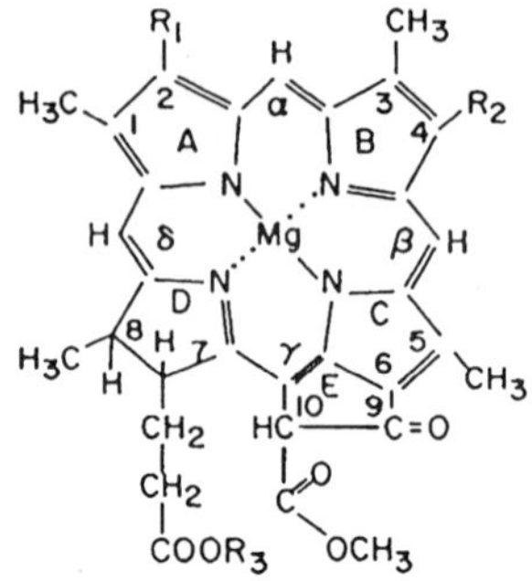


**Chlorophyll *a* (Chl *a*)**

| | R1 | R2 | R3 | Compound |
|---|---|---|---|---|
| I | Vinyl | Vinyl | Phytol | DV Chl *a* |
| II | Vinyl | Ethyl | Phytol | MV Chl *a* |
| III | Vinyl | Vinyl | LCFA | Other DV Chl *a* |
| IV | Vinyl | Ethyl | LCFA | Other MV Chl *a* |

### *10.1.1 Chlorophyll a Formation by Esterification of Chlorophyllide a with Geranylgeraniol in Etiolated Tissues*

In etiolated tissues subjected to a light treatment, formation of Chl *a* by esterification of Chlide *a* involves a complex set of reactions. Initially, it was observed that treatment of etiolated bean leaves with 1 min of light followed by dark incubation resulted in the transient appearance of putative Chlide *a*-geranylgeraniol (GG) which was followed by the formation of Chl *a*-phytol (Ogawa 1975). Subsequently etiolated wheat seedlings treated with herbicides then exposed to light followed by darkness, resulted in the accumulation of Chl *a*-GG and Chl *a*-dihydroxyGG (DHGG) (Rudiger et al. 1976). This was followed by the demonstration of Chlide *a* esterification with GG in a cell-free system from maize shoots (Rudiger et al. 1976).

Further work dealing with the identification of various esterified Chlide *a* in etiolated tissues subjected to a brief light treatment followed by dark incubation, led to the proposal that during phytylation, Chlide *a* is first esterified with GG to yield Chl *a*-GG, which is reduced stepwise to Chl *a*-DHGG, to -Chl a-tetrahydroGG (THGG) and finally to Chl a-hexahydroGG, i.e. Chl a-phytol (Schoch 1978).

The above hypothesis was confirmed in cell-free systems from various etiolated plant tissues. It was demonstrated that in irradiated etioplast-membrane fractions prepared form oat seedlings, [1-$^{3}$H]-GG and its monophosphate were incorporated into Chl *a* only in the presence of exogenous ATP, whereas incorporation of activated [1-$^{3}$H]-GG pyrophosphate (GG-PP) did not require ATP (Rudiger et al. 1980). In order to distinguish this enzymatic activity from chlorophyllase it was named Chl synthetase. Conversion of Chl-GG in vitro to Chl *a*-phytol by hydrogenation required the addition of exogenous NADPH. NADH was not a cofactor (Benz et al. 1980). Enzymic hydrogenation of Chl-GG to Chl a-phytol was inhibited by anaerobiosis (Schoch et al. 1980). Substrate specificity investigations indicated that Chl synthetase requires a chlorin derivative that contains Mg as a central metal ion. A hydrogenated ring D was mandatory since Pchlide *a* with a double bond at position 7–8 of the macrocycle was not a substrate (Benz and Rudiger 1981; Helfrich and Rudiger 1992). However, direct esterification of endogenous Chlide *a* with exogenous phytol in the presence of added ATP, and Mg was also observed in etiolated tissues which led to the proposal that the conversion of Chlide *a* to Chl *a* may follow different biosynthetic routes having different substrate and cofactor requirements, depending on the stage of plastid development (Daniell and Rebeiz 1984). Subsequently it was determined that in oat etioplasts, the relative substrate specificities for GG-PP, Phytol-PP and farnesyl-PP amounted to 6, 3, and 1 respectively (Rudiger 1993).

Chlorophyll synthetase is present mainly in the prothylakoid and prolamellar body of etioplasts (Rudiger 1993). Prolamellar body disaggregation and Chlide *a* esterification appear to be closely related phenomena. It appears that Chlide *a* formed in the prolamellar body can migrate with Pchlide-oxidoreductase to the prothylakoid membranes during light-dependent dissociation of prolamellar bodies (Rudiger 1993).

### *10.1.2 Preferential Chlorophyll a Formation by Esterification of Chlorophyllide a with Phytol in Green Tissues*

Although illumination of etiolated tissues with white light leads to a slow decrease in Chl synthetase activity (Rudiger 1993), the synthetase activity does not disappear completely, and some activity is still observed in mature chloroplasts (Soll and Schultz 1981). In spinach chloroplasts, the relative substrate specificity for Chlide *a* esterification with exogenous GGPP and PhyPP were 1 and 4 respectively *a* (Soll and Schultz 1981).

In *Arabidopsis thaliana* a nuclear encoded gene, *G4*, was identified which exhibited homology to the product of the *Rhodobacter capsulatus bchG* locus which is involved in the esterification of bacteriochlorophyllide with GG (Gaubier et al. 1995). The relationship between gene *G4* and *bchG* was confirmed by isolation and sequencing of a corresponding full length cDNA. The gene appears to consist of 14 exons, some of which were very short. Southern and Northern analyses showed that *G4* is a single copy gene and its transcripts were only detected in green or greening tissues.

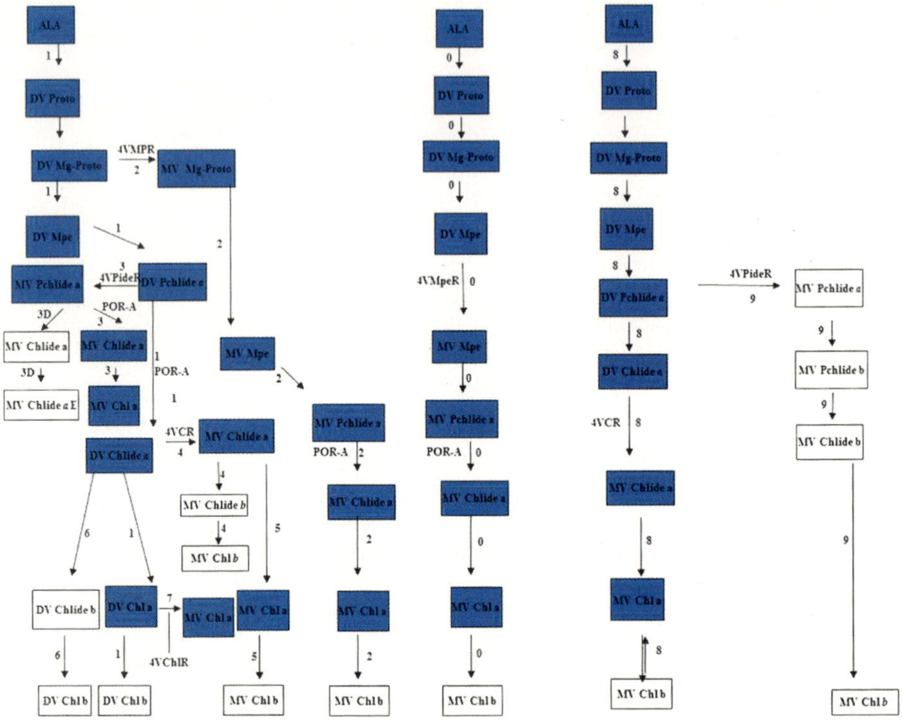


Fig. 10.2 MV Chl *a* biosynthesis via routes 2, 3, 7, 0, and 8 in DDV-LDV-LDDV plant species (Adapted from Fig. 6.3 of Chap. 6)

### 10.1.3 *Biosynthetic Heterogeneity of MV Chlorophyll a in DDV-LDV-LDDV Plants via Routes 2, 3, 5, 7, 0, and 8*

In DDV-LDV-LDDV plant species, MV Chl *a* can be formed via five different biosynthetic routes which are discussed below (Fig. 10.2).

#### 10.1.3.1 Biosynthesis of MV Chlorophyll *a* in DDV-LDV-LDDV Plants via a Combination of routes 1 and 2 in Greening DDV-LDV-LDDV Plants During the Initial Phases of the Photoperiod

In DDV-LDV-LDDV plants growing under photoperiodic conditions, at the onset of illumination, during the first few dark cycles of the photoperiod when t-LW-Pchlide a accumulation is noticeable (Cohen et al. 1977; Cohen and Rebeiz 1978), upon exposure to light, MV Pchlide *a* is photoconverted to MV Chlide *a* by PORA. Most probably, conversion of the nascent MV Chlide *a* to MV Chl *a* via route 2 takes place via Chl *a-GG* followed by stepwise hydrogenation to Chl *a*-Phytol as described later.

#### 10.1.3.2 Biosynthesis of MV Chlorophyll *a* in DDV-LDV-LDDV Plants via a Combination of Routes 1 and 3 in Etiolated DDV-LDV-LDDV Plants After Exposure to Light

During prolonged dark-incubation of DDV-LDV-LDDV plants, small amounts of MV Pchlide *a* are formed via route 3. Upon exposure to light, MV Pchlide *a* is photoconverted to MV Chlide *a* by PORA. Most probably, conversion of the nascent MV Chlide *a* to MV Chl a via route 3 takes place via Chl a-GG followed by stepwise hydrogenation to Chl *a*-Phytol. Since during the initial phases of greening of etiolated tissues, most of the Chl consists of MV Chl *a*, located in PSI and PSII (Akoyunoglou et al. 1981; Alberte et al. 1972), we propose that the MV Chl *a* formed via route 3 is destined to PSI and/or PSII Chl-protein complexes.

#### 10.1.3.3 Biosynthesis of MV Chlorophyll *a* in DDV-LDV-LDDV Plants via a Combination of Routes 1, 4, and 5 in Etiolated Plants After Exposure to Light, and in Greening Plants During the Initial Dark Phases of the Photoperiod

Etiolated DDV-LDV-LDDV plants accumulate mostly DV Pchlide *a*. Upon exposure to light, DV Pchlide *a* is converted to DV Chlide *a* by PORA via route 1. Then DV Chlide *a* is converted to MV Chlide *a* via route 4 by 4-vinyl reduction catalyzed by 4VCR, and finally MV Chlide a is converted to MV Chl *a* via route 5. Since in etiolated tissues, MV Chlide *a* conversion to MV Chl *a* takes place via esterification with GG (Schoch 1978), it is our guess that, conversion of the nascent MV Chlide *a* to MV Chl *a* via route 5 takes place via Chl *a*-GG followed by stepwise hydrogenation to Chl *a*-Phytol as described in Sect. 1.1. This route is also most probably functional in DDV-LDV-LDDV plants growing under photoperiodic conditions, at the onset of illumination, during the first few dark cycles of the photoperiod when t-LW-Pchlide a accumulation is substantial (Cohen et al. 1977; Cohen and Rebeiz 1978). Since during the initial phases of greening of etiolated tissues, most of the Chl consists of MV Chl *a*, located in PSI and PSII (Akoyunoglou et al. 1981; Alberte et al. 1972), we propose that the MV Chl *a* formed via route 3 is destined to PSI and/or PSII Chl-protein complexes.

#### 10.1.3.4 Biosynthesis of MV Chlorophyll *a* in DDV-LDV-LDDV Plants via Route a Combination of Routes 1 and 7 in Etiolated Plants After Exposure to Light, and in Greening Plants During the Initial Dark Phases of the Photoperiod

In etiolated DDV-LDV-LDDV plants exposed to light, DV Pchlide *a* is converted to DV Chlide *a* by PORA via route 1. Then DV Chlide *a* is rapidly converted to MV

Chlide *a* by 4VCR. We have observed that within 10 s the nascent DV Chlide *a* is rapidly esterified to DV Chl *a* and during the following 30 s a decrease in DV Chl *a* is accompanied by a stoichiometric increase in MV Chl *a* as depicted in route 7 (Adra and Rebeiz 1998). Recently Wang et al. have described an enzyme that converts DV Chl *a* to MV Chl *a* in rice (Wang et al. 2010). The size of the MV Chl *a* pool formed from transient DV Chl *a* is rather small, and the nature of the LCFA at position 7 of the macrocycle is unknown. This route may also be functional in DDV-LDV-LDDV plants growing under photoperiodic conditions, at the onset of illumination, during the first few dark cycles of the photoperiod when t-LW-Pchlide *a* accumulation is substantial (Cohen et al. 1977; Cohen and Rebeiz 1978).

It is presently acknowledged that during the initial phases of greening of etiolated tissues, most of the Chl consists of MV Chl *a*, located in PSI and PSII (Akoyunoglou et al. 1981; Alberte et al. 1972). As a consequence we propose that the MV Chl *a* formed via route 7 is destined to PSI and/or PSII Chl-protein complexes.

#### 10.1.3.5 Biosynthesis of MV Chlorophyll *a* in DDV-LDV-LDDV Plants via Route 0

This route is called for by the conversion of DV Mpe to MV Pchlide *a* as described for cucumber etiochloroplasts in Sect. IA1c of Chap. 7. The nascent MV Pchlide *a* can be converted to MV Chlide *a* probably by POR A and the Chlide *a* to MV Chl *a* either via stepwise Hydrogenation of GG as described in Sect. 1.1 or via direct esterification with phytol (Daniell and Rebeiz 1984).

#### 10.1.3.6 Biosynthesis of MV Chlorophyll *a* in DDV-LDV-LDDV Plants via Route 8 in Greening DDV-LDV-LDDV Plants During the Light Phases of the Photoperiod

In photoperiodically-grown DDV-LDV-LDDV plants, during the light phase of the photoperiod, MV Chlide *a* is formed from regenerated DV Pchlide *a* via DV Chlide *a*, by a reaction catalyzed by PORB as discussed in Sect. IIIB in Chap. 9. Based on the prevalence of direct phytylation of MV Chlide *a* in green tissues (Soll and Schultz 1981), it is our guess that in route 8, most of the conversion of nascent MV Chlide *a* to MV Chl *a* proceeds by direct esterification of MV Chlide *a* with phytol. Since in nature, most of the Chl *a* is formed in the light, the size of the MV Chl *a* pool formed via route 8 is most probably very substantial.

Furthermore, since under continuous illumination, most of the synthesized Chl consists of MV Chl *a*, and *b* located in antenna Chl-protein complexes, (Akoyunoglou et al. 1981; Alberte et al. 1972), we propose that the MV Chl a formed via route 8 is destined to LHCII and other antenna Chl-protein complexes.

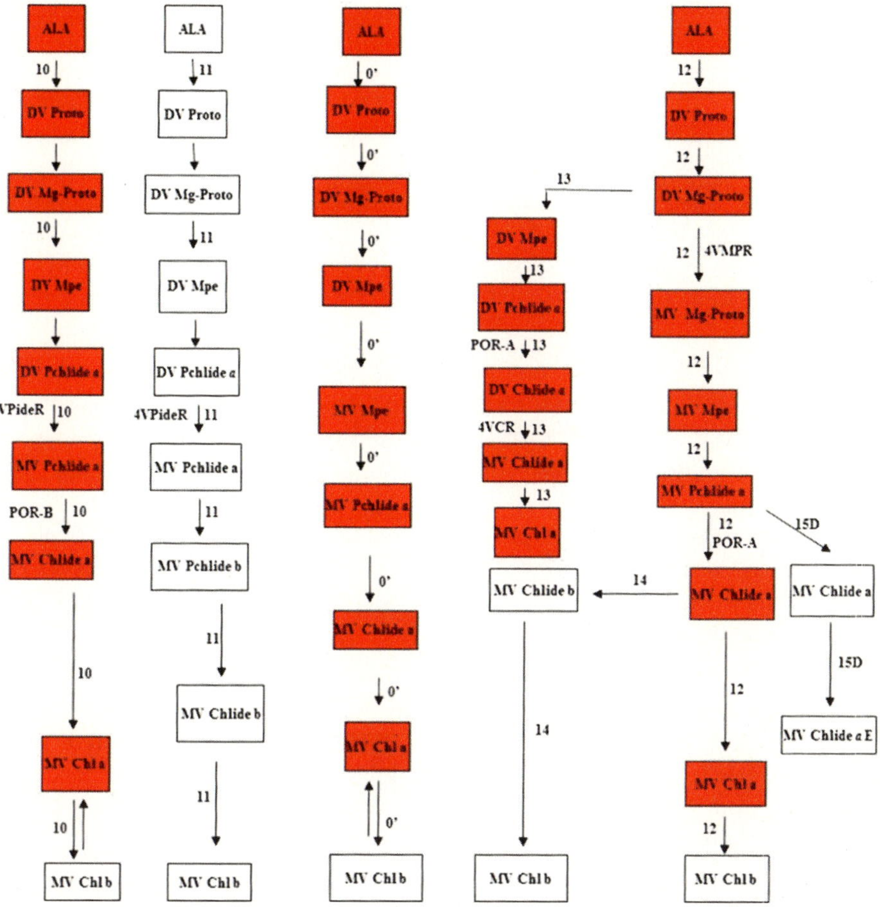


**Fig. 10.3** MV Chl *a* biosynthesis via routes 10, 0′, 13 and 12 in DMV-LDV-LDMV plant species (Adapted from Fig. 6.4 of Chap. 6)

## 10.1.4 Biosynthetic Heterogeneity of MV Chlorophyll *a* in DMV-LDV-LDMV Plants

In DMV-LDV-LDMV plant species, MV Chl *a* can be formed via four different biosynthetic routes which are discussed below (Fig. 10.3).

### 10.1.4.1 Biosynthesis of MV Chlorophyll *a* via Route 10 in Greening DMV-LDV-LDMV Plants During the Light Phases of the Photoperiod

In photoperiodically-grown DMV-LDV-LDMV plants, during the light phase of the photoperiod, MV Chlide *a* is formed from regenerated MV Pchlide *a*, a reaction catalyzed by PORB as discussed in Sect. IIIB of Chap. 9. Based on the prevalence of direct phytylation of MV Chlide *a* in green(ing) tissues (Soll and Schultz 1981), it is our guess that most of the conversion of nascent MV Chlide *a* to MV Chl *a* in route 10, proceeds by direct esterification of MV Chlide *a* with phytol. Since most of the Chl *a* is formed in the light, the size of the MV Chl *a* pool formed via route 7 is most probably very substantial.

Moreover, since under continuous illumination, most of the synthesized Chl consists of MV Chl *a*, and *b* located in antenna Chl-protein complexes, (Akoyunoglou et al. 1981; Alberte et al. 1972), we propose that the MV Chl *a* formed via route 10 is destined to LHCII and other antenna Chl-protein complexes.

#### 10.1.4.2 Biosynthesis of MV Chlorophyll *a* via Route 0′ in Greening DMV-LDV-LDMV Plants During the Light Phases of the Photoperiod

In biosynthetic route 10, DV Mpe is converted to MV Mpe by 4VMpeR (Kolossov and Rebeiz 2001). Then MV Mpe is converted to MV Chlide *a* probably via POR-B and the latter to Chl *a* probably by direct phytylation (Daniell and Rebeiz 1984; Soll and Schultz 1981).

#### 10.1.4.3 Biosynthesis of MV Chlorophyll *a* via Route 12 in Etiolated DMV-LDV-LDMV Plants After Exposure to Light, and in Greening DMV-LDV-LDMV Plants During the Initial Dark Phases of the Photoperiod

Etiolated DMV-LDV-LDMV plants accumulate mostly MV Pchlide *a*. Upon exposure to light, MV Pchlide *a* is converted to MV Chlide *a* by PORA as described in section IIIB of Chap. 9. Since in etiolated tissues, MV Chlide *a* conversion to MV Chl *a* takes place via esterification with GG (Schoch 1978), it is our guess that, conversion of the nascent MV Chlide *a* to MV Chl *a* via route 12 takes place via Chl *a*-GG followed by stepwise hydrogenation to Chl *a*-Phytol as described in Sect. 1.1. This route is also most probably functional in DMV-LDV-LDMV plants growing under photoperiodic conditions, at the onset of illumination, during the first few dark cycles of the photoperiod when t-LW-Pchlide a accumulation is substantial (Cohen et al. 1977; Cohen and Rebeiz 1978).

Since during the initial phases of greening of etiolated tissues, most of the Chl consists of MV Chl *a*, located in PSI and PSII (Akoyunoglou et al. 1981; Alberte et al. 1972), we propose that the MV Chl *a* formed via route 12 is destined to PSI and/or PSII Chl-protein complexes.

#### 10.1.4.4 Biosynthesis of MV Chlorophyll *a* via Route 13 in Etiolated DMV-LDV-LDMV Plants After Exposure to Light, and in Greening DMV-LDV-LDMV Plants During the Initial Dark Phases of the Photoperiod

Etiolated DMV-LDV-LDMV plants accumulate mostly MV Pchlide *a*. In addition in some plant species such as corn, they accumulate smaller yet significant amounts of DV Pchlide *a*. Upon exposure to light, the small amounts of DV Pchlide *a* are converted to DV Chlide *a* by POR-A as described in Sect. IIIA of Chap. 9. Thus In etiolated DMV-LDV-LDMV plants subjected to illumination small amounts of MV Chlide *a* are formed from DV Chlide a by 4-vinyl reduction as discussed for DDV-LDV-LDDV plants in section IV A of Chap. 9. Since in etiolated tissues, MV Chlide *a* conversion to MV Chl *a* takes place via esterification with GG (Schoch 1978), it is our guess that, conversion of the nascent MV Chlide *a* to MV Chl *a* via route 13 takes place via Chl *a*-GG followed by stepwise hydrogenation to Chl *a*-Phytol as described in Sect. 1.1. This route is also most probably functional in DDV-LDV-LDDV plants growing under photoperiodic conditions, at the onset of illumination, during the first few dark cycles of the photoperiod when t-LW-Pchlide a accumulation is substantial (Cohen et al. 1977; Cohen and Rebeiz 1978).

Since during the initial phases of greening of etiolated tissues, most of the Chl consists of MV Chl *a*, located in PSI and PSII (Akoyunoglou et al. 1981; Alberte et al. 1972), we propose that the MV Chl *a* formed via route 13 is destined to PSI and/or PSII Chl-protein complexes.

### *10.1.5 Biosynthetic Heterogeneity of DV Chlorophyll a*

In normal higher plants DV Chl *a* is a transient intermediate during MV Chl *a* formation. Under certain circumstances however, DV Chl *a* is the main Chl *a* that accumulates and participates in photosynthesis. This heterogeneity is discussed below.

#### 10.1.5.1 Transient Formation of DV Chlorophyll *a* via Route 1 in Etiolated DDV-LDV-LDDV Plants After Exposure to Light, and in Greening DDV-LDV-LDDV Plants During the Initial Dark Phases of the Photoperiod

In etiolated DDV-LDV-LDDV tissues, it has been repeatedly observed that when the mixed MV-DV Pchlide *a* is photoconverted into a mixed MV-DV Chlide *a* pool by a 2.5 ms light pulse, some of the nascent DV Chlide *a* is rapidly converted to DV Chl *a* during the first 30 s of dark incubation (Rebeiz et al. 1983). More recently we have observed that within 10 s the nascent DV Chlide *a* is rapidly esterified to DV Chl *a*. In the ensuing 30 s a decrease in DV Chl a is accompanied by a stochiometric rise in MV Chl *a* (Adra and Rebeiz 1998). Recently this reaction has also been reported in rice (Wang et al. 2010). The nature of the long chain fatty acid (LCFA) of the nascent DV Chl *a* at position 7 of the macrocycle is unknown. This route is also most probably functional in DDV-LDV-LDDV plants growing under photoperiodic conditions, at the onset of illumination, during the first few dark cycles of the photoperiod when t-LW-Pchlide a accumulation is substantial (Cohen et al. 1977; Cohen and Rebeiz 1978).

#### 10.1.5.2 Biosynthesis of DV Chlorophyll *a* via Route 1 in the Nec 7 Corn Mutant and in Picoplankton of the Euphotic Zone of the World Tropical and Temperate Oceans, and the Mediterranean Sea

The major fate of DV Chlide *a* resides in its conversion to MV Chlide *a* and MV Chl *a* (vide supra). However under certain circumstances, DV Chlide *a* is massively converted to DV Chl *a* by esterification. For example in the Nec 7 corn mutant (Bazzaz 1981), the major fate of DV Chlide *a* is its conversion to DV Chl *a* (Rebeiz et al. 1983, 2003). So is the case in the prochlorophyte picoplankton of the subtropical waters of the North Atlantic as well as in the picoplankton of the euphotic zone of the world tropical and temperate oceans, and the Mediterranean sea, where DV Chl *a* and *b* are the predominant Chl species (Chisholm et al. 1988, 1992; Goerike and Repeta 1992; Veldhuis and Kraay 1990). The nature of the LCFA at position 7 of the macrocycle and the details of esterification are unknown.

## References

Adra AN, Rebeiz CA (1998) Chloroplast biogenesis 81. Transient formation of divinyl chlorophyll *a* following a 2.5 ms light flash treatment of etiolated cucumber cotyledons. Photochem Photobiol 68:852–856

Akoyunoglou G, Tsakiris S, Argyroudi-Akoyunoglou JH (1981) Independent growth of the photosystem I and II units. The role of the light-harvesting pigment-protein complexes. In: Akoyunoglou G (ed) Photosynthesis V. Chloroplast development. Balaban International Science Services, Philadelphia, pp 523–533

Alberte RS, Thornber JP, Naylor AW (1972) Time of appearance of photosystem I and II in chloroplast of greening jack bean leaves. J Exp Bot 23(77):1060–1069

Bazzaz MB (1981) New chlorophyll chromophores isolated from a chlorophyll deficient mutant of maize. Photobiochem Photobiophys 2:199–207

Benz J, Rudiger W (1981) Chlorophyll biosynthesis: various chlorophyllides as exogenous substrates for chlorophyll synthetase. Z Naturforsch 36c:51–57

Benz J, Wolf C, Rudiger W (1980) Chlorophyll biosynthesis: hydrogenation of geranylgeraniol. Plant Sci Lett 19:225–230

Chisholm S, Olson RJ, Zettler ER et al (1988) A novel free-living prochlorophyte abundant in the oceanic euphotic zone. Nature 334:340–343

Chisholm SW, Frankel S, Goerike R et al (1992) *Prochlorococcus marinus* nov. gen. sp.: an oxyphototrophic marine prokaryote containing divinyl chlorophyll *a* and *b*. Arch Microbiol 157:297–300

Cohen CE, Rebeiz CA (1978) Chloroplast biogenesis 22. Contribution of short wavelength and long wavelength protochlorophyll species to the greening of higher plants. Plant Physiol 61:824–829

Cohen CE, Bazzaz MB, Fullet SE et al (1977) Chloroplast biogenesis XX. Accumulation of porphyrin and phorbin pigments in cucumber cotyledons during photoperiodic greening. Plant Physiol 60:743–746

Daniell H, Rebeiz CA (1984) Bioengineering of photosynthetic membranes: requirement of magnesium for the conversion of chlorophyllide a to chlorophyll a during the greening of etiochloroplasts in vitro. Biotech Bioeng 26:481–487

Gaubier P, Wu HJ, Laudie MD et al (1995) A chlorophyll synthetase gene from *Arabidopsis thaliana*. Mol Gen Genet 249:58–64

Goerike R, Repeta D (1992) The pigments of *Prochlorococcus marinus*. The presence of divinyl-chlorophyll a and b in a marine prochlorophyte. Limnol Oceanogr 37:425–433

Helfrich M, Rudiger W (1992) Various metallopheophorbides as substrates for chlorophyll synthetase. Z Naturforsch 47c:231–238

Kolossov VL, Rebeiz CA (2001) Chloroplast biogenesis 84. Solubilization and partial purification of membrane-bound [4-vinyl] chlorophyllide *a* reductase from etiolated barley leaves. Anal Biochem 295:214–219

Ogawa T (1975) An intermediate in the phytylation of chlorophyllide *a* in vivo. Plant Cell Physiol 16:199–202

Rebeiz CA, Wu SM, Kuhadje M et al (1983) Chlorophyll *a* biosynthetic routes and chlorophyll *a* chemical heterogeneity. Mol Cell Biochem 58:97–125

Rebeiz CA, Kolossov VL, Briskin D et al (2003) Chloroplast biogenesis: chlorophyll biosynthetic heterogeneity, multiple biosynthetic routes and biological spin-offs. In: Nalwa HS (ed) Handbook of photochemistry and photobiology. American Scientific Publishers, Los Angeles, pp 183–248

Rudiger W (1993) Esterification of chlorphyllide and its implication for thylakoids development. In: Sundqvist C, Ryberg M (eds) Pigment-protein complexes in plastids: synthesis and assembly. Academic, New York, pp 219–240

Rudiger W, Benz J, Lempert U et al (1976) Inhibition of phytol accumulation with herbicides: geranylgraniol and dihydrogranylgeraniol-containing chlorophyll from wheat seedlings. Z Pflanzenphysiol 80:131–143

Rudiger W, Benz J, Guthoff C (1980) Detection and characterization of activity of chlorophyll synthetase in etioplast membranes. Eur J Biochem 109:193–200

Schoch S (1978) The esterification of chlorophyllide *a* in greening bean leaves. Z Naturforsch 33c:712–714

Schoch S, Hehlein C, Rudiger W (1980) Influence of anaerobiosis on chlorophyll biosynthesis in greening oat seedlings (Avena sativa L.). Plant Physiol 66:576–579

Soll J, Schultz G (1981) Phytol synthesis from geranylgeraniol in spinach chloroplasts. Biochem Biophys Res Commun 99:907–912

Veldhuis MJW, Kraay GW (1990) Vertical distribution of pigment composition of a picoplankton prochlorophyte in the subtropical north Atlantic: a combined study of pigments and flow cytometry. Mar Ecol Prog Ser 68:121–127

Wang P, Gao J, Wan C et al (2010) Divinyl chlorophyll(ide) *a* can be converted to monovinyl chlorophyll(ide) a by a divinyl reductase in rice. Plant Physiol 153:994–1003

# Chapter 11
# The Fully Esterified Chlorophyll *a* Biosynthetic Routes: Reactions Between Mg-Protoporphyrin IX Diester and Chl *a*

*It takes excellence to recognize excellence, while mediocrity breeds mediocrity.*

## 11.1 The Mg-Proto Diester Pool

The Mg-proto diester (Mpde) pool (Fig. 11.1) consists of the first metabolic intermediates of the fully esterified Chl *a* biosynthetic route (Fig. 11.2).

The fully esterified Chl *a* pathway is populated by tetrapyrroles with a methyl propionate residue at position 6 of the macrocycle and a propionic acid residue at position 7 which is esterified with one of several different long chain fatty alcohols (LCFAs) (Rebeiz et al. 2003) (Fig. 11.1). The two sets of reactions depicted in Fig. 11.2, deal with the least understood phases of the intermediary metabolism of Chl *a*. In our opinion, the unjustified neglect of this facet of Chl *a* biosynthesis is caused by several factors, among which (a) occurrence of the metabolic intermediates, often in very small amounts, (b) slow reaction rates, (c) analytical difficulties, and (d) early misconceptions that ruled out the role of Pchlide ester as a metabolic intermediate in Chl biosynthesis (Rebeiz et al. 2003).

### *11.1.1 Heterogeneity of the Mg-Proto Diester Pool*

A fully esterified Mpde pool was first detected in etiolated cucumber cotyledons incubated overnight with ALA and Dpy in darkness (McCarthy et al. 1981). The novel pool exhibited the chromatographic properties of a fully esterified metalloporphyrin and the spectrofluorometric properties of Mg-Proto. Chemical derivatization coupled to spectrofluorometric and chromatographic analysis identified it as

C.A. Rebeiz, *Chlorophyll Biosynthesis and Technological Applications*, 
DOI 10.1007/978-94-007-7134-5_11, © Springer Science+Business Media Dordrecht 2014

**Fig. 11.1** The Mg-Proto diester ester (Mpde) pool

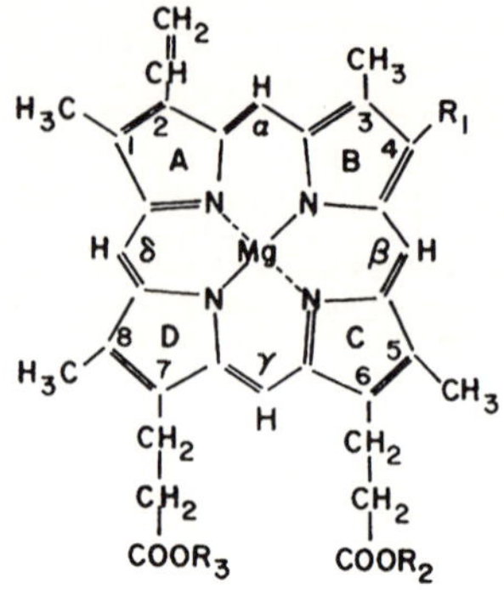


**Mg-Protoporphyrin IX diester (Mpde)**

| | R1 | R2 | R3 | Compound |
|---|---|---|---|---|
| I | Vinyl | Methyl Propionate | Long chain fatty alcohol | DV Mpde |
| II | Ethyl | Methyl Propionate | Long chain fatty alcohol | MV Mpde |

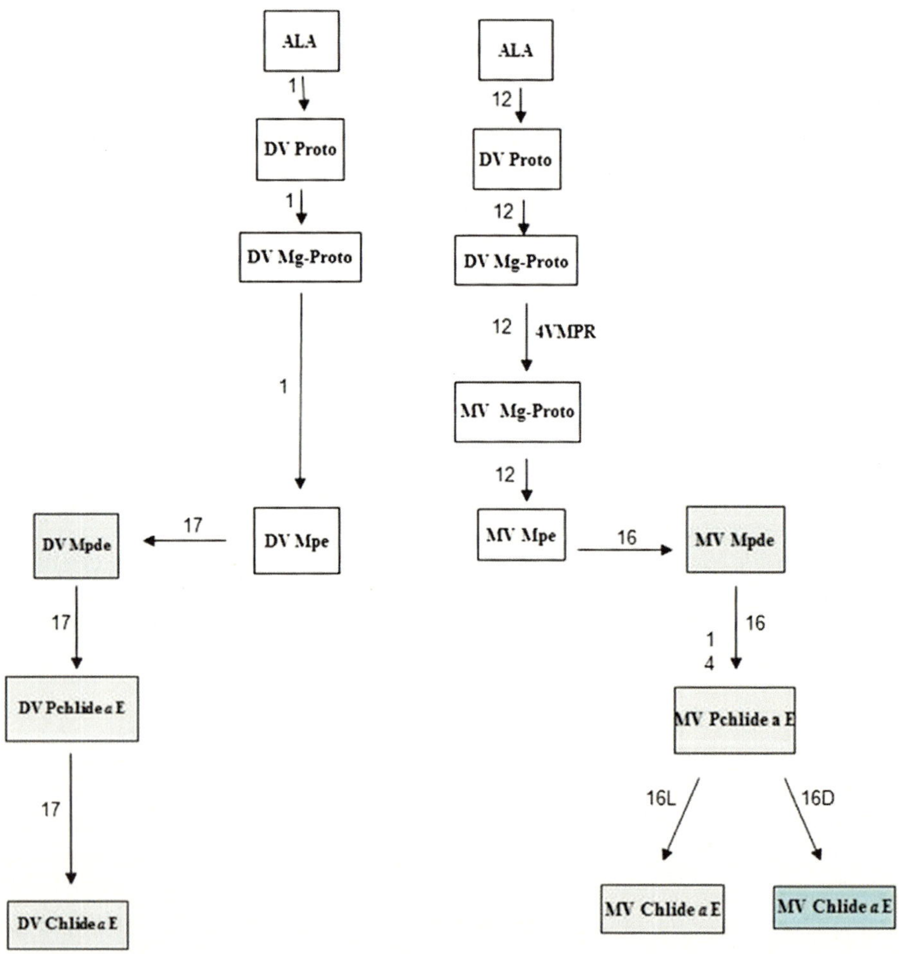


**Fig. 11.2** The fully esterified Chl *a* biosynthetic routes

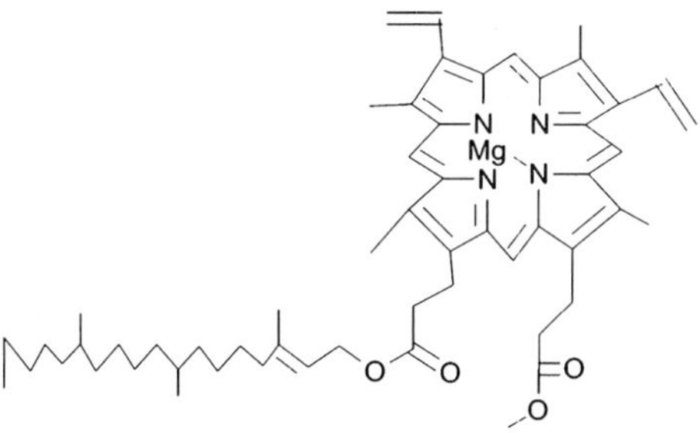


**Fig. 11.3** DV Mg-Proto diester (DV Mpde)

Mg-Proto diester (Mpde). Mpde was also detected in dark-grown Euglena gracilis and in etiolated cucumber cotyledons incubated in darkness with ALA, in the presence and absence of added Dpy. Upon detection of Mpde, it was suggested to be a metabolic precursor of the fully esterified, heterogeneous, Pchlide *a* ester pool (Belanger and Rebeiz 1982; McCarthy et al. 1981).

The Mpde pool exhibits a well pronounced DV-MV chemical heterogeneity (Belanger and Rebeiz 1982). For example in etiolated cucumber cotyledons incubated in darkness with ALA and Dpy, as well as in dark-grown Euglena gracilis, the Mpde pool consisted of DV and MV Mpde components. In general the proportion of DV Mpde was higher than that of MV Mpde, except in *Euglena*.

High-pressure liquid chromatographic analysis has also indicated that the Mpde pool was heterogeneous at position 7 of the macrocycle and consisted of three fully esterified Mg-Protos. Gas-chromatographic /mass spectroscopic analysis of the saponified alcohol fraction of the heterogeneous Mpde pool revealed that the latter consisted of three major long-chain alcohols, none of which was identifiable with known isoprenoids such as, farnesol or phytol (McCarthy et al. 1981). In addition to the heterogeneity of the long chain alcohols esterifying the propionic acid residue at position 7 of the macrocycle, the Mpde pool exhibited a well pronounced DV-MV chemical heterogeneity (see above) (Belanger and Rebeiz 1982).

#### 11.1.1.1 The Divinyl Mpde Pool

Divinyl Mg-Proto diester (Mpde) has two vinyl groups at positions 2 and 4 of the tetrapyrrole macrocycle (Fig. 11.3). The DV nature of the DV component of the Mpde pool of higher plants was determined by chemical derivatization coupled to analytical fluorescence spectroscopy at 77 K (Belanger and Rebeiz 1982).

The biosynthetic origin of DV Mpde is not presently clear, and is tentatively assigned to DV Mpe esterification (Fig. 11.2, route 17). After its detection in cucumber cotyledons and in *Euglena* cultures (Belanger and Rebeiz 1982), DV Mpde, has been proposed as a precursor of fully esterified DV Pchlide *a*, i.e. DV Pchlide *a* ester (DV Pchlide *a* E) (Fig. 11.2, Route 17). A precursor-product relationship between DV Mpde and DV Pchlide *a* E remains to be established and is complicated by the presence of Mpde esterases that convert exogenous Mpde to Mpe (Rebeiz, unpublished).

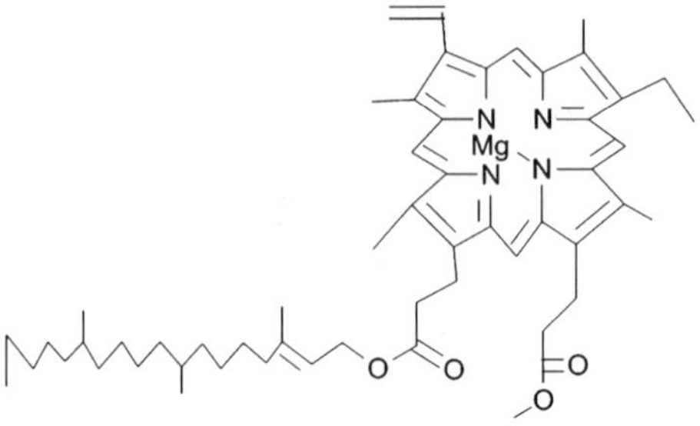


**Fig. 11.4** The MV Mg-Proto diester pool (MV MpeE)

#### 11.1.1.2 The MV Mpde Pool

Monovinyl Mg-Proto diester (MV Mpde) has one vinyl groups at positions 2 and one ethyl group at position 4 of the tetrapyrrole macrocycle (Fig. 11.4). The MV nature of the Mpde pool of higher plants was determined by chemical derivatization coupled to analytical fluorescence spectroscopy at 77 K (Belanger and Rebeiz 1982) (Fig. 11.4).

The biosynthetic origin of MV Mpde is not presently clear, and is tentatively assigned to MV Mpe esterification (Fig. 11.2, route 16). After its detection in cucumber cotyledons and in *Euglena* cultures where it is the main constituent of the Mpde pool (Belanger and Rebeiz 1982), MV Mpde has been proposed as a precursor of MV Pchlide *a* E (Fig. 11.2, route 16). A precursor-product relationship between MV Mpde and DV Pchlide *a* E remains to be established and is complicated by the presence of Mpde esterases that convert exogenous Mpde to Mpe (Rebeiz, unpublished).

### *11.1.2 Pchlide a Ester*

Pchlide *a* E is one of the least understood pools of the Chl biosynthetic pathway and its history is steeped in controversy (Fig. 11.5). Fischer and Oestreicher (1940) synthesized the phytyl ester of Pchlide *a* and showed that it differed from MV Chl *a* by having two fewer hydrogens at position 7 and 8 of the macrocycle. They named this molecule protochlorophyll. Because of the structural similarity between Pchlide *a* phytyl ester and Chl *a*, the erroneous notion evolved that Pchlide *a* phytyl ester was the major immediate photoprecursor of Chl *a* (Smith 1948). When Granick isolated and identified Pchlide *a* from an X-ray *Chlorella* mutant inhibited in its capability to form Chl, he considered it to be the immediate precursor of Pchlide *a* ester. The biological function of Pchlide *a* as the immediate precursor of chlorophyllide (Chlide) *a* was not fully understood till 7 years later (Wolff and Price 1957).

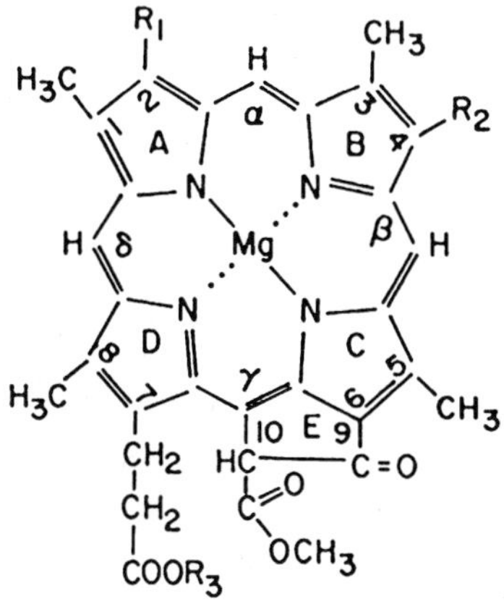


Protochlorophyllide *a* Ester (Pchlide *a* E)

**Fig. 11.5** The Pchlide *a* E pool

| | R1 | R2 | R3 | Compound |
|---|---|---|---|---|
| I | Vinyl | Vinyl | LCFAH | DV Pchlide *a* E |
| II | Vinyl | Ethyl | LCFAH | MV Pchlide *a* E |

#### 11.1.2.1 Biosynthesis of Pchlide *a* Ester (Pchlide *a* E)

Because of the structural similarity between Pchlide *a* and Pchlide *a* E it was convenient to propose that Pchlide *a* was the immediate precursor of Pchlide *a* E (Granick 1950). However, as early as 1970, precursor-product relationship studies in vivo, between the biosynthesis of $^{14}$C-Pchlide *a* and $^{14}$C-Pchlide *a* E failed to establish a precursor-product relationship between these two tetrapyrroles. Instead, the results indicated that Pchlide *a* and Pchlide *a* E were most probably formed in parallel from a common precursor (Rebeiz et al. 1970). These studies were confirmed by in vitro investigations which also failed to establish precursor product relationships between Pchlide *a* and Pchlide *a* E (Ellsworth and Nowak 1973; Mattheis and Rebeiz 1977). Later on, more rigorous precursor-product relationship studies between Pchlide *a* and Pchlide *a* E were carried out (McCarthy et al. 1982). Comparison of the ratio of $^{14}$C-ALA and various $^{14}$C-tetrapyrrole substrates incorporation into $^{14}$ C-Pchlide *a* and $^{14}$C-Pchlide *a* E in vitro, allowed the determination of which exogenous $^{14}$C-tetrapyrrole substrate was the most likely common precursor of Pchlide *a* and Pchlide *a* E. On the basis of these studies, it was proposed that Pchlide *a* was formed via an acidic (monocarboxylic) biosynthetic route while Pchlide *a* E was formed via a fully esterified route. It was also proposed that the two routes are weakly linked at the level of Mg-Proto, Mpe and Pchlide *a* by porphyrin ester synthetases (McCarthy et al. 1982).

## Chemical Heterogeneity of the Long Chain Fatty Alcohols at Position 7 of the Pchlide *a* E Macrocycle

As early as 1958 various researchers started questioning the assumed phytol nature of the long chain fatty alcohol that esterified the propionic acid residue at position 7 of the Pchlide *a* ester macrocycle (Rebeiz and Castelfranco 1973). For example gas chromatographic analysis of the hydrolyzed fatty alcohol fraction of Pchlide *a* ester of etiolated cucumber cotyledons failed to detect any phytol (McCarthy et al. 1981; Rebeiz and Castelfranco 1973). On the other hand, the Pchlide *a* ester of etiolated barley leaves was shown to contain geranylgeraniol (GG) instead of phytol (Liljenberg 1974). The inner seed coat of Cucurbitaceae, a rich source of Pchlide *a* ester was shown to contain a large number of Pchlide *a* esters esterified with different long chain alcohols. The latter consisted of farnesol and all possible C20 alcohols including GG and phytol (Shioi and Sasa 1982). Roots of etiolated wheat, accumulated large amounts of MV Pchlide *a* esters and lesser amounts of MV Pchlide *a*. The alcohol moieties of the four accumulated Pchlide *a* esters consisted of GG, dihydrogeranylgraniol (GHGG), tetrahydrogeranylgraniol (THGG) and phytol (Mc Ewen and Lindsten 1992).

## Monovinyl (MV) and Divinyl (DV) Chemical Heterogeneity at Position 4 of the Pchlide *a* E Macrocycle

The Occurrence of DV Pchlide *a* ester in higher plants was first reported in the inner seed coat of *Cucurbita pepo* (pumpkin) (Jones 1966), and was confirmed by (Houssier and Sauer 1969) (Fig. 11.6). The search for the occurrence of DV Pchlide *a* ester in other higher plant tissues was however unsuccessful till 11 years later (Belanger and Rebeiz 1980a, b). Using sensitive spectrofluorometric techniques, it was possible to show that etiolated cucumber cotyledons, a DDV-LDV-LDDV plant species incubated in darkness with ALA, accumulated mainly MV Pchlide a E and detectable, yet small amounts of DV Pchlide *a* E. The two Pchlide *a* E were separated by chromatography on thin layers of polyethylene and were characterized by fluorescence emission and excitation spectra at room temperature and 77 °K. However, these studies were not extended with rigor to other plant species such as wheat, corn and barley. In other words, it is not certain at this stage whether small amounts of DV Pchlide *a* E also occur in DMV-LMV-LDMV plant species such as johnsongrass and DMV-LDV-LDMV plant species such as barley, wheat and corn.

## Biosynthetic Heterogeneity of Pchlide *a* E

Because of (a) the DV-MV nature of the Mpde and Pchlide *a* E pools, and (b) because of the structural similarities between Mpde and Pchlide *a* E, one fully

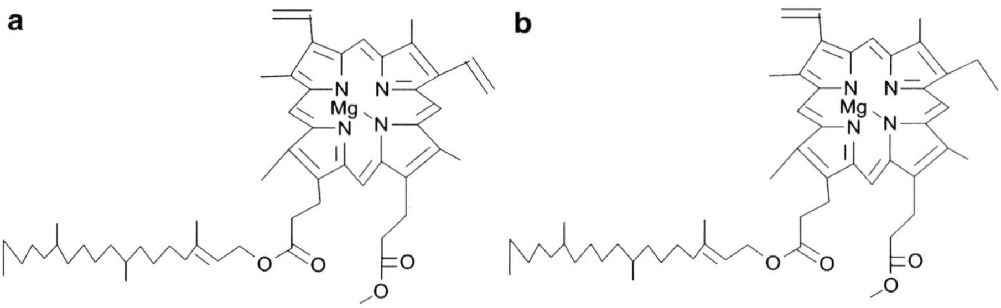


**Fig. 11.6** (**a**) DV and (**b**) MV Pchlide *a* esters

esterified DV biosynthetic route (Fig. 11.2 route 17) and one fully esterified MV biosynthetic route (Fig. 11.2, route 16) are considered to contribute to the formation of Pchlide *a* E. It should be emphasized however, that this hypothesis is based on detection of putative intermediates and structural similarities, but has not been confirmed by demonstration of precursor-product relationships between the Mpde and Pchlide *a* E components of the two putative routes.

#### 11.1.2.2 Metabolism of Pchlide *a* E

Biosynthetically and functionally, the Pchlide *a* ester pool is the least understood pool of the Chl biosynthetic pathway. The full extent of its biological function is still unclear and is surrounded by controversy. At one time, on the basis of its structural similarity to MV Chl *a*, it was assumed to be the immediate major photoprecursor of MV Chl *a* (Granick 1950; Koski 1950; Smith 1948). This hypothesis lost its appeal when (Wolff and Price 1957) demonstrated that the major immediate precursors of MV Chl *a* were MV Pchlide *a* and MV Chlide *a*. Thus for a while Pchlide *a* ester floated as a tetrapyrrole pool without any defined function. However by 1973, several laboratories had reported that Pchlide *a* ester was probably partially photoconvertible to Chl *a* [for a review of this early work, see Rebeiz and Castelfranco (1973)]. Some post-1973 work about the photoconvertibility of Pchlide *a* ester to Chl *a* is described below.

Also, the contribution of Pchlide a E to the greening process is suggested by the pattern of Pchlide *a* E formation under natural photoperiodic greening conditions. Since Pchlide *a* E was observed to accumulate noticeably during the first four dark cycles of the photoperiod, it was suggested by Cohen *et al.* that it may well contribute to Chl *a* biosynthesis and accumulation at the onset of light (i.e. at dawn) during the first few days of photoperiodic greening (Cohen et al. 1977). Although the level of Pchlide *a* E dropped after the fourth dark cycle, it was always detectable in most green plants during all stages of greening during the light phase of the photoperiod (Rebeiz, unpublished).

## Photoconversion of MV Pchlide *a* E to MV Chlide *a* E

Addition of two trans-hydrogens across the 7–8 position of the MV Pchlide *a* ester macrocycle would result in the conversion of MV Pchlide *a* ester to MV Chl *a*. Several laboratories have reported such a reaction in higher plants (Belanger and Rebeiz 1980a, b; Lancer et al. 1976; Liljenberg 1974; Rebeiz and Castelfranco 1973), and lower plant (Kotzabasis and Senger 1989; Sasa and Sugahara 1976). Since other researchers have not been able to detect the photoconversion of Pchlide *a* ester, Rudiger and Schoch suggested that such discrepancies may be due to age of seedlings or the very rapid esterification of Chlide *a* to Chl *a* during the light treatment (Rudiger and Schoch 1991). The latter possibility is unlikely as the photoconversion of Pchlide *a* ester has been also observed at temperatures of −15 to 2 °C (Liljenberg 1974; Rebeiz and Castelfranco 1973). In our opinion, failure to observe the photoconversion of Pchlide *a* E to Chl *a* stems from two considerations: (a) The photoconversion is only partial and very small amounts of Chl *a* are formed, (b) detection of such small amounts of Chl *a* depends a great deal on the sensitivity of the used instrumentation. We have recently reexamined the photoconversion of Pchlide *a* E to Chlide *a* E in isolated cucumber etioplasts. Reaction products were determined by HPLC coupled to high resolution spectrofluorometric detection. It was possible to show that isolated etioplasts of barley and corn subjected to a 2.5 ms flash of light at room temperature followed by immediate precipitation with ammoniacal acetone at various temperatures resulted in the detection of several Chlide a esters (Fig. 11.8B). However, illumination of frozen etioplasts at −18 C did not photoreduce the Pchlide *a* E pool (Adra 1998). These results confirmed the partial photoconvertibility of Pchlide *a* E at room temperatures, but raised the possibility that the enzyme responsible for (photo)reduction of Pchlide *a* E was much more sensitive to low temperatures than conventional Pchlide *a* oxidoreductases (Fig. 11.7).

## Light-Independent Conversion of MV Pchlide *a* E to MV Chl *a* E

On the basis of spectrophotometric and spectrofluorometric analysis, it is presently assumed that the fully esterified tetrapyrrole pool of etiolated tissues consists exclusively of Pchlide *a* E. Recently, it was conjectured that should small amounts of other fully esterified tetrapyrroles be present, their detection would be obscured by the presence of the much larger amounts of Pchlide *a* E. To test this hypothesis, HPLC analysis of etiolated tissues extracts followed by on line spectrofluorometric

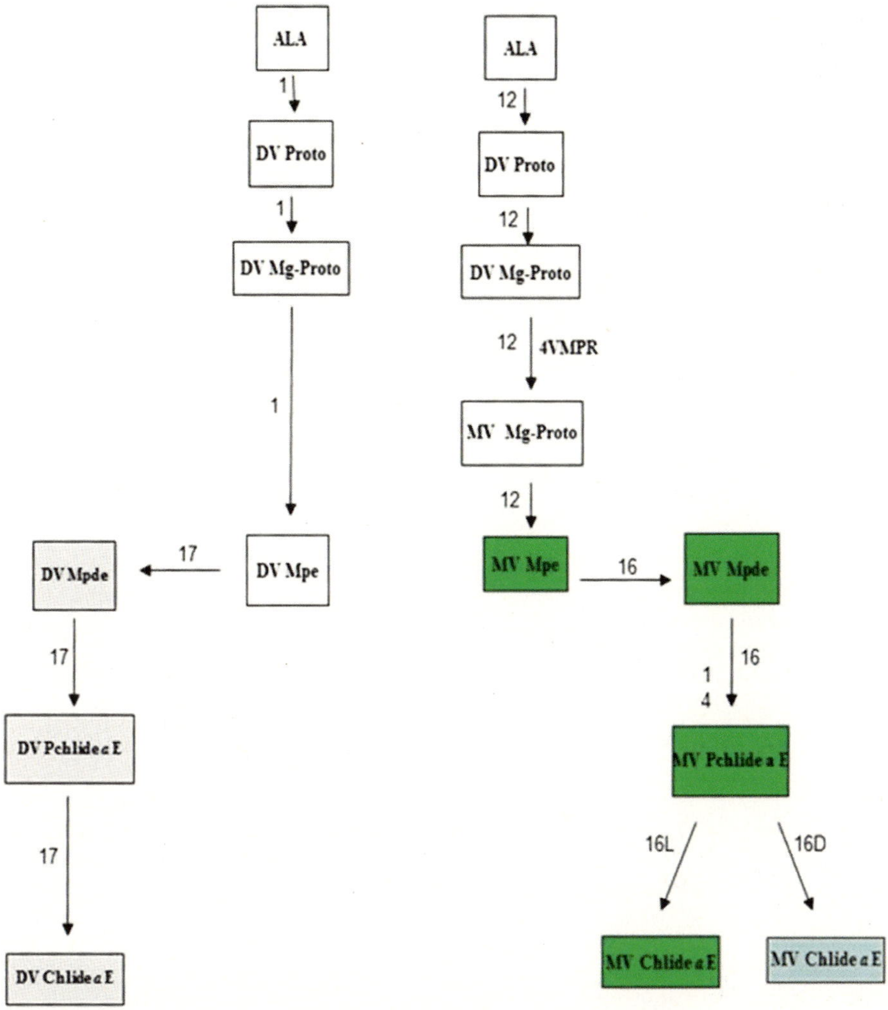


**Fig. 11.7 Biosynthetic route 16, which is responsible for the formation of MV Pchlide *a* ester from MV MpeE**. Routes 16, is highlighted in *green* (Adapted from Rebeiz et al. 2003)

monitoring of all eluting peaks was performed. As expected, high resolution spectrofluorometric analysis of the fully esterified tetrapyrrole pools of etiolated barley and corn detected only MV Pchlide *a* E. However HPLC analysis revealed that the fully esterified Pchlide *a* pools of corn and barley consisted of several fully esterified tetrapyrrole components (Rebeiz et al. 2003). On-line spectrofluorometric analysis of the fully esterified components indicated that they consisted of several different Pchlide *a* E as well as a very small amounts of Chlide *a* E (Fig. 11.8A).

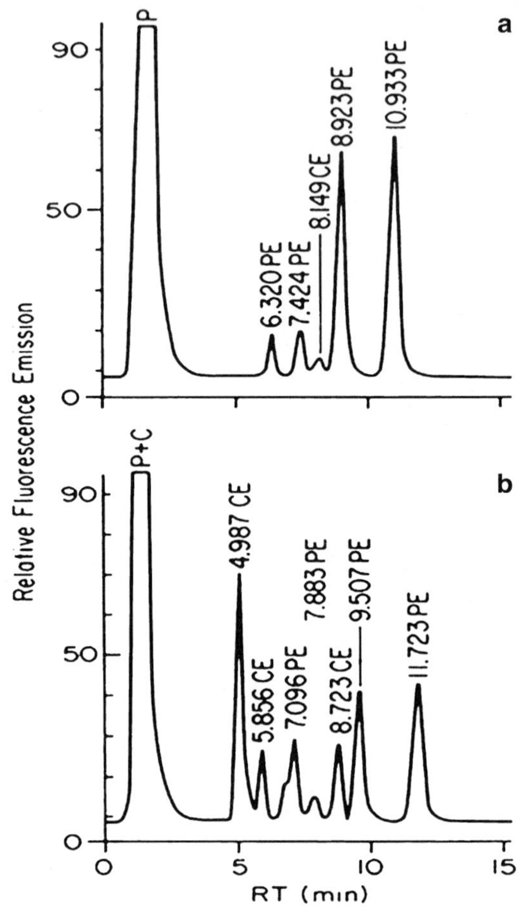


**Fig. 11.8 Elution profile of etiolated corn leaves following a 2.5 ms flash of light at room temperature followed by immediate precipitation with ammoniacal acetone.** Separations were performed on a PE Pecosphere 3 × 3C, C-18 reversed phase, 4 × 0.5 cm column. Elution was with an isocratic, solvent system that consisted of H2O: acetone:methanol (5:20:75 v/v/v. P Pchlide *a*, PE Pchlide *a* ester, CE Chlide *a* ester, RT retention time) (Adapted from Rebeiz et al. 2003)

Formation of the latter implied the involvement of a light-independent Chlide *a* E biosynthetic step in higher plants during dark germination which is depicted by biosynthetic route 16D (Fig. 11.8). Recently, the detection of Chlide *a* E has also been reported by others in etiolated plant tissues (Skribanek et al. 2000) (Fig. 11.9).

Photoconversion of DV Pchlide *a* E to DV Chlide *a* E

In Fig. 11.9, (see below) the photoconversion of DV Pchlide *a* ester to DV Chl *a* is assigned to a fully esterified DV Chl *a* biosynthetic route. This assignment is based on the detection of DV Chl *a* formation immediately following a 47 ms actinic white light treatment of etiolated cucumber cotyledons, at room temperature

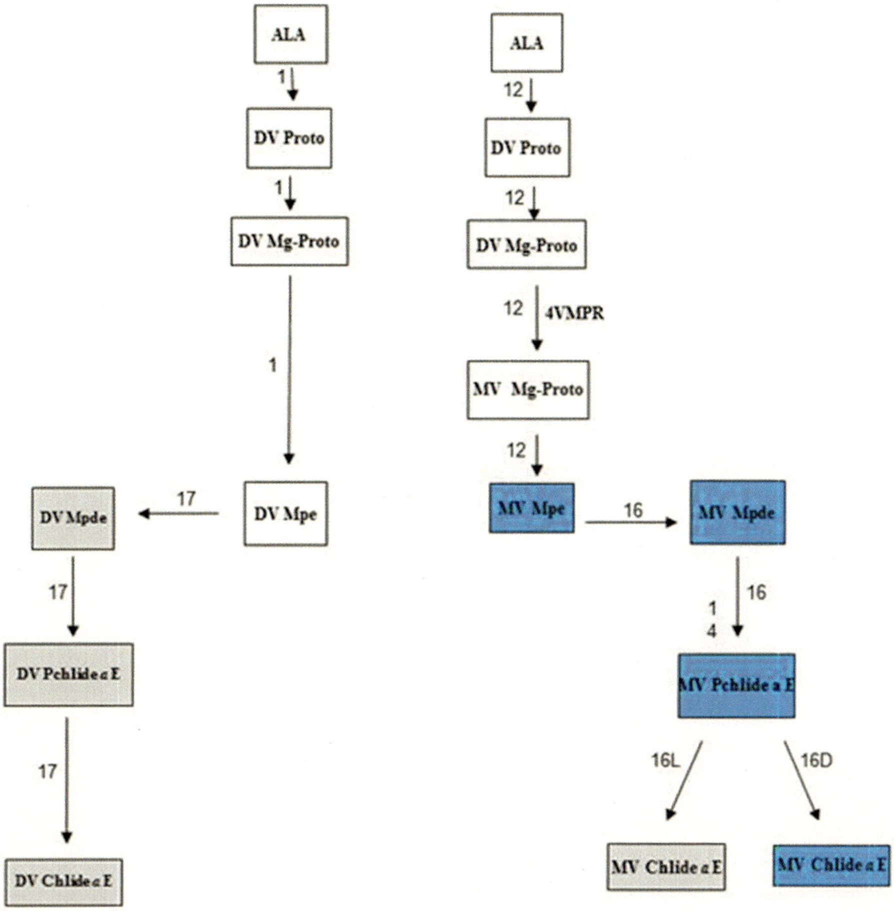


**Fig. 11.9 Biosynthetic route 16D, which is responsible for the formation of MV Chlide *a* ester from MV Pchlide *a* E in darkness**. Routes 16D, is highlighted in *blue* (Adapted from Rebeiz et al. 2003)

(Belanger and Rebeiz 1980a, b). It was assumed that the small amounts of DV Chl *a* were a consequence of the photoconversion of small amounts of DV Pchlide *a* ester. It has since come to our attention, that at room temperature, in-vivo, conversion of newly formed DV Chlide *a* to DV Chl *a* is extremely rapid (Adra 1998). As a consequence the possible photoconversion of DV Pchlide *a* ester to DV Chl *a* should be re-confirmed with isolated plastids at subzero temperatures (Fig. 11.10).

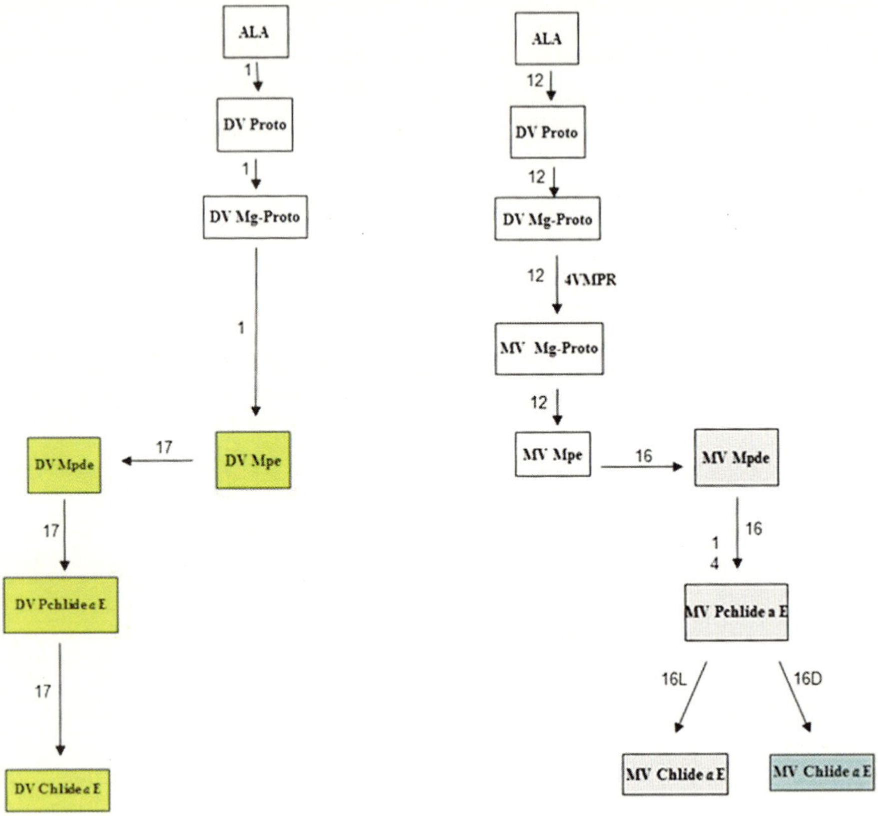


**Fig. 11.10 Biosynthetic route 17, which is supposedly responsible for the formation of DV Chlide *a* ester from DV Pchlide *a* by photoconversion**. Routes 17, is highlighted in *yellow* (Adapted from Rebeiz et al. 2003)

## References

Adra AN (1998) Development of a cell-free system for the study of the terminal stages of the fully esterified chlorophyll *a* biosynthetic routes. MS thesis, University of Illinois, Urbana-Champaign, p 73

Belanger FC, Rebeiz CA (1980a) Chloroplast biogenesis: detection of divinylprotochlorophyllide ester in higher plants. J Biol Chem 19:4875–4883

Belanger FC, Rebeiz CA (1980b) Chloroplast biogenesis 30. Chlorophyll(ide) (E459 F675) and chlorophyll(ide) (E449 F675). The first detectable products of divinyl and monovinyl protochlorophyll photoreduction. Plant Sci Lett 18:343–350

Belanger FC, Rebeiz CA (1982) Chloroplast biogenesis: detection of monovinyl magnesium protoporphyrin monoester and other monovinyl magnesium porphyrins in higher plants. J Biol Chem 257:1360–1371

Cohen CE, Bazzaz MB, Fullet SE, Rebeiz CA (1977) Chloroplast biogenesis XX. Accumulation of porphyrin and phorbin pigments in cucumber cotyledons during photoperiodic greening. Plant Physiol 60:743–746

Ellsworth RK, Nowak CA (1973) The inability of crude homogenates of etiolated wheat seedlings containing protochlorophyllase to convert $^{14}$C-protochlorophyllide to $^{14}$C-protochlorophyll. Photosynthetica 7:246–251

Fischer H, Oestreicher A (1940) Uber Protochlorophyll und Vinyl porphine. Ein Beitrag zur Kenntnis der Oxo-Reaktion. Z Physiol Chem 262:243

Granick S (1950) Magnesium vinyl pheoporphyrin a5, another intermediate in the biological synthesis of chlorophyll. J Biol Chem 183:713–730

Houssier C, Sauer K (1969) Optical properties of the protochlorphyll pigments. I. Isolation, characterization and infrared spectra. Biochem Biophys Acta 172:476–491

Jones OTG (1966) A protein-protochlorophyll complex obtained from inner seed coats of *Cucurbita pepo*. Biochem J 101:153–160

Koski VM (1950) Chlorophyll formation in seedlings of *Zea Mays* L. Arch Biochem 29:339–343

Kotzabasis K, Schuring M-P, Senger H (1989) Occurrence of protochlorophyll and its phototransformation in mutant C-2A′ of *Scenedesmus obliquus*. Physiol Plant 75:221–226

Lancer HA, Cohen CE, Schiff JA (1976) Changing ratios of phototransformable protochlorophyll and protochlorophyllide of bean seedlings developing in the dark. Plant Physiol 57:369–374

Liljenberg C (1974) Characterization and properties of a protochlorophyllide ester in leaves of dark grown barley with geranylgeraniol as esterifying alcohol. Physiol Plant 32:208–213

Mattheis JR, Rebeiz CA (1977) Chloroplast biogenesis XVII. Metabolism of protochlorophyllide and protochlorophyllide ester in developing chloroplasts. Arch Biochem Biophys 184:189–196

Mc Ewen B, Lindsten A (1992) Characterization of protochlorophyllide and protochlorophyllide ester in roots of dark-grown plants. Physiol Plant 84:343–350

McCarthy SA, Belanger FC, Rebeiz CA (1981) Chloroplast biogenesis: detection of a magnesium protoporphyrin diester pool in plants. Biochemistry 20:5080–5087

McCarthy SA, Mattheis JR, Rebeiz CA (1982) Chloroplast biogenesis: biosynthesis of protochlorophyll(ide) via acidic and fully esterified biosynthetic branches in higher plants. Biochemistry 21:242–247

Rebeiz CA, Castelfranco PA (1973) Protochlorophyll and chlorophyll biosynthesis in cell-free systems from higher plants. Annu Rev Plant Physiol 24:129–172

Rebeiz CA, Yaghi M, Abou-Haidar M, Castelfranco PA (1970) Protochlorophyll biosynthesis in cucumber (*Cucumis sativus* L.) cotyledons. Plant Physiol 46:57–63

Rebeiz CA, Kolossov VL, Briskin D, Gawienowski M (2003) Chloroplast biogenesis: chlorophyll biosynthetic heterogeneity, multiple biosynthetic routes and biological spin-offs. In: Nalwa HS (ed) Handbook of photochemistry and photobiology, vol 4. American Scientific Publishers, Los Angeles, pp 183–248

Rudiger W, Schoch S (1991) The last steps of chlorophyll biosynthesis. In: Scheer H (ed) Chlorophylls. Academic, New York, pp 451–464

Sasa T, Sugahara K (1976) Photoconversion of protochlorphyll to chlorophyll *a* in a mutant of *Chlorella regularis*. Plant Cell Physiol 17:273–279

Shioi Y, Sasa T (1982) Separation of protochlorophylls esterified with different alcohols from inner seed coats of three cucurbitaceae. Plant Cell Physiol 23:1315–1321

Skribanek A, Apatini D, Inaoka M, Boddi B (2000) Protochlorophyllide and chlorophyll forms in etiolated plant tissues. J Photochem Photobiol B 55:172–177

Smith JHC (1948) Protochlorophyll, precursor of chlorophyll. Arch Biochem 19:449–454

Wolff JB, Price L (1957) Terminal steps of chlorophyll a biosynthesis in higher plants. Arch Biochem Biophys 72:293–301

# Chapter 12
# The Chlorophyll *b* Biosynthetic Pathway: Novel Metabolic Intermediates

*Imagination is more important than knowledge (Albert Einstein).*

## 12.1 Prologue

The demonstration of metabolic pathways is a multistep process. It involves at least three stages: (a) the detection and characterization of metabolic intermediates, (b) the demonstration of precursor-product relationships between putative intermediates, and (c) purification and characterization of enzymes involved in the metabolic interconversions. These criteria will be applied in our evaluation of the experimental evidence that supports the operation of a multibranched Chl *b* biosynthetic pathway in green (ing) plants.

During the past two decades, several putative metabolic intermediates of the Chl *b* biosynthetic pathway have been detected in higher and lower plants. These intermediates are discussed below.

## 12.2 Monovinyl Protochlorophyllide *b* (MV Pchlide *b*)

The discovery of protochlorophyllide (Pchlide) *b* in higher plants was reported by Shedbalkar et al. (1991). It was first observed as a fluorescent compound at 77 K in diethyl ether, with Soret excitation and red emission maxima at 463 and 643 nm respectively. These fluorescence properties were identical to those of synthetic Pchlide phytyl ester *b*. The chemical structure of the latter was confirmed by proton nuclear magnetic resonance, fast atom bombardment mass spectroscopic analysis and chemical derivatization coupled to electronic spectroscopic analysis (Shedbalkar et al. 1991). MV Pchlide *b* differed from MV Pchlide *a* by the presence

C.A. Rebeiz, *Chlorophyll Biosynthesis and Technological Applications*,

DOI 10.1007/978-94-007-7134-5_12, 

Fig. 12.1 MV Pchlide *b*

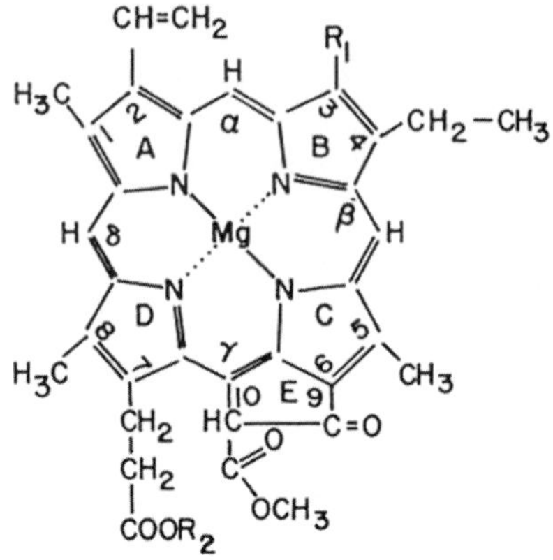


**Protochlorophyllide *b* (Pchlide *b*)**

| R1 | R2 | Compound |
|---|---|---|
| Formyl | H | MV Pchlide *b* |

of a formyl instead of a methyl group at position 3 of the macrocycle 9 (Fig. 12.1). The trivial name MV Pchlide *b* was proposed to differentiate it from MV Pchlide *a*.

Determination of the Amount of MV Pchlide *b* either in the Presence of MV Chl (ide) *a* and *b*, or in the presence of MV Pchlide *a* was achieved by combined spectrofluorometric analysis at room temperature and 77 K (Ioannides et al. 1997). In green cucumber seedlings grown under a 14-h light/10-h dark photoperiod, the amount of MV Pchlide *b* ranged from about 400 to 800 nmoles per 100 mg proteins. MV Pchlide *b* was detectable in green tissues but not in etiolated tissues or during the early phases of greening of etiolated tissues (Kolossov and Rebeiz 2004).

### 12.2.1 Arguments Related to the Spectral Properties of Synthetic Putative Pchlide b

In a recent review Rudiger proposed that the synthetic and natural MV Pchlides *b* described by (Shedbalkar et al. 1991), do not correspond to authentic Pchlide *b* (Rudiger 2003). It was argued that this is because the synthetic MV Pchlide *b* prepared by Shedbalkar et al., exhibits different absorbance and mass spectroscopic properties than the putative Pchlide *b* prepared by Schoch et al. (1995). While the synthetic MV Pchlide *b* prepared by Shedbalkar et al., exhibited a typical Pchlide spectrum with band I and band II maxima at 632 and 582 nm and a band II:I ratio of 0.45 in acetone at room temperature, Schoch et al. putative Pchlide *b* exhibited a quasi-oxorhodo, protopheophytin type spectrum with band I and II maxima at 622 and 578 nm respectively and a band II:I ratio of about 1.57. As a consequence Rudiger goes on to propose that contrary to the assertions of Shedbalkar et al., MV Pchlide *b* does not really occur in green plants (Rudiger 2003).

### 12.2.2 Rebuttal of Above Claims

The absorbance properties of the tetrapyrrole synthesized by Schoch et al. (1995), were assigned to Pchlide *b* on the basis of (a) their similarity to the absorbance properties of Chl $c_3$ reported by Jeffrey and Wright (1987), and (b) the structural similarities of MV Pchlide *b* and Chl $c_3$ both of which contain a porphin macrocycle having a double bond at position 7–8 of the macrocycle and a formyl group at position 3. In this comparison the authors overlooked the fact that in addition to the rhodofying formyl group at position 3 (a rhodofying group increases the absorbance intensity of band III, in porphins and band II in phorbins), Chl $c_3$ also contains a second rhodofying group (Propyl acrylic group) at position 7 of the porphin macrocycle (Jeffrey and Wright 1987) which is not present in Pchlide *b*. Instead the latter contains an etio propyl group at that position. The authors obviously overlooked the well-known spectroscopic fact that two rhodofying groups on diagonally opposite pyrrole rings as occurs in Chl $c_{3,}$ enhance each other rhodofying effect (Falk 1964) and result in the quasi-oxorhodo type spectrum observed for Chl $c_3$. In an oxorhodo spectrum, the ratio of the absorbance intensities of bands II:I increases significantly (Falk 1964). For example in diethyl ether at room temperature, the ratio of bands II:I in Pchlide *a* which lacks a rhodofying group at position 7, is 0.54, it is 1.17 in Chl $c_2$ which has a rhodofying group at position 7, and 3.79 in Chl $c_3$ which has two rhodofying groups on diagonally opposite pyrrole rings . The rhodofying effects of the formyl group at position 3 of the macrocycle is usually insignificant as observed for Chl *b* in comparison to Chl *a*. In our opinion the spectral properties assigned by Schoch et al. to their synthetic putative Pchlide *b* correspond very closely to the spectral properties of a pheoporphyrin, that is to a metal-free Pchlide, also referred to as protopheophytin (Houssier and Sauer 1969). For example the ratio of bands II:I in protopheophytin *a* in diethyl ether is 1,67; it is 1.57 in the putative Pchlide *b* spectrum reported by Schoch et al. in acetone at room temperature (Schoch et al. 1995). Loss of Mg may have occurred during the preparation of Pchlide *b* in the presence of excess 2,3-dichloro-5,6-dicyanobenzoquinone (DDQ) at room temperature, instead of under the milder DDQ oxidation conditions (4°C), recommended by Shedbalkar et al. (1991).

As for the mass spectroscopic data reported by Shedbalkar et al., it corresponds to that of a tetrapyrrole axially coordinated to one molecule of methanol the solvent in which the tetrapyrrole was dissolved prior to FAB mass spectroscopic analysis. Axial coordination to lewis bases such as methanol is a very well documented phenomenon particularly in the *b* tetrapyrrole series (Belanger and Rebeiz 1984; Rebeiz and Belanger 1984). As for the failure to detect the 7–8 double bond by mass spectroscopy it was due to the liability of that bond in Pchlide *b*, as discussed in (Shedbalkar et al. 1991). Furthermore, Schoch et al. (1995) completely ignored the extensive and detailed absorbance, fluorescence and NMR evidence that led to the assignment of a Pchlide *b* structure to the synthetic compound by Shedbalkar et al. (1991).

Finally, the proposal that Pchlide *b* does not occur in nature goes counter to the recent finding of Xu et al. (2002). These authors reported the accumulation of a putative Pchlide *b* in a triple *Synechocystis* mutant (PS I-less/*chl*$L^{-}$/*lhcb*$^{+}$/cao$^{+}$) grown under light activated heterogeneous growth (LAHG) conditions that also led to the accumulation of pheophorbide *b* (*i.e.* demetalated Chlide *b*). The putative Pchlide *b* exhibited an absorbance spectrum very similar to that reported by Schoch et al. (1995) for their synthetic putative Pchlide *b*. In this case too, it is very probable that the accumulated putative Pchlide *b* reported by Xu et al. (2002), was demetalated along with the reported demetalation of Chlide *b*.

Although the mass spectroscopic data reported by Schoch et al. (1995) and Xu et al. (2002) is compatible with the molecular weight of MV Pchlide *b*, it is very possible that this is due to the vagary of mass spectroscopy caused by the formation of a protopheophytin *a* adduct having the same molecular weight as MV Pchlide *b*.

## 12.3 Divinyl Protochlorophyllide *b* (DV Pchlide *b*)

So far it has not been possible to detect DV Pchlide *b* in plants (Fig. 12.2).

## 12.4 Monovinyl Chlorophyllide *b* (MV Chlide *b*)

Monovinyl chlorophyllide *b* was first detected in greening (Duggan and Rebeiz 1981, 1982) and green higher plant tissues (Aronoff 1981). The pool of MV Chlide *b* exhibited the spectrofluorometric properties of MV Chl *b* in diethyl ether at 298 and 77 K, but had the chromatographic mobility and solubility of a monocarboxylic phorbin. The presence of a free carboxylic group and a formyl group was demonstrated by methylation with diazomethane and conversion to a Chlide *b* oxime upon reaction with hydroxylamine (Duggan and Rebeiz 1982). The concentration of Chlide *b* in green tissues was in the same range as that of MV Pchlide *a* and MV Chlide *a*. It was estimated that less than 15 % of the Chlide *b* pool could have arisen from chlorophyllase activity in vitro as confirmed by the extent of hydrolysis of $^{14}$C-labeled MV Chl *b* added to green tissues just before pigment extraction (Duggan and Rebeiz 1982) (Fig. 12.3).

## 12.5 Divinyl Chlorophyllide *b* (DV Chlide *b*)

DV Chlide *b* has so far been detected only in the Nec 2 maize mutant (that used to be known as the ON 8147 mutant) (C. A. Rebeiz, unpublished). This mutation is a lethal mutation, the leaves are pale yellow, and accumulate only DV Chl *a* and *b* (Bazzaz 1981). Nec2 maize leaves accumulate DV Chlide *b* to the extent of about 1.00 nmoles per gram of fresh leaves (C. A. Rebeiz, unpublished) (Fig. 12.4).

Fig. 12.2 DV Pchlide *b*

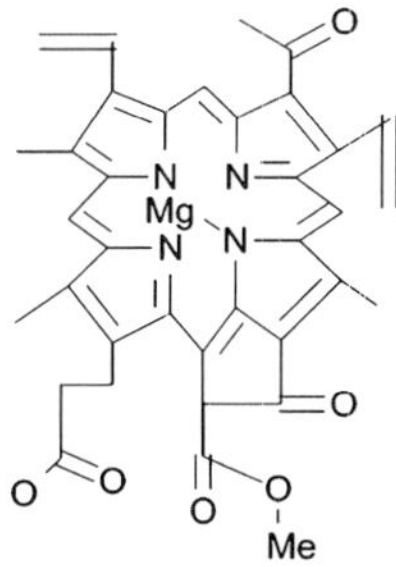


Fig. 12.3 MV Chlide *b*

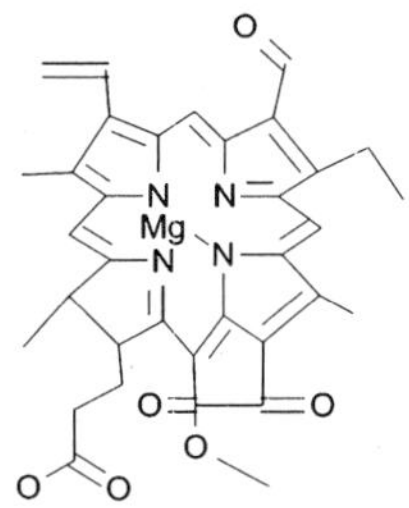


Fig. 12.4 DV Chlide *b*

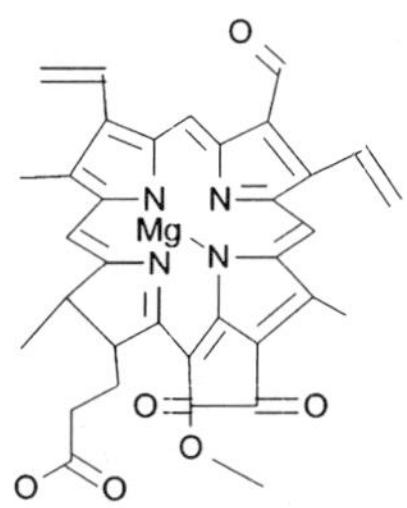


DV Chlide *b* may also be present in the prochlorophyte picoplankton of the subtropical waters of the North Atlantic as well as in the picoplankton of the euphotic zone of the world tropical and temperate oceans, and the Mediterranean sea, where DV Chl *a* and *b* are the predominant Chl species (Chisholm et al. 1990, 1992; Goerike and Repeta 1992; Veldhuis and Kraay 1990).

DV Chlide *b* exhibits the same electronic spectroscopic properties as DV Chl *b* (see below) but differs from the latter by its solubility in more polar organic solvents and its chromatographic mobility.

## 12.6 DV Chl *b*

The possible occurrence of DV Chl *b* in green plants was predicted after the discovery of DV Chl *a* (Rebeiz et al. 1980). It was next detected in the Nec 2 maize mutant (ex-ON 8147) by electronic spectroscopy (Bazzaz 1981). Its chemical structure was

**Fig. 12.5** DV Chl *b*

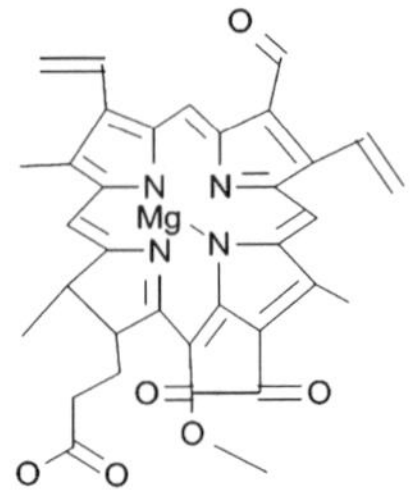

ascertained by fast atom mass spectroscopic (Brereton et al. 1983), and $^1$H NMR analysis (Wu and Rebeiz 1985). It accumulates to the extent of about 100 nmoles per gram fresh weight of Nec 2 leaves (Rebeiz, Unpublished) (Fig. 12.5).

DV Chl *b* also accumulates in the prochlorophyte picoplankton of the subtropical waters of the North Atlantic as well as in the picoplankton of the euphotic zone of the world tropical and temperate oceans, and the Mediterranean sea, where DV Chl *a* and *b* are the predominant Chl species (Chisholm et al. 1990, 1992; Goerike and Repeta 1992; Veldhuis and Kraay 1990).

## References

Aronoff S (1981) Chlorophyllide *b*. Biochem Biophys Res Commun 102:108–112

Bazzaz MB (1981) New chlorophyll chromophores isolated from a chlorophyll deficient mutant of maize. Photobiochem Photobiophys 2:199–207

Belanger FC, Rebeiz CA (1984) Chloroplast biogenesis 47: spectroscopic study of net spectral shifts induced by ligand coordination in metalated tetrapyrroles. Spectrochim Acta 40A:807–827

Brereton RG, Bazzaz MB (1983) Positive and negative fast atom bombardment mass spectroscopic studies on chlorophylls: structure of 4-vinyl-4-desethyl chlorophyll b. Tetrahedron Lett 24:5775–5778

Chisholm SW, Olson RJ, Zettler ER, Goericke R, Waterbury JB, Welschmeyer NA (1990) A novel free-living prochlorophyte abundant in the oceanic euphotic zone. Nature 334:340–343

Chisholm SW, Frankel S, Goerike R, Olson R, Palenic R, Urbach B, Waterbury J, Zettler E (1992) Prochlorococcus marinus nov.gen. sp.: an oxyphototropic marine prokaryote containing divinyl chlorophyll a and b. Arch Microbiol 157:297–300

Duggan JX, Rebeiz CA (1981) Detection of a naturally occurring chlorophyllide b pool in higher plants. Plant Physiol 67(suppl):267

Duggan JX, Rebeiz CA (1982) Chloroplast biogenesis 38. Quantitative detection of a chlorophyllide *b* pool in higher plants. Biochim Biophys Acta 714:248–260

Falk JE (1964) Porphyrins and metalloporphyrins. Elsevier, Amsterdam

Goerike R, Repeta D (1992) The pigments of Prochlorococcus marinus. The presence of divinyl-chlorophyll a and b in a marine prochlorophyte. Limnol Oceanogr 37:425–433

Houssier C, Sauer K (1969) Optical properties of the protochlorophyll pigments II. Electronic absorption, fluorescence and circular dichroism spectra. Biochim Biophys Acta 172:492–502

Ioannides MI, Shedbalkar VP, Rebeiz CA (1997) Quantitative determination of 2-monovinyl protochlorphyll(ide) b by spectrofluorometry. Anal Biochem 249:241–244

Jeffrey SW, Wright SW (1987) A new spectrally distinct component in preparations of chlorophyll *c* from micro-alga (Prymnesiophyceae). Biochim Biophys Acta 894:180–188

Kolossov VL, Rebeiz CA (2004) Chloroplast biogenesis 88. Protochlorophyllide b occurs in green but not in etiolated plants. J Biol Chem 278:49675–49678

Rebeiz CA, Belanger FC (1984) Chloroplast biogenesis 46: calculation of net spectral shifts induced by axial ligand coordination in metalated tetrapyrroles. Spectrochim Acta 40A:793–806

Rebeiz CA, Belanger FC, Freyssinet G, Saab DG (1980) Chloroplast biogenesis. XXIX. The occurrence of several novel chlorophyll a and b chromophores in plants. Biochim Biophys Acta 590:234–247

Rudiger W (2003) The last steps of chlorophyll synthesis. In: Kadish KM, Smith KM (eds) Chlorophylls and bilins: biosynthesis, synthesis, and degradation, vol 13. Elsevier, New York, pp 71–108

Schoch S, Helfrich M, Wiktorsson B et al (1995) Photoreduction of zinc protopheophorbide *b* with NADPH-protochlorophyllide oxidoreductase from etiolated wheat (*Triticum aestivum*). Eur J Biochem 229:229–298

Shedbalkar VP, Ioannides IM, Rebeiz CA (1991) Chloroplast biogenesis. Detection of monovinyl protochlorophyll(ide) b in plants. J Biol Chem 266:17151–17157

Veldhuis MJW, Kraay GW (1990) Vertical distribution of pigment composition of a picoplanktonic prochlorophyte in the subtropical north Atlantic: a combined study of pigments and flow cytometry. Mar Ecol Prog Ser 68:121–127

Wu SM, Rebeiz CA (1985) Chloroplast biogenesis. Molecular structure of chlorophyll *b* (E489 F666). J Biol Chem 260:3632–3634

Xu H, Vavilin D, Vermass W (2002) The presence of chlorophyll b in synechocystis sp. PCC 6803 disturbs tetrapyrrole biosynthesis and enhances chlorophyll degradation. J Biol Chem 277:42726–42732

# Chapter 13
# The Chl *b* Biosynthetic Pathway: Intermediary Metabolism

*The time continuum consists of the past present and future. Understanding the past is essential for a better understanding of the present and future. Technology can help solve the problems of the present. However understanding the forces that control and regulate the physical and biological components of the universe is mandatory for the development of future technologies (C. A. Rebeiz).*

## 13.1 Introduction

### *13.1.1 Determination of Precursor-Product Relationships In Vivo*

In discussing the Chl *b* biosynthetic pathway, use will be made of kinetic analysis of precursor-product relationships in vivo. In 1988, equations were derived to investigate possible precursor-product relationships in branched, and interconnected pathways (Rebeiz et al. 1988; Tripathy and Rebeiz 1988). It was shown that for any two compounds **A** and **B**, formed from a common precursor **P** such as ALA, and having a possible direct precursor-product relationship between them, for any number of time intervals $t_1$ to $t_2$, the following equation describes the relationship between the specific radioactivity of compound **A**, possible radiolabel incorporation from compound **A** into compound **B**, and the net synthesis of compound **B** from compound **A** (Rebeiz et al. 1988):

$$\mathbf{QB}_2 = ({}^*\mathbf{A}_1 + {}^*\mathbf{A}_2)/2) \cdot ({}^*\mathbf{B}_2) \quad (13.1)$$

C.A. Rebeiz, *Chlorophyll Biosynthesis and Technological Applications*,
DOI 10.1007/978-94-007-7134-5_13, © Springer Science+Business Media Dordrecht 2014

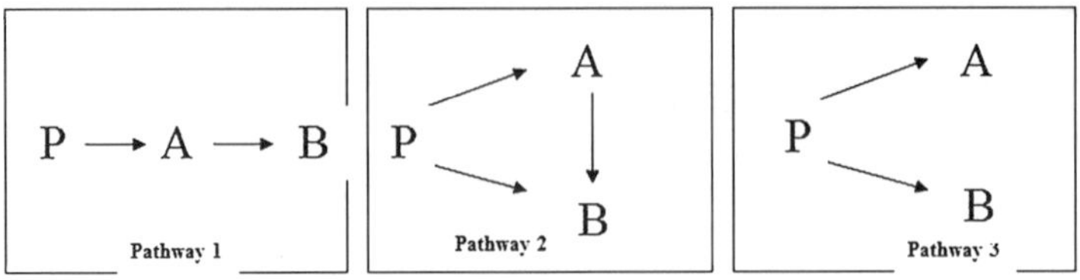


**Fig. 13.1** The three possible irreversible precursor-product relationships between two precursors (*P*, *A*) and one end product (*B*) (Adapted from Rebeiz et al. 1988)

where:

$\mathbf{QB}_2$ = amount of radiolabel incorporated into compound B during time interval $t_1$–$t_2$;

$^*\mathbf{A}_1$, $^*\mathbf{A}_2$ = specific radioactivity of compound **A** at the beginning and end of time interval $t_1$–$t_2$ respectively.

$^*\mathbf{B}_2$ = amount of compound **B** synthesized during time interval $t_1$–$t_2$.

By comparing expected radiolabel incorporation into compound **B**, as calculated from Eq. 13.1, with experimentally determined incorporations into compound **B**, it is possible to tell whether compound **B** was formed exclusively from compound **A** or not. If compound **B** is formed exclusively from compound **A**, then within the range of experimental error, the theoretical and experimental radiolabel incorporations into compound **B** should be identical or reasonably similar. On the other hand, if compound **B** is not formed from compound **A**, or is partially formed from compound **A**, then the calculated and experimental radiolabel incorporations into compound **B** will be different. The difference between the calculated and experimental values may then depend, among other things, on the extent of partial contribution of compound **A** to the synthesis of compound **B**.

If comparison of calculated and experimental results indicates that compound **B** is not totally formed from compound **A** via pathway 1 (Fig. 13.1), then the question arises as to whether compound **B** is formed via pathway 2 or 3. Furthermore if compound **B** is found to be formed via pathway 2, then the contribution of compound **A** to the formation of compound **B** needs to be assessed. The determination of whether compound **B** is formed via pathway 2 or 3 can be achieved from conventional in vitro investigations of precursor product relationship between compound **A** and compound **B**. In other words, if pathway 2 is found to be operational then the contribution of compound **A** to the formation of compound **B** can be assessed from Eq. 13.2 (Rebeiz et al. 1988):

$$\%\,\text{Conversion} = 100 - [(|\text{Exp} - \mathbf{QB}_X|/\text{Exp})100] \quad (13.2)$$

where:

% Conversion = maximum possible percent conversion of compound **A** to compound **B** during any time interval X,

Exp = actual $^{14}$C-incorporation into compound **B** by the end of time interval X, as determined experimentally,

$Q\mathbf{B}_X$ = theoretical $^{14}C$-incorporation into compound **B** by the end of time interval X, as calculated from Eq. 13.1,

$|Exp - Q\mathbf{B}_X|$ = absolute difference between the experimental and theoretical $^{14}C$-incorporation of precursor **P** into compound **B** during time interval X.

### *13.1.2 Source of Oxygen During the Formation of the Formyl Group of Chl b*

Mass spectra of [7-hydroxymethyl]-chlorophyll *b* extracted from leaves greened in the presence of either $^{18}O_2$ or $H_2{}^{18}O_2$ revealed that $^{18}O$ was incorporated only from molecular oxygen into the 7-formyl group of Chl *b* (Porra et al. 1993, 1994). The high enrichment using $^{18}O_2$, and the absence of labeling by $H_2{}^{18}O_2$, demonstrated that molecular oxygen is the sole precursor of the 7-formyl oxygen of chlorophyll *b* in greening maize leaves. This in turn suggested that a mono-oxygenase is involved in the oxidation of the methyl group to a formyl.

## 13.2 The Chl *b* Biosynthetic Pathway

As was mentioned in the previous chapter, the demonstration of metabolic pathways is a multistep process that involves at least three stages: (a) the detection and characterization of metabolic intermediates, (b) the demonstration of precursor-product relationships between putative intermediates, and (c) purification and characterization of enzymes involved in the metabolic interconversions. In this chapter I will mainly emphasize the demonstration of precursor-product relationships between various intermediates.

### *13.2.1 Chlorophyllide b (Chlide b)*

Chlorophyllides *b* (Chlides *b*) are the immediate precursors of the *b* chlorophylls. Chlides *b* are chemically and biochemically heterogeneous. Chemical heterogeneity consists in MV and DV substitutions at position 4 of the macrocycle, while biochemical heterogeneity consists of multiple biosynthetic routes involving MV and DV Chlides *a* and MV Pchlide *b* precursors (Fig. 13.2).

#### 13.2.1.1 Monovinyl Chlorophyllides *b*

The biosynthetic heterogeneity of Chlides *b* is manifested by their biosynthesis via multiple biosynthetic routes in DV and MV plant species as described below.

Fig. 13.2 Monovinyl (MV) and Divinyl (DV) Chlorophyllide *b*

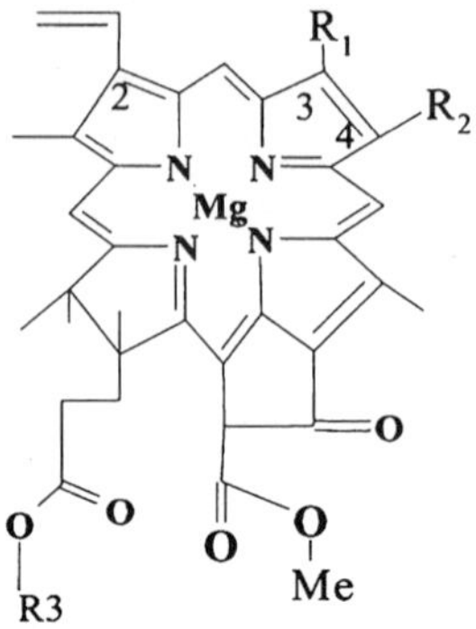


**Chlorophyllide *b* (Chlide *b*)**

| | R1 | R2 | R3 | Compound |
|---|---|---|---|---|
| I | Formyl | Vinyl | H | DV Chlide *b* |
| II | Formyl | Ethyl | H | MV Chlide *b* |

## Metabolism of MV Chlide *b*

MV Chlide *b* was simultaneously detected in greening (Duggan and Rebeiz. 1981, 1982) and green higher plant tissues (Aronoff 1981). It was proposed as a logical immediate precursor of MV Chl *b*. Subsequently the conversion of exogenous MV Chlide *b* to MV Chl *b* in etiolated oat was reported (Benz and Rudiger 1981).

The pool of MV Chlide *b* exhibited the spectrofluorometric properties of MV Chl *b* in diethyl ether at 298 and 77 K, but had the chromatographic mobility and solubility of a monocarboxylic phorbin. The presence of a free carboxylic group and a formyl group was demonstrated by methylation with diazomethane and conversion to a Chlide *b* oxime upon reaction with hydroxylamine (Duggan and Rebeiz 1982). The concentration of Chlide *b* in green tissues was in the same range as that of MV Pchlide *a*, and MV Chlide *a*. It was estimated that less than 15 % of the Chlide *b* pool could have arisen by hydrolysis of phytol at position 7 of the macrocycle via chlorophyllase activity in vitro. This was confirmed by the extent of hydrolysis of $^{14}$C-labeled MV Chl *b* added to green tissues just before pigment extraction (Duggan and Rebeiz 1982).

The source of oxygen of the formyl group at position 3 of the macrocycle has been investigated by Porra et al. (1993, 1994). Mass spectra of [7-hydroxymethyl]-Chl *b* extracted from leaves greened in the presence of either $^{18}O_2$ or $H_2{}^{18}O_2$ revealed that $^{18}O$ was incorporated only from molecular oxygen into the 3-formyl group of Chl *b*. The high enrichment using $^{18}O_2$, and the absence of labeling by $H_2{}^{18}O_2$, suggested that molecular oxygen is the sole precursor of the 3-formyl oxygen of Chl(ide) *b* in greening maize leaves. This in turn suggested that a mono-oxygenase is involved in the oxidation of the methyl group to a formyl.

The biosynthesis of MV Chlide *b* is highly heterogeneous (*vide infra*). In the elucidation of this biosynthetic heterogeneity, extensive use was made of kinetic analysis of precursor-product relationships in vivo. For that purpose, equations were derived to investigate possible precursor-product relationships in branched and interconnected pathways (Rebeiz et al. 1988; Rebeiz 2002; Tripathy and Rebeiz 1988). It was shown that for any two compounds A and B, formed from a common precursor "P" such as ALA, and having a possible direct precursor-product relationship between them, for any number of time intervals t1 to t2, an equation can be derived that describes the relationship between (a) the specific radioactivity of compound "A", and the possible radiolabel incorporation from compound "A" into compound "B", and (b) the possible net synthesis of compound "B" from compound "A" (Rebeiz et al. 1988).

*Formation of MV Chlide b from DV Chlide a and MV Chlide a via Route 4 in Greening DDV-LDV-LDDV Plant Species*

Conversion of DV Pchlide *a* to MV Chlide *b* via route 4 *i.e.* via DV Chlide *a*, and MV Chlide *a*, in greening DDV-LDV-LDDV plants is supported by the conversion of exogenous ALA to MV Chlide *b* and Chl *b* in etiochloroplasts prepared from etiolated cucumber cotyledons subjected to 4 h of illumination prior to etiochloroplast isolation (Kolossov et al. 1999). In such systems ALA is converted mainly to DV Pchlide *a* (Tripathy and Rebeiz 1986) which is readily convertible to MV Chlide *a* via DV Chlide *a* (Fig. 13.3).

*Formation of MV Chlide* b *from MV Chlide* a *via Route 14 in Greening DMV-LDV-LDMV Plants*

Conversion of MV Chlide *a* to MV Chlide *b* via route 14 in DMV-LDV-LDMV plants is supported by precursor-product relationships analysis in vivo between MV Chlide *a* and MV Chl *b* in greening corn seedlings, (Rebeiz et al. 1999). After 5 h of greening of etiolated corn seedlings, about 27–36 % of the MV Chl *b* was formed from MV Chlide *a*. Under these conditions MV Chlide *a* is formed in turn from MV Pchlide *a* (Fig. 13.4).

Formation of MV Chlide *b* from MV Pchlide *b* in Greening Plants

Two MV Chlide b pools are putatively formed form MV Pchlide *b* via routes 9, and 11. These two putative biosynthetic routes are discussed below.

*Possible Formation of MV Chlide b from MV Pchlide b in Green DDV-LDV-LDDV Plants via Biosynthetic Route 9*

The operation of this hypothetical route is suggested by (a) the formation of MV Pchlide *b* in green cucumber seedlings, and (b) the structural relationship between

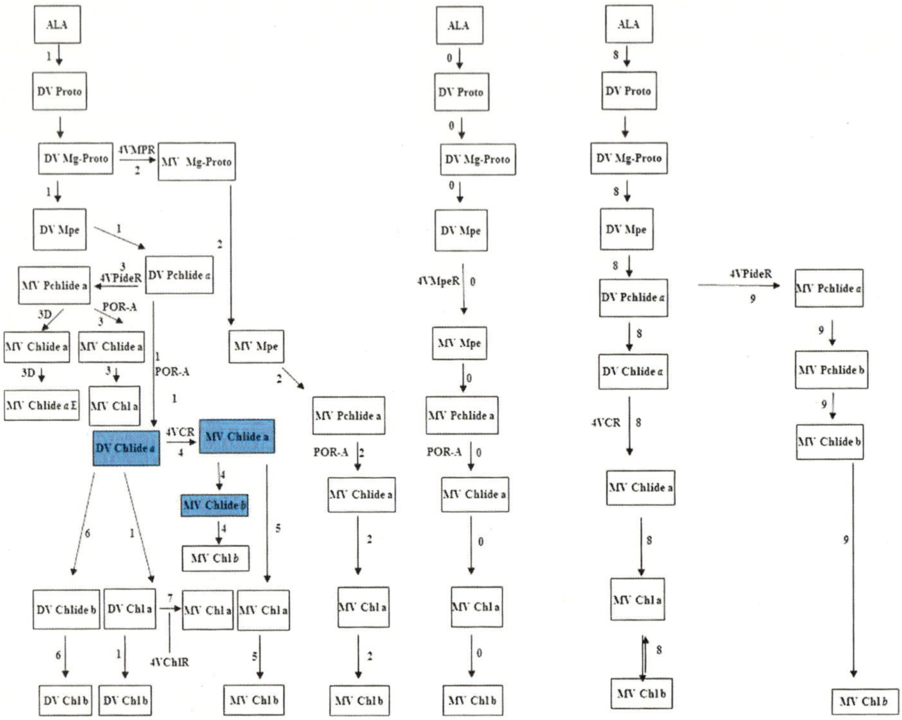


**Fig. 13.3** MV Chlide *b* biosynthesis via route 4 in DDV-LDV-LDDV plant species (Adapted from Fig. 6.3 of Chap. 6)

MV Pchlide *a* and MV Pchlide *b*. The first traces of MV Pchlide *b* in greening etiolated cucumber cotyledons are detected after 14 h of greening (Ioannides 1993). In photoperiodically grown green seedlings, higher amounts of MV Pchlide *b* are detected (Ioannides et al. 1997; Kolossov and Rebeiz 2003). To ascertain the operation of biosynthetic route 9, in DDV-LDV-LDMV plants, precursor product relationships between MV Pchlide *a* and MV Pchlide *b* and between MV Pchlide *b* and MV Chlide *b* need to be established (Fig. 13.5).

### *Possible Formation of MV Chlide b from MV Pchlide b in Green DMV-LDV-LDMV Plants via Biosynthetic Route 11*

The operation of this hypothetical route is suggested by (a) the formation of MV Pchlide *b* in green photoperiodically-grown barley (Kolossov and Rebeiz 2003), and wheat seedlings (Rebeiz, unpublished), (b) the structural relationship between MV Pchlide *a* and MV Pchlide *b*, and (c) the reported photoconversion of MV Pchlide

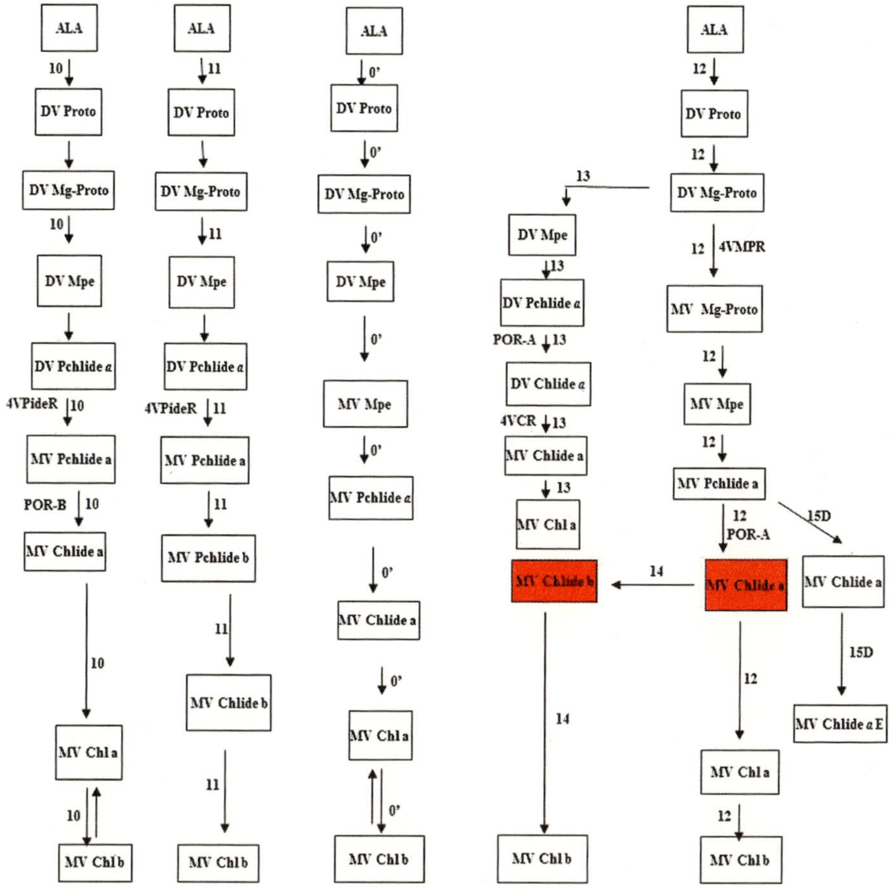


**Fig. 13.4** MV Chlide *b* biosynthesis via route 14 in DMV-LDV-LDMV plant species (Adapted from Fig. 6.4 of Chap. 6)

*b* to MV Chlide *b* (Klement et al. 1999). To ascertain the operation of biosynthetic route 11, in DMV-LDDV-DLMV plants, precursor product relationships between MV Pchlide *a* and MV Pchlide *b* need to be established (Fig. 13.6).

### Formation of DV Chlide *b* via Route 6 in DDV-LDV-LDDV Plants

The DV Chlide *b* pool is putatively formed from DV Chlide *a* via biosynthetic route 6. In higher plants DV Chlide *b* has so far been detected only in the Nec 2 maize mutant (that used to be known as the ON 8,147 mutant) (C. A. Rebeiz, unpublished). This mutation is a lethal mutation, the leaves are pale yellow, and accumulate only DV Chl *a* and *b* (Bazzaz 1981). Nec2 maize leaves accumulate DV Chlide

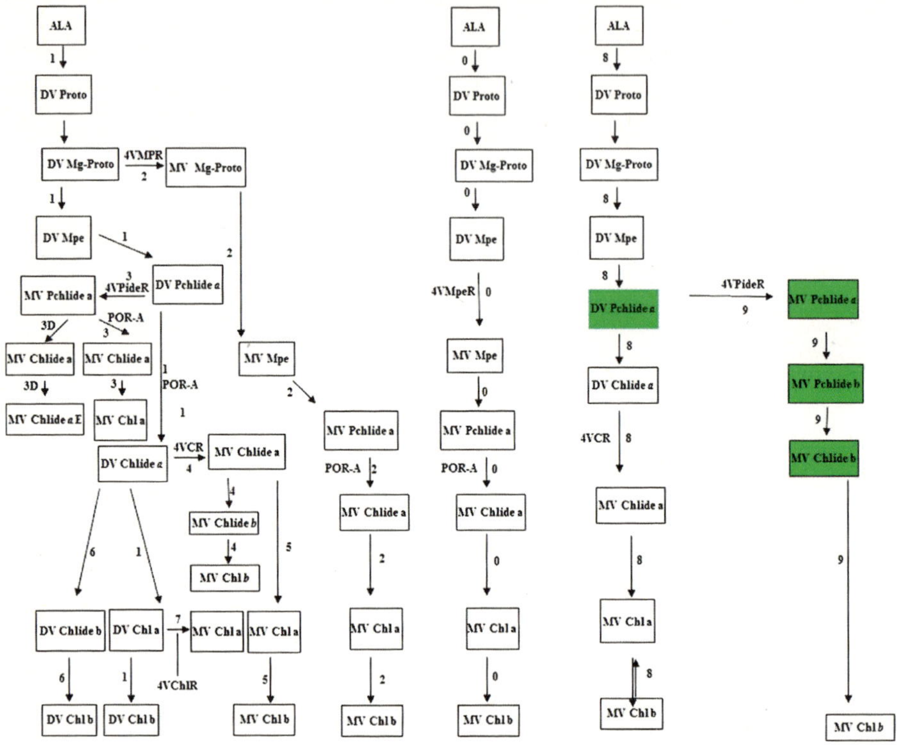


**Fig. 13.5** MV Chlide *b* biosynthesis via route 9 in DDV-LDV-LDDV plant species (Adapted from Fig. 6.3 of Chap. 6)

*b* to the extent of about 1.00 nmoles per gram of fresh leaves (C. A. Rebeiz, unpublished). DV Chlide *b* may also be present in the prochlorophyte picoplankton of the subtropical waters of the North Atlantic as well as in the picoplankton of the euphotic zone of the world tropical and temperate oceans, and the Mediterranean sea, where DV Chl *a* and *b* are the predominant Chl species (Chisholm et al. 1988, 1992; Goerike and Repeta 1992; Veldhuis and Kraay 1990). DV Chlide *b* exhibits the same electronic spectroscopic properties as DV Chl *b* (see below) but differs from the latter by its solubility in organic solvents and its chromatographic mobility (Fig. 13.7).

Conversion of DV Pchlide *a* to DV Chlide *b* via DV Chlide *a*, is suggested by the absence of DV Pchlide *b* occurrence in the Nec 2 corn mutant which forms and accumulate only DV Chl *a* and DV Chl *b*. The establishment of precursor- product relationships in-vivo and in-vitro is required however, to validate the operation of route 6 in the Nec 2 corn mutant and the picoplankton of the euphotic zone of the world tropical and temperate oceans, and the Mediterranean sea.

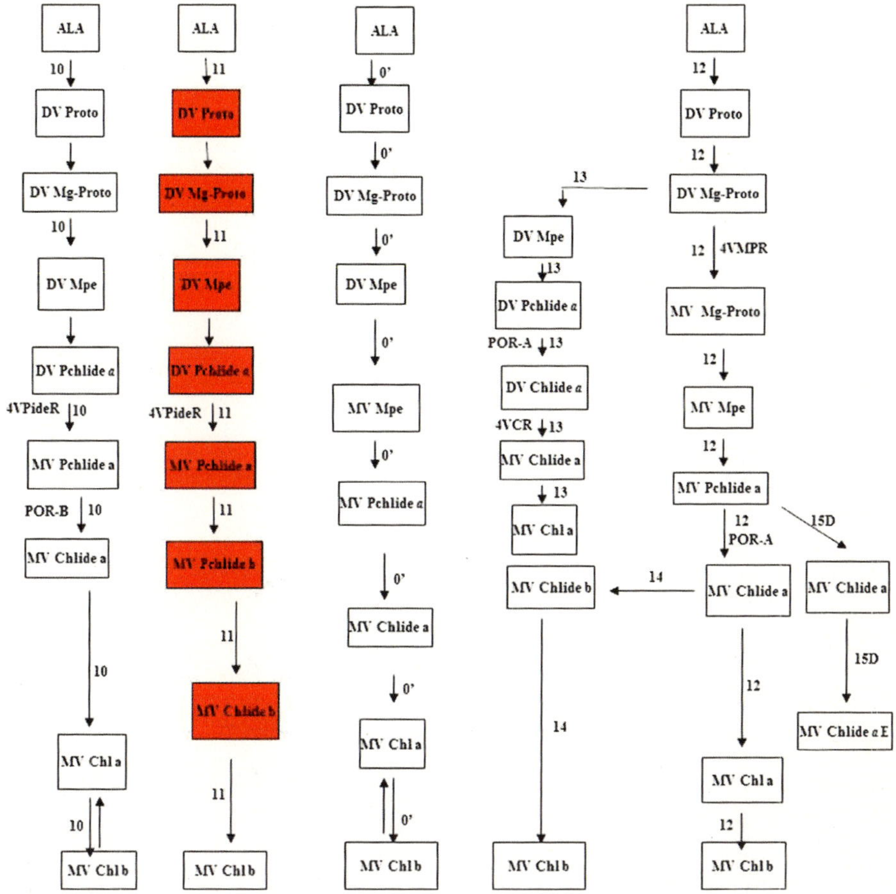


**Fig. 13.6** MV Chlide *b* biosynthesis via route 11 in DMV-LDV-LDMV plant species (Adapted from Fig. 6.4 of Chap. 6)

## *13.2.2 Chlorophyll b*

### 13.2.2.1 Chlorophyll *b* Biosynthetic Heterogeneity

The biosynthetic heterogeneity of Chl *b* is more complex than that of Chl *a* since it is based on the biosynthetic heterogeneity of Chl *a*, of MV Pchlide *b* and of MV Chlide *b*. As a consequence 12 different Chl *b* pools destined to different Chl-protein complexes appear to be formed during greening (Fig. 13.8).

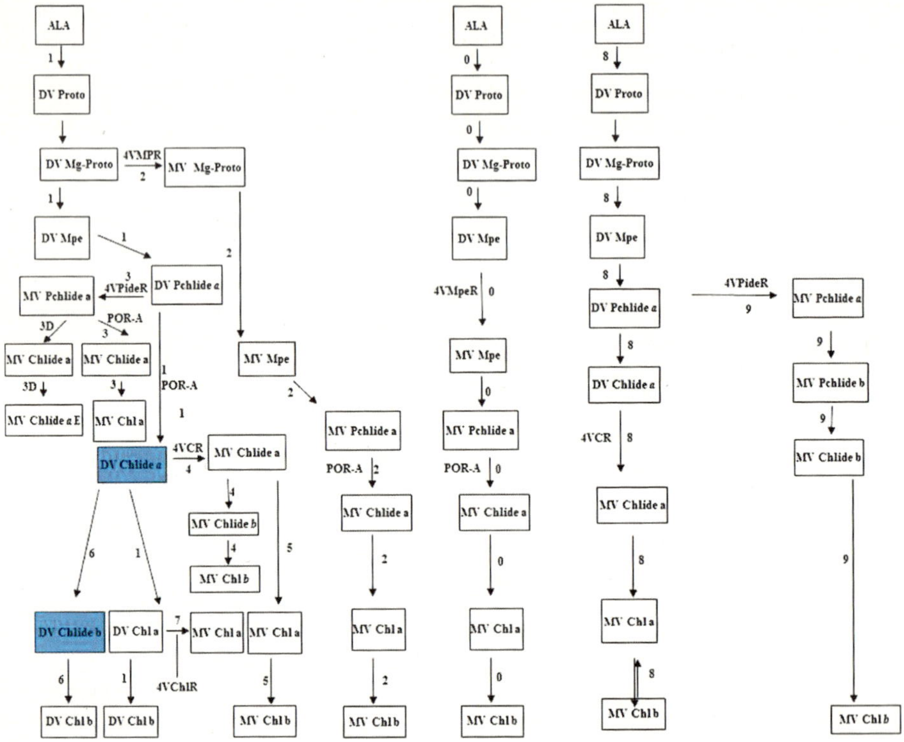


**Fig. 13.7** DV Chlide *b* biosynthesis via route 6 in DDV-LDV-LDDV plant species (Adapted from Fig. 6.3 of Chap. 6)

**Fig. 13.8** Monovinyl (MV) and Divinyl (DV) Chlorophyll *b*

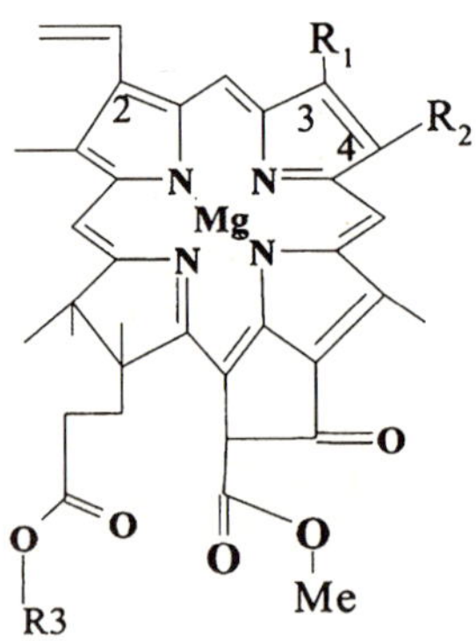


**Chlorophyll *b* (Chl *b*)**

| | R1 | R2 | R3 | Compound |
|---|---|---|---|---|
| I | Formyl | Vinyl | Phytol | DV Chl *b* |
| II | Formyl | Ethyl | Phytol | MV Chl *b* |

### Biosynthetic Heterogeneity of MV Chlorophyll *b* in DDV-LDV-LDDV Plant Species

Six different pools of MV Chl *b* are formed from MV Chlide *b*, MV Chl *a* and MV Pchlide *b*. The biosynthesis of these pools is discussed below.

#### *Formation of MV Chl b from MV Chlide a and MV Chl a via Routes 2, and 5 and from MV Chlide b via Route 4 in Etiolated DDV-LDV-LDDV Plants Subjected to Illumination*

We have repeatedly observed that in DDV-LDV-LDDV plant tissues such as etiolated cucumber cotyledons subjected to a brief light illumination then returned to darkness, MV Chl *b* accumulation as monitored by sensitive fluorescence techniques (Rebeiz 2002) is observed after 15–30 min of dark-incubation. By using less sensitive spectrophotometric techniques, it is observed that etiolated cucumber cotyledons subjected to continuous illumination, start accumulating measurable amounts of MV Chl *b* after 2 h of illumination (Rebeiz 1967). We propose that the biosynthesis of this MV Chl *b* can originate in routes 2, 4 and 5, as described below.

In route 2, MV Chl *b* would be formed from MV Chl *a* which is formed in turn from MV Pchlide *a* and MV Chlide *a* (Fig. 13.9).

In route 5, MV Chl *b* would be formed from MV Chl *a* which is formed in turn from DV Chlide *a*, and MV Chlide *a* (Fig. 13.9). In both cases the conversion of MV Chl *a* to MV Chl *b* is substantiated by favorable precursor-product relationships between these two tetrapyrroles in DDV-LDV-LDDV plants (Rebeiz et al. 1999).

In route 4, MV Chl *b* would be formed by esterification of nascent MV Chlide *b* which is formed in turn from DV Chlide *a* and MV Chlide *a* (Fig. 13.9). Esterification of MV Chlide *b* with GG followed by stepwise hydrogenation of MV Chl b-GG is strongly suggested by the detection of MV Chlide *b*-GG, Chlide *b*-DHGG, MV Chlide *b*-THGG, and MV Chl *b*-phytol in greening etiolated cucumber cotyledons (Shio and Sasa 1983).

Since during the initial phases of greening of etiolated tissues, most of the Chl consists of MV Chl, located in PSI and PSII (Akoyunoglou 1978; Akoyunoglou et al. 1981; Alberte et al. 1972), we propose that the MV Chl *b* formed via routes 2, 4, and 5 is destined to PSI and/or PSII inner antenna Chl-protein complexes. It is not certain at this stage whether the MV Chl *b* formed via routes 2, and 5 during the very early phases of greening is convertible to MV Chl *a* or not as has been reported for later stages of greening (Ohtsuka et al. 1996).

#### *Biosynthesis of MV Chl b from MV Chl a via Route 0 in Greening DDV-LDV-LDDV Plants During the Light Phases of the Photoperiod*

The operation of this biosynthetic route depends on the Detection of DV Mg-Proto reductase in DDV-LDV-LDDV plant species (Fig. 13.10).

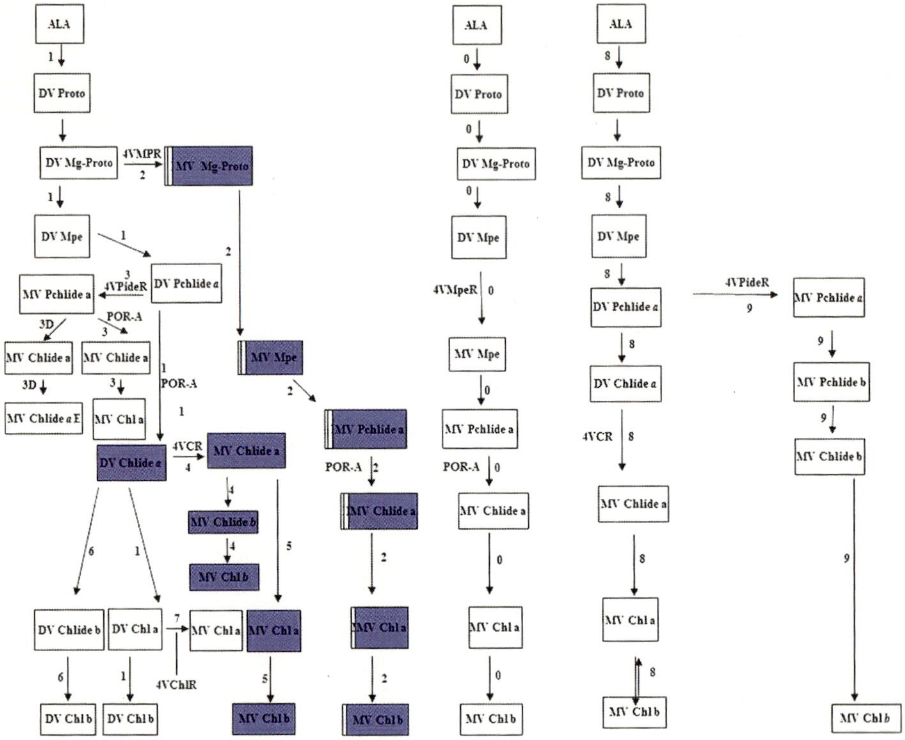

**Fig. 13.9** MV Chlide *b* biosynthesis via routes 2, 4 and 5 in DDV-LDV-LDDV plant species at the beginning of illumination of etiolated plant species (Adapted from Fig. 6.3 of Chap. 6)

*Biosynthesis of MV Chl b from MV Chl a via Route 8 in Greening DDV-LDV-LDDV Plants During the Light Phases of the Photoperiod*

Most of the MV Chl *b* accumulation in DDV-LDV-LDDV plants takes place during the light phases of the photoperiod. In route 8, MV Chl *b* is formed from MV Chl *a* which is formed in turn from DV Chlide *a* and MV Chlide *a* (Fig. 13.11). In this case too, conversion of MV Chl *a* to MV Chl *b* is supported by favorable precursor product relationship between MV Chl *a* and MV Chl *b* after 7 and 8 h of illumination of DDV-LDV-LDDV plant tissues (Rebeiz et al. 1999).

Since under continuous illumination, most of the synthesized Chl consists of MV Chl *a*, and *b* located in antenna Chl-protein complexes, (Akoyunoglou 1978; Akoyunoglou et al. 1981; Alberte et al. 1972), we propose that the MV Chl *b* formed via route 8 is destined to LHCII and other outer antenna Chl-protein complexes.

It is presently acknowledged that under certain conditions, MV Chl *b* is converted to MV Chl *a* in green cucumber cotyledons (Ohtsuka et al. 1996). These authors proposed that during the light phase of the photoperiod, the photosynthetic apparatus is reorganized during acclimation to various light environments. This

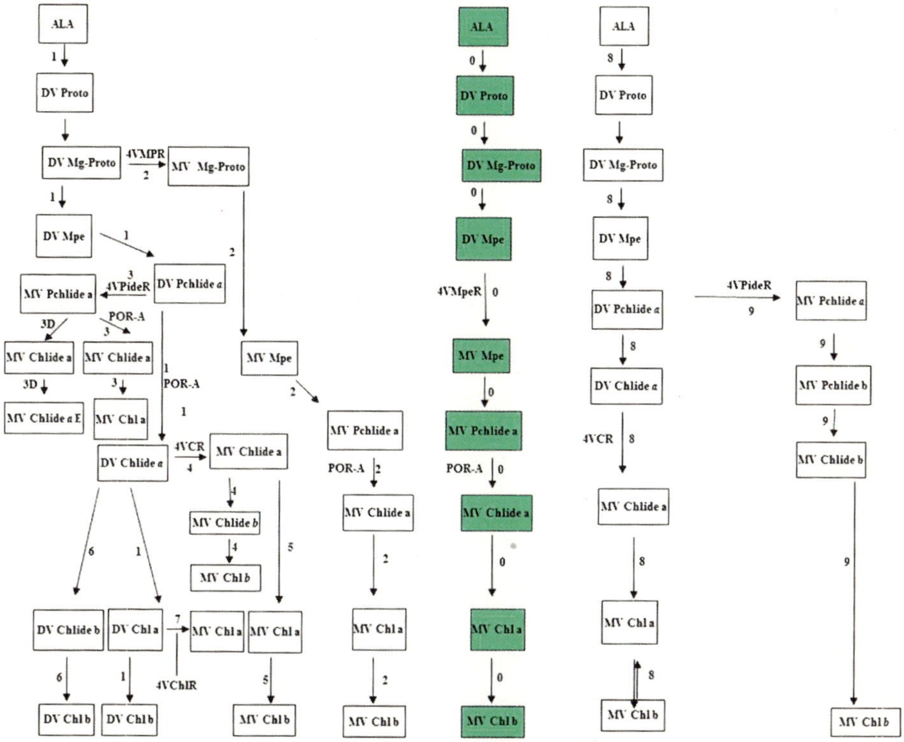


**Fig. 13.10** Proposed MV Chlide *b* biosynthesis via routes 0 in DDV-LDV-LDDV plant species in greening plant species (Adapted from Fig. 6.3 of Chap. 6)

reorganization involves release of MV Chl *b* from the light-harvesting Chl *a*/*b* protein complex of PSII. The released MV Chl *b* is then converted to MV Chl *a* by a Chl *b* formyl reductase. The nascent MV Chl *a* is then used for the formation of core complexes of PSI and PSII. On the basis of these results we propose that MV Chl *b* formed via route 8, is convertible to MV Chl *a* in green DDV-LDV-LDDV plant tissues.

*Formation of Chl b from MV Pchlide b via Route 9 in Greening DDV-LDV-LDDV Plants During the Light Phases of the Photoperiod*

In route 9, MV Chl *b* is formed from MV Chlide *b* which is formed in turn from MV Pchlide *a* and MV Pchlide *b* in green DDV-LDV-LDDV plants (Fig. 13.12). The conversion of MV Pchlide *b* to MV Chlide *b* is suggested by the detection of MV Pchlide *b* (Ioannides et al. 1997; Kolossov and Rebeiz 2003) and MV Chlide *b* (Duggan and Rebeiz 1982) in green(ing) cucumber cotyledons. A specific precursor-product relationship remains to be established however between MV Pchlide *b* and MV Chlide *b* in DDV-LDV-LDDV plant tissues. As in the case

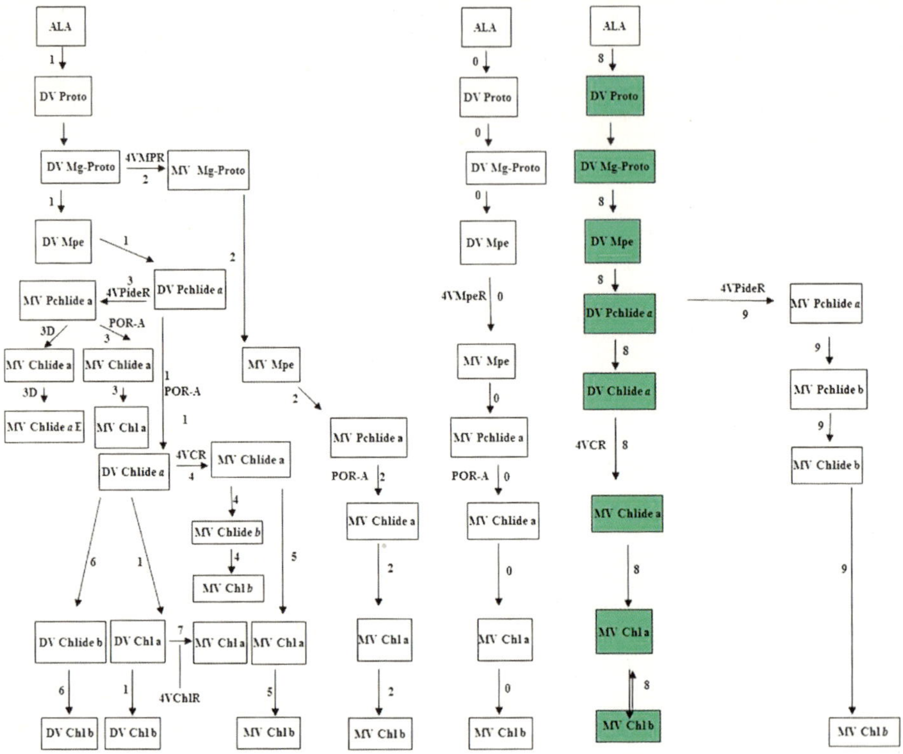

Fig. 13.11 Proposed MV Chlide *b* biosynthesis via routes 8 in DDV-LDV-LDDV plant species. During the light phases of the photoperiod (Adapted from Fig. 6.3 of Chap. 6)

with other Chl biosynthetic routes that operate during the light phases of the photoperiod, it is our guess that the putative MV Chl *b* formed via route 9 is destined to LHCII and other outer antenna Chl-protein complexes (Fig. 13.12).

## Biosynthetic Heterogeneity of MV Chlorophyll *b* in DMV-LDV-LDMV Plant Species

Monovinyl chlorophyll *b* is formed via five different biosynthetic routes in DMV-LDV-LDMV Plant species. These routes are discussed below.

### *Biosynthesis of MV Chl b from MV Chl a via Route 10 in Greening DMV-LDV-LDMV Plants During the Light Phases of the Photoperiod*

As was observed in DDV-LDV-LDDV plants, in DMV-LDV-LDMV plants, most of the MV Chl *b* accumulation takes place during the light phases of the photoperiod. In route 10, MV Chl *b* is formed from MV Chl *a* which is formed in

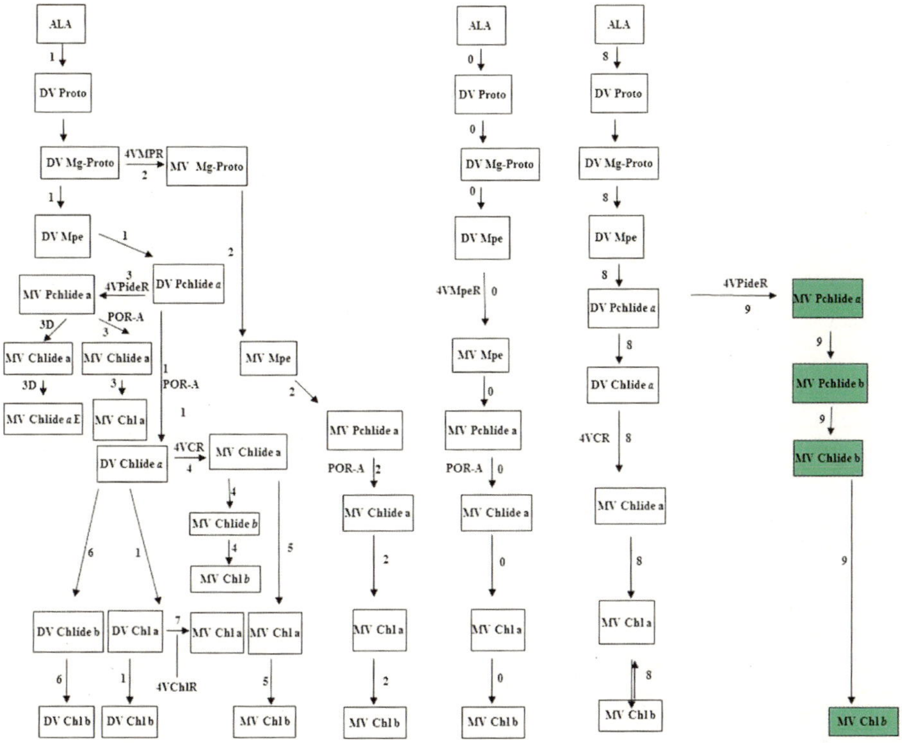


**Fig. 13.12** Proposed MV Chlide *b* biosynthesis via route 9 in DDV-LDV-LDDV plant species. During the light phases of the photoperiod. (Adapted from Fig. 6.4 of Chap. 6)

turn from MV Chlide *a* and MV Pchlide *a* (Fig. 13.13). In this case too conversion of MV Chl *a* to MV Chl *b* is supported by the favorable precursor product relationship between MV Chlide *a* and MV Chl *b* in DMV-LDV-LDMV plant tissues after 7 and 8 h of illumination (Rebeiz et al. 1999). At this stage, we have no reason to argue against a certain degree of direct conversion of MV Chlide *a* to MV Chl *a*-phytol in green DMV-LDV-LDMV plants as was observed in spinach chloroplasts (Soll et al. 1983), and greening cucumber etiochloroplasts (Daniell and Rebeiz 1984).

Since under continuous illumination, most of the synthesized Chl consists of MV Chl *a*, and *b* located in antenna Chl-protein complexes, (Akoyunoglou 1978; Akoyunoglou et al. 1981; Alberte et al. 1972), we propose that the MV Chl *a* formed via route 10 is destined to LHCII and other antenna Chl-protein complexes. We also propose that MV Chl *b* formed via route 10, is convertible to MV Chl *a* as the photosynthetic apparatus is reorganized during acclimation to various light environments (Ohtsuka et al. 1996).

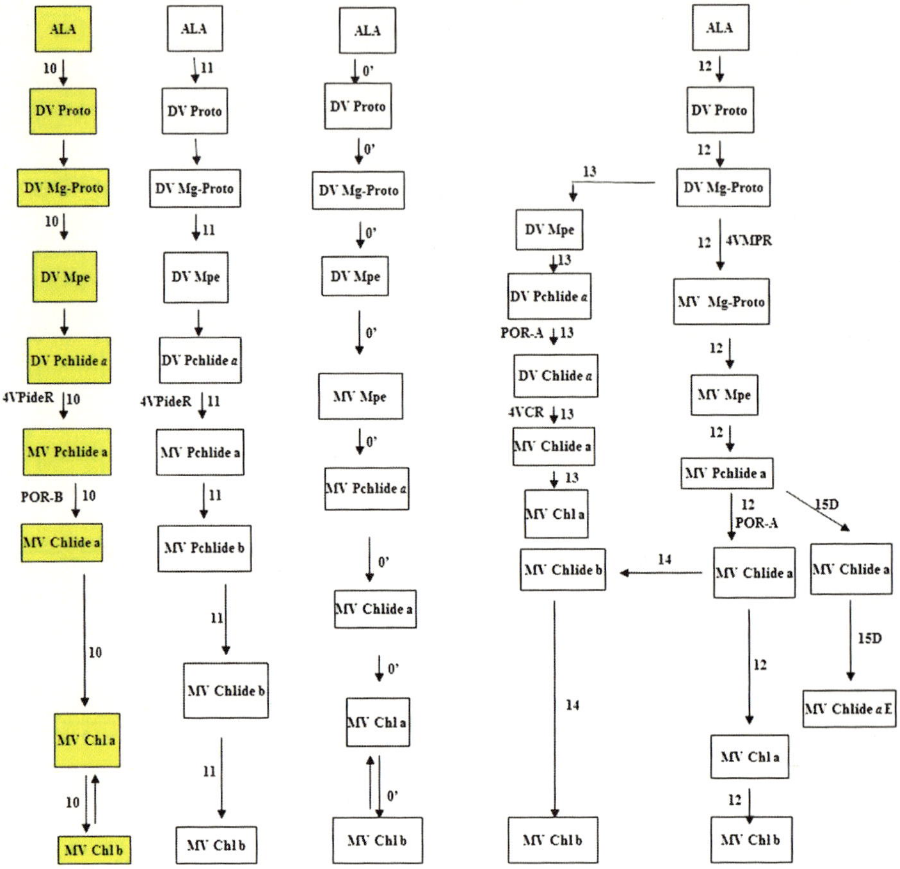


**Fig. 13.13** Biosynthesis of MV Chl *b* via route 10 in greening DMV-LDV-LDMV plants during the light phases of the photoperiod (Adapted from Fig. 6.4 of Chap. 6)

### *Biosynthesis of MV Chl b from MV Pchlide b via Route 11 in Greening DMV-LDV-LDMV Plants During the Light Phases of the Photoperiod*

In route 11, MV Chl *b* is formed from MV Chlide *b* which in turns is formed from MV Pchlide *a* and MV Pchlide *b* in green DMV-LDV-LDMV plants. The conversion of MV Pchlide *b* to MV Chlide *b* is suggested by the detection of MV Pchlide *b* in barley, a DMV-LDV-LDMV plant (Kolossov and Rebeiz 2003), and the lack of precursor product relationship between MV Chlide *a* and MV Chlide *b* in greening corn seedlings during lengthy (15 h) illuminations (Ioannides 1993). This observation argues against the formation of MV Chl *b* from MV Chlide *a* and MV Chlide *b* during lengthy light phases of the photoperiod. A specific precursor-product relationship remains to be established however between MV Pchlide *b* and MV Chlide *b* and MV Chl *b* in green DMV-LDV-LDMV plant tissues. As in the case of

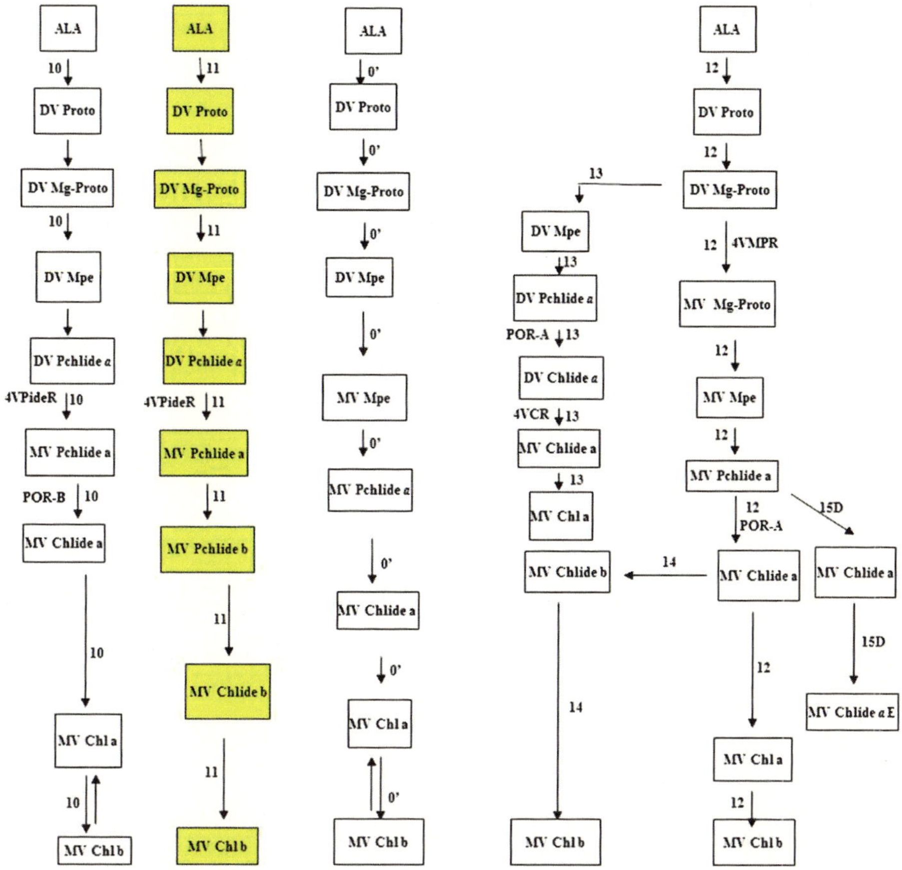


**Fig. 13.14** Biosynthesis of MV Chl *b* via route 11 in greening DMV-LDV-LDMV plants during the light phases of the photoperiod (Adapted from Fig. 6.4 of Chap. 6)

other biosynthetic routes that operate during the light phases of the photoperiod, our guess would be that the putative MV Chl *b* formed via route 11 is destined to LHCII and other antenna Chl-protein complexes (Fig. 13.14).

*Biosynthesis of MV Chl b from MV Pchlide b via Route 0′ in Greening DMV-LDV-LDMV Plants During the Light Phases of the Photoperiod*

The Operation of biosynthetic route 0′ in DMV-LDV-LDMV plants during photoperiodic greening is justified by the detection and solubilization of 4-Vinyl Mpe reductase (4VMpeR) in greening barley etiochloroplasts (Kolossov and Rebeiz 2010). Such etiochloroplasts can actively convert MV Mpe to MV Pchlide *a* (Tripathy and Rebeiz 1986) (Fig. 13.15).

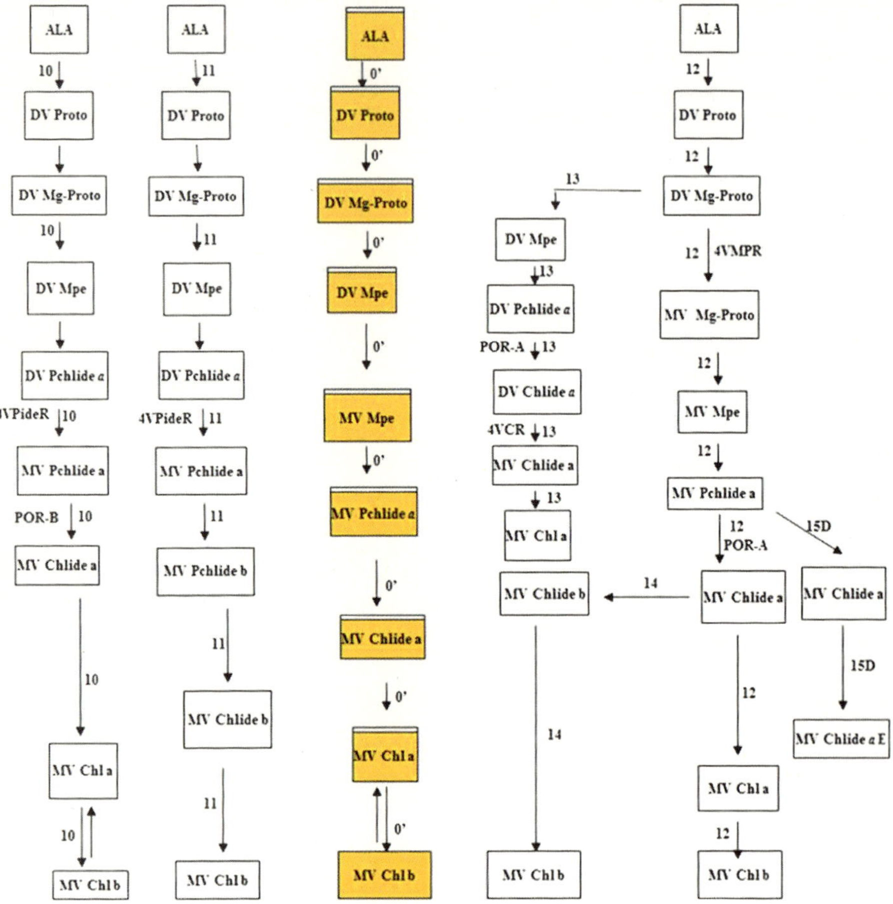


**Fig. 13.15** Biosynthesis of MV Chl *b* via route 0′ in greening DMV-LDV-LDMV plants during the light phases of the photoperiod (Adapted from Fig. 6.4 of Chap. 6)

In biosynthetic route 0′, DV Mpe is converted to MV Mpe by 4VMpeR (Kolossov and Rebeiz 2001). Then MV Mpe is converted to MV Chlide *a* probably via POR-B and the latter to Chl *a* probably by direct phytylation (Daniell and Rebeiz 1984; Soll and Schultz 1981). MV Chl *a* is then proposed to be converted reversibly to Chl *b* as proposed by Ohtsuka et al. (1996 #214).

### *Biosynthesis of MV Chl b from MV Chl a via Route 12 and from MV Chlide a via Route 14 in Etiolated DMV-LDV-LDMV Plants After Exposure to Light, and in Greening DMV-LDV-LDMV Plants During the Initial Dark Phases of the Photoperiod*

We have repeatedly observed that in DMV-LDV-LDMV plant tissues such as etiolated barley seedlings subjected to brief illumination then returned to darkness,

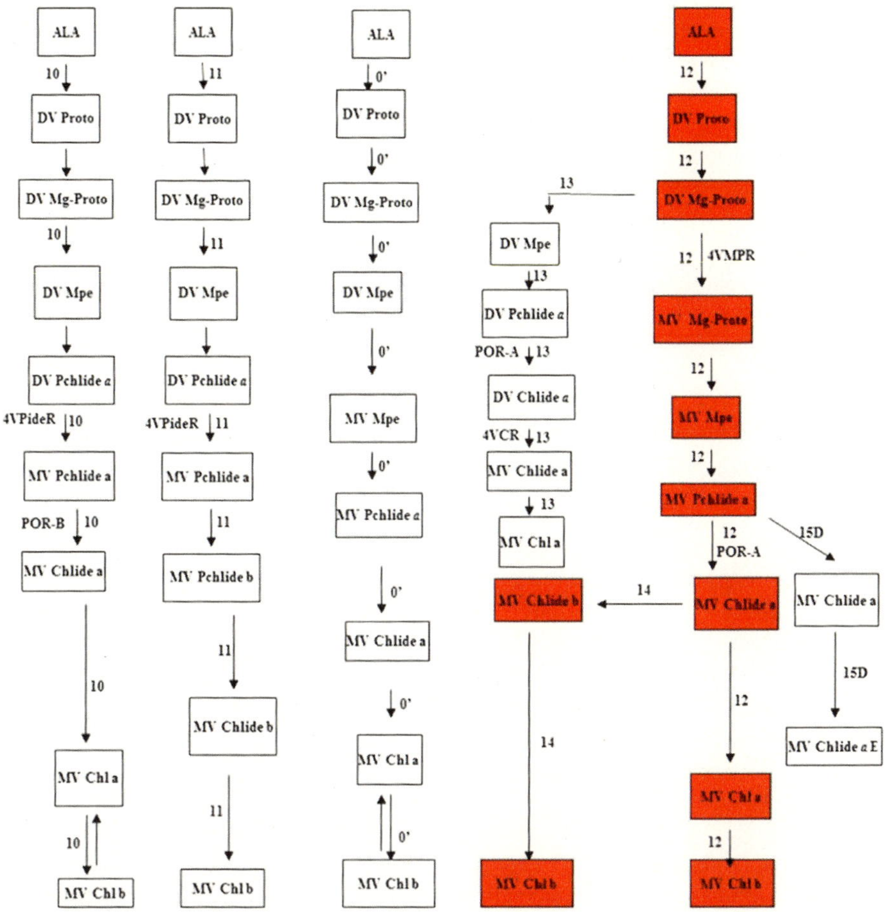


**Fig. 13.16** Biosynthesis of MV Chl *b* via routes 12 and 14 in greening DMV-LDV-LDMV plants during the light phases of the photoperiod (Adapted from Fig. 6.4 of Chap. 6)

MV Chl *b* accumulation, monitored by sensitive fluorescence techniques (Rebeiz 2002) is observed after 15–30 min of dark-incubation. We propose that the biosynthesis of this MV Chl *b* originates in routes 12 and 14.

In route 12, MV Chl *b* would be formed from MV Chl *a* which is formed in turn from MV Pchlide *a* and MV Chlide *a* (Fig. 13.16). Under the same conditions, conversion of MV Chl *a* to MV Chl *b* by route 12 appeared to be low, and amounted to a maximum of 5 % after 7 h of greening (Rebeiz et al. 1999). These results indicated that under these conditions, the fate of the nascent MV Chlide *a* was conversion to MV Chl *b* by way of MV Chlide *b* rather than conversion to MV Chl *b* via MV Chl *a*. After 7 h of greening, the rate of MV Chl *b* formation from MV Chlide *a* and *b* increased to 56–68 % (Rebeiz et al. 1999).

Since during the initial phases of greening of etiolated tissues, most of the Chl consists of MV Chl, located in PSI and PSII (Akoyunoglou 1978; Akoyunoglou et al. 1981; Alberte et al. 1972), we propose that the MV Chl *b* formed via routes

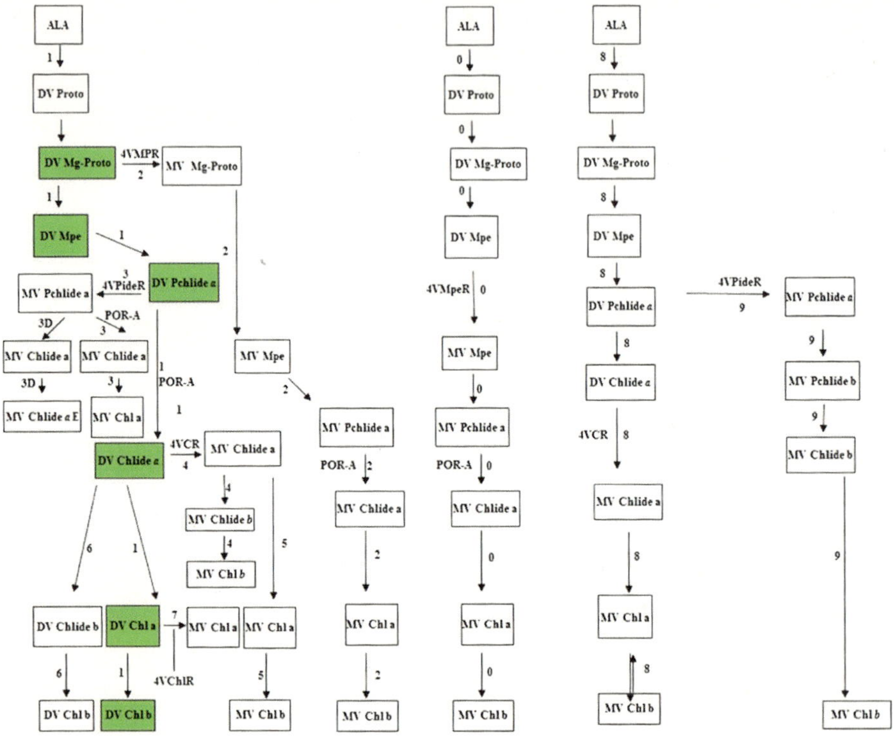


**Fig. 13.17** Biosynthesis of MV Chl *b* via route 1 in the Nec 7 corn mutant and in picoplanktons (Adapted from Fig. 6.3 of Chap. 6)

12, and 14 is destined to PSI and/or PSII inner antenna Chl-protein complexes. It is not certain at this stage whether the MV Chl *b* formed via routes 12, 14 during the very early phases of greening is convertible to MV Chl *a* or not as has been reported for later stages of greening (Ito et al. 1994, 1996).

### 13.2.2.2 Biosynthesis of DV Chlorophyll *b*

The occurrence of DV Chl *b* has not been observed in normal higher plants. However, DV Chl *b* is the major Chl *b* that accumulates in the Nec 2 corn mutant and in the prochlorophyte picoplankton of the subtropical waters of the North Atlantic as well as in the picoplankton of the euphotic zone of the world tropical and temperate oceans, and the Mediterranean sea (Chisholm et al. 1988, 1992; Goerike and Repeta 1992).

Conversion of DV Chl *a* to DV Chl *b* via Route 1

Conversion of DV Chl *a* to DV Chl *b* via DV Pchlide *a* and DV Chlide *a* (Fig. 13.17) probably takes place in the Nec 2 corn mutant and in the prochlorophyte

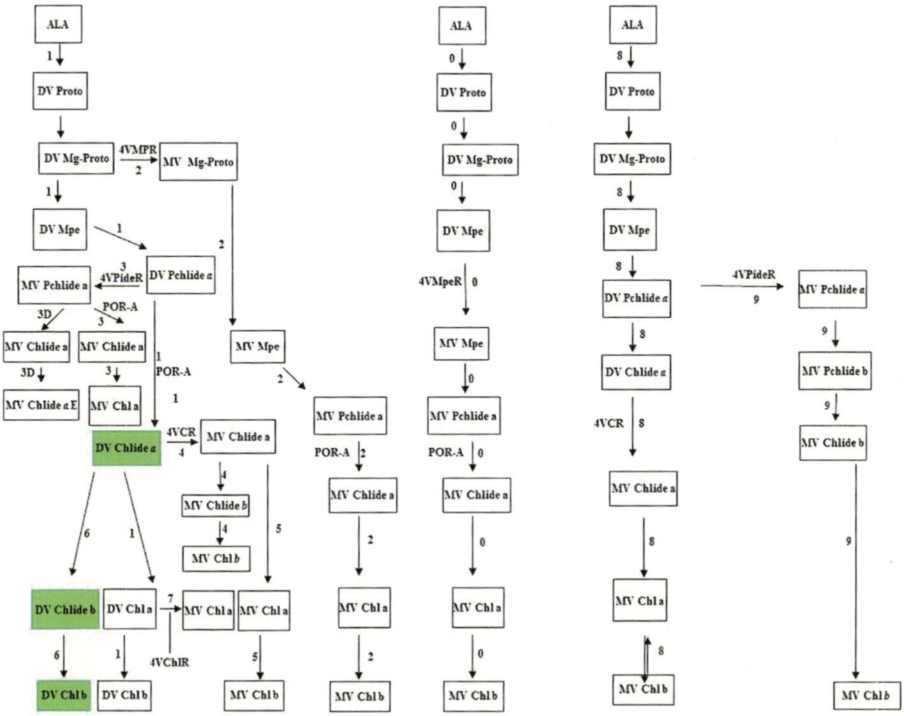

Fig. 13.18 Biosynthesis of MV Chl *b* via route 6 in the Nec 2 corn mutant and in picoplanktons (Adapted from Fig. 6.3 of Chap. 6)

picoplankton (see above) which form and accumulate only DV Chl *a* and DV Chl *b*. Precursor- product relationships between DV Chl *a* and DV Chl *b* in vivo and in vitro is required however, to validate this hypothesis.

### Conversion of DV Chlide *b* to DV Chl *b* via Route 6

Conversion of DV Chlide *b* to DV Chl *b* via route 6 (Fig. 13.18) may also take place in the Nec 2 corn mutant and in the prochlorophyte picoplankton (Chisholm et al. 1988, 1992; Goerike and Repeta 1992) which form and accumulate only DV Chl *a* and DV Chl *b*. Precursor- product relationships between DV Chlide *b* and DV Chl *b* in these organisms in vivo and in vitro is required however, to validate this hypothesis.

## References

Akoyunoglou G (1978) Growth of the PS II unit by fluorescence measurements, the photoinduced absorbance change at 518 nm and photochemical activity. In: Akoyunoglou G, Argyroudi-Akoyunoglou JH (eds) Chloroplast development. Elsevier, New York, p 2

Akoyunoglou G, Tsakiris S, Argyroudi-Akoyunoglou JH (1981) Independent growth of the photosystem I and II units. The role of the light-harvesting pigment-protein complexes. In: Akoyunoglou G (ed) Photosynthesis V. Chloroplast development. Balaban International Science Services, Philadelphia, pp 523–533

Alberte RS, Thornber JP, Naylor AW (1972) Time of appearance of photosystem I and II in chloroplast of greening Jack bean leaves. J Exp Bot 23(77):1060–1069

Aronoff S (1981) Chlorophyllide *b*. Biochem Biophys Res Commun 102:108–112

Bazzaz MB (1981) New chlorophyll chromophores isolated from a chlorophyll deficient mutant of maize. Photobiochem Photobiophys 2:199–207

Benz J, Rudiger W (1981) Chlorophyll biosynthesis: various chlorophyllides as exogenous substrates for chlorophyll synthetase. Z Naturforsch 36c:51–57

Chisholm S, Olson RJ, Zettler ER et al (1988) A novel free-living prochlorophyte abundant in the oceanic euphotic zone. Nature 334:340–343

Chisholm SW, Frankel S, Goerike R et al (1992) *Prochlorococcus marinus* nov. gen. sp.: an oxyphototrophic marine prokaryote containing divinyl chlorophyll *a* and *b*. Arch Mikrobiol 157:297–300

Daniell H, Rebeiz CA (1984) Bioengineering of photosynthetic membranes: requirement of magnesium for the conversion of chlorophyllide a to chlorophyll a during the greening of etiochloroplasts in vitro. Biotechnol Bioeng 26:481–487

Duggan JX, Rebeiz CA (1981) Detection of a naturally occurring chlorophyllide *b* pool in higher plants. Plant Physiol 67(suppl):267

Duggan JX, Rebeiz CA (1982) Chloroplast biogenesis 38. Quantitative detection of a chlorophyllide b pool in higher plants. Biochim Biophys Acta 714:248–260

Goerike R, Repeta D (1992) The pigments of *Prochlorococcus marinus*. The presence of divinyl-chlorophyll a and b in a marine prochlorophyte. Limnol Oceanogr 37:425–433

Ioannides IM (1993) Intermediate metabolism of the chlorophyll *b* biosynthetic pathway. Nat Res Env Sci, Urbana, p 132

Ioannides ML, Shedbalkar VP, Rebeiz CA (1997) Quantitative determination of 2-monovinyl protochlorophyllide *b* by spectrofluorometry. Anal Biochem 249:241–244

Ito H, Takaichi S, Tsuji H et al (1994) Properties of synthesis of chlorophyll *a* from chlorophyll *b* in cucumber etioplasts. J Biol Chem 269:22034–22038

Ito H, Ohtsuka T, Ayumi T (1996) Conversion of chlorophyll *b* to chlorophyll *a* via 7-hydroxymethyl chlorophyll. J Biol Chem 271:1475–1479

Klement H, Helfrich M, Oster U et al (1999) Pigment-free NADPH: protochlorophyllide oxidoreductase from *Avena sativa* L: purification and substrate specificity. Eur J Biochem 265 (3):862–874

Kolossov VL, Rebeiz CA (2003) Chloroplast biogenesis 88. Protochlorophyllide *b* occurs in green but not in etiolated plants. J Biol Chem 278(50):49675–49678

Kolossov VL, Rebeiz CA (2010) Evidence for various 4-vinyl reductase activities in higher plants. In: Rebeiz CA, Benning C, Bohnert HJ et al (eds) The chloroplast: basics and applications. Springer, Dordrecht/London, pp 25–38

Kolossov V, Ioannides IM, Kulur S et al (1999) Chloroplast biogenesis 82: development of a cell-free system capable of the net synthesis of chlorophyll(ide) *b*. Photosynthetica 36:253–258

Ohtsuka T, Ito H, Ayumi T (1996) Conversion of chlorophyll *b* to chlorophyll *a* and the assembly of chlorophyll with apoproteins by isolated chloroplasts. Plant Physiol 113:137–147

Porra RJ, Schafer W, Cmiel E, Katheder I, Scheer H (1993) Derivation of the formyl-group oxygen of chlorophyll *b* from molecular oxygen in greening leaves of higher plants (*Zea mays*). FEBS 323:31–34

Porra RJ, Schafer W, Cmiel E et al (1994) The derivation of the formyl-group oxygen of chlorophyll *b* in higher plants from molecular oxygen. Achievement of high enrichment of the 7-formyl-groupoxygen from $^{18}O_2$ in greening maize leaves. Eur J Biochem 219:671–679

Rebeiz CA (1967) Studies on chlorophyll biosynthesis in etiolated excised cotyledons of germinating cucumber at different stages of seedling development. Magon Serie Scientifique 13:1–21

Rebeiz CA (2002) Analysis of intermediates and end products of the chlorophyll biosynthetic pathway. Heme chlorophyll and bilins. In: Smith A, Witty M (eds) Methods and protocols. Humana Press, Totowa, pp 111–155
Rebeiz CA, Mayasich JM, Tripathy BC (1988) Chloroplast biogenesis 61: kinetic analysis of precursor-product relationships in complex biosynthetic pathways. J Theor Biol 133:319–326
Rebeiz CA, Ioannides IM, Kolossov V et al (1999) Chloroplast biogenesis 80. Proposal of a unified multibranched chlorophyll a/b biosynthetic pathway. Photosynthetica 36:117–128
Shio Y, Sasa T (1983) Esterification of chlorophyllide *b* in higher plants. Biochem Biophys Acta 756:127–131
Soll J, Schultz G (1981) Phytol synthesis from geranylgeraniol in spinach chloroplasts. Biochem Biophys Res Commun 99:907–912
Soll J, Schultz G, Rudiger W et al (1983) Hydrogenation of grenylgeraniol: two pathways exist in spinach chloroplasts. Plant Physiol 71:849–854
Tripathy BC, Rebeiz CA (1986) Chloroplast biogenesis. Demonstration of the monovinyl and divinyl monocarboxylic routes of chlorophyll biosynthesis in higher plants. J Biol Chem 261:13556–13564
Tripathy BC, Rebeiz CA (1988) Chloroplast biogenesis 60: conversion of divinyl protochlorophyllide to monovinyl protochlorophyllide in green(ing) barley, a dark monovinyl light divinyl plant species. Plant Physiol 87:89–94
Veldhuis MJW, Kraay GW (1990) Vertical distribution of pigment composition of a picoplankton prochlorophyte in the subtropical north Atlantic: a combined study of pigments and flow cytometry. Mar Ecol Prog Ser 68:121–127

# Chapter 14
# Relationship of Chlorophyll Biosynthetic Heterogeneity to the Greening Group Affiliation of Plants

*If scientific reasoning were limited to the logical processes of arithmetic, we should not get very far in our understanding of the physical world (Vannevar Bush).*

## 14.1 Introduction

As we have described at great length in the first 13 chapters, biosynthetic heterogeneity refers to the biosynthesis of a particular metabolite by an organelle, tissue or organism via multiple biosynthetic routes (Rebeiz et al. 2003). It has been well documented in delta-aminolevulinic acid (ALA), chlorophyll (Chl) *a* and vitamin $B_{12}$ biosynthesis (Arigoni 1994; Rebeiz et al. 1994; Scott 1994). As described in previous chapters, It has been demonstrated that in green plants, Chl *a* and Chl *b* are formed via parallel biosynthetic routes, namely (a) DV Chl *a* biosynthetic routes, (b) MV routes and (c) mixed DV-MV routes (Kolossov and Rebeiz 2010). Intermediates of the DV carboxylic route consist of dicarboxylic and monocarboxylic tetrapyrroles with vinyl groups at positions 2 and 4 of the macrocycle, such as DV protoporphyrin IX (Proto), DV Mg-Proto, DV Mg-Proto monomethyl ester (Mpe), DV Pchlide *a*, and DV Chlide *a*. The MV carboxylic routes involve dicarboxylic and monocarboxylic tetrapyrroles including, MV Mg-Proto, MV Mpe, MV Pchlide *a*, and MV Chlide *a*, which have one vinyl and one ethyl group at positions 2 and 4 of the macrocycle, respectively. The mixed DV-MV carboxylic routes involve monocarboxylic tetrapyrroles such as DV and MV Pchlide *a* and Chlide *a*.

C.A. Rebeiz, *Chlorophyll Biosynthesis and Technological Applications*,
DOI 10.1007/978-94-007-7134-5_14, 

In higher plants, the end product of the Chl *a* biosynthetic heterogeneity is invariably MV Chl *a* and *b*, with the only known exception of a lethal maize mutant (Bazzaz 1981) which forms only DV Chl *a* and *b*. However, in the prochlorophyte picoplankton of the subtropical waters of the North Atlantic as well as in the picoplankton of the euphotic zone of the world tropical and temperate oceans, and the Mediterranean sea, DV Chl *a* and *b* are the predominant Chl species (Chisholm et al. 1988, 1992; Goerike and Repeta 1992; Shioi and Sasa 1983; Veldhuis and Kraay 1990). It has been proposed that in green plants, the multiplicity of Chl *a* biosynthetic routes, produces different pools of MV Chl *a*, complexed to different pigment-proteins at specific sites of the photosynthetic membranes (Rebeiz et al. 1983, 1994, 2003, 2005).

The DV and MV Chl *a* biosynthetic routes are linked at the level of DV Mg-Proto (Kim and Rebeiz 1996), DV Mpe (Kolossov and Rebeiz 2010), DV Pchlide *a* (Tripathy and Rebeiz 1988), DV Chlide *a* (Parham and Rebeiz 1992, 1995), and DV Chl *a* (Adra and Rebeiz 1998) by [4-vinyl] reductase(s) that convert the 4-vinyl group at position 4 to ethyl, thus converting a DV tetrapyrrole to a MV tetrapyrrole.

It is relevant to point out that Whyte and Griffiths (1993) have interpreted the accumulation of DV and MV Pchlide *a* in term of a dual pathway with a single vinyl reductase of broad specificity. In their scheme the major route converts DV Pchlide *a* to MV Chl *a* via DV Chlide *a*, and MV Chlide *a*. The minor route consists of the formation of MV Chlide *a*, and by inference of MV Chl *a* via DV Pchlide *a*, which is converted to MV Pchlide *a* by the non-specific vinyl reductase. This hypothesis is not compatible with the following observations: (a) It has been demonstrated that during DV and MV Pchlide *a* biosynthesis, only a fraction of the MV Pchlide *a* pool can be formed by reduction of DV Pchlide *a* (Tripathy and Rebeiz 1988), and (b) In *Rhodobacter capsulatus* in which the bchJ gene which codes for DV Pchlide *a* reductase (4VpideR) has been deleted, in addition to the accumulation of DV Pchlide *a*, accumulation of MV Mg-Proto, its monoester (precursors of Pchlide *a*), and MV Pchlide *a* have been observed (Suzuki and Bauer 1995). This in turn indicates that at least one separate [4-vinyl] reductase exists which acts prior to DV Pchlide *a* and DV Chlide *a* vinyl reduction. This enzyme would be responsible for the accumulation of MV Mg protoporphyrins in plants (Belanger and Rebeiz 1982), as well as for MV Pchlide *a* formation in the absence of 4VpideR. Finally Significant evidence indicates the existence of multiple vinyl-reductases in green plants (Kolossov and Rebeiz 2010).

Very recently the Chl *a* biosynthetic heterogeneity has been extended to the level of Pchlide *a* photoreduction, by the discovery of a second Pchlide *a* oxidoreductase (POR), in addition to the conventional POR of etiolated tissues (Armstrong et al. 1995; Holtorf et al. 1995). One, POR-A, is the conventional photoenzyme, which occurs in etiolated tissues but disappears during greening. The second, POR-B is present throughout angiosperm development. Armstrong et al. (1995) have suggested that POR-A performs a specialized function restricted to the initial stages of greening, while POR-B is involved in maintaining Chl levels throughout angiosperm development.

## 14.2 Greening Group Affiliation of Green Plants: Discovery of the Divinyl (DV) and Monovinyl (MV) Greening Groups of Plants

Prior to 1985 it was assumed that the greening process was uniform across all green plants *i.e.* all green plants formed Chl via the same Chl biosynthetic route(s). Several observations made in the early 1980s, helped dispel, this misconception as described below.

The first observation related to multiple greening groups of plants was made in 1985. By then the Chl biosynthetic heterogeneity was already established (Rebeiz et al. 1983). It was realized that different species of plants greened differently by using different Chl Biosynthetic routes. Thus, On the basis of the DV-MV biosynthetic heterogeneity, green plants were initially classified into three different greening groups depending upon MV or DV Pchlide *a* accumulation during the dark and light phases of the photoperiod (Carey and Rebeiz 1985; Ioannides et al. 1994; Shioi and Takamiya 1992). Dark Divinyl-Light Divinyl (DDV-LDV) plants, such as *Pogostemon cablin*, cucumber (*Cucumis sativus*), velvetleaf (*Abutilon theophrastii*), common morningglory (*Ipomea purpurea*), prickly sida (*Sida spinosa*), and mustard (*Brassica nigra*), accumulated mainly DV Pchlide *a* at night and in daytime the analyzed Pchlide *a* pool consisted mainly of DV Pchlide *a*. All representative primitive plant species, including algae, bryophytes, ferns, and gymnosperms, fell into this greening group. This led Ioannides et al. (1994) to propose that this greening group is evolutionary ancestral (Table 14.1).

Dark Monovinyl-Light Monovinyl (DMV-LMV) plants, such as barnyardgrass (*Echinochloa crus-galli*) and johnsongrass (*Sorghum halepense*), accumulated mainly MV Pchlide *a* at night and in daytime the Pchlide *a* consisted mainly of MV Pchlide *a* (Table 14.1). This greening group comprised a small number of plants, and evolutionary studies suggested that it is derived (Ioannides et al. 1994). Finally, it was observed that Dark Monovinyl-Light Divinyl (DMV-LDV) plants such as French bean (*Phaseolus vulgaris L.*), corn (*Zea mays*), wheat (*Triticum aestivum*), wild oat (*Avena fatua*), barley (*Hordeum vulgare*), soybean (*Glycine max*), lambsquarter (*Chenopodium album*), jimsonweed (*Datura strmonium*), red-root pigweed (*Amaranthus retroflexus*), cocklebur (*Xanthium stumarium*), etc., accumulated MV Pchlide *a* at night. In daytime the Pchlide *a* pool consisted mainly of DV Pchlide *a*. This greening group comprised by far the largest number of surveyed plant species, and evolutionary studies suggested that it was evolutionary intermediate (Ioannides et al. 1994). Plant species of major agronomic importance belonged to this group.

Table 14.1 extracted from Ioannides et al. (1994), depicts a brief survey of various plant species belonging to the three greening groups mentioned above.

**Table 14.1** Plant material examined and greening group

| Scientific name | Family | Subclass | DDV-LDV | DMV-LDV | DMV-LMV | $C_3$–$C_4$ |
|---|---|---|---|---|---|---|
| **Non-flowering plants** | | | | | | |
| *Chara rusbyana* | Characeae | | 0–0 | | | |
| *Nitellopsis obtusa* | Characeae | | 0–0 | | | |
| *Marchantia polymorha* | Marchantiaceae | | 0–0 | | | |
| *Cystopteris bulbifera* | Aspleniaceae | | 0–0 | | | $C_3$ |
| *Lycopodium clavatum* | Lycopodiaceae | | 0–0 | | | $C_3$ |
| *Ophioglossum vulgatum* | Ophioglossaceae | | 0–0.06 | | | |
| *Selaginella rupestris* | Selaginellaceae | | 0.22–0 | | | $C_3$ |
| *Equisetum arvense* | Equisetaceae | | 0.01–0.04 | | | $C_3$ |
| *Pinus strobus* | Pinaceae | | 0–0.4 | | | $C_3$ |
| *Pseudotsuga douglasii* | Pinaceae | | 0.14–0.3 | | | |
| *Taxus canadensis* | Taxaceae | | 0–0.1 | | | $C_3$ |
| *Zamia ottonis* | Cycadaceae | | 0.07–0.04 | | | |
| *Cycas revoluta* | Cycadaceae | | 0–0 | | | |
| *Cycas circinalis* | Cycadaceae | | 0.03–0.01 | | | |
| *Ephedra* sp. | Ephdraceae | | 0.11–0.25 | | | |
| *Gnetum leyboldii* | Gnetaceae | | 0.07–0.04 | | | |
| *Ginkgo biloba* | Ginkgoaceae | | 0.43–0.41 | | | |
| *Welwitschia mirabilis* | Welcitchiaceae | | 0–0.09 | | | |
| **Angiosperms** | | | | | | |
| *Liriodendron tulipifera* | Magnoliaceae | Magnoliidae | | 1.87–0.74 | | $C_3$ |
| *Magnolia acuminata* | Magnoliaceae | Magnoliidae | | 4.8–0.06 | | $C_3$ |
| *Sassafras albidum* | Lauraceae | Magnoliidae | | 6.01–0.31 | | $C_3$ |
| *Peperomia obtusifolia* 'Variegata' | Piperaceae | Magnoliidae | | 1.82–0 | | |
| *Asarum canadense* | Aristolochiaceae | Magnoliidae | | 14.8–0.21 | | |
| *Illicium anisatum* | Illiciaceae | Magnoliidae | | 1.06–0.28 | | |
| *Cercidiphyllum japonicum* | Cercidiphyllaceae | Hamamelidae | 0.66–0 | | | |

(continued)

**Table 14.1** (continued)

| Scientific name | Family | Subclass | DDV-LDV | DMV-LDV | DMV-LMV | $C_3$–$C_4$ |
|---|---|---|---|---|---|---|
| *Hamamelis virginiana* | Hamamelidaceae | Hamamelidae | | 2.42–0.07 | | $C_3$ |
| *Myrica pensylvanica* | Myricaceae | Hamamelidae | 0.06–0.05 | | | |
| *Fagus grandifolia* | Fagaceae | Hamamelidae | 0.13–0.19 | | | $C_3$ |
| *Fagus sylvatica* | Fagaceae | Hamamelidae | 0.10–0 | | | |
| *Quercus alba* | Fagaceae | Hamamelidae | | 4.93–0.07 | | $C_3$ |
| *Casuarina equisetifolia* | Casuarinaceae | Hamamelidae | | 4.87–0.01 | | |
| *Chenopodium album* | Chenopodiaceae | Caryophyllidae | | 49–0.49 | | $C_3$ |
| *Amaranthus retroflexus* | Amaranthaceae | Caryophyllidae | | 6.4–0.55 | | $C_4$ |
| *Rumex acetosella* | Polygonaceae | Caryophyllidae | 0.01–0 | | | $C_3$ |
| *Plumbago auriculata* | Plumbaginaceae | Caryophyllidae | | 2.71–0 | | |
| *Abutilon theophrastii* | Malvaceae | Dilleniidae | 0.3–0 | | | $C_3$ |
| *Gossypium hirsutum* | Malvaceae | Dilleniidae | | 10.2–0.2 | | $C_3$ |
| *Sida spinosa* | Malvaceae | Dilleniidae | 0.25–0 | | | $C_3$ |
| *Viola affinis* | Violaceae | Dilleniidae | 0.02–0 | | | $C_3$ |
| *Cucumis sativus* | Cucurbitaceae | Dilleniidae | 0.9–0.2 | | | $C_3$ |
| *Salix nigra* | Salicaceae | Dilleniidae | 0.16–0 | | | $C_3$ |
| *Polanisia dodecandra* | Capparaceae | Dilleniidae | | 2.13–0.09 | | $C_3$ |
| *Brassica nigra* | Brassicaceae | Dilleniidae | 0.3–0.06 | | | $C_3$ |
| *Prunus padus* | Rosaceae | Rosidae | | 3.96–0.23 | | $C_3$ |
| *Glycine max* | Fabaceae | Rosidae | | 31.4–0.35 | | $C_3$ |
| *Melilotus alba* | Fabaceae | Rosidae | | 1.9–0 | | $C_3$ |
| *Melilotus officinalis* | Fabaceae | Rosidae | | 2.9–0 | | $C_3$ |
| *Phaseolus vulgaris* | Fabaceae | Rosidae | | 12.7–0.4 | | $C_3$ |
| *Pisum sativum* | Fabaceae | Rosidae | 0.54–0 | | | |
| *Cornus florida* | Cornaceae | Rosidae | | 4.89–0.22 | | |
| *Euonymus alata* | Celastraceae | Rosidae | 0.69–0 | | | $C_3$ |
| *Euphorbia milii* | Euphorbiaceae | Rosidae | | 3.08–0.23 | | |
| *Manihot dulcis* | Euphorbiaceae | Rosidae | | | 1.1–5.3 | |
| *Reinwardtia indica* | Linaceae | Rosidae | | 4.37–0.15 | | |
| *Pelargonium Xhortorum* | Geraniaceae | Rosidae | 0.07–0.01 | | | |

(continued)

**Table 14.1** (continued)

| Scientific name | Family | Subclass | DDV-LDV | DMV-LDV | DMV-LMV | $C_3$–$C_4$ |
|---|---|---|---|---|---|---|
| *Apocynum cannabinum* | Apocynaceae | Asteridae | | 7.6–0.08 | | $C_3$ |
| *Datura stramonium* | Solanaceae | Asteridae | | 9.15–0.24 | | $C_3$ |
| *Lycopersicum esculentum* | Solanaceae | Asteridae | | 11.6–0.1 | | $C_3$ |
| *Solanum cornutum* | Solanaceae | Asteridae | | 6.6–0.5 | | $C_3$ |
| *Convolvulus arvensis* | Convolvulaceae | Asteridae | | 15.8–0.06 | | $C_3$ |
| *Ipomoea hederacea* | Convolvulaceae | Asteridae | 0.74–0 | | | |
| *Ipomoea purpurea* | Convolvulaceae | Asteridae | 0.86–0 | | | |
| *Glechoma hederacea* | Lamiaceae | Asteridae | 0.49–0.27 | | | |
| *Pogostemon cablin* | Lamiaceae | Asteridae | 0–0 | | | |
| *Plantago lanceolata* | Plantaginaceae | Asteridae | | 2.3–0.6 | | $C_3$ |
| *Cymbalaria muralis* | Scrophulariaceae | Asteridae | 0.01–0 | | | |
| *Campanula cochleariifolia* 'Fragilis Jewel' | Campanulacease | Asteridae | | 1.44–0.08 | | |
| *Galium aparine* | Rubiaceae | Asteridae | 0.94–0 | | | $C_3$ |
| *Pentas lanceolata* | Rubiaceae | Asteridae | | 1.33–0 | | |
| *Ambrosia artemisiifolia* | Asteraceae | Asteridae | | 27–0.49 | | $C_3$ |
| *Ambrosia trifida* | Asteraceae | Asteridae | | 31.7–0.12 | | $C_3$ |
| *Cirsium discolor* | Asteraceae | Asteridae | 0.14–0.05 | | | |
| *Matricaria chamomilla* | Asteraceae | Asteridae | 0.55–0.04 | | | $C_3$ |
| *Taraxacum officinale* | Asteraceae | Asteridae | | 11.8–0.4 | | $C_3$ |
| *Xanthium strumarium* | Asteraceae | Asteridae | | 15.8–0.1 | | |
| *Sagittaria latifolia* | Alismataceae | Alismatidae | | 3.39–0.08 | | $C_3$ |
| *Limnobium spongia* | Hydrocharitaceae | Alismatidae | | 1.65–0.04 | | |
| *Lilaea subulata* | Juncaginaceae | Alismatidae | | 15.36–0.15 | | |
| *Phoenix canariensis* | Arecaceae | Arecidae | 0.39–0 | | | |

(continued)

**Table 14.1** (continued)

| Scientific name | Family | Subclass | DDV-LDV | DMV-LDV | DMV-LMV | $C_3$–$C_4$ |
|---|---|---|---|---|---|---|
| *Philodendron domesticum* | Araceae | Arecidae | 0.59–0.20 | | | |
| *Tradescantia ohiensis* | Commelinaceae | Commelinidae | | 14–0.11 | | |
| *Alopecurus pratensis* | Poaceae | Commelinidae | | 9.21–0.13 | | $C_3$ |
| *Arundinaria gigantea* | Poaceae | Commelinidae | | 7.9–0.3 | | |
| *Avena fatua* | Poaceae | Commelinidae | | 15.8–0.03 | | $C_3$ |
| *Digitaria sanguinalis* | Poaceae | Commelinidae | | 7.4–0 | | $C_3$ |
| *Echinochloa crus-galli* | Poaceae | Commelinidae | | | 8.4–1.02 | |
| *Hordeum vulgare* | Poaceae | Commelinidae | | 2.4–0.01 | | $C_3$ |
| *Panicum miliaceum* | Poaceae | Commelinidae | | 5.5–0.7 | | $C_4$ |
| *Poa pratensis* | Poaceae | Commelinidae | | 8.7–0.5 | | $C_3$ |
| *Setaria faberi* | Poaceae | Commelinidae | | 16–0.99 | | $C_4$ |
| *Setaria viridis* | Poaceae | Commelinidae | | 4.9–0.5 | | $C_4$ |
| *Sorghum halepense* | Poaceae | Commelinidae | | | 6.9–4.2 | $C_4$ |
| *Triticum aestivum* | Poaceae | Commelinidae | | 4.72–0 | | $C_3$ |
| *Zea mays* | Poaceae | Commelinidae | | 3.5–0.11 | | $C_4$ |
| *Typha latifolia* | Typhaceae | Commelinidae | | 1.77–0.21 | | $C_3$ |
| *Neoregelia chlorosticta* | Bromeliaceae | Zingiberidae | 0.5–0 | | | |
| *Costus elatus* | Zingiberaceae | Zingiberidae | 0.67–0 | | | |
| *Allium cepa* | Liliaceae | Liliidae | | 3.4–0 | | |
| *Paphiopedilum incabuena* | Orchidaceae | Liliidae | 0–0 | | | |

Non-flowering plants are arranged by major groups; angiosperms are arranged by subclass of Cronquist (1981) and then by family. Ratios of monovinyl to divinyl protochlorophyllide for the three greening groups are given in columns labelled DDV-LDV, DMV-LDV and DMV-LMV Information on $C_3$–$C_4$ photo-synthesis is from Downton (1975) (Adapted from Ionannides et al. 1994)

## 14.3 Discovery of the Dark-Light Greening Group of Plants

When the DDV-LDV, DMV-LDV, and DMV-LMV greening groups (Table 14.1) were first described (Carey et al. 1985). It was believed that the DV Pchlide *a* pool that was detected in the light was a manifestation of the Functional DV Chl Biosynthetic route in the light. Then the question arose as to whether the Pchlide *a* detected in the light was the manifestation of an overly active DV Pchlide *a* biosynthetic route or the results of a sluggish DV Pchlide *a* biosynthetic route coupled to an active MV Pchlide *a* biosynthetic route coupled to a rapid photoconversion of the nascent MV Pchlide *a* to MV Chlide *a*. This issue was investigated as described below.

### 14.3.1 The Dark-Monovinyl/Light-Divinyl/Light-Dark Monovinyl/Greening Group of Plants

In order to determine whether the Pchlide *a* detected in the light was the manifestation of an overly active DV Pchlide *a* biosynthetic route or the results of a sluggish DV Pchlide *a* biosynthetic route coupled to an active MV Pchlide *a* biosynthetic route and a rapid photoconversion of the nascent MV Pchlide *a* to MV Chlide *a*, the following experimental strategy was devised.

Corn seedlings that exhibited DV Pchlide *a* accumulation in the light, were transferred to darkness for various periods of time after 8 h of light during the light phase of the photoperiod, and the MV and DV Pchlide *a* content of the Pchlide *a* pool were determined immediately after transfer to darkness and after various period of time in darkness.

As shown in Fig. 14.1, immediately after transfer to darkness, the rate of MV Pchlide *a* started to increase sharply while the content of DV Pchlide *a* content decreased significantly indicating a sluggish DV Pchlide *a* biosynthetic route.

Consequently this greening group of plants was called a Dark Monovinyl/Light-Divinyl/Light-dark Monovinyl/Greening group.

### 14.3.2 The Dark-Divinyl/Light-Divinyl/Light-Dark Divinyl Greening Group of Plants

The functional Pchlide *a* biosynthetic route was investigated in the same manner in Dark Divinyl-Light Divinyl plants. After 8 h into the light phase of the photoperiod, cucumber seedlings were moved to darkness and the DV and MV Pchlide *a* content of the green cotyledons was investigated after various times in darkness.

As shown in Fig. 14.2, the rate of DV Pchlide *a* biosynthesis increased sharply while the rate of MV Pchlide formation dropped continuously. This in turn strongly suggested that in these plants the functional Pchlide *a* biosynthetic route was a DV route.

This greening group of plants was designated as a Dark Divinyl/Light-Divinyl/Light-dark Divinyl/Greening group.

### 14.3.3 The Dark Monovinyl/Light-Monovinyl/Light-Dark Monovinyl Greening Group of Plants

DMV-LMV monocotyledonous plants such as johnsongrass differ from monocotyledonous DMV-LDV plants such as oat and barley by the amounts of DV and MV

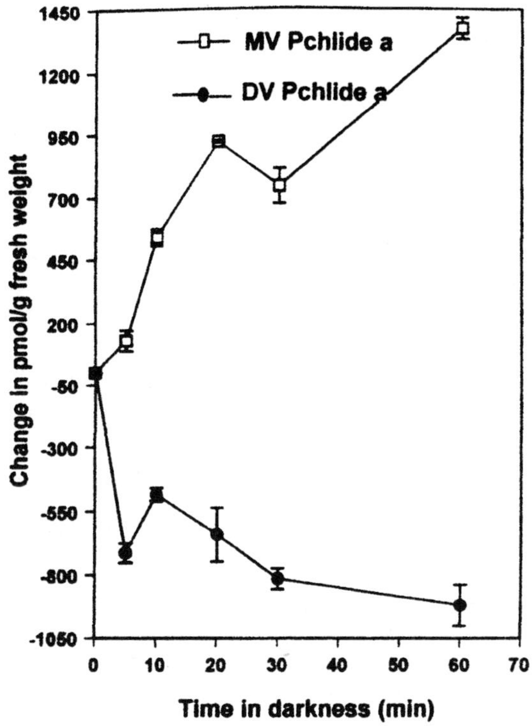


**Fig. 14.1** Changes in the DV and MV Pchlide *a* content in 6-day-old photoperiodically grown corn leaves after transfer from light to darkness. Transfer from light to darkness was carried out after 8 h in the light (Adapted from Abd-el-Magid et al. 1997)

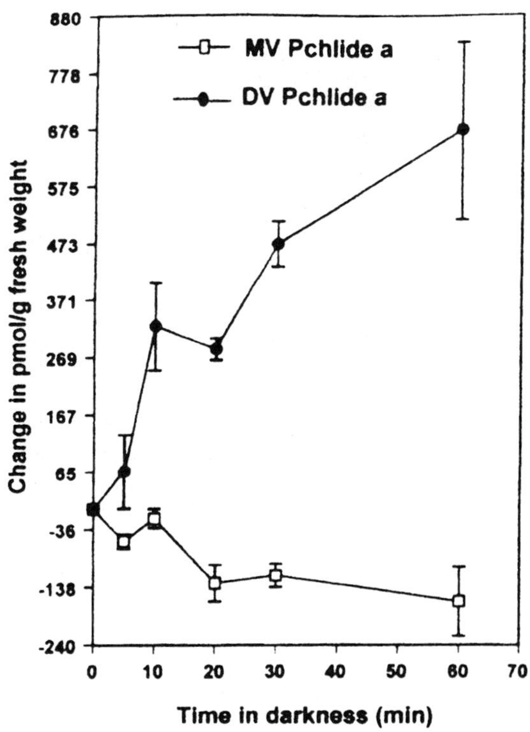


**Fig. 14.2** Changes in the DV and MV Pchlide *a* content in 6-day-old photoperiodically-grown cucumber cotyledons after transfer from light to darkness. Transfer from light to darkness was carried out after 8 h in the light (Adapted from Abd-el-Magid et al. 1997)

Pchlide *a* that accumulate in daylight. While monocotyledonous DMV-LMV plants accumulate mainly MV Pchlide *a* in daylight, under the same conditions, monocotyledonous DM-LDV plants accumulate mainly DV Pchlide *a*. The possible relationship of this difference between the two greening groups of plants to the rates of DV and MV Pchlide *a* biosynthesis in daylight was therefore investigated by determining the light-dark-rate of MV and DV regeneration during the light phase of the photoperiod.

As reported in Table 14.2, in all DMV-LDV and DMV-LMV species that were examined, the observed rates of MV Pchlide *a* regeneration upon returning the plants from light to darkness were much higher than those of DV Pchlide *a* during the first 5 min of dark-Pchlide *a* regeneration and thereafter. The same was also true after 30 and 60 min of dark-Pchlide *a* regeneration (Table 14.2).

These results strongly suggested that in both DMV-LMV and DMV-LDV monocotyledonous plants, the MV Chl *a* biosynthetic route is the dominant route in daylight. As a consequence DMV-LMV plants such as Jhonsongrass are and DMV-LDV were assigned to a DMV-LMV-LDMV greening group.

Figure 14.3 summarizes the greening group affiliation of plants and take into account the Dar-Light Chl biosynthetic routes involvement in the process.

If the DV/MV Pchlide *a* ratio for a plant species is larger than 1, for a particular phase of the photoperiod, then the plant is considered to be DV for that phase of the photoperiod. If the ratio is less than one, the plant is considered to be MV for that phase of the photoperiod. Plants that are DV during the dark phase of the photoperiod are referred to as DDV. Plants that are MV during the dark phase of the photoperiod are referred to as DMV. Likewise Plants that are DV during the light phase of the photoperiod are referred to as LDV. Plants that are MV during the light phase of the photoperiod are referred to as LMV. Plants that accumulate DV Pchlide *a* immediately after returning them from light to darkness are referred to as LDDV. Plants that accumulate MV Pchlide *a* immediately after returning them from light to darkness are referred to as LDMV. As a consequence for plants can be classified as DDV/LDV/LDDV, DMV/LDV/LDMV or DMV/LMV/LDMV.

## 14.4 Biological Significance of the Greening Group Affiliation of Green Plants

Natural selection has often produced multiple (bio) chemical and physical ways of conveying the same message. Genetic redundancy achieves the same purpose. It is therefore logical to thing of the Chl biosynthetic heterogeneity as a hedge against lethal mutations.

It is also possible that via natural selection, Chl biosynthetic heterogeneity has imparted an evolutionary advantage to higher plants. As was just described, the preferential operation of the MV or DV Chl *a* monocarboxylic biosynthetic routes during the formation of Chl *a* in higher plants is a species-dependent phenomenon with evolutionary significance (Abd-El-Mageed et al. 1997; Ioannides et al. 1994).

**Table 14.2** Rates of DV and MV Pchlide *a* regeneration upon returning various plant species from light to darkness

| Plant species | MV/DV Pchlide ratio | | Pchlide regenerated after the indicated times of transfer from light to darkness (pmol/g fresh weight) | | | | | | Greening | |
|---|---|---|---|---|---|---|---|---|---|---|
| | End of darkness | In the light | 5 min DV | 5 min MV | 30 min DV | 30 min MV | 60 min DV | 60 min MV | Group | Subgroup |
| Johnson-grass | 35.28 ± 7.78 | 0.34 ± 0.04 | 107 ± 50 | 650 ± 87 | 137 ± 15 | 1,173 ± 38 | 127 ± 29 | 1,320 ± 130 | DMV-LMV | LDMV |
| Corn | 63.91 ± 34 | 0.39 ± 0.10 | −713 ± 46 | 130 ± 50 | −783 ± 100 | 750 ± 85 | −920 ± 101 | 1,373 ± 51 | DMV-LDV | LDMV |
| Rice | 11.02 ± 2 | 0.17 ± 0.05 | −157 ± 6 | 13 ± 6 | −147 ± 8 | 40 ± 10 | −133 ± 12 | 193 ± 21 | DMV-LDV | LDMV |
| Sorghum | 12.93 ± 1.43 | 0.12 ± 0.09 | 20 ± 44 | 217 ± 6 | −6 ± 21 | 453 ± 35 | −27 ± 15 | 530 ± 53 | DMV-LDV | DLMV |
| Wild proso | 3.59 ± 1.5 | 0.11 ± 0.05 | −140 ± 60 | 127 ± 47 | −147 ± 45 | 287 ± 6 | −20 ± 3 | 487 ± 45 | DMV-LDV | LDMV |
| Wheat | 3.83 ± 1.1 | 0.25 ± 0.09 | −490 ± 101 | −100 ± 96 | 13.3 ± 205 | 836 ± 162 | 67 ± 200 | 1,117 ± 125 | DMV-LDV | LDMV |
| Oat | 3.92 ± 0.66 | 0.14 ± 0.03 | −80 ± 50 | 180 ± 36 | −80 ± 50 | 180 ± 36 | −230 ± 56 | 263 ± 15 | DMV-LDV | LDMV |
| Barley | 16.25 ± 1.87 | 0.21 ± 0.02 | −333 ± 106 | 150 ± 36 | −387 ± 78 | 747 ± 50 | −380 ± 78 | 927 ± 139 | DMV-LDV | LDMV |
| Quackgrass | 9.09 ± 0.75 | 0.56 ± 0.07 | 213 ± 23 | 493 ± 76 | 197 ± 21 | 790 ± 70 | 210 ± 35 | 1,013 ± 35 | DMV-LDV | LDMV |
| Canary-grass | 2.77 ± 0.56 | 0.13 ± 0.0.2 | −13 ± 70 | 40 ± 35 | 263 ± 45 | 1,127 ± 58 | 367 ± 85 | 1,693 ± 143 | DMV-LDV | LDMV |
| Tomato | 7.61 ± 0.36 | 0.28 ± 0.04 | −260 ± 30 | −100 ± 30 | −263 ± 15 | 70 ± 10 | −293 ± 60 | 250 ± 30 | DMV-LDV | LDMV |
| French bean | 5.58 ± 3.49 | 0.04 ± 0.05 | | | 150 ± 147 | 451 ± 77 | 110 ± 204 | 571 ± 105 | DMV-LDV | LDMV |
| Cucumber | 0.15 ± 0.03 | 0.07 ± 0.02 | −40 ± 65 | 63 ± 36 | 457 ± 50 | −6 ± 1 | 147 ± 66 | 21 ± 13 | DDV-LDV | LDDV |
| Pogostemom | 0.24 ± 0.04 | 0.01 ± 0.01 | | | 21 ± 81 | 3 ± 3 | 37 ± 75 | 14 ± 8 | DDV-LDV | LDDV |

All values are means of three replicate

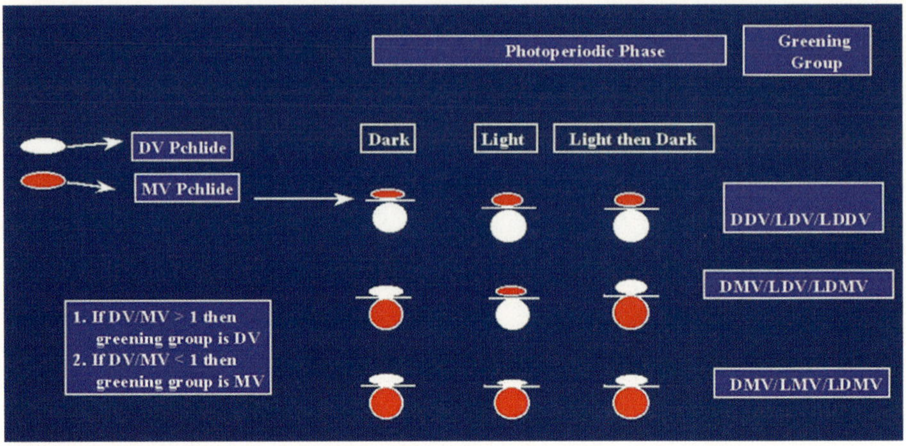


**Fig. 14.3** Greening group affiliation of green plants

Accordingly, evolution under domestication, as manifested by selection for higher plant yields, may favor either one of the two routes. It was shown for example (Fasoula et al. 1996), that in spring wheat (*Triticum aestivum* L.) and corn (*Zea mays* L.), plants that use preferentially the more evolved MV Chl *a* biosynthetic route at night, selection for yield does not favor the DV route, whereas in certain cases the MV route is enhanced.

## References

Abd-El-Mageed HA, El Sahhar KF, Robertson KR et al (1997) Chloroplast biogenesis 77. Two novel monovinyl and divinyl light-dark greening groups of plants and their relationship to the chlorophyll *a* biosynthetic heterogeneity of green plants. Photochem Photobiol 66:89–96

Adra AN, Rebeiz CA (1998) Chloroplast biogenesis 81. Transient formation of divinyl chlorophyll *a* following a 2.5 ms light flash treatment of etiolated cucumber cotyledons. Photochem Photobiol 68:852–856

Arigoni D (1994) Summing up. In: Chadwick DJ, Ackrill K (eds) The biosynthesis of the tetrapyrrole pigments. Wiley, New York, pp 285–308

Armstrong GA, Runge S, Frick G et al (1995) Identification of NADPH:protochlorophyllide oxidoreductases a and B: a branched pathway for light-dependent chlorophyll biosynthesis in *Arabidopsis thaliana*. Plant Physiol 108:1505–1517

Bazzaz MB (1981) New chlorophyll chromophores isolated from a chlorophyll deficient mutant of maize. Photobiochem Photobiophys 2:199–207

Belanger FC, Rebeiz CA (1982) Chloroplast biogenesis: detection of monovinyl magnesium protoporphyrin monoester and other monovinyl magnesium porphyrins in higher plants. J Biol Chem 257:1360–1371

Carey EE, Rebeiz CA (1985) Chloroplast biogenesis 49. Difference among angiosperms in the biosynthesis and accumulation of monovinyl and divinyl protochlorophyllide during photoperiodic greening. Plant Physiol 79:1–6

Carey EE, Tripathy BC, Rebeiz CA (1985) Chloroplast biogenesis 51. Modulation of monovinyl and divinyl protochlorophyllide biosynthesis by light and darkness in vitro. Plant Physiol 79:1059–1063

Chisholm S, Olson RJ, Zettler ER et al (1988) A novel free-living prochlorophyte abundant in the oceanic euphotic zone. Nature 334:340–343

Chisholm SW, Frankel S, Goerike R et al (1992) *Prochlorococcus marinus* nov. gen. sp.: an oxyphototrophic marine prokaryote containing divinyl chlorophyll *a* and *b*. Arch Mikrobiol 157:297–300

Fasoula DA, Smyth C, A RC (1996) Relationship of the monovinyl protochlorophyllide *a* content to plant yield. In: Pessarakli M (ed) Handbook of photosynthesis. CRC press, Boca Raton, pp 671–679

Goerike R, Repeta D (1992) The pigments of *Prochlorococcus marinus*. The presence of divinyl-chlorophyll a and b in a marine prochlorophyte. Limnol Oceanogr 37:425–433

Holtorf R, Reinbothe S, Reinbothe C et al (1995) Two routes of chlorophyllide synthesis that are differentially regulated by light in barley (*Hordeum vulgare* L.). Proc Natl Acad Sci U S A 92:3254–3258

Ioannides IM, Fasoula DM, R RK et al (1994) An evolutionary study of chlorophyll biosynthetic heterogeneity in green plants. Biochem Syst Ecol 22:211–220

Kim JS, Rebeiz CA (1996) Origin of the chlorophyll *a* biosynthetic heterogeneity in higher plants. J Biochem Mol Biol 29:327–334

Kolossov VL, Rebeiz CA (2010) Evidence for various 4-vinyl reductase activities in higher plants. In: Rebeiz CA, Benning C, Bohnert HJ et al (eds) The chloroplast: basics and applications. Springer, Dordrecht, pp 25–38

Parham R, Rebeiz CA (1992) Chloroplast biogenesis: [4-vinyl] chlorophyllide *a* reductase is a divinyl chlorophyllide *a*-specific NADPH-dependent enzyme. Biochemistry 31:8460–8464

Parham R, Rebeiz CA (1995) Chloroplast biogenesis 72: a [4-vinyl] chlorophyllide *a* reductase assay using divinyl chlorophyllide *a* as an exogenous substrate. Anal Biochem 231:164–169

Rebeiz CA, Wu SM, Kuhadje M et al (1983) Chlorophyll *a* biosynthetic routes and chlorophyll *a* chemical heterogeneity. Mol Cell Biochem 58:97–125

Rebeiz CA, Parham R, Fasoula DA et al (1994) Chlorophyll biosynthetic heterogeneity. In: Chadwick DJ, Ackrill K (eds) The biosynthesis of the tetrapyrrole pigments. Wiley, New York, pp 177–193

Rebeiz CA, Kolossov VL, Briskin D et al (2003) Chloroplast biogenesis: chlorophyll biosynthetic heterogeneity, multiple biosynthetic routes and biological spin-offs. In: Nalwa HS (ed) Handbook of photochemistry and photobiology, vol 4. American Scientific Publishers, Los Angeles, pp 183–248

Rebeiz CA, Kopetz KJ, Kolossov VL (2005) Chloroplast biogenesis: probing the relationship between chlorophyll biosynthetic routes and the topography of chloroplast biogenesis by resonance excitation energy transfer determinations. In: Pessarkli M (ed) Handbook of photosynthesis, 2nd edn. Marcel Dekker, Inc, New York, Revised and Expanded

Scott AI (1994) Recent studies of the enzymically controlled steps in B12 biosynthesis. In: Chadwick DJ, Ackrill K (eds) The biosynthesis of the tetrapyrrole pigments. Wiley, New York, pp 285–308

Shioi Y, Sasa T (1983) Formation and degradation of protochlorophylls in etiolated and greening cotyledons of cucumber. Plant Cell Physiol 24:835–840

Shioi Y, Takamiya KI (1992) Monovinyl and divinyl protochlorophyllide pools in etiolated tissues of higher plants. Plant Physiol 100:1291–1295

Suzuki JY, Bauer CE (1995) Altered monovinyl and divinyl protochlorophyllide pools in bchJ mutants of rhodobacter capsulatus. Possible monovinyl substrate discrimination of light-independent protochlorophyllide reductase. J Biol Chem 270:3732–3740

Tripathy BC, Rebeiz CA (1988) Chloroplast biogenesis 60. Conversion of divinyl protochlorophyllide to monovinyl protochlorophyllide in green(ing) barley, a dark monovinyl/light divinyl plant species. Plant Physiol 87:89–94

Veldhuis MJW, Kraay GW (1990) Vertical distribution of pigment composition of a picoplankton prochlorophyte in the subtropical north Atlantic: a combined study of pigments and flow cytometry. Mar Ecol Prog Ser 68:121–127

Whyte BJ, Griffiths TW (1993) 8-Vinyl reduction and chlorophyll *a* biosynthesis in higher plants. Biochem J 291:939–944

# Chapter 15
# Relationship of Chlorophyll Biosynthesis to the Assembly of Chlorophyll-Protein Complexes

*The greatest use of life is to spend it for something that will outlast it (William James).*

## 15.1 Introduction

Chlorophyll biosynthetic heterogeneity (Rebeiz et al. 2003a) refers (a) either to spatial biosynthetic heterogeneity, (b) to chemical biosynthetic heterogeneity, or (c) to a combination of spatial and chemical biosynthetic heterogeneities. Spatial biosynthetic heterogeneity refers to the biosynthesis of an anabolic tetrapyrrole or end product by identical sets of enzymes, at several different locations of the thylakoid membranes. On the other hand, chemical biosynthetic heterogeneity refers to the biosynthesis of an anabolic tetrapyrrole or end product at several different locations of the thylakoid membranes, via different biosynthetic routes, each involving at least one different enzyme. Figures 6.3, 6.4 and 6.5 in Chap. 6 organize all known Chl biosynthetic reactions into a logical scheme made up of multiple biosynthetic routes. Each route consists of one or more biosynthetic reactions that has been discussed in some details in previous chapters.

## 15.2 Relationship of Chlorophyll Biosynthetic Heterogeneity to Thylakoid Membrane Biogenesis

During the past few years a systematic research effort has been undertaken to explore the relationship of Chl biosynthetic heterogeneity to the assembly of thylakoid membranes. In a departure from a conventional discussion of the Chl biosynthetic pathway as a standalone entity, this section will focus on a

C.A. Rebeiz, *Chlorophyll Biosynthesis and Technological Applications*, 
DOI 10.1007/978-94-007-7134-5_15, 

discussion of Chl biosynthetic heterogeneity in the context of a photosynthetic unit made up of PSI, PSII and light harvesting Chl-protein complexes (LHCs) (see below).

### 15.2.1 Chlorophyll Biosynthesis-Thylakoid Membrane Biogenesis Working Models

In 1999, Rebeiz and coworkers proposed that the unified multibranched Chl *a*/*b* biosynthetic pathway might be visualized as a template of Chl-protein biosynthesis centers, upon which the assembly of PSI, PSII and light harvesting Chl-protein complexes (LHC) takes place (Rebeiz et al. 1999). Then in an extension of the working hypothesis described above, we have recently proposed three Chl-thylakoid apoprotein biosynthesis models (Kolossov et al. 2003; Rebeiz et al. 2003b, 2004). These models took into account the structure and dimensions of the photosynthetic unit (PSU) (Allen and Forsberg 2001; Anderson 2002; Bassi et al. 1990; Staehelin 2003), the biochemical heterogeneity of the Chl biosynthetic pathway (Rebeiz et al. 1994), and the biosynthetic and structural complexity of thylakoid membranes (Sundqvist and Ryberg 1993). Within a PSU, the three putative Chl-apoprotein thylakoid biosynthesis models are referred to as: (a) the single branched biosynthetic pathway (SBP)-single location model, (b) the SBP-multilocation model and (c) the multibranched biosynthetic pathway (MBP)-sublocation model.

#### 15.2.1.1 Single-Branched Pathway (SBP)-Single Location Model

Within the PSU, the SBP-single location model Fig. 15.1 was considered to accommodate only one Chl-apoprotein thylakoid biosynthesis center and no Chl-apoprotein thylakoid biosynthesis sub-centers. Within the Chl-apoprotein thylakoid biosynthesis center, Chl *a* and *b* were considered to be formed via the conventional single-branched Chl biosynthetic pathway Described below in Fig. 15.2, at a single location accessible to all Chl-binding apoproteins. An apoprotein moved to that location in the unfolded state, picked up a complement of MV Chl *a* and/or MV Chl *b*, and underwent appropriate folding. Then the folded Chl-apoprotein complex moved at random from the Chl biosynthesis location to a specific PSI, PSII, or LHC location within the Chl-apoprotein biosynthesis center.

#### 15.2.1.2 Single-Branched Pathway (SBP)-Multilocation Model

In the SBP-multilocation model Fig. 15.3, every location within the photosynthetic unit is considered to be a Chl-apoprotein thylakoid biosynthesis subcenter. In every Chl-apoprotein biosynthesis subcenter, a complete single-branched Chl *a*/*b* biosynthetic pathway is active. Association of Chl *a* and/or Chl *b* with specific PSI, PSII, or LHC apoproteins at any location is random. In every Chl-apoprotein biosynthesis

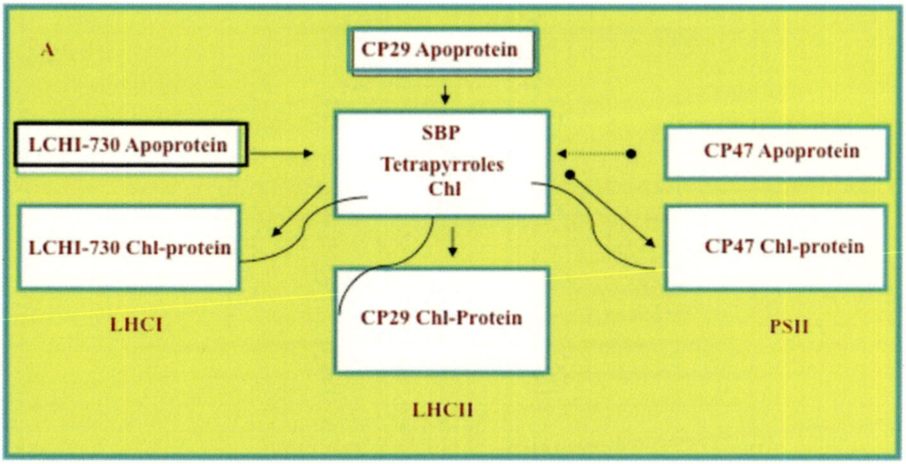


**Fig. 15.1** Schematics of the SBP-single location model in a PSU. As an example, the functionality of the model was illustrated with the use of three apoproteins namely CP29, LCHI-730 and CP47. Abbreviations: *SBP* single-branched Chl biosynthetic pathway, *PSII* photosystem II, LHCII, the major light-harvesting Chl-protein complex of PSII, LHCI, one of the LHC antennae of PSI, CP47 and CP29, two PSII antennae, LHCI-730, the LHC antenna of PSI. Curved lines indicate putative energy transfer between tetrapyrroles and a Chl-protein complex (Adapted from Rebeiz et al. 2003b)

subcenter, distances separating metabolic tetrapyrroles from the Chl-protein complexes are shorter than in the SBP-single-location model.

#### 15.2.1.3 Multibranched-Pathway (MBP)-Sublocation Model

In the MBP-sublocation model (Fig. 15.4), the unified multi-branched Chl *a*/*b* biosynthetic pathway is visualized as the template of a Chl-protein biosynthesis center where the assembly of PSI, PSII and LHC take place (Kolossov et al. 2003; Rebeiz et al. 2003b). The multiple Chl biosynthetic routes are visualized, individually or in groups of one or several adjacent routes, as Chl-apoprotein thylakoid biosynthesis subcenters earmarked for the coordinated assembly of individual Chl-apoprotein complexes. Apoproteins destined to some of the biosynthesis subcenters may possess specific signals for specific Chl biosynthetic enzymes peculiar to that subcenter, such as 4-vinyl reductases, formyl synthetases or Chl *a* and Chl *b* synthetases. Once an apoprotein, formed in the cytoplasm or in the plastid, reaches its biosynthesis subcenter destination and its signal is split off, it binds nascent Chl formed via one or more biosynthetic routes. During Chl binding, the apoprotein folds properly and act at that location, while folding or after folding, as a template for the assembly of other apoproteins. In this case too, shorter distances separate the accumulated tetrapyrroles from the Chl-protein complexes.

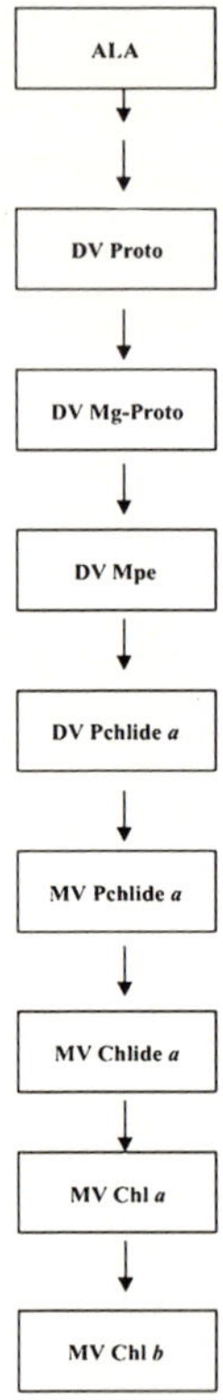


**Fig. 15.2** The single branched Chl biosynthetic pathway proposed by Granick (1950) and modified by Wolfe and Price (1957) and Jones (1963)

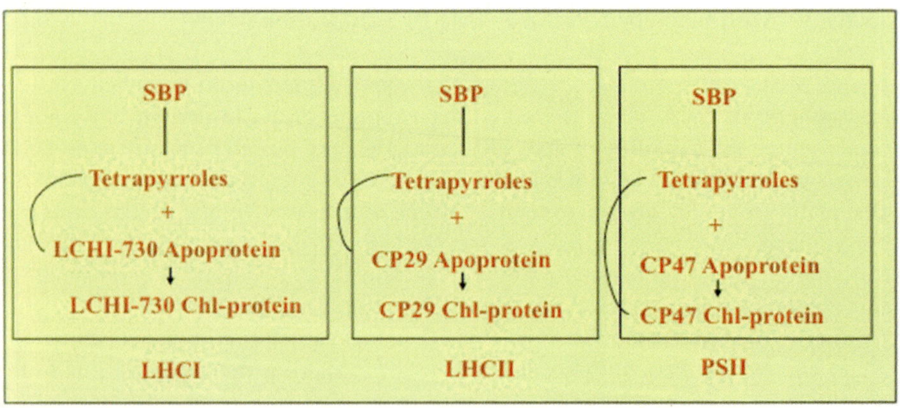


**Fig. 15.3** Schematics of the SBP-multilocation model in a PSU. All abbreviations and conventions are as in Fig. 15.1

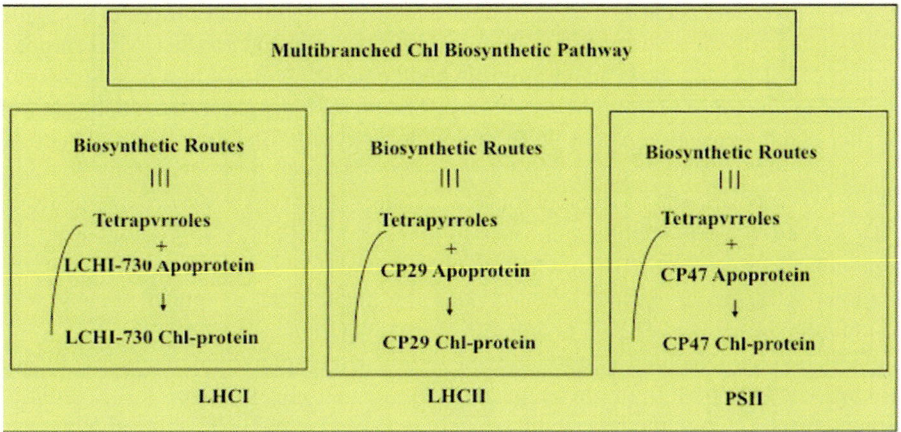


**Fig. 15.4** Schematics of the MBP-sublocation model in a PSU. All abbreviations and conventions are as in Fig. 15.1

## 15.3 Resonance Excitation Energy Transfer from Metabolic Tetrapyrroles to Various Chl-Protein Complexes Indicate that Resonance Excitation Energy Transfer Takes Place from Multiple Heterogeneous Sites

Fluorescence resonance energy transfer involves the transfer of excited state energy from an excited donor "D*" to an unexcited acceptor "A" (Calvert and Pitts 1967; Lakowicz 1999; Turro 1965). The transfer is the result of dipole-dipole interaction between donor and acceptor and does not involve the exchange of a photon. The rate of energy transfer depends upon (a) the extent of overlap of the emission spectrum of the donor and the absorption spectrum of the acceptor, (b) the relative orientation of the donor and acceptor transition dipoles, and (c) the distance between donor and acceptor molecules. As soon as the excited donor "D*" and unexcited acceptor "A" states are coupled by an appropriate interaction, they become degenerate if there is an excited state of the acceptor "A", which requires exactly the same excitation energy available in "D*". When such a condition exists, excitation of one of the degenerate states leads to a finite probability that the same excitation will appear in the other degenerate state (Turro 1965). This probability increases with time but is inversely proportional to the sixth power of the fixed distance separating the centers of the donor and acceptor molecules. It has been estimated that dipole-dipole energy transfer between donor and acceptor molecules may occur up to a separation distance of 50–100 Å (Calvert and Pitts 1967).

Resonance excitation energy transfer from three tetrapyrrole donors to the Chl *a* of Chl-protein complexes were monitored, namely: from protoporphyrin IX (Proto), divinyl (DV) Mg-Proto and its methyl ester and monovinyl (MV) and DV Pchlide *a*. DV Proto is a common precursor of heme and Chl. It is the immediate precursor of

DV Mg-Proto. As such, it is an early intermediate along the Chl biosynthetic chain. Biosynthetically, it is several steps removed from the Chl end product. Mg-Proto is a mixed MV-DV, dicarboxylic tetrapyrrole pool, consisting of DV and MV Mg-Proto. It is the precursor of DV and MV Pchlide *a*. The protochlorophyll(ide) [(Pchl(ide)] of higher plants consists of about 95 % protochlorophyllide (Pchlide) *a* and about 5 % Pchlide *a* ester (Pchlide *a* E). The latter is esterified with long chain fatty alcohols (LCFAs) at position 7 of the macrocycle. While Pchlide *a* ester consists mainly of MV Pchlide *a* ester, Pchlide *a* consists of DV and MV Pchlide *a*. The latter are the immediate precursors of DV and MV chlorophyllide (Chlide) *a*. Accumulation of the various tetrapyrrole donors was induced by incubation of green tissues with δ-aminolevulinic acid (ALA) and/or 2,2′-dipyridyl (Rebeiz et al. 1988).

The task of selecting appropriate Chl *a*-protein acceptors was facilitated by the fluorescence properties of green plastids. At 77 K, emission spectra of isolated chloroplasts exhibit maxima at 683–686 nm (~F685), 693–696 nm (~F695), and 735–740 nm (~F735). It is believed that the fluorescence emitted at ~F685 nm arises from the Chl *a* of LHCII, the major thylakoid LHC antenna, and LHCI-680, one of the LHC antennae of PSI (Bassi et al. 1990). That emitted at ~F695 nm is believed to originate mainly from the Chl *a* of CP47 and CP29, two PSII antennae (Bassi et al. 1990). That emitted at ~F735 nm is believed to originate primarily from the Chl *a* of LHCI-730, a PSI antenna (Bassi et al. 1990). Since these emission maxima are readily observed in the fluorescence emission spectra of green tissues and are associated with definite thylakoid Chl *a*-protein complexes, it was conjectured that they would constitute a meaningful resource for monitoring excitation resonance energy transfer between anabolic tetrapyrroles and representative Chl *a*-protein complexes. To monitor the possible occurrence of resonance excitation energy transfer between the accumulated anabolic tetrapyrroles and Chl *a*-protein complexes, excitation spectra were recorded at 77 K at the respective emission maxima of the selected Chl *a* acceptors, namely at ~685, ~695, and ~735 nm. It was conjectured that if excitation resonance energy transfers were to be observed between the tetrapyrrole donors and the selected Chl *a* acceptors, definite excitation maxima would be observed. These excitation maxima would correspond to absorbance maxima of the various tetrapyrrole donors, and would represent the peaks of the excitation resonance energy transfer bands.

Pronounced excitation resonance energy transfer bands from Proto (Table 15.1), Mp(e), and Pchl(ide) *a* to Chl *a* ~F685, ~F695, and ~F735 were detected (Table 6.1, Chap. 6). Assignment of *in situ* excitation maxima to various metabolic tetrapyrroles was unambiguous except for a few cases at the short wavelength and long wavelength extremes of excitation bands. Contrary to previous believes, it was surprising to observe a significant diversity in various intra-membrane environments of Proto, Mp(e), and Pchl(ide) *a*. This diversity was manifested by a differential donation of resonance excitation energy transfer to the different Chl *a*-apoprotein complexes from multiple Proto, Mp(e) and Pchl(ide) a sites, and is highly compatible with biosynthetic heterogeneity of the Chl biosynthetic pathway. Thus, the multi-branched Chl biosynthetic pathways reported in Figs. 6.3, 6.4 and 6.5 Chap. 6, account for the existence of multiple Proto, Mp(e) and Pchl(ide) *a* donor sites by depicting multiple Biosynthetic routes that originate in multiple ALA, Proto, Mg-Proto and Pchlide *a* sites.

**Table 15.1** Mapping of excitation resonance energy transfer maxima to Chl *a* F686, Chl *a* F694 and Chl *a*F738–742 *in situ*

| Plant species | Major donor | Undil donor conc (pmoles/ml suspension) | Dil donor conc (pmoles/ml suspension) | Excitation resonance energy maxima to: Chl *a* F686 | Chl *a* F694 (nm) | Chl *a* F738 | Conc ALA (nM) | Conc Dpy (nM) | Incub (h) |
|---|---|---|---|---|---|---|---|---|---|
| Cucumber | Proto | 1,620 | 54 | 397p, 402p, 410p, 415p | 390s, 400p, 409p | 390s, 395s, 408p, 417p | 4.5 | 3.7 | 6 |
| Cucumber | Proto | 1,242 | 83 | 387p, 402p, 412p | 392p, 406p | 388p, 399p, 403p, 410p, 415p | 20 | 4 | 6 |
| Cucumber | Proto | 1,374 | 92 | 390p, 399p, 405p, 412p | 399p, 409p, 412s | 399p, 400p, 416p | 20 | 0 | 6 |
| Cucumber | Proto | 5,640 | 376 | 395p, 404s, 411p, 416p | 395p,406p, 414p | 393p, 400s, 407p | 20 | 16 | 6 |
| Cucumber | Proto | 3,138 | 1,046 | 402s, 411p | 404p, 410s, 416p, | 399s, 405s, 411p | 20 | 0 | 12 |
| Barley | Proto | 390 | 13 | 391, 398s, 404s, 411p | 389s, 395p, 406p, 414p | 390s, 393p, 400s, 406p, 412p, 416s, | 4.5 | 3.7 | 6 |
| Barley | Proto | 1,492 | 61 | 389p, 396s, 404s, 410p, 412p | 396p, 406p, 412p | 389s, 395p, 406s, 410p, | 20 | 16 | 6 |
| Barley | Proto | 966 | 64 | 395s, 400p, 405s, 413p | 389p, 397s, 403p, 412p | 388s, 393p, 400s, 406p, 412p | 20 | 0 | 6 |
| Barley | Proto | 1,015 | 68 | 389p, 396p, 412p, 413s | 389p, 398p, 409p, | 396s, 400p, 412p, 414s | 20 | 4 | 6 |

Peak (p) and shoulders (s) of excitation resonance energy transfer from Proto to various Chl-protein complexes are interpreted as transfer from different environments to the Chl-protein complexes

*Undil* donor concentration before dilution, *Dil* donor concentration after dilution, *s* shoulder, *p* peak

Adapted from Kolossov et al. (2003)

## 15.4 Incompatibility of the Single-Branched Pathway (SBP)-Single Location Model with Resonance Excitation Energy Transfer from Anabolic Tetrapyrroles to Various Chl-Protein Complexes

Since resonance energy transfer is insignificant at distances larger than 100 Å (Calvert and Pitts 1967) the detection of pronounced resonance excitation energy transfer from Proto, Mp(e), and Pchl(ide) *a* to Chl *a* ~F685, ~F695, and ~F735 (Table 6.1, Chap 6) indicates that these anabolic tetrapyrroles are within distances of 100 Å or less of the Chl *a* acceptors. This in turn is incompatible with the functionality of the SBP-single location Chl-thylakoid biogenesis model. Indeed, it can be estimated from published data that the size of the PSU that includes the two PS, LHC, as well as the CF0-CF1 ATP synthase is about 130 × 450 Å (Bassi et al. 1990). Most PSU models depict a central Cyt $b_6$ complex flanked on one side by PSI and coupling factor CF1, and on the other side by PSII and LHCPII. With this configuration, the shortest distance between the single-branched Chl biosynthetic pathway and PSI, PSII, and LHCII, in the SBP-single location model would be achieved if the SBP occupied a central location within the PSU. In that case it can be calculated from the PSU model proposed by Bassi et al. (1990) that the core of PSII including CP29, would be located about 126 Å away from the SBP. On the other hand, LHCI-730 would be located about 159 Å on the other side of the SBP. The centers of the inner and outer halves of LHCII surrounding the PSII core would be located about 156 (outer half) and 82 (inner half) Å from the SBP. The detection of pronounced excitation resonance energy transfer from Proto, Mp(e) and Pchl (ide) *a* to Chl *a* ~F685, ~F695, and ~F735 indicates that these anabolic tetrapyrroles are within distances of 100 Å or less of the Chl *a* acceptors. In view of the above considerations it was concluded that the detection of resonance excitation energy transfer between anabolic tetrapyrroles and Chl *a* of the various thylakoid Chl-protein complexes was not compatible with the functionality of the SBP-single location Chl-thylakoid biogenesis model.

## 15.5 Compatibility of the Multibranched Pathway (MBP)-Multilocation Model with Resonance Excitation Energy Transfer from Anabolic Tetrapyrroles to Various Chl-Protein Complexes

Further calculations of resonance excitation energy transfer rates, and distances separating tetrapyrrole donors from Chl *a* acceptors and other considerations favored the operation of the MBP-sublocation Chl biosynthesis-thylakoid biogenesis model as described below.

**Table 15.2** Calculated distances R, that Separate Proto, Mp(e) and Pchlide *a* Donors from Chl *a*-Protein complexes acceptors in barley and cucumber chloroplasts at 77 K *in situ*

| | Proto | | Mp(e) | | | |
|---|---|---|---|---|---|---|
| | Barley | cucumber | Barley | Cucumber | MV Pchlide | DV Pchlide |
| Chl *a* species | R (Å) | | | | *a* barley | *a* cucumber |
| Chl *a* F685 (LHCI-680 + outer half of LHCII) | 38.24 | 29.6 | 35.06 | 38.16 | 37.31 | 34.79 |
| Chl *a* F695 (CP47) + CP29) | 40.6 | 29.48 | 34.62 | 40.91 | 38.97 | 38.14 |
| Chl *a* F735 (LHCI-730) | 22.34 | 16.26 | 19.07 | 23.41 | 23.06 | 21.84 |

*Note*: The distances "R" were determined from $[(R^6)^{1/6}\ cm]10^8\ Å.cm^{-1}$. The $R^6$ values were calculated in Table 9 of Kopetz et al. 2004 (Adapted from Kopetz et al. 2004)

First analytical tools were developed to calculate the distances separating Proto, Mp(e), and DV and MV Pchlide *a* from Chl *a* acceptors (Table 15.2) (Kopetz et al. 2004). The calculated distances were next compared to current concepts of the photosynthetic unit structure (Allen and Forsberg 2001; Anderson 2002; Staehelin 2003) and the Chl-thylakoid biogenesis models proposed in Figs. 15.1, 15.2, and 15.3 (see above). The calculated distances separating Proto, Mp(e) and DV and MV Pchlide *a* from various Chl *a* acceptors *in situ* are reported in Table 15.2.

The early concept of a PSU consisting of about 500 antennas Chl per reaction center has evolved into two pigment systems each with its own reaction center and antennas Chl (Allen and Forsberg 2001; Anderson 2002; Staehelin 2003). The early visualization of the two photosystems consisted of various pigment-protein complexes arrayed into a linear PSU (the continuous array model), about 450 Å in length and 130 Å in width (Bassi et al. 1990). Within the PSU, the LHCII was considered shared between the two photosystems. More recent models favored the concept of a laterally heterogeneous PSU (Allen and Forsberg 2001; Anderson 2002; Staehelin 2003). In this model LHCII was considered to shuttle between PSI and PSII upon phosphorylation and dephosphorylation (Allen and Forsberg 2001). Furthermore while PSII is mainly (but not exclusively) located in appressed thylakoid domains, PSI is located in non-appressed stroma thylakoids, grana margins, and end membranes (Anderson 2002; Staehelin 2003). The calculated distances separating Proto, Mp(e) and DV and MV Pchlide *a* from various Chl *a* acceptors in situ are reported in Table 15.2.

Distances separating anabolic tetrapyrroles from various Chl-protein complexes ranged from a low of 16.26 Å for Proto-Chl *a* separation in cumber, to a high of 40.91 Å for Proto-Chl *a*-F695 separation in barley (Table 15.2). The magnitude of these distances is certainly compatible with the observation of intense resonance excitation energy transfer reported in Kolossov et al. (2003).

In cucumber, a DDV-LDDV plant species (Abd-El-Mageed et al. 1997), the distances that separate Proto were shorter than those that separate Mp(e) and DV Pchlide *a* from the Chl *a* species (Table 15.2). Since Proto is an earlier intermediate of Chl *a* biosynthesis than Mp(e) and Pchlide *a*, it indicates that in cucumber, the Chl *a*-protein biosynthesis subcenter is a highly folded entity, where linear distances between intermediates and end products bear little meaning (see discussion). On the other hands, in barley, a DMV-LDMV plant species (Abd-El-Mageed et al. 1997) distances separating Proto from various Chl *a* acceptors were generally longer than those separating Mp(e) and MV Pchlide a from the Chl *a* acceptors (Table 15.2). This in turn suggests that the tetrapyrrole-protein complex folding in cucumber (DV subcenters) is different than in barley (MV subcenters).

On the other hands the shorter distances separating anabolic tetrapyrroles from Chl-protein complexes (Table 15.2) are compatible with the SBP-multilocation and MBP-sublocation models. Since overwhelming experimental evidence argues against the operation of a single-branched Chl biosynthetic pathway in plants (Rebeiz et al. 2003b) that leaves us with the MBP-sublocation model alternative. In this model, the unified multibranched Chl *a*/*b* biosynthetic pathway, is visualized as the template of a Chl-protein biosynthesis center where the assembly of PSI, PSII and LHC takes place (Rebeiz et al. 1999). The multiple Chl biosynthetic routes are visualized, individually or in groups of one or several adjacent routes, as Chl-apoprotein biosynthesis subcenters earmarked for the coordinated assembly of individual Chl-apoprotein complexes. Apoproteins destined to some of the subcenters may possess specific polypeptide signals for specific Chl biosynthetic enzymes peculiar to that subcenter, such as 4-vinyl reductases, formyl synthetases or Chl *a* and Chl *b* synthetases. Once an apoprotein formed in the cytoplasm or in the plastid reaches its subcenter destination and its signal is split off, it binds nascent Chl formed via one or more biosynthetic routes, as well as carotenoids. During pigment binding, the apoprotein folds properly and acts at that location, while folding or after folding, as a template for the assembly of other pigment-proteins. This model is certainly compatible with the lateral heterogeneity of the PSU and can account for the observed resonance excitation energy transfer and the short distances separating anabolic tetrapyrroles from Chl-protein complexes in the distinct PSI, PSII and shuttling LHCII entities that compose the PSU.

In all cases, it was observed that while distances separating metabolic tetrapyrroles from Chl *a* E670F685 and Chl *a* E677F695 were in the same range, those separating Chl *a* E704F735 from the anabolic tetrapyrroles were much shorter (Table 15.2). As may be recalled, it is believed that the fluorescence emitted at F685 nm arises from the Chl *a* of the light-harvesting Chl-protein complexes (LHCII and LHCI-680), that emitted at F695 nm originates mainly from the PSII antenna Chl *a* (CP47 and/or CP29), while that emitted at F735 nm originates primarily from the PSI antenna Chl *a* (LHCI-730) (Bassi et al. 1990). This in turn suggests that in the Chl *a*-protein biosynthesis subcenters, protein folding is such that the PSI antenna Chl *a* (LHCI-730) is much closer to the terminal steps of anabolic tetrapyrrole biosynthesis than the LHCII and LHCI-680 Chl-protein complexes or the CP47 and/or CP29 PSII antenna Chl *a* complexes.

# References


Abd-El-Mageed HA, El Sahhar KF, Robertson KR et al (1997) Chloroplast biogenesis 77: two novel monovinyl and divinyl light-dark greening groups of plants and their relationship to the chlorophyll a biosynthetic heterogeneity of green plants. Photochem Photobiol 66:89–96

Allen JF, Forsberg J (2001) Molecular recognition in thylakoid structure and function. Trends Plant Sci 6:317–326

Anderson JM (2002) Changing concepts about the distribution of photosystem I and II between grana-appressed and stroma-exposed thylakoid membranes. Photosynth Res 73:157–164

Bassi R, Rigoni F, Giacometti GM (1990) Chlorophyll binding proteins with antenna function in higher plants and green algae. Photochem Photobiol 52:1187–1206

Calvert JG, Pitts JN (1967) Photochemistry. Wiley, New York

Granick S (1950) Magnesium vinyl pheoporphyrin $a_5$, another intermediate in the biological synthesis of chlorophyll. J Biol Chem 183:713–730

Jones OTG (1963) Magnesium 2,4-divinyl phaeoporphyrin $a_5$ monomethyl ester, a protochlorophyll-like pigment produced by *Rhodopseudomonas spheroides*. Biochem J 89:182–189

Kolossov VL, Kopetz KJ, Rebeiz CA (2003) Chloroplast biogenesis 87: evidence of resonance excitation energy transfer between tetrapyrrole intermediates of the chlorophyll biosynthetic pathway and chlorophyll *a*. Photochem Photobiol 78:184–196

Kopetz KJ, Kolossov VL, Rebeiz CA (2004) Chloroplast biogenesis 89: development of analytical tools for probing the biosynthetic topography of photosynthetic membranes by determination of resonance excitation energy transfer distances separating metabolic tetrapyrrole donors from chlorophyll *a* acceptors. Anal Biochem 329:207–219

Lakowicz JR (1999) Principles of fluorescence spectroscopy. Kluwer Academic/Plenum Press, New York

Rebeiz CA, Montazer-Zouhoor A, Mayasich JM et al (1988) Photodynamic herbicides: recent developments and molecular basis of selectivity. Crit Rev Plant Sci 6:385–434

Rebeiz CA, Parham R, Fasoula DA et al (1994) Chlorophyll biosynthetic heterogeneity. In: Chadwick DJ, Ackrill K (eds) The biosynthesis of the tetrapyrrole pigments. Wiley, New York, pp 177–193

Rebeiz CA, Ioannides IM, Kolossov V et al (1999) Chloroplast biogenesis 80. Proposal of a unified multibranched chlorophyll a/b biosynthetic pathway. Photosynthetica 36:117–128

Rebeiz CA, Kolossov VI, Briskin D et al (2003a) Chloroplast biogenesis 86: chlorophyll biosynthetic heterogeneity, multiple biosynthetic routes and biotechnological spin-offs. In: Nalwa N (ed) Handbook of photochemistry and photobiology. American Scientific Publisher, Los Angeles, pp 183–248

Rebeiz CA, Kolossov VL, Briskin D et al (2003b) Chloroplast biogenesis: chlorophyll biosynthetic heterogeneity, multiple biosynthetic routes and biological spin-offs. In: Nalwa HS (ed) Handbook of photochemistry and photobiology, vol 4. American Scientific Publishers, Los Angeles, pp 183–248

Rebeiz CA, Kolossov VL, Kopetz KK (2004) Chloroplast bioengineering: photosynthetic efficiency, modulation of the photosynthetic unit size, and the agriculture of the future. In: Nelson DW (ed) Agricultural applications in green chemistry, vol 887. American Chemical Society, Washington, DC, pp 81–105

Staehelin LA (2003) Chloroplast structure: from chlorophyll granules to supra-molecular architecture of thylakoid membranes. Photosynth Res 76:185–196

Sundqvist C, Ryberg M (eds) (1993) Pigment-protein complexes in plastids: synthesis and assembly. Academic, New York

Turro NJ (1965) Molecular photochemistry. Benjamin, London

Wolff JB, Price L (1957) Terminal steps of chlorophyll *a* biosynthesis in higher plants. Arch Biochem Biophys 72:293–301

# Chapter 16
# The Chlorophyll Biosynthetic Heterogeneity and Chloroplast Bioengineering

*You can have brilliant ideas, but if you can't get them across, your ideas won't get you anywhere. (Lee Iacocca)*

## 16.1 Introduction

By the year 2030, the world population may increase significantly and top the nine billion benchmark. This is particularly significant since worldwide there has been a decline in cereal yield that is causing the annual rate of increase in yield to fall below the rate of population increase. Furthermore it will be difficult to increase the land area under cultivation without serious environmental complications. As a consequence the increased demand for food and fiber will have to be met by higher agricultural plant productivity.

Since plant productivity depends on photosynthetic efficiency, there is hope that agricultural productivity can be significantly increased by alteration of the photosynthetic unit size (Rebeiz et al. 2003a). Indeed, on the basis of recent advances in the understanding of the chemistry and biochemistry of the greening process and significant advances in molecular biology, we believe that alteration of the PSU size has become a realistic possibility.

Life in the biosphere is carbon based. All molecules needed for life are made up of a carbon skeleton which is complemented by organic elements such as O, H, N, and inorganic elements such as K, P, Ca, Fe, etc. Carbon, O and H of organic compounds originate in $CO_2$ and $H_2O$. The carbon skeleton is assembled via the process of photosynthesis that essentially converts solar energy into chemical energy. Nitrogen originates in NH3 and inorganic elements originate in the rocks of the biosphere and are incorporated into the carbon skeleton via enzymatic reactions. Chemical energy consists of the covalent bond energy embedded into

C.A. Rebeiz, *Chlorophyll Biosynthesis and Technological Applications*, 
DOI 10.1007/978-94-007-7134-5_16, © Springer Science+Business Media Dordrecht 2014

the carbon-carbon skeleton as well as the high energy bonds of ATP and NADPH which are formed during the process of photosynthesis.

The carbon cycle essentially describes how photosynthesis supports organic life in the biosphere. The carbon skeleton formed via the process of photosynthesis is converted into the simple and complex food consumed by organic life. The needed energy for enzymatic inter conversions and biosynthetic processes is provided by ATP and NADPH. The organic matter of dead biota is converted in turn into $CO_2$, $H_2O$, and inorganic elements by bacterial activity. Then the carbon cycle repeats itself all over again.

At issue then, is whether agricultural productivity at today's levels of photosynthetic efficiency is efficient enough to feed a growing world population. This issue will be explored below.

## 16.2 Relationship of Agricultural Productivity to Photosynthetic Efficiency

Since plants form food by conversion of solar energy, $CO_2$, and $H_2O$ into chemical energy via the process of photosynthesis, it ensues that agricultural productivity depends in turn upon photosynthetic efficiency. Let us therefore briefly describe the components of photosynthetic efficiency.

Photosynthetic efficiency is controlled by intrinsic and extrinsic factors (Lien and San Pietro 1975). Extrinsic factors include the availability of water, $CO_2$, inorganic nutrients, ambient temperature, and the metabolic and developmental state of the plant. The most important intrinsic factor is the efficiency of the photosynthetic electron transport system (PETS). The latter is driven by two photochemical reactions that take place in membrane-bound photosystem I (PSI), and PSII chlorophyll (Chl)-protein complexes.

### *16.2.1 The Primary Photochemical Acts of Photosystem I (PSI) and PSII*

Conversion of solar energy into chemical energy is the results of two photochemical acts that take place in PSI and PSII. The primary photochemical act of PSII is initiated by the absorption of light by antenna Chl *a* and *b*. The absorbed photons are conveyed to special Chls in the PSII reaction center. There, the light energy is used to generate a strong oxidant $Z^+$ that liberates oxygen from water. It also generates a weak reductant $Q^-$ that together with plastoquinone electron acceptor pools serve as temporary storage of the electrons extracted from water. The primary photochemical act of PSI is also initiated by the absorption of light by antenna Chl *a* and *b*, and in this case too, the absorbed photons are conveyed to special Chls in

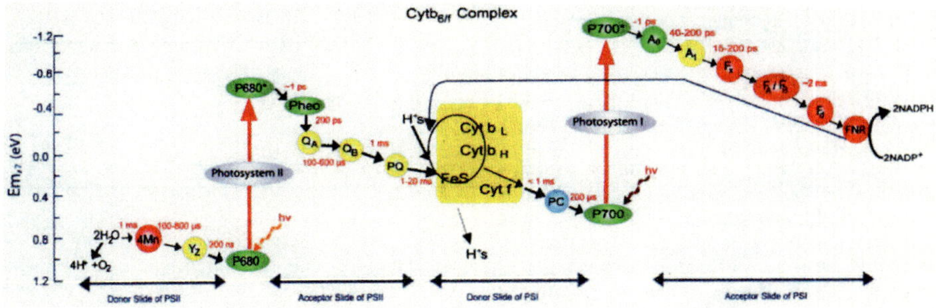


**Fig. 16.1** The Z-scheme of oxygenic photosynthesis for electron transfer from water to oxidized nicotinamide adenine dinucleotide phosphate (NADP). The symbols are: Mn Mn cluster, Y Tyrosine-161 on the D1 protein, p680 a pair of Chls the reaction center (RC) Chls of PSII having one of its absorption bands at 689 nm, P680* excited P680, Pheo the primary electron acceptor of PS II, $Q_A$ the primary plastoquinone electron acceptor of PS II, $Q_B$ secondary plastoquinone electron acceptor of PS II, PQ plastoquinone pool, FeS Rieske iron sulfur protein, Cyt f cytochrome *f*, CytbII high potential cytochrome *b*6, PC plastocyanin, P700 a pair of Chl *a* and *a′* the RC Chls of PS I, P700* excited P700, $A_0$ primary electron acceptor of PS I a Chl monomer, $A_1$ secondary electron acceptor of PS I vitamin K, $F_x$ $F_A$ and $F_B$ three different iron sulfur centers, Fd ferredoxin, and FNR ferredoxinNADP reductase. Approximate estimated times = various are also noted on the figure. A *circular path* in the Cyt b6*f* complex symbolizes the existence of a cyclic flow around PS I under certain conditions (Reproduced from Govindjee (2004); colored version is from Satoh et al. (2005))

the PSI reaction center. There the light energy generates a weak oxidant P700* which receives electrons from the plastoquinone pools via cytochrome f and plastocyanin. It also generates a strong reductant $A_0$ that donates electrons to $NADP^+$ via a series of electron carriers. As a consequence, $NADP^+$ is converted it to NADPH. The photochemical acts of PSII and PSI, and the flow of electrons between PSII and PSI are depicted in Fig. 16.1.

During electron and proton flow, energy rich ATP and NADPH are formed. The energy of NAPDH and ATP is used for the enzymatic conversion of $CO_2$ into carbohydrates. The latter are in turn converted into a variety of organic molecules. In summary the efficiency of food formation by green plants depends to a great extent on the efficiency of NADPH and ATP formation, and that in turn depends on the efficiency of the PETS.

The rest of this chapter will therefore be devoted to a discussion of the efficiency of the PETS and possible alterations in the circuitry of the chloroplast that may lead to a higher efficiency of the PETS and higher plant productivity under field conditions.

### *16.2.2 Theoretical Maximal Energy Conversion Efficiency of the PETS of Green Plants*

This discussion is essentially extracted from a 1975 RANN report (Lien and San Pietro 1975).

At the maximal quantum efficiency of one, two photons are required to move one electron across the potential difference of about 1.25 V between $Z^+$ and $A_0$. The maximal efficiency of the photochemical reactions leading to the formation of $Z^+$ and $A_0$ is then given by

$$E = 1.25\,eV/2\,h\nu \tag{16.1}$$

Where,
E = efficiency of PETS
eV = Energy units in electron volts
$\nu$ = Energy of the absorbed photon in eV

Since the red 680 nm photons absorbed by PSI and PSII have an energy of 1.83 eV, it ensues from Eq. 1 that

$$E = 1.25\,eV/2 * 1.83\,eV = 0.34\,eV \tag{16.2}$$

Therefore under red light, the absolute maximal efficiency of the PETS is

$$(0.34\,eV/1.25\,eV) * 100 = 27\,\% \tag{16.3}$$

Under natural white light, although the Chl concentration in photosynthetic membranes is high enough to result in the near total absorption of all incident photosynthetically active photons between 400 and 700 nm. These photons represent only about 44.5 % of the total incident solar radiation, under normal weather conditions. Therefore under these conditions, the possible overall maximal energy conversion efficiency amounts to:

$$(27\,\% * 44.5\,\%)/100 = 12\,\% \tag{16.4}$$

### *16.2.3 Actual Energy Conversion Efficiency of the PETS of Green Plants Under Field Conditions*

However, under field conditions, the average net photosynthetic efficiency results in a net agricultural productivity in the range of 2–8 tons of dry organic matter per acre per year (Lien and San Pietro 1975). This corresponds to a solar conversion efficiency of 0.1–0.4 % of the total average incident radiation. Therefore the discrepancy between the 12 % maximal theoretical efficiency of the PETS, and the agricultural photosynthetic efficiency observed under field conditions ranges from

$$(12\,\%/0.4\,\%) * 100 = 3000\,\% \tag{16.5}$$

to

$$(12\,\%/0.1\,\%) * 100 = 12000\,\% \tag{16.6}$$

## 16.3 Molecular Basis of the Discrepancy Between the Theoretical Maximal Efficiency of the PETS and the Actual Solar Conversion Efficiency of Photosynthesis Under Field Conditions

The discrepancy between the 12 % theoretical maximal efficiency of the PETS and the actual 0.1–0.4 % solar conversion efficiency of photosynthesis observed under field conditions can be attributed to (a) factors extrinsic to the PETS, and (b) to intrinsic rate limitations of the PETS (Lien and San Pietro 1975).

These factors will be examined below.

### *16.3.1 Contribution of Extrinsic PETS Parameters to the Discrepancy Between the Theoretical Photosynthetic Efficiency of 12 % and the Actual Photosynthetic Field Efficiency of 0.1–0.4 %*

Extrinsic factors such as ambient weather conditions, availability of water, $CO_2$, and inorganic nutrients, as well as the metabolic and developmental state of the plant directly affect photosynthetic efficiency under field conditions. Some of those factors are under human control while others are not. They do contribute nevertheless, to variations in photosynthetic efficiency under field conditions.

The rest of this discussion will focus however upon the impact of intrinsic factors that affect the PETS and photosynthetic efficiency.

### *16.3.2 Contribution of Intrinsic PETS Parameters to the Discrepancy Between the Theoretical Photosynthetic Efficiency of 12 % and the Actual Photosynthetic Field Efficiency of 0.1–0.4 %*

The 12 % theoretical efficiency of the PETS assumes that under natural conditions, PSI and PSII operate at a maximal quantum efficiency of ONE. In other words, it is assumed that every absorbed photon is completely converted into energy without losses (Lien and San Pietro 1975).

For a photosynthetic unit (PSU) size of 200 i.e. for 200 light harvesting Chl molecules per reaction center (RC), under the moderate light intensities of a shady sky (about 1/10 of full sunlight), each RC would receive about 200 photons per second (s) (Lien and San Pietro 1975). In other words, each RC would receive about 200 hits or excitons per s. In order to maintain a quantum efficiency of ONE, the slowest dark reaction of the entire PS must have a turnover rate of 200 per s (Lien and San Pietro 1975).

Under full sunlight, which is about tenfold higher than in the shade, the turnover rate of the limiting dark reaction would be 200*10 = 2,000 per s. This turnover rate corresponds to a rate of $O_2$ evolution of about 9,000 μmol of $O_2$ per mg Chl per hour (h). Yet, under saturating light intensities, and other optimal conditions, the maximal rate of $O_2$ evolution observed during a Hill reaction, which results in the oxidation of $H_2O$ and the release of $O_2$, rarely exceeds 5–10 % of the above value. In other words, that rate is equal to the optimal rate of $O_2$ evolution observed in the shade (Lien and San Pietro 1975).

Furthermore extensive kinetic studies have demonstrated that the rate limiting steps of the PETS do not reside in the initial photochemical reactions that take place in the RC, but reside within the redox-carriers, i.e. the electron transport chains connecting PSII to PSI. The discrepancy between the capacity of the photon gathering apparatus, i.e. the antenna Chl-protein complexes and the capacity of the rate-limiting dark reactions has been named the antenna/PS Chl mismatch (Lien and San Pietro 1975).

### *16.3.3 Impact of the Antenna/PS Chl Mismatch*

The first and most important effect of the antenna/PS Chl mismatch is one of reduced quantum conversion efficiencies at light intensities above shade levels. The second effect relates to the photodestructive effects of the excess photons collected by antenna Chl but not used in the initial photochemical acts. The energy of these unused photons leads to serious photodestruction of the PETS that must be repaired at a cost (Lien and San Pietro 1975).

## 16.4 Correction of the Antenna/Photosystem Chlorophyll Mismatch

Early on, the possible correction of the antenna/PS mismatch attracted the interest and curiosity of the photosynthesis community. It was suggested that one way of correcting the mismatch was to reduce the size of the PSU, which may be achieved by growing plants with chloroplasts having less antenna and more RC Chl per unit thylakoid area (Lien and San Pietro 1975). Research performed in the early 1970s failed however in its effort to alter significantly the PSU size in algal cell cultures (Lien and San Pietro 1975). Now, on the basis of deeper understanding of the chemistry and biochemistry of the greening process, which was achieved during the past 40 years, we have reason to believe that alteration of the PSU has become a realistic possibility.

## 16.5 What Kind of Scientific Knowledge Is Needed to Bioengineer a Reduction in PSU Size

Thorough and integrated anabolic and catabolic knowledge in the following fields of research is needed for successful research aimed at the bioengineering of a reduced PSU size. That include:

(a) Chls,
(b) lipids,
(c) carotenoids,
(d) plastoquinones,
(e) chloroplast apoproteins,
(f) assembly of pigment-protein complexes.

Because of space limitations, the remainder of this discussion will focus on the Chl, and apoprotein components of chloroplasts as well as on the assembly of Chl-protein complexes.

### *16.5.1 State of the Art in Our Understanding of Chl Biosynthesis*

During the past 30 years, it has become apparent that contrary to previous beliefs, the Chl biosynthetic pathway, is not a simple single-branched pathway, but a complex multibranched pathway that consist of about 15 carboxylic and two fully esterified biosynthetic routes (Chap. 6, Figs. 6.3, 6.4 and 6.5). The single and multibranched carboxylic pathways are briefly discussed below. Because of the importance of Chl-apoprotein assembly for the bioengineering of smaller photosynthetic units, these topics will also be discussed in this chapter which will create some necessary redundancy with Chap. 15.

#### 16.5.1.1 The Single-Branched Chl Biosynthetic Pathway Does Not Account for the Formation of All the Chl in Green Plants

The single-branched Chl biosynthetic pathway is depicted below in Fig. 16.2. It consists of a linear sequence of biochemical reactions which convert divinyl (DV) protoporphyrin IX (Proto) to monovinyl (MV) Chl *a* via DV Mg-Proto, DV Mg-Proto monomethyl ester (Mpe), DV protochlorophyllide *a* (Pchlide *a*), MV Pchlide *a*, and MV Chlorophyllide *a* (Chlide *a*). The salient features of this pathway are (a) the assumption that DV Pchlide *a* does not accumulate in higher plants, but is a transient metabolite which is rapidly converted to MV Chl *a* via MV Pchlide *a*, and (b) that the formation and accumulation of MV tetrapyrroles between Proto and Mpe and DV tetrapyrroles between Pchlide *a* and Chl *a* does not take place

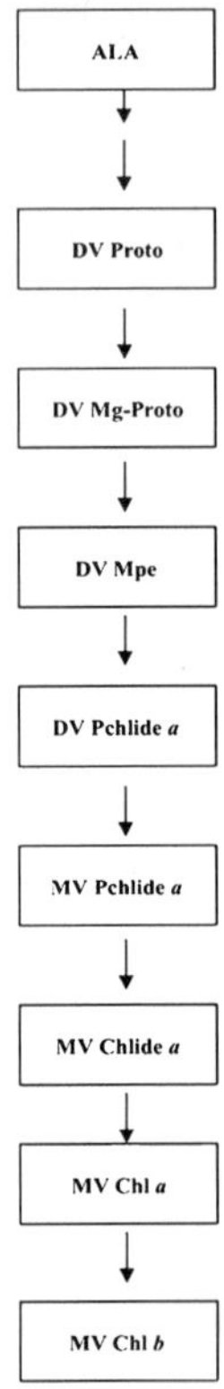


**Fig. 16.2** The single branched Chl biosynthetic pathway Proposed by (Granick 1950) and modified by (Wolff and Price 1957) and (Jones 1963)

(Rebeiz et al. 1994). All in all, experimental evidence gathered over the past 33 years indicates that only a small fraction of the total Chl of green plants is formed via this pathway (Rebeiz et al. 1994, 2003a).

#### 16.5.1.2 The Chl of Green Plants Is Formed via a Multibranched Biosynthetic Pathway

Our understanding of the Chl biosynthetic pathway has changed dramatically since the 1963 seminal review of Smith and French (1963). Several factors have contributed to this phenomenon, among which: (a) development of systems capable of Chl and thylakoid membrane biosynthesis in organello and in vitro, (Daniell and Rebeiz 1982a, b; Kolossov et al. 1999; Rebeiz and Castelfranco 1971a, b; Rebeiz et al. 1984), (b) powerful analytical techniques that allowed the qualitative and quantitative determination of various intermediates of the pathway (Rebeiz 2002), (c) recognition that the greening process proceeds differently in etiolated and green tissues, in darkness and in the light, (Carey and Rebeiz 1985), and in plants belonging to different greening groups (Abd-El-Mageed et al. 1997; Ioannides et al. 1994), and (d) recognition of the probability that the structural and functional complexity of thylakoid membranes is rooted in a multibranched, heterogeneous Chl biosynthetic pathway (Rebeiz et al. 1999).

It was also proposed that Chlorophyll biosynthetic heterogeneity referred either (a) to spatial biosynthetic heterogeneity, (b) to chemical biosynthetic heterogeneity, or (c) to a combination of spatial and chemical biosynthetic heterogeneities (Rebeiz et al. 2003b). Spatial biosynthetic heterogeneity was defined as the biosynthesis of an anabolic tetrapyrrole or end product by identical sets of enzymes, at several different locations of the thylakoid membrane. On the other hand, chemical biosynthetic heterogeneity was defined as the biosynthesis of an anabolic tetrapyrrole or end product at several different locations of the thylakoid membrane, via different biosynthetic routes, each involving at least one different enzyme. Figures 6.3, 6.4 and 6.5 of Chap. 6 organizes all known biosynthetic reactions into a logical scheme made up of various different biosynthetic routes.

## *16.5.2 Thylakoid Apoprotein Biosynthesis*

The biosynthesis of thylakoid apoproteins is a very complex phenomenon. Some apoproteins are coded for by nuclear DNA, are translated on cytoplasmic ribosomes and are transported to developing chloroplasts. Other apoproteins are coded for by plastid DNA and are translated on chloroplast ribosomes. A detailed discussion of chloroplast apoprotein biosynthesis is beyond the scope of this discussion. The reader is referred to reference (Sundqvist and Ryberg 1993) for a comprehensive discussion of this topic. For the purpose of this discussion it suffices to say that a PSU is an extremely complex structure that consists of many highly folded thylakoid and soluble proteins as well as membrane-bound pigment protein complexes having different functions in the light and dark steps of photosynthesis. An early visualization of a linear model of a PSU in the unfolded state is depicted in Fig. 16.3.

### 16.5.2.1 Assembly of Chl-Protein Complexes

Success in the bioengineering of smaller PSUs resides in a thorough understanding of how the Chl and thylakoid apoprotein biosynthetic pathways are coordinated to generate a specific functional Chl-protein complex. It is known for example that an apoprotein formed in the cytoplasm or in the chloroplast has to bind Chl molecules, has to fold properly, and has to wind up in the right place on the thylakoid. This process has to take place in order for the Chl-apoprotein to become a functional Chl-protein complex having a specific role in photosynthesis. What is unknown however is how an apoprotein formed in the cytoplasm or in the chloroplast becomes associated with Chl to become a specific Chl-protein complex of PSI, PSII or a light harvesting Chl-protein complex, having a specific function in photosynthesis.

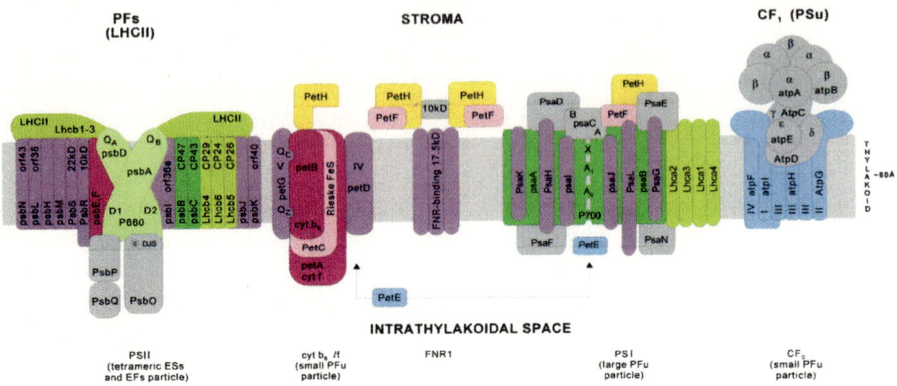


**Fig. 16.3** Schematics of a linear model of a PSU in the unfolded state (Reproduced from von Wettstein et al. 1995)

As mentioned in Chap. 15, we have recently examined three possible models for that scenario, which have been referred to as: (a) the single-branched Chl biosynthetic pathway (SBP)-single location model, (b) the SBP-multilocation model, and (c) the multi-branched Chl biosynthetic pathway (MBP)-sublocation model. The models take into account the dimension of the PSU (Bassi et al. 1990), the biochemical heterogeneity of the Chl biosynthetic pathway (Rebeiz et al. 1994, 2003a) and the biosynthetic and structural complexity of the thylakoid and the Chl.

## Assembly of Chl-Protein Complexes: The SBP-Single Location Model

The SBP-single location model is depicted schematically below, in Fig. 16.4, which has been reproduced from Fig. 15.1. As mentioned previously, within the PSU, this model accommodates only one Chl-apoprotein biosynthesis center and no Chl-apoprotein biosynthesis subcenters. Within the Chl-apoprotein biosynthesis center, Chl *a* and *b* are formed via a single-branched Chl biosynthetic pathway (Fig. 16.2) at a location accessible to all Chl-binding apoproteins. The latter will have to access that location in the unfolded state, pick up a complement of MV Chl *a* and/or MV Chl *b*, and undergo appropriate folding. Then the folded Chl-apoprotein complex has to move from the central location to a specific PSI, PSII, or Chl *a*/*b* LHC-protein location within the Chl-apoprotein biosynthesis center over distances of up to 225 Å (Kolossov et al. 2003; Kopetz et al. 2004). In this model, it is unlikely to observe resonance energy transfer between metabolic tetrapyrroles and some of the Chl-apoprotein complexes located at distances longer than 100 Å. This is because resonance excitation energy transfer takes place only over distances shorter than 100 Å (Calvert and Pitts 1967).

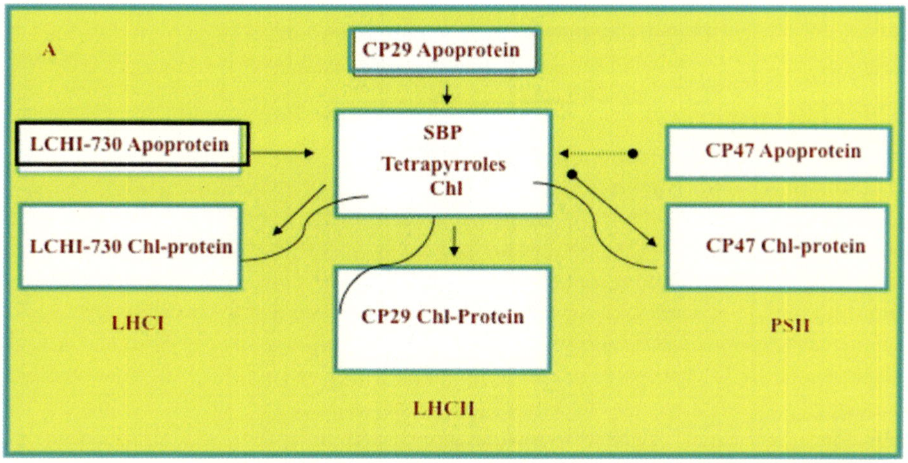


**Fig. 16.4** Schematics of the SBP-single location model in a PSU. As an example, the functionality of the model was illustrated with the use of three apoproteins namely CP29, LCHI-730 and CP47. Abbreviations: *SBP* single-branched Chl biosynthetic pathway, *PSII* photosystem II, *LHCII* the major light-harvesting Chl-protein complex of PSII, LHCI, one of the LHC antennae of PSI, CP47 and CP29, two PSII antennae, LHCI-730, the LHC antenna of PSI. *Curved lines* indicate putative energy transfer between tetrapyrroles and a Chl-protein complex (Adapted from Kopetz et al. 2004)

### Assembly of Chl-Protein Complexes: The SBP-Multilocation Location Model

The SBP-Multilocation model is depicted schematically below, in Fig. 16.5 which is lifted from Chap. 15.

In this model, every biosynthetic location within the PSU is considered to be a Chl-apoprotein thylakoid biosynthesis center. In every Chl-apoprotein biosynthesis location, a complete single-branched Chl *a*/*b* biosynthetic pathway (Fig. 16.2) is active. Association of Chl *a* and/or Chl *b* with specific PSI, PSII, or LHC apoproteins at any location is random. In every Chl-apoprotein biosynthesis center, distances separating metabolic tetrapyrroles from the Chl-protein complexes are shorter than in the SBP-single-location model.

### Assembly of Chl-Protein Complexes: The MBP-Sublocation Model

The SBP-sublocation model is depicted schematically below, in Fig. 16.6 which has been lifted from Chap. 15. In this model, the unified multibranched Chl *a*/*b* biosynthetic pathway, (Rebeiz et al. 2003a, b) is visualized as the template of a Chl-protein biosynthesis center where the assembly of PSI, PSII and LHC takes place. The multiple Chl biosynthetic routes are visualized, individually or in groups of one or several adjacent routes, as Chl-apoprotein biosynthesis subcenters

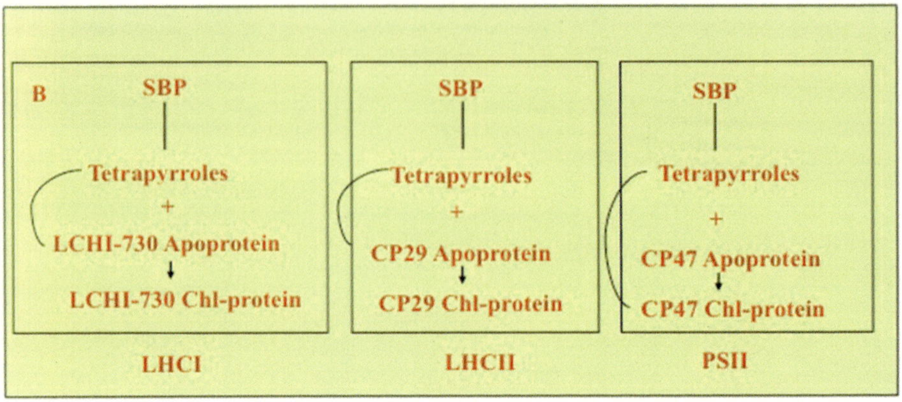


**Fig. 16.5** Schematics of the SBP-Multilocation model in a PSU. All abbreviations and conventions are as in Fig. 16.4 (Adapted from Kopetz et al. 2004)

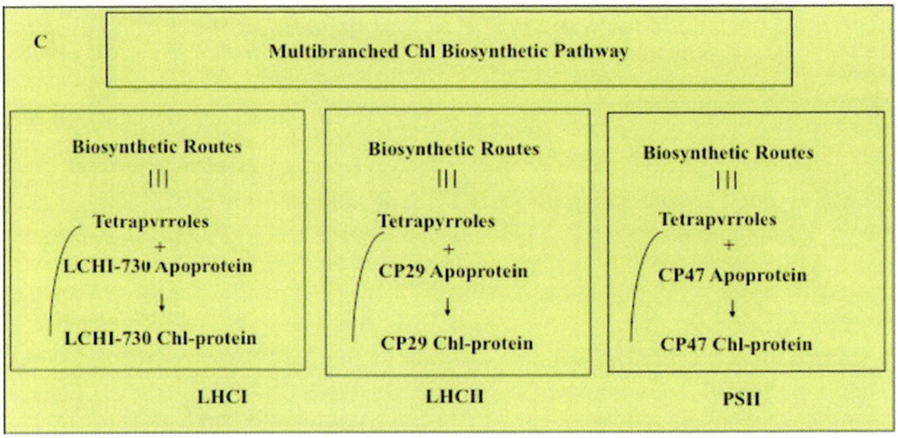


**Fig. 16.6** Schematics of the MBP-sublocation model in a PSU. All abbreviations and conventions are as in Fig. 16.4 (Adapted from Kopetz et al. 2004)

earmarked for the coordinated assembly of individual Chl-apoprotein complexes. Apoproteins destined to some of the subcenters may possess specific polypeptide signals for specific Chl biosynthetic enzymes peculiar to that subcenter, such as 4-vinyl reductases, formyl synthetases or Chl *a* and Chl *b* synthetases. Once an apoprotein formed in the cytoplasm or in the plastid reaches its subcenter destination and its signal is split off, it binds nascent Chl formed via one or more biosynthetic routes, as well as carotenoids. During pigment binding, the apoprotein folds properly and acts at that location, while folding or after folding, as a template for the assembly of other pigment-proteins.

Because of the shorter distances separating the accumulated tetrapyrroles from Chl-protein complexes, within each subcenter, resonance excitation energy transfer between various metabolic tetrapyrroles and Chl is readily observed. In this model, both MV and DV Mp(e) may be present in some pigment-protein complexes, in particular if more than one Chl biosynthetic route is involved in the Chl formation of a particular Chl-protein complex.

#### 16.5.2.2 Which Chl-Thylakoid Apoprotein Assembly Model Is Favored by Experimental Evidence?

We tested the compatibility of the three aforementioned models by resonance excitation energy transfer between anabolic tetrapyrrole intermediates of the Chl biosynthetic pathway and various thylakoid Chl-protein complexes, in order to determine which Chl-thylakoid apoprotein assembly model is likely to be functional during thylakoid membrane formation.

Resonance excitation energy transfer from three tetrapyrrole donors to the Chl *a* of various Chl-protein complexes were monitored, namely: from Proto, Mp (e) and MV and DV Pchlide *a*. DV Proto is a common precursor of heme and Chl. It is the immediate precursor of DV Mg-Proto. As such, it is an early intermediate along the Chl biosynthetic chain. Biosynthetically, it is several steps removed from the Chl end product (Rebeiz et al. 2003b). Mg-Proto is a mixed MV-DV, dicarboxylic tetrapyrrole pool, consisting of DV and MV Mg-Proto (Rebeiz et al. 2003b). It is the precursor of DV and MV Pchlide *a*. The [(Pchl(ide)] of higher plants consists of about 95 % Pchlide *a* and about 5 % Pchlide *a* ester . The latter is esterified with long chain fatty alcohols at position 7 of the macrocycle. While Pchlide *a* ester consists mainly of MV Pchlide a ester, Pchlide *a* consists of DV and MV Pchlide *a*. The latter are the immediate precursors of DV and MV Chlide *a* (Rebeiz et al. 2003b).

Accumulation of the various tetrapyrrole donors was induced by incubation of green tissues with δ-aminolevulinic acid (ALA) and/or 2,2′-dipyridyl (Rebeiz et al. 1988). The task of selecting appropriate Chl *a*-protein acceptors was facilitated by the fluorescence properties of green plastids. At 77 K, emission spectra of isolated chloroplasts exhibit maxima at 683–686 nm (~F685), 693–696 nm (~F695), and 735–740 nm (~F735) (see Chap. 15). Since these emission maxima are readily observed in the fluorescence emission spectra of green tissues and are associated with definite thylakoid Chl *a*-protein complexes, it was conjectured that they would constitute a meaningful resource for monitoring excitation resonance energy transfer between anabolic tetrapyrroles and representative Chl *a*-protein complexes.

To monitor the possible occurrence of resonance excitation energy transfer between the accumulated anabolic tetrapyrroles and Chl *a*-protein complexes, excitation spectra were recorded at 77 K at the respective emission maxima of the selected Chl *a* acceptors, namely at ~685, ~695, and ~735 nm. It was conjectured that if resonance excitation energy transfers were to be observed between the tetrapyrrole donors and the selected Chl *a* acceptors, definite excitation maxima would be observed. These excitation maxima would correspond to absorbance maxima of the various tetrapyrrole donors, and would correspond to the peaks of the resonance excitation energy transfer bands.

## The SBP-Single Location Model Is Not Compatible with Resonance Excitation Energy Transfer Between Anabolic Tetrapyrrole Donors and chl *a*-Proteins Acceptors in Chloroplasts

The compatibility of the SBP-single location model with the formation of Chl *a*-thylakoid proteins was tested by monitoring resonance excitation energy transfer between anabolic tetrapyrrole intermediates of the Chl biosynthetic pathway and various thylakoid Chl *a*-protein complexes. Pronounced resonance excitation energy transfer bands from Proto, Mp(e), and Pchl(ide) a to Chl *a* ~F685, ~F695, and ~F735 were detected (Chap. 6, Table 6.1).

Assignment of *in situ* resonance excitation energy transfer maxima to various metabolic tetrapyrroles was unambiguous except for a few cases at the short wavelength and long wavelength extremes of excitation bands. It was surprising to observe a significant diversity in the various intra-membrane environments of Proto, Mp(e), and Pchl(ide) *a* (Kolossov and Rebeiz 2003). A differential donation of resonance excitation energy transfer from multiple Proto, Mp(e) and Pchl(ide) *a* sites to different Chl *a*-apoprotein complexes, expressed this diversity which was strongly compatible with the biosynthetic heterogeneity of the Chl biosynthetic pathway.

Since resonance excitation energy transfer is insignificant at distances larger than 100 Å (Calvert and Pitts 1967), the detection of pronounced resonance excitation energy transfer from Proto, Mp(e), and Pchl(ide) *a* to Chl *a* ~ F685, ~F695, and ~F735 (Chap. 6, Table 6.1) indicated that these anabolic tetrapyrroles were within distances of 100 Å or less of the Chl *a* acceptors. This was incompatible with the functionality of the SBP-single location Chl-thylakoid biogenesis model as detailed below.

As mentioned in Chap. 15, the early concept of a PSU consisting of about 500 antenna Chl per reaction center has evolved into two pigment systems each with its own reaction center and antenna Chl (Allen and Forsberg 2001; Anderson 2002; Staehelin 2003). The early visualization of the two photosystems consisted of various pigment-protein complexes arrayed into a linear PSU (the continuous array model), about 450 Å in length and 130 Å in width (Bassi et al. 1990). In the PSU, the LHCII was depicted as being shared between the two photosystems. More recent models however, favor the concept of a laterally heterogeneous PSU (Allen and Forsberg 2001; Anderson 2002; Staehelin 2003). In this model LHCII shuttles between PSI and PSII upon phosphorylation and dephosphorylation (Allen and Forsberg 2001). Furthermore while PSII is mainly (but not exclusively) located in oppressed thylakoid domains, PSI is located in non-appressed stroma thylakoids, grana margins, and end membranes (Anderson 2002; Staehelin 2003).

The SBP-single location model is incompatible with the linear continuous array model, and the laterally heterogeneous PSU model. The SBP-single location model calls for Chl *a* and *b* to be formed via a single-branched Chl biosynthetic pathway at a location accessible to all Chl-binding apoproteins. The latter will have to access that location in the unfolded state, pick up a complement of MV Chl *a* and/or MV Chl *b*, and undergo appropriate folding. Then the folded Chl-apoprotein complex has to move from the central location to a specific PSI, PSII, or Chl a/b LHC-protein location within the Chl-apoprotein biosynthesis center over distances of up to 225 Å in the linear continuous array model, or over larger distances, in the laterally heterogeneous model, to become part of PSI, PSII or LHCII. If this were the case, then no resonance excitation energy transfer would be observed between anabolic tetrapyrroles and the various Chl-protein complexes, and the distances separating the anabolic tetrapyrroles from the various Chl-protein complexes would be much larger than the values reported in Table 15.2 of Chap. 15.

#### Incompatibility of the SBP-MLM Model with Experimental Data

The shorter distances separating anabolic tetrapyrroles from Chl-protein complexes reported in Table 6.1 of Chap. 6 are compatible with the SBP-multilocation and MBP-sublocation models. However, overwhelming experimental evidence argues against the operation of a single-branched Chl biosynthetic pathway in plants (Rebeiz et al. 2003a).

#### Compatibility of the MBP-SUBLM Model with Experimental Data

The various considerations discussed above leaves the MBP-sublocation model (MBPSUBLM) as a viable working hypothesis. In this model the unified multi-branched Chl *a*/*b* biosynthetic pathway, is visualized as the template of a Chl-protein biosynthesis center where the assembly of PSI, PSII and LHC takes place (Rebeiz et al. 1999, 2004). The multiple Chl biosynthetic routes are visualized, individually or in groups of one or several adjacent routes, as Chl-apoprotein biosynthesis subcenters earmarked for the coordinated assembly of individual Chl-apoprotein complexes. Apoproteins destined to some of the subcenters may possess specific polypeptide signals for specific Chl biosynthetic enzymes peculiar to that subcenter, such as 4-vinyl reductases, formyl synthetases or Chl *a* and Chl *b* synthetases. Once an apoprotein formed in the cytoplasm or in the plastid reaches its subcenter destination and its signal is split off, it binds nascent Chl formed via one or more biosynthetic routes, as well as carotenoids. During pigment binding, the apoprotein folds properly and acts at that location, while folding or after folding, as a template for the assembly of other pigment-proteins. This model is compatible with the lateral heterogeneity of the PSU and can account for the observed resonance excitation energy transfers Table 6.1 of Chap. 6, and the short distances separating anabolic tetrapyrroles from Chl-protein complexes in the distinct PSI, PSII and shuttling LHCII entities that compose the PSU.

## 16.6 Guidelines and Suggestions to Bioengineer Plants with Smaller Photosynthetic Unit Size

The compatibility of the MBP-sublocation model of Chl-thylakoid protein assembly has opened the way for testing the hypothesis of whether certain Chl biosynthetic routes are indeed involved in the formation of specific Chl-protein complexes. Below are outlined some guidelines and suggestions for investigating this issue.

The experimental strategy involves a two-pronged experimental approach. In a first attempt, a variety of higher and lower plant mutants that lack specific Chl-protein complexes could be used to determine which specific Chl biosynthetic route(s) is/are missing from the mutant Chl biosynthetic pathway. In this manner it may be possible to link a particular Chl biosynthetic route to a specific Chl-protein complex formation.

Likewise in the second approach functional PS I, and PS II particles as well as LHCII preparations could be isolated from wild types and mutants using mild detergents and the putative Chl biosynthetic routes associated with a particular preparation could be determined. In this manner it may be possible to link particular Chl biosynthetic routes to the lateral heterogeneity of the PSU.

### *16.6.1 Selection of Mutants*

A literature search of higher and lower plant mutants deficient in specific Chl-protein complexes revealed a rather large number of such mutants. Final selection of specific mutants for specific studies will therefore depend on the nature of the missing Chl-protein complexes and availability of plant material. Below are listed some of the candidate mutants

#### 16.6.1.1 Mutants of Higher Plants Other Than Arabidopsis

- *chlorina–f2 viridis-m*$^{29}$ *viridis-n*$^{34}$ *and viridis-zd*$^{69}$ of Barley (Henry et al. 1983; Machold et al. 1979; Preiss and Thornber 1995; White and Green 1987).
- Chl *b*-less barley mutant (Bellemare et al. 1982; Mullet et al. 1980).
- *viridis-zb*63 and *viridis-h*15 of barley (Hiller et al. 1980).
- Qy/+ hcf3/hcf3 of maize (Polacco 1984).
- *hcf**-3 nuclear maize mutant (Leto et al. 1985).
- hcf1-2-3-6-19- 38-42-44-50-101-102-103-104-108-111 in maize (Miles 1994).
- U374 mutant of sweet clover (Markwell et al. 1985).
- C*ab4Bst*EII, C*ab4.23,* and C*ab4.3* mutants of tomato (Huang et al. 1992).

#### 16.6.1.2 Arabidopsis Mutants

A large number of mutants with defects in or elimination of chlorophyll/protein complexes in *Arabidopsis thaliana* has already been identified (H. Bohnert, personal communication). Also some interesting mutants can be obtained from the US or European seed banks.

#### 16.6.1.3 Lower Plant Mutants

- *Y*-1 mutant of *Chlamydomonas reinhardtii* (Gershoni et al. 1982; Hoober 1990; Ish-Shalom and Ohad 1983).
- Gr1BSL, G1BU, and O4BSL of *Euglena gracilis* (Cunningham and Schiff 1986)
- PS I-less/*apc*E$^{-}$ *Synechocystis* sp. PCC 6803 mutant (Shen and Vermaas 1994)

### *16.6.2 Preparation of Photosynthetic Particles*

A large volume of literature dealing with the preparation of various photosynthetic particles is readily available. The most recent review of various procedures for the preparation of PS I, PS II, LHCII, and a variety of smaller Chl-protein complexes has been reported in (Paulsen and Scmid 2002). Procedures described there can be complemented by original standalone procedures available in the photosynthesis literature.

### *16.6.3 Determination of Biosynthetic Routes Functional in a Specific Mutant or Photosynthetic Particle*

Partial and full biosynthetic routes that are functional in various mutants and isolated photosynthetic particles and complexes can be determined by various techniques described in (Rebeiz 2002, 2003b). These techniques have been developed over a period of three and half decades and are routinely used on daily basis by several scientists. Cold and $^{14}$C-substrated can be prepared and used as described by (Rebeiz 2002). Wild types and mutants can be light or dark adapted in order to poise them in the DV or MV modes (Carey et al. 1985) prior to subplastidic particle isolation. Single or multistep reaction sequences can be executed by feeding appropriate substrates in well defined cofactor media capable of supporting nearly all the reactions described in Fig. 16.3.

### *16.6.4 Epilogue*

Future research dealing with the bioengineering of smaller PSU sizes will have to use as a working hypothesis the MBP-sublocation Chl *a*-thylakoid protein biosynthesis model. First the researcher will have to deal with the determination of which Chl biosynthetic routes gives rise to PSI, PSII and LHCII Chl-protein complexes. The greening process may then be manipulated to bioengineer genetically modified plants with a smaller PSU, i.e. with more PSU units having fewer antenna Chl per unit thylakoid area. Nevertheless this type of agriculture using genetically modified plants with smaller PSU sizes and higher photosynthetic conversion efficiencies will still be at the mercy of extrinsic factors and weather uncertainties.

In our opinion the ultimate agriculture of the future should consist of bioreactors populated with bioengineered, highly efficient photosynthetic membranes, with a small PSU size and operating at efficiencies that approach the 12 % maximal theoretical efficiency of the PETS that may be observed under white light, or the 27 % maximal theoretical efficiency that may be achieved under red light. Such conditions may be set up during space travel, in large space stations, or in human colonies established on the moon or on Mars (Rebeiz et al. 1982). The photosynthetic product may well be a short chain carbohydrate such as glycerol that can be converted into food fiber and energy.

In the meanwhile, let us not forget that a journey of 10,000 miles starts with the first step.

## References

Abd-El-Mageed HA, El Sahhar KF, Robertson KR et al (1997) Chloroplast biogenesis 77. Two novel monovinyl and divinyl light–dark greening groups of plants and their relationship to the chlorophyll *a* biosynthetic heterogeneity of green plants. Photochem Photobiol 66:89–96

Allen JF, Forsberg J (2001) Molecular recognition in thylakoid structure and function. Trends Plant Sci 6:317–326

Anderson JM (2002) Changing concepts about the distribution of photosystem I and II between grana-appressed and stroma-exposed thylakoid membranes. Photosynth Res 73:157–164

Bassi R, Rigoni F, Giacometti GM (1990) Chlorophyll binding proteins with antenna function in higher plants and green algae. Photochem Photobiol 52:1187–1206

Bellemare G, Bartlett SG, Chua NH (1982) Biosynthesis of chlorophyll *a*/*b*-binding polypeptides in wild type and the Chlorina f2 mutant of barley. J Biol Chem 257(13):7762–7767

Calvert JG, Pitts JN (1967) Photochemistry. Wiley, New York

Carey EE, Rebeiz CA (1985) Chloroplast biogenesis 49. Difference among angiosperms in the biosynthesis and accumulation of monovinyl and divinyl protochlorophyllide during photoperiodic greening. Plant Physiol 79:1–6

Carey EE, Tripathy BC, Rebeiz CA (1985) Chloroplast biogenesis 51. Modulation of monovinyl and divinyl protochlorophyllide biosynthesis by light and darkness in vitro. Plant Physiol 79:1059–1063

Cunningham FXJ, Schiff J (1986) Chlorophyll-protein complexes from *Euglena gracilis* and mutants deficient in chlorophyll *b*. Plant Physiol 80:231–238

Daniell H, Rebeiz CA (1982a) Chloroplast culture VIII. A new effect of kinetin in enhancing the synthesis and accumulation of protochlorophyllide in vitro. Biochem Biophys Res Commun 104:837–843

Daniell H, Rebeiz CA (1982b) Chloroplast culture IX. Chlorophyll(ide) *a* biosynthesis in vitro at rates higher than in vivo. Biochem Biophys Res Commun 106:466–470

Gershoni JM, Shochat S, Malkin S et al (1982) Functional organization of the chlorophyll-containing complexes of *Chlamydomonas reinhardi*. Plant Physiol 70:637–644

Govindjee (2004) Chlorophyll *a* fluorescence: a bit of basics and history. In: Papageorgiou GC, Govindjee (eds) Chlorophyll *a* fluorescence. A signature of photosynthesis, vol 19. Springer, Dordrecht, pp 1–41

Granick S (1950) Magnesium vinyl pheoporphyrin $a_5$, another intermediate in the biological synthesis of chlorophyll. J Biol Chem 183:713–730

Henry LEA, Dalgaard mikkelsen J, Lindberg Moller B (1983) Pigment and acyl lipid composition of photosystem I and II vesicles and of photosynthetic mutants in barley. Carlsberg Res Commun 48:131–148

Hiller RG, Lindberg Moller B, Hoyer-Hansen G (1980) Characterization of six putative photosystem I mutants in barley. Carlsberg Res Commun 45:315–328

Hoober KJ (1990) Accumulation of chlorophyll *a*/*b*-binding polypeptides in *Chlamydomonas reinhardi y-1* in the light or dark at 38 C. Evidence for proteolytic control. Plant Physiol 92:419–426

Huang L, Adam Z, Hoffman NE (1992) Deletion mutants of chlorophyll *a*/*b* binding proteins are efficiently imported into chloroplasts but do not integrate into thylakoid membranes. Plant Physiol 99:247–255

Ioannides IM, Fasoula DM, R. RK (1994) An evolutionary study of chlorophyll biosynthetic heterogeneity in green plants. Biochem Syst Ecol 22:211–220

Ish-Shalom D, Ohad I (1983) Organization of chlorophyll-protein complexes of photosystem I in *Chlamydomonas reinhardi*. Biochem Biophys Acta 722:498–507

Jones OTG (1963) Magnesium 2,4-divinyl phaeoporphyrin $a_5$ monomethyl ester, a protochlorophyll-like pigment produced by *Rhodopseudomonas spheroides*. Biochem J 89:182–189

Kolossov VL, Rebeiz CA (2003) Chloroplast biogenesis 88. Protochlorophyllide *b* occurs in green but not in etiolated plants. J Biol Chem 278(50):49675–49678

Kolossov VL, Ioannides IM, Kulur S et al (1999) Chloroplast biogenesis 82: development of a cell-free system capable of the net synthesis of chlorophyll(ide) *b*. Photosynthetica 36:253–258

Kolossov VL, Kopetz KJ, Rebeiz CA (2003) Chloroplast biogenesis 87: evidence of resonance excitation energy transfer between tetrapyrrole intermediates of the chlorophyll biosynthetic pathway and chlorophyll *a*. Photochem Photobiol 78:184–196

Kopetz KJ, Kolossov VL, Rebeiz CA (2004) Chloroplast biogenesis 89: development of analytical tools for probing the biosynthetic topography of photosynthetic membranes by determination of resonance excitation energy transfer distances separating metabolic tetrapyrrole donors from chlorophyll *a* acceptors. Anal Biochem 329:207–219

Leto KJ, Bell E, McIntosh L (1985) Nuclear mutation leads to an accelerated turnover of chloroplast-encoded 48 kd polypeptides in thylakoids lacking photosystem II. EMBO J 4 (7):1645–1653

Lien S, San Pietro A (1975) An inquiry into the biophotolysis of water to produce hydrogen. Indiana University, Bloomington, p 50

Machold O, Simpson DJ, Lindberg Moller B (1979) Chlorophyll-proteins of thylakoids from wild-type and mutants of barley (*Hordeum vulgare* L.). Carlsberg Res Commun 44:235–254

Markwell J, Webber AN, Lake B (1985) Mutants of sweetclover (*Melilotus alba*) lacking chlorophyll b. Plant Physiol 77:948–951

Miles D (1994) The role of high chlorophyll fluorescence photosynthesis mutants in the analysis of chloroplast thylakoid membrane assembly and function. Maydica 39:35–45

Mullet JE, Burke JJ, Arntzen CJ (1980) A developmental study of photosystem I peripheral chlorophyll proteins. Plant Physiol 65:823–827

Paulsen H, Scmid VHR (2002) Analysis and reconstitution of chlorophyll-proteins. In: Smith AG, Witty M (eds) Heme, chlorophyll, and bilins. Methods and protocols. Humana Press, Totowa, pp 235–253

Polacco ML (1984) Chl (A/B) light harvesting complex assembly in maize: genetic evidence that it may compete with PSII for Chl. Curr Top Plant Biochem Physiol 3:167

Preiss S, Thornber PJ (1995) Stability of the apoproteins of light-harvesting complex I and II during biogenesis of thylakoids in the chlorophyll *b*-less barely mutant *Chlorina f2*. Plant Physiol 107:709–717

Rebeiz CA (2002) Analysis of intermediates and end products of the chlorophyll biosynthetic pathway. In: Smith A, Witty M (eds) Heme chlorophyll and bilins, *methods and protocols*. Humana Press, Totowa, pp 111–155

Rebeiz CA, Castelfranco P (1971a) Protochlorophyll biosynthesis in a cell-free system from higher plants. Plant Physiol 47:24–32

Rebeiz CA, Castelfranco P (1971b) Chlorophyll biosynthesis in a cell-free system from higher plants. Plant Physiol 47:33–37

Rebeiz CA, Daniell H, Mattheis JR (1982) Chloroplast bioengineering: the greening of chloroplasts in vitro. In: Scott CD (ed) Biotechnology bioengineering symposium, vol 12. John Wiley, New York, pp 414–439

Rebeiz CA, Montazer-Zouhoor A, Daniell H (1984) Chloroplast culture X: thylakoid assembly in vitro. Isr J Bot 33:225–235

Rebeiz CA, Juvik JA, Rebeiz CC (1988) Porphyric insecticides 1. Concept and phenomenology. Pesticide Biochem Physiol 30:11–27

Rebeiz CA, Parham R, Fasoula DA et al (1994) Chlorophyll biosynthetic heterogeneity. In: Chadwick DJ, Ackrill K (eds) The biosynthesis of the tetrapyrrole pigments. Wiley, New York, pp 177–193

Rebeiz CA, Ioannides IM, Kolossov V et al (1999) Chloroplast biogenesis 80. Proposal of a unified multibranched chlorophyll a/b biosynthetic pathway. Photosynthetica 36:117–128

Rebeiz CA, Kolossov VI, Briskin D et al (2003a) Chloroplast biogenesis 86: chlorophyll biosynthetic heterogeneity, multiple biosynthetic routes and biotechnological spin-offs. In: Nalwa N (ed) Handbook of photochemistry and photobiology. American Scientific Publishers, Los Angeles, pp 183–248

Rebeiz CA, Kolossov VL, Briskin D et al (2003b) Chloroplast biogenesis: chlorophyll biosynthetic heterogeneity, multiple biosynthetic routes and biological spin-offs. In: Nalwa HS (ed) Handbook of photochemistry and photobiology, vol 4. American Scientific Publishers, Los Angeles, pp 183–248

Rebeiz CA, Kolossov VL, Kopetz KK (2004) Chloroplast bioengineering: photosynthetic efficiency, modulation of the photosynthetic unit size, and the agriculture of the future. In: Nelson DW (ed) Agricultural applications in green chemistry, vol 887. American Chemical Society, Washington, DC, pp 81–105

Satoh K, Wydrzynski T, Govindgee (2005) Introduction to photosynthesis. In: Wydrzynyski T, Satoh K (eds) Photosystem II: the light-driven water: plastoquinone oxidoreductase. Springer, Dordrecht, pp 11–22

Shen G, Vermaas WFJ (1994) Mutation of chlorophyll ligands in the chlorophyll-binding CP47 protein as studied in a *synechocystis* sp. PCC 6803 photosystem I-less background. Biochemistry 33:7379–7388

Smith JHC, French CS (1963) The major accessory pigment in photosynthesis. Annu Rev Plant Physiol 14:181–224

Staehelin LA (2003) Chloroplast structure: from chlorophyll granules to supra-molecular architecture of thylakoid membranes. Photosynth Res 76:185–196

Sundqvist C, Ryberg M (eds) (1993) Pigment-protein complexes in plastids: synthesis and assembly. Academic, New York

von Wettstein D, Gough S, Kannangara CG (1995) Chlorophyll biosynthesis. Plant Cell 7:1039–1057

White MJ, Green BR (1987) Polypeptides belonging to each of the three major chlorophyll *a* + *b* protein complexes are present in a chlorophyll-*b*-less barley mutant. Eur J Biochem 165:531–535

Wolff JB, Price L (1957) Terminal steps of chlorophyll *a* biosynthesis in higher plants. Arch Biochem Biophys 72:293–301

# Chapter 17
# Photodynamic Herbicides

*It is the habit of mediocre minds to condemn all that is beyond their grasp (La Rochefoucaud)*

## 17.1 Prologue

In 1982, after having researched the chemistry and biochemistry of the greening process for 18 years, it was felt that enough was known about this important biological phenomenon to translate it into biotechnological developments. In looking for a handle on the problem we opted for the development of photodynamic herbicides. That decision was prompted by two considerations: (a) the size and importance of the herbicide industry, and (b) the interesting photosensitizing properties of tetrapyrroles. Indeed, in 1974 it was reported that etiolated cucumber cotyledons incubated with ALA for 16 h in darkness accumulated, as expected, exogenous tetrapyrroles. It was noticed that upon exposure of the etiolated cotyledons to light in order to study the lag phase of Chl Biosynthesis (Rebeiz 1967), the etiolated tissue underwent visible damage that was attributed to the accumulated tetrapyrroles (Castelfranco et al. 1974). However, at the time it was not known whether green tissues incubated with ALA would accumulate exogenous tetrapyrroles, since it was not known how active the Chl biosynthetic pathway was in green tissues. This information was needed in order to develop a tetrapyrrole-dependent photosensitizing herbicidal technology for green plants.

## 17.2 Chlorophyll Biosynthesis Is Indeed Very Active in Green Tissues

The key to the success of the paper chemistry photodynamic herbicide tetrapyrrole hypothesis mentioned above resided therefore in determining whether green plants could be induced to accumulate enough tetrapyrroles to trigger a damaging photosensitization reaction in the green plant tissues.

C.A. Rebeiz, *Chlorophyll Biosynthesis and Technological Applications*,
DOI 10.1007/978-94-007-7134-5_17, © Springer Science+Business Media Dordrecht 2014

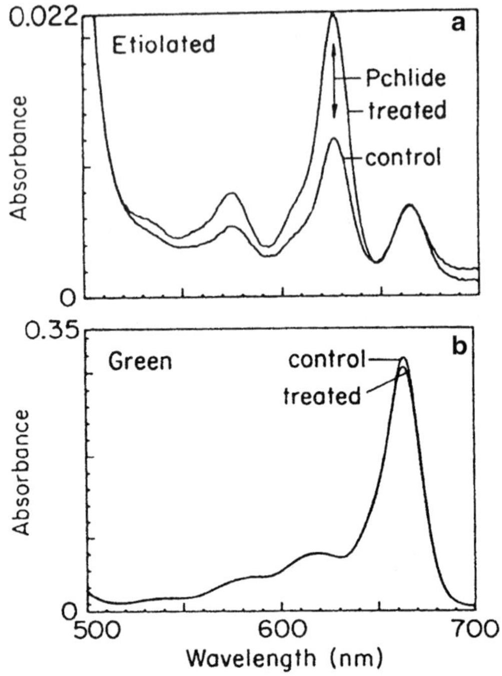


**Fig. 17.1** Room temperature absorption spectrum of (**a**) acetone extract of etiolated control and etiolated ALA-treated seedlings and (**b**) green control and green ALA-treated seedlings. Cucumber seeds were grown for 5 days either in darkness (etiolated seedlings) or under a 14 h light/10 h dark photoperiod (green seedlings) at 28 °C. Etiolated and green control seedlings were sprayed in darkness under a green safelight with water: 0.1 % Tween 80 (99:1 v/v), adjusted to PH 3.5. Treated etiolated and green seedlings were sprayed in darkness with 20 mM ALA dissolved in water: 0.1 % Tween 80 (99:1 v/v) and similarly adjusted to PH 3.5. After spraying, the seedlings were wrapped in aluminum foil to induce maximum penetration of the spray and were incubated in darkness for a period of 17 h to allow for the biosynthesis and accumulation of tetrapyrroles. At the end of dark incubation 3 g of tissue was homogenized under subdued laboratory light (about 10-foot candles) in 15 mL of acetone: 0.1 N $NH_4OH$ pH (90:10 v/v). The homogenate was centrifuged at 18,000 rpm for 10 min at 1 °C to separate the acetone extract from cell debris. The acetonic supernatant containing the pigments was decanted and used for spectrophotometric analysis (Reproduced from Rebeiz 1991)

In 1982 the prevailing conventional wisdom stated that this was impossible. It was believed that green plants that had accumulated all the Chl required for photosynthesis did not need to synthesize more Chl and had lost their Chl-making capabilities (Perkins and Roberts 1960; Virgin 1961; Wickliff and Aronoff 1963). This dogma was reinforced by observations indicating that when etiolated plants were treated with ALA and analyzed by absorption spectrophotometry, substantial protochlorophyllide (Pchlide) accumulation was observed in the treated plants, over and beyond control plants that were treated with solvent only (Fig. 17.1a). δ-Aminolevulinic acid (ALA) is the precursor of all tetrapyrroles in living cells, and etiolated plants are notorious for their vigorous Chl biosynthetic capabilities in the light and their propensity for forming Pchlide when incubated with ALA in

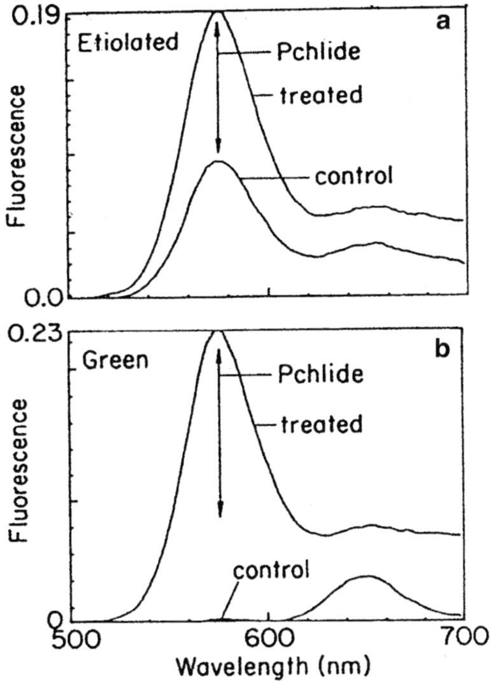


**Fig. 17.2** Room-temperature fluorescence emission spectrum of (**a**) the hexane-extracted acetone extract of etiolated control and etiolated ALA-treated seedlings and (**b**) green control and ALA-treated seedlings. The acetone extract containing various pigments was extracted with one volume of hexane, then with a third volume of hexane to remove the Chl. The hexane-extracted acetone residue containing monocarboxylic tetrapyrroles such as Pchlide and dicarboxylic tetrapyrroles such as protoporphyrin IX (if present) was subjected to high resolution spectrofluorometric analysis according to methods described by Rebeiz (2002)

darkness (Sisler and Klein 1963). Protochlorophyllide, a Mg-porphyrin, is the immediate precursor of chlorophyllide which upon esterification is converted to Chl. However when green plants were treated in a similar manner, and their acetone extract was analyzed by absorbance spectroscopy, tetrapyrrole accumulation was not detected (Fig. 17.1b). These results reinforced the notion that (a) in green plants the Chl biosynthetic pathway was not highly functional, and (b) that the tetrapyrrole-dependent photodynamic herbicide (TDPH) hypothesis was not likely to succeed. At that stage it was conjectured that the lack of observed tetrapyrrole accumulation in green tissues treated with ALA might be due to inadequate analytical techniques than lack of appropriate metabolic activity. To test this hypothesis, acetone extracts of control and ALA-treated etiolated and green cucumber cotyledons that were previously analyzed by absorbance spectroscopy, were processed and analyzed by fluorescence spectroscopy as described in (Rebeiz 2002). Surprisingly it was observed that green tissues that had been treated with ALA had actually accumulated more Pchlide than the etiolated tissues (Fig. 17.2a, b). This tetrapyrrole accumulation was not detectable however with classical sample preparation and absorbance spectroscopy. When a sample preparation was used that eliminated the Chl from the green extracts prior to analysis by room temperature fluorescence spectroscopy (Rebeiz 2002), a true picture of the tetrapyrrole profile emerged that indicated that green tissues were actually more active than etiolated tissues at converting exogenous ALA to tetrapyrroles Fig. 17.2. This in turn suggested that the TDPH hypothesis may after all be reduced to practice.

## 17.3 Photodynamic Herbicides: Concept and Phenomenology

Tetrapyrrole-dependent photodynamic herbicides (TDPH) are compounds that force green plants to accumulate undesirable amounts of metabolic intermediates of the chlorophyll (Chl) and heme metabolic pathways, namely tetrapyrroles (Duke and Rebeiz 1994; Rebeiz et al. 1984b, 1987, 1988b, 1991, 1994; Rebeiz 1991; Reddy and Rebeiz 1994). In the light the accumulated tetrapyrroles photosensitize the formation of singlet oxygen that kills the treated plants by oxidation of their cellular membranes. Tetrapyrrole-dependent photodynamic herbicides usually consist of a 5-carbon amino acid, ALA, the precursor of all tetrapyrroles in plant and animal cells, and one of several chemicals referred to as modulators. Delta-aminolevulinic acid and the modulators act in concert. The amino acid serves as a building block of tetrapyrrole accumulation, while the modulator alters quantitatively and qualitatively the pattern of tetrapyrrole accumulation (Amindari et al. 1995). The tetrapyrrole-dependent connotation is meant to differentiate between this class of photodynamic herbicides and other light activated herbicides such as paraquat, that are not dependent on tetrapyrrole metabolism for herbicidal activity. During the past several years, the scope of TDPH research has expanded considerably, as some established herbicides and a plethora of new compounds that act via the TDPH phenomenon have been discovered. However commercialization of ALA-dependent photodynamic herbicides has not yet been achieved. The rest of this chapter will therefore attempt to lay down the foundations for such an undertaking by future researchers in academia and industry.

## 17.4 Photodynamic Effects of Metabolic Tetrapyrroles on Isolated Chloroplasts

While delta-aminolevulinic acid (ALA)-dependent photodynamic destruction of insect and animal tissues is mainly photosensitized by protoporphyrin IX (Proto), additional Mg-containing tetrapyrroles are involved in the photodynamic destruction of plant tissues. To gain better understanding of the destructive photodynamic effects of these plant tetrapyrroles, the effects of divinyl (DV) Proto, DV Mg-Proto and its monomethyl ester and DV and monovinyl (MV) protochlorophyllides (Pchlides) on isolated chloroplasts was compared. Incubation of isolated cucumber chloroplasts with tetrapyrroles, in the light, exhibited various effects on the pigments and pigment-protein complexes of the plastids. These effects are described below. The state of pigment- protein complexes was monitored by analysis of pigment content and by spectrofluorometry of isolated chloroplasts at 77 K (Amindari et al. 1995).

### *17.4.1 Fluorescence Properties of Chl and Freshly Isolated Chloroplast at 77 K*

Fluorescence spectroscopy at 77 K has been used extensively to probe the effects of various metabolic tetrapyrroles on the state of organization of the chloroplast membranes. It was deemed important therefore, to discuss the 77 K fluorescence properties of Chl and freshly isolated chloroplasts before proceeding with a discussion of experimental results.

#### 17.4.1.1 Fluorescence Properties of Chlorophyll at 77 K in Organic Solvents

Most of the light energy absorbed by Chls dissolved in organic solvents, is dissipated as fluorescence. At the temperature of liquid $N_2$ (77 K) MV Chl *a* dissolved in diethyl ether coordinates to two solvent molecules (i.e. the central Mg atom becomes hexacoordinated by axial coordination to two Lewis bases) (Belanger and Rebeiz 1984; Rebeiz and Belanger 1984). It exhibits a major red emission maximum at 674 nm [Qy (0′-0) transition], a minor maximum at 725 nm [Qy (0′-1) transition], and Soret excitation maxima at 447 nm [By, Bx (0-0′) transition] (Belanger and Rebeiz 1984; Rebeiz and Belanger 1984). It also exhibits, a 422 nm (eta $\eta_1$ transition) and a 400 nm (eta $\eta_2$ transition) (Weiss 1975, 1978). Under the same conditions, MV Chl *b* is also hexacoordinated and exhibits a major red emission maximum at 659 nm [Qy (0′-0) transition], a minor maximum at 722 nm [Qy (0′-1) transition], and Soret excitation maxima at 475 nm [By (0-0′) transition], 449 nm ($\eta_1$ transition) and 427 nm ($\eta_2$ transition) (Duggan and Rebeiz 1982). The eta ($\eta$) transitions are forbidden in unsubstituted porphyrins (Weiss 1975, 1978), but become allowed in reduced porphyrins or when there is a conjugated carbonyl substituent as in the Chls (Weiss 1978).

#### 17.4.1.2 Fluorescence Properties of Chloroplasts at 77 K

In the chloroplast, MV Chl *a* and *b* are non-covalently associated with various thylakoid polypeptides. This special pigment-protein environment changes drastically the population and energy levels of various electronic transitions and results in different spectroscopic properties than in ether. As a consequence the spectroscopic properties of a given Chl-polypeptide complex, depends on the specific Chl-protein interactions within the complex. This picture is complicated further by the fact that not all Chl-protein complexes are capable of fluorescence. Depending on the structural proximity of various complexes, some Chl-polypeptides transfer their excitation energy to other fluorescing complexes, instead of emitting their excitation energy as fluorescence. These non-fluorescing Chl-polypeptides may become fluorescent only when their structural relationship to other Chl-polypeptides is disrupted. For example Chl *b* does not fluoresce in healthy thylakoid membranes because it transfers its excitation energy to Chl *a*. It becomes fluorescent when its structural organization is disrupted.

A fraction of the light energy absorbed by chloroplast membranes is converted to chemical energy via the process of photosynthesis. Another fraction of that energy is dissipated via several mechanisms including fluorescence. As mentioned in Chap. 16, At 77 K, freshly isolated chloroplasts exhibit a deceptively simple three banded fluorescence emission spectrum with emission maxima at 683–686 nm (F686), 693–696 nm (F696) and 735–740 nm (F740) (Bassi et al. 1990; Butler and Kilajima 1975). It is believed that the fluorescence emitted at F686 nm arises from the Chl *a* of the light-harvesting Chl-protein complexes (LHCII and LHCI-680), that emitted at F696 nm originates mainly from the Photosystem (PS) II antenna Chl *a* (CP47 and/or CP29). That emitted at F740 nm originates primarily from the PS I antenna Chl *a* (LHCI-730) (Bassi et al. 1990; Butler and Kilajima 1975). Under the same experimental conditions, each fluorescence excitation spectrum recorded at emission wavelengths of 685 (LHCII and LHCI-680), 695 (CP47 and/or CP29) or 740 nm (LHCI-730) exhibits four excitation bands with maxima at 415–417, 440 nm, 475 nm and 485 nm. The excitation band with a maximum at 415–417 nm is probably caused by the $\eta 1$ transition of Chl *a*, while the 440 nm band corresponds to the bulk of light absorption by Chl *a* in the Soret region. The excitation bands with maxima at 475 and 485 nm are excitation energy transfer bands and correspond to light absorbed by Chl *b* and carotenoids in the Soret region. In healthy chloroplasts the photons absorbed at these wavelength by Chl *b* and by carotenoids, are transferred to Chl *a* where they are converted to chemical energy or wasted as Chl *a* fluorescence.

As mentioned above, this simple picture of the fluorescence properties of thylakoid membranes is rather deceptive, since thylakoid membranes contain several Chl *a* and *b*-binding polypeptides which may not fluorescence until their structural organization is disrupted. In this context, the ratio of emission at 739–740 nm relative to that at 685 nm (F740/F686), as well as F740/F696 have been used to determine changes in the relative distribution of excitation energy between PSI and PSII which is mediated mainly by LHCII (Hipkins 1986). The magnitude and blue shift of these fluorescence ratios have also been used to study the onset of chloroplast degradation that disrupts the normal distribution of excitation energy between the photosystems and results in a steady decrease in the F740/F696 and F740/F686 fluorescence emission ratios (Rebeiz and Bazzaz 1978). Furthermore disorganization of the chloroplast structure results in a blue shift of the emission and excitation maxima to shorter wavelength and eventual disappearance of the emission peaks between 680 and 740 nm, and the excitation bands between 470 and 490 nm. With this introduction to Chl and Chloroplast fluorescence the effects of exogenous tetrapyrroles on isolated chloroplasts will now be discussed.

### 17.4.2 Effects of Exogenous Tetrapyrroles on Isolated Chloroplasts

Only one of the five exogenous tetrapyrroles failed to trigger chloroplast destruction in the light, namely divinyl (DV) Mg-Protoporphyrin IX (Mg-Proto). Esterification of DV Mg-Proto to yield DV Mg-Proto monomethyl ester (Mpe) rendered this tetrapyrrole extremely destructive. While overall destructive effects were manifested by Chl *a* and *b* disappearance and the appearance of Chl degradation products, such as chlorophyllide *a*, and *b* and pheophytin and pheophorbide *a*, more specific effects on the pigment-protein complexes became evident from in organello 77 K fluorescence spectroscopy. DV Proto, an early intermediate in Chl *a* biosynthesis, affected the photosystem (PS) II antenna Chl *a* pigment-protein complexes, but had no effect on the PS I antenna complex and the Chl *a*/*b* light harvesting antenna complex (LHCII). On the other hand DV Mpe and DV Pchlide *a*, destroyed completely all the thylakoid pigment-protein complexes. As for DV-Pchlide *a*, it exhibited its strongest effect on the disorganization of the PS I antenna LHCI-730 complex. Altogether these results indicate that individual tetrapyrroles have distinct and different disruptive effects on the structure of thylakoid membranes in the light. Specific effects appear to be related to the position of particular tetrapyrrole in the Chl *a* biosynthetic chain and its electrostatic properties (Amindari et al. 1995)

#### 17.4.2.1 Effect of Exogenous DV Proto on Photodynamic Damage in Isolated Cucumber Chloroplasts

As described in Chaps. 5 and 7, DV Proto is the precursor of DV Mg-Proto. In its native state DV Proto is loosely bound to the plastid membranes (Smith and Rebeiz 1979). It is formed by oxidation of DV protoporphyrinogen IX (Protogen) by protoporphyrinogen oxidase (Protox). DV Protogen is the hexahydro reduction product of DV Proto. It is a highly mobile metabolite. It moves readily from one cellular compartment to another where it is rapidly converted to DV Proto by Protox. Indeed, diphenyl ethers belong to a family of potent herbicides that act via Protox inhibition in the chloroplast. Protogen that can no longer be converted to Proto in the chloroplast, diffuses out of the chloroplast to various subcellular compartments. There, it causes considerable photodynamic damage after conversion to DV Proto by Protoxes, that are resistant to inhibition by diphenyl ethers. A large number of photodynamic herbicide modulators also result in the accumulation of DV Proto in the chloroplast, when plants are treated with modulators and ALA (Rebeiz et al. 1988b, 1990, 1991, 1994).

After 2 h of incubation of isolated chloroplasts with Proto in the light, about 90 % of the added Proto disappeared (Rebeiz et al. 1984a, b). Proto photosensitization exerted negligible effects, however, on other pigments. Except for a very modest increase in Chlide *a* and *b* content, Proto had essentially no effects on other pigment pools a (Amindari et al. 1995).

Further insight into the effects of DV Proto photosensitization on isolated chloroplasts was derived from 77 K spectroscopy. Proto photosensitization appeared to affect mainly the structure of CP47 and/or CP29. CP47 is an internal Chl *a* antenna that is in direct contact with PSII. In that polypeptide, 14 conserved histidine residues bind 9-10 Chl *a* molecules by coordination of the central Mg atom to the imidazole N. CP29 is a minor PSII antenna that consists of a 31 KDa polypeptide. It is located close to the PSII reaction center and is confined to grana membranes (Bassi et al. 1990). Each 31 KDa polypeptide binds about 4–12 Chls with an *a*/*b* ratio of 2.8–3.0. The disruption of these pigment-protein complexes by exogenous Proto in the light is based on the following observations (a) reduction in the magnitude of the emission at 695 nm which originates in CP47 and/or CP29, and (b) near disappearance of the 440 nm excitation maximum, blue-shift and considerable reduction in the magnitude of the Chl *b*-carotenoids energy transfer bands at 470–490 nm, in the excitation spectrum of CP47 and/or CP29 (recorded at F695 nm). This in turn was further indication of structural disruption of the CP complexes, (c) appearance of a 674 nm emission peak which is identical to the 674 nm [$Q_y$ (0′-0)] transition of hexacoordinated Chl *a* at 77°K (Belanger and Rebeiz 1984), and (d) appearance of a pronounced excitation maximum at 411 nm. This Soret excitation maximum is that of DV Proto. It is detectable in the excitation spectrum recorded at F696 nm. This is because Proto under these conditions has been able to transfer its excitation energy to CP47 and/or CP29. This in turn indicated that the Proto substrate had positioned itself close enough to CP47 and/or CP29 to cause efficient excitation energy transfer to these complexes. The confinement of the effects of DV Proto to CP47 and/or CP29 was further indicated by (a) the lack of effects on the 685 nm emission maximum (of LHCII and LHCI-680 complexes), the 739 nm emission maximum (of LHCI-730 complex), and (b) lack of effects on the 440 nm Chl *a* Soret excitation maximum and the Chl *b*-carotenoids energy transfer bands of the LHCII, LHCI-680 and LHCI-730 complexes, in excitation spectra recorded at (F685 and F739 nm).

#### 17.4.2.2 Effect of Exogenous DV Mg-Proto on Photodynamic Damage in Isolated Cucumber Chloroplasts

As discussed in Chap. 7, DV Mg-Proto is the precursor of Mpe. In its native state it is bound to the plastid membranes (Smith and Rebeiz 1979). After 2 h of incubation with isolated chloroplasts in the light, 87 % of the added Mg-Proto disappeared probably as a consequence of photodestruction (Rebeiz et al. 1984a, b). In the light DV Mg-Proto had no significant effects on the pigment pools of incubated chloroplasts, except for a modest increase in the Chlide *a* and *b* content.

After 2 h of incubation in the absence of added DV Mg-Proto, the 77 K fluorescence emission and excitation profiles of the incubated chloroplasts, were indistinguishable from those of freshly isolated chloroplasts (Amindari et al. 1995). Likewise, after incubation with 7,289 nmol of Mg-Proto per 100 mg plastid protein for 2 h in the light, the 77 K fluorescence emission profile, and the excitation profile recorded at F739 nm (for LHCI-730 complex) were also indistinguishable from those of freshly isolated chloroplasts. However, slight differences in the low temperature fluorescence excitation profile recorded at F685 (for LHCII and the LHCI-680 complexes) and at F696 nm (for the CP47 and/or CP29 complexes), became apparent. It consisted in the appearance of a 418 nm excitation peak, a 430 nm excitation shoulder, and a 2–3 nm blue shift of the remaining excitation maxima to 437, 473 and 483 nm respectively. The 2–3 nm blue shifts of these peaks indicated a slight disorganization of the LHCII and the LHCI-680 light-harvesting, and the CP47 and/or CP29 Chl-protein complexes. The 418 nm excitation maximum corresponds to the Soret excitation of pentacoordinated Mg- Proto in a semi-aqueous environment, such as aqueous acetone, at room temperature (Belanger and Rebeiz 1984; Hinchigeri et al. 1984; Rebeiz and Belanger 1984). Nevertheless in this environment DV Mg-Proto was able to transfer its excitation energy to the LHCII, LHCI-680 and the CP47 and/or CP29 Chl-protein complexes, as evidenced by the presence of the 418 nm peak in the excitation spectra recorded at an emission maximum of 685 nm and 695 nm. Because of the slight disorganization of the LHCII and the LHCI-680 light-harvesting, and the CP47 and/or CP29 Chl-protein complexes it was conjectured that the 430 nm excitation shoulder may be a degradation product of these complexes (Amindari et al. 1995).

#### 17.4.2.3 Effect of Exogenous DV Mpe on Photodynamic Damage in Isolated Cucumber Chloroplasts

As discussed in Chap. 7, DV Mpe is the precursor of DV Pchlide *a*. It differs from DV Mg-Proto by methyl esterification of the propionic acid residue at position six of the macrocycle. In its native state Mpe is bound to the plastid membranes (Smith and Rebeiz 1979). Membrane-bound Mpe exhibits an emission maximum at 598–600 nm and a Soret excitation maximum at 424–425 nm, at room temperature (Rebeiz et al. 1975).

After 2 h of incubation with isolated chloroplasts in the light, 79 % of the added Mg-Proto disappeared probably as a consequence of photodestruction (Rebeiz et al. 1984a, b). It resulted in the destruction of MV Chl *a* and *b* and the formation of MV Chlide *a*, probably by hydrolysis of MV Chl *a* (Amindari et al. 1995)

In three replicates, the 77 K fluorescence emission and excitation profiles after 2 h of incubation in the absence of added DV Mpe, were indistinguishable from 0 h controls. After incubation with 6,698 nmol of DV Mpe per 100 mg plastid protein for 2 h in the light, the 77 K fluorescence emission and excitation profiles underwent dramatic changes (Amindari et al. 1995). Essentially the organized structure of the chloroplast was completely destroyed. This was evidenced by disappearance of the normal three-banded emission chloroplast profile and the appearance of only one fluorescence emission maximum. This in turn indicated the complete disorganization of the LHCII, LHCI-680, CP47 and/or CP29, and LHCI-730 complexes. Excitation at 472 nm *i.e.* close to the Soret excitation maximum of MV Chl *b*, elicited a Chl *b* emission maximum at 660 nm. Appearance of Chl *b* fluorescence in vivo or in organello is usually an indication of a certain degree of disruption of the structural relationship of Chl *a* and *b* in the thylakoid membranes. Indeed, Chl *b* in healthy thylakoids, does not fluoresce but transfers its excitation energy to Chl *a*.

Further evidence of disorganization of the aforementioned complexes, was evidenced by the state of the 77 K fluorescence excitation spectra. In three excitation spectra, recorded at F685, 696 and 739 nm, the normal three-banded fluorescence excitation profile with maxima at 440, 475 and 485 nm was replaced by one Soret excitation maximum at 443 nm (Amindari et al. 1995). This Soret excitation maximum corresponds to MV Chl *a* coordinated to two small ligands such as pyridine at room temperature. The Soret excitation maximum at 424 nm is that of membrane bound DV Mpe which transfers its excitation energy to the remnants of the LHCII, LHCI-680, Chl *a* CP47 and/or CP29, and LHCI-730 complexes, as evidenced by the sloping tail between 685 and 740 nm (Amindari et al. 1995).

#### 17.4.2.4 Effect of Exogenous DV Pchlide *a* on Photodynamic Damage in Isolated Cucumber Chloroplasts

As discussed in Chap. 8, Divinyl Pchlide *a* is the precursor of DV Chlide *a*. It differs from DV Mpe by the presence of a fifth ring, the cyclopentanone ring, at position six and δ of the macrocycle. In its native state it is bound to the plastid membranes (Smith and Rebeiz 1979). Membrane-bound DV Pchlide *a* exhibits emission maxima between 629 and 658 nm, depending on its state of aggregation and the stage of greening of the tissue (Cohen and Rebeiz 1978).

After 2 h of incubation with isolated chloroplasts in the light, 20 % of the added DV Pchlide *a* disappeared probably as a consequence of photodestruction (Rebeiz et al. 1984a). Added DV Pchlide caused considerable destruction of MV Chl *a* and *b*, which was accompanied by the formation of significant amounts of chlorophyllide (Chlide) *a* and *b* probably by hydrolysis of the long chain fatty acid at position seven of the macrocycle of the corresponding Chls. In three replicates, the 77 K fluorescence emission and excitation profiles after 2 h of incubation in the absence of added DV Pchlide *a*, were indistinguishable from 0 h controls. However after incubation with 5,493 nmol of DV Pchlide *a* per 100 mg plastid protein for 2 h in the light, the 77 K fluorescence emission and excitation profiles underwent profound changes. Essentially the organized structure of the chloroplast was profoundly disrupted. The 740 nm fluorescence emission maximum decreased considerably in magnitude thus indicating disruption of the LHCI-730 protein-pigment complex (Amindari et al. 1995). The fluorescence emission maximum at 696 nm disappeared completely, thus indicating the complete disorganization of the CP47 and/or CP29 complex. The 686 nm fluorescence emission peak underwent a 3 nm blue shift which also indicated a certain degree of disorganization of the LHCII and LHCI-680 complexes (Amindari et al. 1995).

Evidence of strong disorganization of the aforementioned complexes was provided by changes in the 77 K fluorescence excitation spectra. In excitation spectra recorded at F685, 696 and 739 nm, the normal three-banded fluorescence excitation profile with maxima at 416, 440, 475 and 485 nm disappeared and was replaced by one Soret excitation maximum at 447 nm and an excitation shoulder at 421 nm (Amindari et al. 1995). The Soret excitation maximum at 447 nm belongs to MV Chl *a* hexacoordinated to a small ligand such as diethyl ether at 77 K (Belanger and Rebeiz 1984). The Soret excitation maximum at 421 nm probably corresponds to the $\eta_1$ transition of Chl *a*.

#### 17.4.2.5 Effect of Exogenous MV Pchlide *a* on Photodynamic Damage in Isolated Cucumber Chloroplasts

As discussed in Chap. 9, MV Pchlide *a* is the precursor of MV Chlide *a*. It differs from DV Pchlide *a* by the presence of an ethyl instead of a vinyl group at position two of the macrocycle. In its native state it is bound to the plastid membranes (Smith and Rebeiz 1979). Membrane-bound MV Pchlide *a* exhibits emission maxima between 629 and 658 nm, depending on its state of aggregation and the stage of greening of the tissue (Cohen and Rebeiz 1978).

After 2 h of incubation with isolated chloroplasts in the light, 12 % of the added MV Pchlide *a* disappeared probably as a consequence of photodestruction (Rebeiz et al. 1984a, b). During incubation in the light, MV Pchlide *a* exerted negligible effects on the pigment pools of the chloroplast. Nevertheless it did cause disruption of various pigment-protein complexes as evidenced by 77 K fluorescence spectroscopy.

In one of three replicates the 77 K fluorescence emission and excitation profiles after 2 h of incubation in the absence of added MV Pchlide *a*, were indistinguishable from those of the 0 h control. In the other two replicates, the amplitudes of the 694 emission became larger than that of the 685 nm emission latter was split into two emissions, and became red shifted by about 8 nm (Amindari et al. 1995).

After incubation with 5,997 nmol of MV Pchlide *a* per 100 mg plastid protein for 2 h in the light, the 77 K fluorescence emission and excitation profiles underwent profound changes. Essentially the organized structure of the chloroplast was strongly disrupted. The 740 nm fluorescence emission disappeared and was replaced by a long wavelength emission at 747 nm, thus indicating disruption of the LHCI-730 protein-pigment complex (Amindari et al. 1995). The fluorescence emission maxima at 686 and 696 nm also disappeared and were replaced by a single emission maximum at 690 nm. This also indicated a certain disorganization of the LHCII, LHCI-680 and CP47 and/or CP29 complexes. Excitation at 472 nm *i.e.* close to the Soret excitation maximum of MV Chl *b*, elicited a Chl *b* emission maximum at 660 nm. Appearance of Chl *b* fluorescence is usually an indication of a certain degree of disruption of the structural relationship of Chl *a* and *b* in the thylakoid membranes, as Chl *b* in healthy thylakoids, does not fluoresce but transfers its excitation energy to Chl *a*. The 77 K fluorescence excitation recorded at F685, 696 and 739 nm, were similar. They exhibited excitation maxima at 418, 443, and 476 nm and a 485 nm excitation shoulder (Amindari et al. 1995). The presence of the 476 and 485 nm excitation bands indicated that the disruption of the Chl *b*-carotenoid association was not as complete as in the case of DV Pchlide *a*. The Soret excitation maximum at 443 nm is equivalent to that of MV Chl *a* coordinated to two small ligands such as pyridine at room temperature (Belanger and Rebeiz 1984). The 418 nm excitation peak probably corresponds to the $\eta_1$ transition of Chl *a*.

## 17.5 Molecular and Plant Tissue Bases of Tetrapyrrole-Dependent Photodynamic Herbicide Selectivity

Originally photodynamic herbicides were assumed to be non-selective in their mode of action. Further experimentation under controlled laboratory and field conditions indicated that various ALA and modulator combinations exhibited a significant degree of photodynamic herbicidal selectivity. This selectivity appeared to be rooted (a) in the different tetrapyrrole accumulating capabilities of various plant tissues, (b) in the differential susceptibility of various greening group of plants to the accumulation of various DV and MV tetrapyrroles, and (c) in the differential response of various greening groups of plants to photodynamic herbicide modulators.

### *17.5.1 Dependence of the Differential Photodynamic Herbicidal Susceptibility Upon the Extent of Tetrapyrrole Accumulation by Plant Tissues*

It soon became apparent that different plant tissues were not equally capable of tetrapyrrole accumulation. Since the tetrapyrrole-dependent photodynamic herbicidal (TDPH) phenomenon depended on photosensitization by accumulated tetrapyrroles it was expected that only tissues that accumulate tetrapyrroles will be susceptible to TDPH. This situation was encountered in green soybean seedlings where the stems, leaves and cotyledons, exhibited different susceptibilities toward ALA plus Dpy treatments. The leaves, which accumulated high amounts of tetrapyrroles, were quite susceptible to photodynamic damage while the stems (Rebeiz et al. 1984a, b) and cotyledons (Rebeiz et al. 1988a, b), which were very poor tetrapyrrole accumulators, exhibited resistance to treatment.

### *17.5.2 Dependence of the Differential Photodynamic Herbicidal Susceptibility of Plant Species Upon Greening Group Affiliation of Plant*

On the basis of the photodynamic herbicidal response of cucumber a DDV/LDV plant species and soybean a DMV/LDV species (see Chap. 12), toward various ALA and modulator combinations (Abd-El-Mageed et al. 1997; Rebeiz et al. 1987, 1988b) the following working hypothesis was proposed: (a) that plants poised in the DV tetrapyrrole biosynthetic state are more photodynamically susceptible to the accumulation of MV tetrapyrroles than to the accumulation of DV tetrapyrroles; and (b) that plants poised in the MV tetrapyrrole biosynthetic state are more susceptible to the accumulation of DV tetrapyrroles than to MV tetrapyrroles. This phenomenon rested on the hypothesis that plant species poised in the MV-greening pattern could not cope as well with a massive influx of DV tetrapyrroles. Likewise, plants poised in the DV greening pattern could not cope as effectively with a massive influx of MV tetrapyrroles. Being unable to rapidly metabolize the wrong tetrapyrroles, In the light the latter would linger around and photosensitize the destruction of the host plant before being eventually degraded by light. With the discovery of the Dark-Light (DL) subgroups in 1997 (Abd-El-Mageed et al. 1997), this hypothesis which was proposed in 1990 (Mayasich et al. 1990) did not withstand the rigors of further testing.

## 17.6 Modulation of TDPH Activity

The dependence of TDPH susceptibility upon the nature and amount of accumulated tetrapyrroles suggested that it may be possible to chemically modify the activity of TDPH by appropriate selections of modulators. It was conjectured that this may be achieved with the use of chemicals that may modulate the Chl *a* biosynthetic pathway by forcing ALA-treated plants to accumulate substantial amounts of various types of tetrapyrroles. An initial search led to the identification of 14 chemicals that acted in concert with ALA and that exhibited a definite modulating propensity toward the Chl *a* biosynthetic pathway. These chemicals were therefore designated as TDPH modulators. They were classified into four groups depending on their effects on the Chl *a* biosynthetic pathway.

### *17.6.1 The Four Classes of Modulators*

In order to determine whether a compound acted as an ALA-based photodynamic herbicide modulator (TDPH), the chemical was usually sprayed on a plant with and without ALA, and the treated plant was kept in darkness for several hours for tetrapyrrole accumulation to take place. After dark incubation, the plant tissues were analyzed for tetrapyrrole accumulation before exposure to light and determination of photodynamic damage. Upon exposure to light, tissues that had accumulated tetrapyrroles in darkness exhibited rapid photodynamic damage within the first hour of illumination. The classification of a modulator as an enhancer, inducer or inhibitor was then determined from the pattern of tetrapyrrole accumulation in the presence and absence of ALA and modulator (Rebeiz et al. 1987, 1988b).

Based on their mechanism of action TDPH modulators were classified into four distinct groups (Rebeiz et al. 1987, 1988b): (a) enhancers of ALA conversion to DV Pchlide, which enhanced the conversion of exogenous ALA to DV Pchlide, (b) enhancers of ALA conversion to MV Pchlide, which enhanced the conversion of exogenous ALA to MV Pchlide, (c) inducers of tetrapyrrole accumulation, which induced the plant tissues to form large amounts of endogenous ALA and enhanced the conversion of this endogenous ALA, as well as any exogenously supplied ALA, to tetrapyrroles; and (d) inhibitors of MV Pchlide accumulation, which appear to block the detoxification of DV tetrapyrroles by inhibiting their conversion to MV tetrapyrroles. Of all the aforementioned modulators, only inducers of tetrapyrrole accumulation were capable of exhibiting tetrapyrrole accumulation in the absence of added ALA, since they forced the plant tissue to form high levels of endogenous ALA, which were then converted to tetrapyrroles. The three other classes of modulators did not lead to significant levels of tetrapyrrole accumulation in the absence of added ALA. However in all cases, the use of ALA together with a modulator resulted in enhanced tetrapyrrole accumulation and photodynamic damage over and beyond the levels caused by ALA alone. A more detailed description of the four classes of modulators and of their criteria of classification can be found in Rebeiz et al. 1987 and 1988b.

### 17.6.2 Response of Various Greening Groups of Plants to TDPH Modulators

It is extremely desirable to be able to predict the mode of action of a modulator from its chemical structure. The advantages of such an undertaking become obvious when one takes into account: (a) the amount of time and effort involved in determining experimentally the mode of action of a modulator in a particular plant species, (b) that 79 highly potent modulators, belonging to the four modulator classes have so far been discovered and can be investigated, and (c) that at this rate many more additional modulators may be discovered (see below). As a consequence we have undertaken a research effort aimed at probing: (a) whether different plant species belonging to the same greening group react in a similar manner toward a particular modulator, (b) whether a particular plant species would react the same way toward a group of modulators that belong to the same chemical category, and (c) whether a particular modulator would exhibit the same effects on the chlorophyll biosynthetic pathway in plant species belonging to different greening groups.

Determination of the mode of action of a modulator were performed on cucumber cotyledons, a representative DDV-LDV tissue, on soybean primary leaves, a representative DMV-LDV tissue, and on Johnsongrass, a representative DMV-LMV species. Treatments consisted of spraying the seedlings with 2.5 and 5 mM solutions of ALA with or without various concentrations of modulators ranging from 5 to 30 mM. At the end of 17 h of dark incubation the plants were extracted with ammoniacal acetone under a green safelight and the porphyrin metabolic pool sizes between ALA and Chl were determined from room temperature and 77 K fluorescence spectral analysis (Rebeiz 2002). From a comparison of the relative pool sizes of the treated and control seedlings, the site of action of the various modulators along the various Chl monocarboxylic biosynthetic routes was then determined.

Preliminary results suggested that: (a) a modulator that acted in a certain way on the chlorophyll biosynthetic pathway of one greening group of plants did not necessarily act the same way on plant species belonging to a different greening group, (b) different plant species belonging to the same greening group tended to exhibit similar chlorophyll biosynthetic reactivities toward a given modulator, and (c) modulators that belonged to the same chemical category tended to exhibit the same chlorophyll biosynthetic mode of action toward a particular plant species. The above results suggested that it may be possible to make certain predictions about the mode of action of a modulator on a particular plant species belonging to a particular greening group once the mode of action of the chemical category to which it belongs, had been determined on that particular greening group.

### *17.6.3 Discovery of Novel TDPH Modulators*

Because of the central importance of modulators to the performance of TDPH, considerable time and efforts have been devoted during the past several years, to the discovery of novel TDPH modulators. The experimental strategy used in that successful undertaking used two dimensional and three dimensional computer modeling and resulted in the discovery of several hundred potent TDPH modulators (Rebeiz et al. 1990, 1991, 1994; Reddy and Rebeiz 1994).

#### 17.6.3.1 The Use of 2,2′-Dipyridyl as a TDPH Modulator

The first modulator that was used jointly with ALA for TDPH purposes was 2,2′-dipyridyl (Dpy) (Rebeiz et al. 1984a, b). The decision to use Dpy in concert with ALA was based on the following considerations: (a) In the early 1960s Granick (1961) had demonstrated that etiolated barley leaves incubated with

ALA and Dpy accumulated large amounts of Proto, Pchlide, and small amounts of Mpe, (b) Since the early 1970s, Rebeiz and coworkers have been routinely using ALA plus Dpy treatments to force the accumulation of massive amounts of Pchlide, and Mpe by etiolated plant tissues (Rebeiz et al. 1970), (c) With the use of analytical fluorescence techniques developed by Rebeiz et al. (Rebeiz 2002), Bazzaz and Rebeiz (1978) had demonstrated that the chlorophyll biosynthetic pathway was still very active in green mature tissues. It. was therefore conjectured that in order to achieve the massive accumulation of tetrapyrroles by green mature tissues, for possible photodynamic herbicidal purposes, such tissues should be treated with combinations of ALA and Dpy (Rebeiz et al. 1984a, b).

#### 17.6.3.2 The Search for Other Classes of Compounds Capable of Modulating the Chlorophyll Biosynthesis Pathway

The successful synergistic use of ALA plus Dpy for TDPH purposes (Rebeiz et al. 1984a, b), prompted a search for additional compounds that may affect the chlorophyll biosynthetic pathway. To this end, we undertook a search of the literature for chemicals and biochemicals that have been known to affect, in a general way, chlorophyll and Pchlide biosynthesis. That search identified a total of 14 chemicals that had been implicated in one way or another in chlorophyll, Pchlide and Mpe formation. One of those compound, 8-hydroxyquinoline was selected for its effects on bacteriochlorophyll (Bchl) biosynthesis in *Rhodapseudomonas spheroides*. Four compounds (1,10-phenanthroline, 2-pyridyl aldoxime, 2-pyridyl aldehyde and picolinic acid) had been described by Duggan and Gassman (1974) for their effects on the chlorophyll biosynthetic pathway in etiolated seedlings of red kidney bean, corn and cucumber.

Two other compounds, namely 1,7-phenanthroline and phenanthridine were chosen because of their effect on the chlorophyll biosynthetic pathway in *Chlamydomonas reinhardtii* (Bednarick and Hoober 1985). These compounds are all derived from pyridine. Detailed mode of action studies of these compounds on cucumber seedlings revealed interesting TDPH modulating properties (Rebeiz et al. 1988b). For example, it was noted that the presence of a nitrogen (N) atom in a benzene-like 6-membered ring structure or a phenanthrene-like structure was essential for TDPH activity. Indeed, benzene and phenanthrene, which are identical in chemical structure to pyridine and phenanthroline respectively but lack N atoms in their rings, are not active as TDPH modulators.

#### 17.6.3.3 Selection of Other Pyridine-Based TDPH Modulators

Encouraged by these preliminary results the search continued for additional pyridine derivatives that may have TDPH modulating potential. The following compounds (Rebeiz et al. 1991), were chosen: nicotinic acid because it is a geometrical isomer of picolinic acid; nicotinamide because of its structural

**Table 17.1** The six- and five-membered N-heterocyclic templates

| Template | No. of touted modulators | No. exhibiting activity | No. exhibiting greater than 88 % kill |
|---|---|---|---|
| Nicotinic acid (i.e. Niacin) | 2.′$ | 14 | 7 |
| Nicotinamide | 13 | 12 | 2 |
| Picolinic acid | 14 | 12′ | 3 |
| 2-Pyridine aldehyde | 3 | 1 | 0 |
| 2-Pyridine aldoxime | 2 | 2 | 2 |
| 2-Oxypyridine | 21 | 18 | 11 |
| 4-Oxopyridine | 2 | 2 | 0 |
| Pyridinium ion | 27 | 24 | 14 |
| 2,2′-Bipyridine | 7 | 7 | 2 |
| 2,2′-Dipyridyldisulfide | 4 | 3 | 2 |
| 8-Hydroxyquinoline | 15 | 11 | 4 |
| Quinoline | 58 | 58 | 5 |
| 1,10-Phenanthroline | 14 | 10 | 10 |
| Phenanthridine | 5 | 5 | 5 |
| Pyridazine | 2 | 1 | 0 |
| Quinoxaline | 4 | 4 | 0 |
| Pyrrole | :i:i | 31 | 12 |
| Total | 247 | 215 | 79 |

similarity to nicotinic acid; 2-oxypyridine and 4-oxopyridine because of their structural similarity to pyridine; 2,3-dipyridyl, 2,4-dipyridyl and 4,4′-dipyridyl because they are structural isomers of 2,2′-dipyridyl; 2,2′-dipyridylamine and 2,2′-dipyridyl disulfide because of their structural similarity to 2,2′-dipyridyl; quinoline because of its structural similarity to 5-OH-quinoline; and 4,7-phenanthroline because it is a structural isomer of 1,10-phenanthroline. Pyridinium (Rebeiz et al. 1991) was also selected because of its structural similarity to pyridine, and because a database of herbicidal compounds available from ISI (Philadelphia, PA, USA), listed several pyridinium compounds with general herbicidal properties. Finally 1,2-diazine and 1,4-benzodiazine were also chosen from the ISI database for the same reason as pyridinium. Some of the above compounds are depicted in Table 17.1 depicted above.

#### 17.6.3.4 Discovery of a Pyrrole-Based TDPH Template

Rapid testing for TDPH activity of representative members of the aforementioned chemical categories, that will be referred to hereafter as TDPH templates or templates for short, unraveled interesting TDPH properties. A closer examination of the chemical structure of the various templates revealed that they fell into two separate categories: (a) those that were structurally related to half a tetrapyrrole molecule such as the phenanthrolines (Rebeiz et al. 1991); and (b) those that were structurally related to one quadrant of a tetrapyrrole molecule, i.e. to an individual pyrrole, such as picolinic acid, nicotinic acid and substituted pyridyls (Fig. 17.4). This in turn suggested that this similarity between the chemical structures of the templates and between tetrapyrrole halves or quadrants may be essential to the TDPH activity of the templates, as it may facilitate the binding of the templates to or close to the reaction sites of specific enzymes of the chlorophyll biosynthetic pathway. It was possible to visualize how such enzymes may be fooled by the structural similarities between the templates and those parts of the tetrapyrrole substrates that bind to the chlorophyll biosynthetic enzymes. What was not clear at this stage was the way that such a template-enzyme binding may modulate the activity of chlorophyll biosynthetic enzymes.

The aforementioned template-enzyme binding hypothesis led in turn to the discovery of an important novel template. It was conjectured that substituted 5-membered N-containing rings, i.e. substituted pyrroles (Fig. 17.3) may prove to

| Parent Nucleus | Substituent(s) | Name of Template | Chemical Structure |
|---|---|---|---|
| Pyridine | 5,6–benz | 5,6–Benzpyridine (Quinoline) | 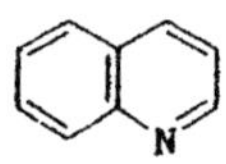 |
| Pyridine | 1–OH,5,6–benz | 8–Hydroxyquinoline | 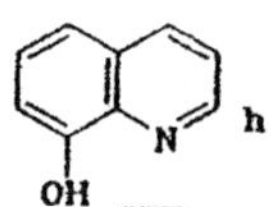 |
| Pyridine | 2,3,7',8'–Quinoline | 1,10–Phenanthroline | 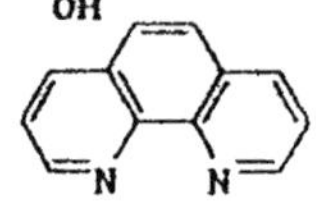 |
| Pyridine | 3,4,5,6–Dibenz | Phenanthridine | 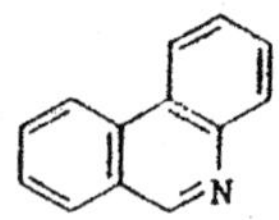 |
| Pyridazine | None | 1,2–diazine | 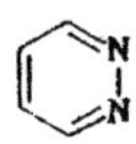 |
| 1,4–diazine | 5,6–benz | 1,4–benzodiazine | 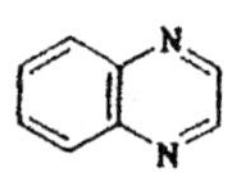 |
| Pyrrole | None | Pyrrole | 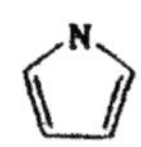 |

**Fig. 17.3** (continued)

| Parent Nucleus | Substituent(s) | Name of Template | Chemical Structure |
|---|---|---|---|
| Pyridine | 3-carboxylic | Nicotinic acid | 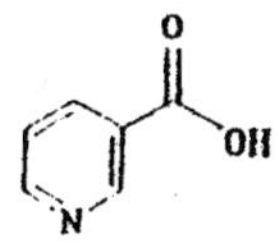 |
| Pyridine | 3-carboxamide | Nicotinamide | 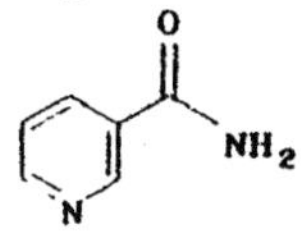 |
| Pyridine | 2-carboxylic | Picolinic acid | 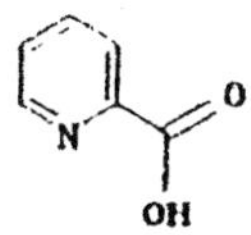 |
| Pyridine | 2-carboxaldehyde | 2-Pyridyl aldehyde | 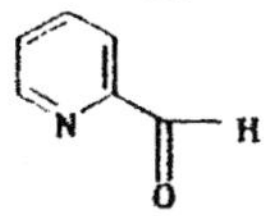 |
| Pyridine | 2-carboxaldoxime | 2-Pyridyl aldoxime | 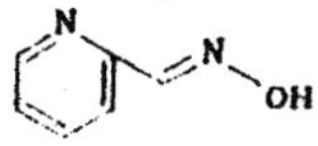 |
| Pyridine | 2-oxy | 2-Oxypyridine | 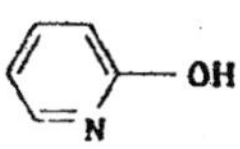 |
| Pyridine | 4-oxo | 4-oxopyridine | 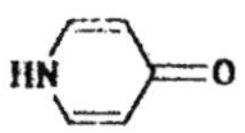 |
| Pyridine | ionized N | Pyridinium ion |  |
| Pyridine | 2,2'-Pyridyl | 2,2'-Bipyridine | 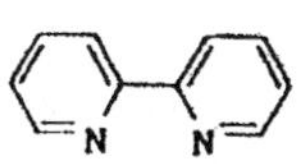 |
| Pyridine | 4,4'-Pyridyl | 4,4'-Bipyridine | 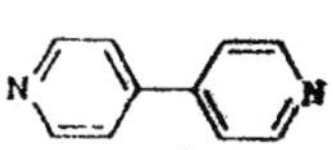 |
| Pyridine | 2,2'-Pyridyl amine | 2,2'-Dipyridylamine | 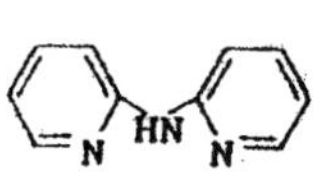 |
| Pyridine | 2,2'-Dithiopyridine | 2,2'-dipyridyl disulfide | 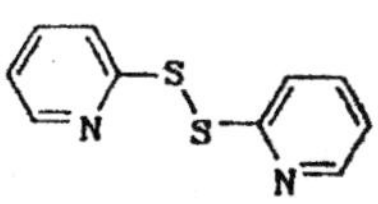 |

**Fig. 17.3** Chemical structures of the 17 photodynamic herbicide templates derived either from pyridine, diazine or pyrrole nuclei

be as good TDPH modulators as substituted 6-membered N-containing rings, and may prove to be as good TDPH modulators as substituted 6-membered N-containing rings to Fig. 17.3. Chemical structures of photodynamic herbicide templates derived from pyridine, diazine or pyrrole nuclei exhibit closure structural similarity to tetrapyrrole quadrants. A rapid evaluation of the TDPH activity of a few representative pyrroles confirmed our suspicions.

#### 17.6.3.5 Search of Chemical Analogues of the Pyridine, Diazine and Pyrrole Templates

Since the 19 TDPH templates depicted in Fig. 17.3 exhibited noticeable TDPH activity it was conjectured that chemical analogues of these templates may also exhibit TDPH activity. Because of limitations in man-power and the prohibitive cost of synthesizing de novo, hundreds of TDPH template analogues, we opted for an alternate strategy which is described below.

(a) Development of a Commercially Available Database of TDPH Analogues

A database of available TDPH template analogues was developed from commercial sources. Commercial catalogues of organic compounds were scrutinized for all N-containing *6* and 5-membered ring compounds. All such compounds were then entered into a ChemBase™ database using ChemBase software. ChemBase is a personal chemical database management system from Molecular Design Limited (San Leandro, CA, USA). The completed database consisted of 2,118 commercially available compounds and was used for analogue searches.

(b) Computer Search of the N-Heterocyclic Database

The computer-aided search for commercially available analogues of the TDPH templates consisted in extracting out of the N-heterocyclic database all analogues of the 19 templates shown in Fig. 17.3. This was achieved by carrying substructure searches of the database, using the 17 templates one at a time. In other words every substructure search consisted in extracting out of the N-heterocyclic database all analogues of that particular template. A typical template substructure search took less than 1 min. The total search yielded 247 analogues belonging to the 17 different templates. The chemicals were purchased from Aldrich (St Louis, MO, USA) and were tested for TDPH activity as described below.

(c) Primary Screening of 247 Analogues for TDPH Activity

In order to determine which of the 247 TDPH template analogues were actually active as TDPH modulators, these compounds were rapidly screened for TDPH activity.

Preliminary screening of the 247 putative modulators was performed under controlled experimental conditions using greenhouse-grown cucumber seedlings (cotyledon stage) a DDV/ LDV plant tissue and whenever possible single representative species of the DMV/LDV and DMV/LMV greening groups. The seedlings were sprayed in the late afternoon with 5 mM ALA plus 20 mM modulator at a rate of 40 gal/acre and an average droplet size of 75 μm. The sprayed plants were wrapped in aluminum foil to maximize penetration of the active ingredients and were placed in darkness at 28 °C to induce tetrapyrrole accumulation. The next day the plants were unwrapped and were exposed to light in the greenhouse. Photodynamic damage was evaluated visually and photographically over a period of 10 days. Modulators that exhibited photodynamic damage of 88–90 % or better were retained for further experimentation. Seventy-nine commercially available compounds exhibited 88–100 %.

Photodynamic kill on cucumber seedlings are reported in Table 17.2. These TDPH modulators belonged to 13 different templates. The 13 TDPH templates and the 79 corresponding analogues that exhibited TDPH activity are reported in Table 17.2. Perhaps the most dramatic spinoff of the modulator search strategy was the discovery of 9 modulators that belonged to the nicotinic acid and nicotinamide templates and which exhibited (88–100 %) photodynamic kill on cucumber when used in concert with ALA. These modulators are simple vitamin derivatives and may herald an era in herbicide design where it may be possible to design totally biodegradable and safe TDPH formulations made up of ALA and a vitamin derivative.

(d) Discovery of Additional 5-Membered N-Heterocyclic Modulators

The successful discovery of several TDPH pyrrole analogs prompted a closer look at various commercially available 5-membered N-heterocyclic compounds. By following the same approach as the one described above namely: (a) development of a commercially available database of 5-membered N-heterocyclic, compounds, (b) computer search of the database for 5-membered N-heterocyclic TDPH analogs of' the selected templates and (c) screening for TDPH: activity of the selected analogs, 75 new modulators that exhibited minimal phytotoxicity on corn and 88 % or better kill on cucumber, pigweed and johnsongrass, were discovered. They are depicted below in Table 17.3.

## 17.7 Effect of TDPH on Plant Tissues Lacking Chlorophyll

In 1988 we described the concept and phenomenology of a new insecticidal system (Rebeiz et al. 1988a). The system consisted of certain modulators of the porphyrin-heme biosynthetic pathway, which when used singly or in combination with ALA, induced the massive accumulation of Proto in treated insects.

The uncontrolled protoporphyrin biosynthesis and accumulation caused death of the treated insects in darkness (dark death) via an unknown mechanism, and in the

**Table 17.2** Primary screening on cucumber seedlings of modulators belonging to 13 different N-heterocyclic templates

| Modulator template | Modulator ID | Modulator | % Death ALA | % Death MOD | % Death A + N! |
|---|---|---|---|---|---|
| 1,10-Phenanthroline | 445 | 2,9-Dimethyl-4,7-diphenyl-l, 10-phen | 75 | 100 | 100 |
| 1,10-Phenanthroline | 1,018 | 3,4,7,8-Tetramethyl-1,10-phen | 75 | 100 | 100 |
| 1,10-Phenanthroline | 271 | 5-Chloro-l,10phen | 50 | 100 | 100 |
| 1,10-Phenanthroline | 454 | 5,6-Dimethyl-l, 10-phen | 50 | 100 | 100 |
| 1,10-Phenanthroline | 738 | 5-Methyl-l,10-phen | 50 | 100 | 100 |
| 1,10-Phenanthroline | 814 | 5-Nitro-l, 10-phen | 50 | 81 | 100 |
| 1,10-Phenanthroline | 453 | 4,7-Dimethyl-l,10-phen | 20 | 100 | 100 |
| 1,10-Phenanthroline | 476 | 4,7-Diphenyl-l,10-phen | 20 | 100 | 100 |
| 1,10-Phenanthroline | 839 | 1,10 Phenanthroline | 20 | 100 | 100 |
| 1,10-Phenanthroline | 1,102 | 4-Methyl-l,10-phen | | 100 | 100 |
| 2,2′-Bipyridine | 446 | 4,4′-Dimethyl-2\2′-dipyridyl | 19 | 78 | 100 |
| 2,2′-Bipyridine | 999 | 2,2:6′,2″-Terpyridine | 19 | 100 | 100 |
| 2,2′-Dipyridyl disulfide | 488 | 2,2′-Dithiobis (pyridine iV-oxide) | 38 | 88 | 100 |
| 2,2′-Dipyridyl disulfide | 489 | 6,6-Dithiodinicotinic acid | 50 | 0 | 94 |
| 2-Oxypyridine | 75 | 5-Amino-2-methoxypyridine | 69 | 0 | 100 |
| 2-Oxypyridine | 427 | 2,3-Dihydroxypyridine | 50 | 0 | 100 |
| 2-Oxypyridine | 598 | 2-Hydroxy-4-methylpyridine | 50 | 0 | 100 |
| 2-Oxypyridine | 648 | Isocarbostyryl | 50 | 13 | 100 |
| 2-Oxypyridine | 55 | 3-Amino-2,6-dimethoxy pyridine, HC1 | 69 | 0 | 97 |
| 2-0xypyridine | 259 | 2-Chloro-6-methoxypyridine | 69 | 0 | 97 |
| 2-Oxypyridine | 301 | 3-Cyano-4,6-dimethyl-2-hydroxypyridine | 50 | 0 | 97 |
| 2-Oxypyridine | 363 | Dibucaine hydrochloride | 50 | 38 | 97 |
| 2-Oxypyridine | 603 | 2-Hydroxy-3-nitropyridine | 50 | 0 | 94 |
| 2-Oxypyridine | 433 | 2,6-Dimethoxypyridine | 50 | 0 | 93 |
| 2-Oxypyridine | 297 | Citrazinic acid | 69 | 9,790 | 88 |
| 2-Pyridine aldoxine | 485 | Di-2-pyridyl ketone oxime | 50 | 25 | 100 |
| 2-Pyridine aldoxine | 853 | Phenyl 2-pyridyl ketoxime | 38 | 25 | 100 |
| 8-Hydroxyquinoline | 606 | 8-Hydroxy-5-nitroquinoline | 56 | 100 | 100 |
| 8-Hydroxyquinoline | 254 | 5-Chloro-8-hydroxy-7-iodoquinoline | 25 | 75 | 100 |
| 8-Hydroxyquinoline | 368 | 5,7-Dichloro-8-hydroxyquinoline | 25 | 83 | 100 |
| 8-Hydroxyquinoline | 361 | 5,7-Dichloro-8-hydroxyquinoline | 25 | 44 | 91 |
| Nicotinic acid | 164 | MBenzyl-i-V-nicotoyl nicotinamide | 44 | 0 | 94 |
| Nicotinic acid | 732 | Ar-Methylnicotinamide | 44 | 0 | 93 |
| Nicotinic acid | 519 | Ethyl 2-methylnicotinate | 50 | 0 | 100 |
| Nicotinic acid | 801 | Nifumic acid | 50 | 93 | 96 |
| Nicotinic acid | 601 | 2-Hydroxynicotinic acid | 50 | 0 | 95 |
| Nicotinic acid | 403 | Diethyl 3.4 pyridine dicarboxylate | 38 | 0 | 93 |

(continued)

**Table 17.2** (continued)

| Modulator template | Modulator ID | Modulator | % Death ALA | % Death MOD | % Death A + N! |
|---|---|---|---|---|---|
| Nicotinic acid | 527 | Ethyl nicotinate | 30 | 0 | 93 |
| Nicotinic acid | 596 | 2-Hydroxy-6-methylpyridine-3-carboxylic acid | 38 | 0 | 91 |
| Phenanthridine | 627 | 4-Hydroxy-7-trifluoromethyl-3-quinolinecarboxy | 44 | 65 | 88 |
| Phenanthridine | 470 | Dimidium bromide monohydrate | 44 | 97 | 100 |
| Phenanthridine | 496 | Ethidium bromide | 44 | 94 | 100 |
| Phenanthridine | 892 | Propidium iodine hydrate | 44 | 94 | 100 |
| Phenanthridine | 836 | Phenanthridine | 44 | 0 | 88 |
| Phenanthridine | 988 | Sanguinarine chloride | 44 | 0 | 88 |
| Picolinic acid | 610 | 3-Hydropicolinic acid | 40 | 88 | 100 |
| Picolinic acid | 867 | Picolinic acid | 40 | 0 | 100 |
| Picolinic acid | 659 | 1-Isoquinalie carboxylic acid | 50 | 0 | 95 |
| Pyridinium | 441 | 2-[4-(Dimethylamino)styryl]-1-ethylpyridinium | 56 | 75 | 100 |
| Pyridinium | 442 | 2-[4-(Dimethylamino)styryl]-1-methylpyridinium | 56 | 56 | 100 |
| Pyridinium | 172 | Berberine hydrochloride hydrate | 38 | 81 | 100 |
| Pyridinium | 162 | Bis-*N*-methyl acridinium nitrate | 38 | 100 | 100 |
| Pyridinium | 237 | 1-(Carboxymethyl)pyridinum chloride | 38 | 0 | 100 |
| Pyridinium | 855 | 5-Phenyl-2-(4-pyridyl)oxazole | 38 | 31 | 100 |
| Pyridinium | 393 | 1,1-Diethyl-2,2-cyanine iodide | 25 | 94 | 100 |
| Pyridinium | 394 | 1,1-Diethyl-2,4-cyanine iodide | 25 | 63 | 100 |
| Pyridinium | 395 | 1,1-Diethyl-4,4-cyanine iodide | 25 | 97 | 100 |
| Pyridinium | 492 | 1-Dodecylpyridinium chloride monohydrate | 25 | 100 | 100 |
| Pyridinium | 776 | Methyl viologen dichloride hydrate | 25 | 100 | 100 |
| Pyridinium | 298 | 2,4,6-Collidine *p*-toluene sulfonate | 25 | 0 | 97 |
| Pyridinium | 513 | 1-Ethyl-3-OH-pridinium bromide | 56 | 0 | 94 |
| Pyridinium | 440 | 4-(Dimethylamino)bromide perbromide | 56 | 0 | 88 |
| Pyrrole | 406P | 3-Ethyl-2-methyl-4,5,6,7-tetrahydroindol | 81 | 0 | 100 |
| Pyrrole | 396P | Ethyl 3,5-dimethyl-2-pyrrolecarboxylate | 63 | 0 | 100 |
| Pyrrole | 814P | Pyrrol [1,2-*a*]quinoxaline | 63 | 12 | 100 |
| Pyrrole | 282P | Diethyl 2,4-dimethylpyrrole-3,5-dicarboxylate | 38 | 50 | 100 |
| Pyrrole | 796P | Pyrrole-2-carboxaldehyde | 38 | 100 | 100 |
| Pyrrole | 431P | 1-Furfurylpyrrole | 63 | 0 | 94 |
| Pyrrole | 613P | Methyl 5-(benzoxycarbonyl)-2,4-dimethyl-3-pyrr | 63 | 0 | 94 |
| Pyrrole | 663P | 1-Methyl-2-pyrrolecarboxaldehyde | 50 | 0 | 94 |

(continued)

**Table 17.2** (continued)

| Modulator template | Modulator ID | Modulator | % Death ALA | % Death MOD | % Death A + N! |
|---|---|---|---|---|---|
| Pyrrole | 309P | 1-(Dimethylamino)pyrrole | 38 | 0 | 94 |
| Pyrrole | 235P | 1-(2-Cyanomethyl)pyrrole | 63 | 0 | 88 |
| Pyrrole | 664P | 1-Methyl-2-pyrrolecarboxylic acid | 63 | 0 | 88 |
| Pyrrole | 183P | *tert*-Butyl 4-acetyl-3,5-dimethyl-2-pyrrolecarb | 31 | 100 | 88 |
| Quinoline | 1,004 | 6-Nitroquinoline | 30 | 30 | 100 |
| Quinoline | 1,005 | 8-Nitroquinoline | 30 | 100 | 100 |
| Quinoline | 819 | 5-Nitroquinoline | 25 | 95 | 100 |
| Quinoline | 840 | 4,7 Phenanthroline | 35 | 40 | 95 |
| Quinoline | 838 | 1,7 Phenanthroline | 25 | 35 | 95 |

Percent death was monitored 10 days after Spraying
*ALA* ALA alone, *Mod* Modulator alone, *A + M* ALA + modulator, *ID* modulator identification number in the database

light (light death) probably via singlet oxygen formation. Protoporphyrin is a transient metabolite, which does not accumulate to any large extent in normal tissues. It is an immediate precursor of protoheme which in turn is the prosthetic group of cytochromes in mitochondria and chloroplasts. It is also the prosthetic group of catalases and peroxidases.

Since non-chlorophyllous plant tissues such as roots consist of cells containing an abundance of mitochondria, which in turn contain cytochromes and presumably an active porphyrin-heme biosynthetic pathway, the issue was raised as to whether some TDPH formulations would be effective against plant roots, in the same manner they were effective against insects. In particular, it was interesting to determine whether plant roots would be susceptible to TDPH dependent dark death since in their natural environment in the soil, roots are usually shielded from light.

Chung and Rebeiz (unpublished) investigated the effects of ALA and four modulators belonging to four different templates on excised and attached cucumber roots. The results of these investigations are summarized below.

### 17.7.1 Effects of TDPH on Excised Cucumber Roots

Excised cucumber roots were incubated overnight in darkness with ALA and modulators. At the end of dark incubation the tissue was analyzed for tetrapyrrole accumulation and was exposed to light for evaluation of photodynamic damage. The latter was evaluated visually, and polarographically by the decrease in oxygen consumption of treated roots as compared to controls. Excised roots incubated with

**Table 17.3** Primary screening of modulators belonging to various 5-membered N-heterocyclic templates

| Template | Mod ID | Modulator | Plant | DTH ALA | DTH Mod | DTH A + M |
|---|---|---|---|---|---|---|
| Thiazole | P00616P | Methyl 3-chlorocarbonyl-L-thiazolidine-4-carboxylate | Pigweed | 30 | 100 | 100 |
| Thiazole | P00616P | Methyl 3-chlorocarbonyl-L-thiazolidine-4-carboxylate | Cucumber | 35 | 20 | 60 |
| Thiazole | P00616P | Methyl 3-chlorocarbonyl-L-thiazolidine-4-carboxylate | Johnsongrass | 30 | 30 | 10 |
| Thiazole | P00616P | Methyl 3-chlorocarbonyl-L-thiazolidine-4-carboxylate | Corn | 0 | 0 | 0 |
| Thiazole | P00733P | (–)-2-Oxo-4-thiazolidine carboxylic acid | Pigweed | 30 | 50 | 100 |
| Thiazole | P00733P | (–)-2-Oxo-4-thiazolidine carboxylic acid | Cucumber | 35 | 10 | 100 |
| Thiazole | P00733P | (–)-2-Oxo-4-thiazolidine carboxylic acid | Corn | 0 | 10 | 5 |
| Thiazole | P00733P | (–)-2-Oxo-4-thiazolidine carboxylic acid | Johnsongrass | 40 | 0 | 20 |
| Thiazole | P00281P | 5-(4-Diethylaminobenzylidene-rhodamine) | Pigweed | 50 | 95 | 90 |
| Thiazole | P00281P | 5-(4-Diethylaminobenzylidene-rhodamine) | Cucumber | 30 | 0 | 25 |
| Thiazole | P00212P | 5-Chloro-2-mercaptobenzothiazole | Johnsongrass | 30 | 80 | 70 |
| Thiazole | P00212P | 5-Chloro-2-mercaptobenzothiazole | Cucumber | 60 | 95 | 100 |
| Thiazole | P00212P | 5-Chloro-2-mercaptobenzothiazole | Pigweed | 70 | 100 | 100 |
| Thiazole | P00212P | 5-Chloro-2-mercaptobenzothiazole | Corn | 0 | 15 | 5 |
| Thiazole | P00304P | 5-(4-Dimethylamino benzylidine) rhodinine | Corn | 0 | 0 | 0 |
| Thiazole | P00304P | 5-(4-Dimethylamino benzylidine) rhodinine | Cucumber | 60 | 0 | 100 |
| Thiazole | P00304P | 5-(4-Dimethylamino benzylidine) rhodinine | Pigweed | 60 | 60 | 100 |
| Thiazole | P00304P | 5-(4-Dimethylamino benzylidine) rhodinine | Johnsongrass | 40 | 60 | 100 |
| Thiazole | P00137P | 4-(4-Biphenyllyi)2-methyl thiazole | Corn | 0 | 0 | 0 |
| Thiazole | P00137P | 4-(4-Biphenyllyi)2-methyl thiazole | Cucumber | 50 | 0 | 100 |
| Thiazole | P00137P | 4-(4-Biphenyllyi)2-methyl thiazole | Pigweed | 70 | 90 | 95 |
| Thiazole | P00137P | 4-(4-Biphenyllyi)2-methyl thiazole | Johnsongrass | 25 | 0 | 100 |
| Thiazole | P00220P | 3-(4-Chlorophenyl)-2-ethyl-2,3,5,6-tetrahydroimidazo...[a] | Corn | 0 | 0 | 0 |
| Thiazole | P00220P | 3-(4-Chlorophenyl)-2-ethyl-2,3,5,6-tetrahydroimidazo...[a] | Cucumber | 60 | 0 | 100 |
| Thiazole | P00220P | 3-(4-Chlorophenyl)-2-ethyl-2,3,5,6-tetrahydroimidazo...[a] | Pigweed | 60 | 60 | 80 |
| Thiazole | P00220P | 3-(4-Chlorophenyl)-2-ethyl-2,3,5,6-tetrahydroimidazo...[a] | Johnsongrass | 40 | 0 | 100 |
| Thiazole | P00288P | 3,3-Diethylthiocarbocyanine iodide | Johnsongrass | 40 | 20 | 100 |

(continued)

**Table 17.3** (continued)

| Template | Mod ID | Modulator | Plant | DTH ALA | DTH Mod | DTH A + M |
|---|---|---|---|---|---|---|
| Thiazole | P00288P | 3,3-Diethylthiocarbocyanine iodide | Cucumber | 60 | 100 | 100 |
| Thiazole | P00288P | 3,3-Diethylthiocarbocyanine iodide | Pigweed | 60 | 80 | 100 |
| Thiazole | P00288P | 3,3-Diethylthiocarbocyanine iodide | Corn | 0 | 10 | 10 |
| Thiazole | P00042P | 2-Amino-6-fluorobenzothiazole | Corn | 0 | 0 | 0 |
| Thiazole | P00042P | 2-Amino-6-fluorobenzothiazole | Cucumber | 50 | 0 | 95 |
| Thiazole | P00042P | 2-Amino-6-fluorobenzothiazole | Pigweed | 70 | 0 | 100 |
| Thiazole | P00042P | 2-Amino-6-fluorobenzothiazole | Johnsongrass | 40 | 60 | 100 |
| Thiazole | P00034P | 2-Amino-5,6-dimethylbenzothiazole | Johnsongrass | 40 | 40 | 100 |
| Thiazole | P00034P | 2-Amino-5,6-dimethylbenzothiazole | Cucumber | 50 | 0 | 100 |
| Thiazole | P00034P | 2-Amino-5,6-dimethylbenzothiazole | Pigweed | 70 | 100 | 100 |
| Thiazole | P00034P | 2-Amino-5,6-dimethylbenzothiazole | Corn | 0 | 0 | 0 |
| Thiazole | P00049P | 2-(4-Aminophenyl)-6-methylbenzothiazole | Johnsongrass | 55 | 0 | 95 |
| Thiazole | P00049P | 2-(4-Aminophenyl)-6-methylbenzothiazole | Cucumber | 65 | 40 | 90 |
| Thiazole | P00049P | 2-(4-Aminophenyl)-6-methylbenzothiazole | Pigweed | 30 | 10 | 100 |
| Thiazole | P00049P | 2-(4-Aminophenyl)-6-methylbenzothiazole | Corn | 0 | 0 | 0 |
| Thiazole | P00178P | 2-Bromothiazole | Johnsongrass | 25 | 0 | 100 |
| Thiazole | P00178P | 2-Bromothiazole | Cucumber | 50 | 0 | 100 |
| Thiazole | P00178P | 2-Bromothiazole | Pigweed | 70 | 20 | 100 |
| Thiazole | P00178P | 2-Bromothiazole | Corn | 0 | 0 | 0 |
| Thiazole | P00064P | (+)6-Aminopenicillanic acid | Johnsongrass | 40 | 0 | 100 |
| Thiazole | P00064P | (+)6-Aminopenicillanic acid | Cucumber | 50 | 0 | 100 |
| Thiazole | P00064P | (+)6-Aminopenicillanic acid | Pigweed | 70 | 40 | 100 |
| Thiazole | P00064P | (+)6-Aminopenicillanic acid | Corn | 0 | 0 | 0 |
| Thiazole | P00061P | 2-Amino-6-nitrobenzothiazole | Corn | 0 | 0 | 0 |
| Thiazole | P00061P | 2-Amino-6-nitrobenzothiazole | Cucumber | 30 | 30 | 40 |
| Thiazole | P00061P | 2-Amino-6-nitrobenzothiazole | Pigweed | 60 | 30 | 30 |
| Thiazole | P00061P | 2-Amino-6-nitrobenzothiazole | Johnsongrass | 20 | 40 | 90 |
| Thiazole | P00021P | 2-Acetylthiazole | Corn | 0 | 0 | 0 |
| Thiazole | P00021P | 2-Acetylthiazole | Johnsongrass | 40 | 0 | 80 |
| Thiazole | P00021P | 2-Acetylthiazole | Cucumber | 50 | 0 | 60 |
| Thiazole | P00021P | 2-Acetylthiazole | Pigweed | 70 | 0 | 90 |
| Thiazole | P00100P | Basic blue 66 | Pigweed | 70 | 50 | 90 |
| Thiazole | P00100P | Basic blue 66 | Cucumber | 50 | 0 | 50 |
| Thiazole | P00100P | Basic blue 66 | Johnsongrass | 25 | 0 | 50 |
| Thiazole | P00100P | Basic blue 66 | Corn | 0 | 0 | 0 |
| Thiazole | P00312P | 3,6-Dimethylbenzothiazole | Corn | 0 | 0 | 0 |
| Thiazole | P00312P | 3,6-Dimethylbenzothiazole | Johnsongrass | 25 | 60 | 100 |

(continued)

**Table 17.3** (continued)

| Template | Mod ID | Modulator | Plant | DTH ALA | DTH Mod | DTH A + M |
|---|---|---|---|---|---|---|
| Thiazole | P00312P | 3,6-Dimethylbenzothiazole | Cucumber | 50 | 50 | 80 |
| Thiazole | P00312P | 3,6-Dimethylbenzothiazole | Pigweed | 60 | 100 | 100 |
| Thiazole | P00347P | 4,5-Dimethylthiazole | Pigweed | 60 | 70 | 100 |
| Thiazole | P00347P | 4,5-Dimethylthiazole | Cucumber | 50 | 0 | 80 |
| Thiazole | P00347P | 4,5-Dimethylthiazole | Johnsongrass | 25 | 0 | 80 |
| Thiazole | P00347P | 4,5-Dimethylthiazole | Corn | 0 | 0 | 0 |
| Thiazole | P00310P | 2-[4-(Dimethylamino)styryl]-3-ethylbenzothiazolium iodide | Corn | 0 | 10 | 20 |
| Thiazole | P00310P | 2-[4-(Dimethylamino)styryl]-3-ethylbenzothiazolium iodide | Johnsongrass | 25 | 60 | 20 |
| Thiazole | P00310P | 2-[4-(Dimethylamino)styryl]-3-ethylbenzothiazolium iodide | Cucumber | 50 | 50 | 95 |
| Thiazole | P00310P | 2-[4-(Dimethylamino)styryl]-3-ethylbenzothiazolium iodide | Pigweed | 60 | 100 | 100 |
| Thiazole | P00862P | 2-(4-thiazolyi)benzimidazole | Cucumber | 60 | 80 | 100 |
| Thiazole | P00862P | 2-(4-thiazolyi)benzimidazole | Pigweed | 70 | 0 | 80 |
| Thiazole | P00862P | 2-(4-thiazolyi)benzimidazole | Johnsongrass | 50 | 0 | 100 |
| Thiazole | P00862P | 2-(4-thiazolyi)benzimidazole | Com | 0 | 0 | 0 |
| Thiazole | P00399P | Ethyl 2-(formylamino)-4-thiazolegloxylate | Cucumber | 60 | 0 | 100 |
| Thiazole | P00399P | Ethyl 2-(formylamino)-4-thiazolegloxylate | Pigweed | 70 | 80 | 80 |
| Thiazole | P00399P | Ethyl 2-(formylamino)-4-thiazolegloxylate | Johnsongrass | 50 | 0 | 100 |
| Thiazole | P00399P | Ethyl 2-(formylamino)-4-thiazolegloxylate | Com | 0 | 0 | 0 |
| Thiazole | P00866P | Thiaflavin T | Cucumber | 100 | 100 | 100 |
| Thiazole | P00866P | Thiaflavin T | Pigweed | 70 | 100 | 100 |
| Thiazole | P00866P | Thiaflavin T | Johnsongrass | 20 | 90 | 90 |
| Thiazole | P00866P | Thiaflavin T | Corn | 0 | 50 | 90 |
| Thiazole | P00388P | Ethy12-amino-alpha-(methoxyimino)-4-thiazole acetate | Pigweed | 70 | 25 | 25 |
| Thiazole | P00388P | Ethy12-amino-alpha-(methoxyimino)-4-thiazole acetate | Johnsongrass | 20 | 20 | 50 |
| Thiazole | P00388P | Ethy12-amino-alpha-(methoxyimino)-4-thiazole acetate | Corn | 0 | 0 | 0 |
| Thiazole | P00388P | Ethy12-amino-alpha-(methoxyimino)-4-thiazole acetate | Cucumber | 100 | 0 | 100 |
| Thiazole | P00903P | 2-(Tritylamino)-alpha-(methoxylimino)-. . .t | Cucumber | 70 | 0 | 30 |
| Thiazole | P00903P | 2-(Tritylamino)-alpha-(methoxylimino)-. . .t | Pigweed | 70 | 80 | 100 |

(continued)

**Table 17.3** (continued)

| Template | Mod ID | Modulator | Plant | DTH ALA | DTH Mod | DTH A + M |
|---|---|---|---|---|---|---|
| Thiazole | P00903P | 2-(Tritylamino)-alpha-(methoxylimino)-. . .f | Johnsongrass | 70 | 0 | 80 |
| Thiazole | P00903P | 2-(Tritylamino)-alpha-(methoxylimino)-. . .f | Corn | 0 | 0 | 0 |
| Thiazole | P00763P | 1-Phenyl-3-(2-thiazolyil-2-thiourea) | Cucumber | 70 | 0 | 40 |
| Thiazole | P00763P | 1-Phenyl-3-(2-thiazolyil-2-thiourea) | Pigweed | 70 | 80 | 100 |
| Thiazole | P00763P | 1-Phenyl-3-(2-thiazolyil-2-thiourea) | Johnsongrass | 80 | 10 | 80 |
| Thiazole | P00763P | 1-Phenyl-3-(2-thiazolyil-2-thiourea) | Corn | 0 | 0 | 0 |
| Thiazole | P00784P | Pscudothiohydintoin | Cucumber | 40 | 0 | 70 |
| Thiazole | P00784P | Pscudothiohydintoin | Pigweed | 10 | 0 | 100 |
| Thiazole | P00784P | Pscudothiohydintoin | Johnsongrass | 80 | 0 | 90 |
| Thiazole | P00784P | Pscudothiohydintoin | Com | 0 | 0 | 0 |
| Tetrazole | P00135P | 3-3′-(4,4′-Biphenylene)bis (2,5-diphenyl-2H-tetrazolium) | Cucumber | 50 | 100 | 100 |
| Tetrazole | P00135P | 3-3′-(4,4′-Biphenylene)bis (2,5-diphenyl-2H-tetrazolium) | Pigweed | 50 | 100 | 100 |
| Tetrazole | P00135P | 3-3′-(4,4′-Biphenylene)bis (2,5-diphenyl-2H-tetrazolium) | Johnsongrass | 15 | 50 | 10 |
| Tetrazole | P00135P | 3-3′-(4,4′-Biphenylene)bis (2,5-diphenyl-2H-tetrazolium) | Com | 0 | 10 | 10 |
| Tetrazole | P00153P | Blue tetrazolim | Cucumber | 20 | 20 | 20 |
| Tetrazole | P00153P | Blue tetrazolim | Pigweed | 0 | 0 | 100 |
| Tetrazole | P00153P | Blue tetrazolim | Johnsongrass | 20 | 0 | 20 |
| Tetrazole | P00153P | Blue tetrazolim | Com | 0 | 10 | 10 |
| Tetrazole | P00901P | 2,3,5-Triphenyl-2H-tetrazolium chloride | Cucumber | 50 | 100 | 100 |
| Tetrazole | P00901P | 2,3,5-Triphenyl-2H-tetrazolium chloride | Pigweed | 50 | 100 | 100 |
| Tetrazole | P00901P | 2,3,5-Triphenyl-2H-tetrazolium chloride | Johnsongrass | 30 | 100 | 50 |
| Tetrazole | P00901P | 2,3,5-Triphenyl-2H-tetrazolium chloride | Com | 0 | 0 | 0 |
| Pyrrolidine | P00233P | (–)-Cotinine | Cucumber | 5 | 0 | 80 |
| Pyrrolidine | P00233P | (–)-Cotinine | Pigweed | 15 | 0 | 85 |
| Pyrrolidine | P00233P | (–)-Cotinine | Johnsongrass | 50 | 0 | 90 |
| Pyrrolidine | P00233P | (–)-Cotinine | Corn | 0 | 0 | 0 |
| Pyrrole | P00183P | Tert-Butyl 4-acetyl-3,5-dimethyl-2-pyrrolecarboxylate’ | Cucumber | 31 | 100 | 88 |
| Pyrrole | P00814P | Pyrrolo [1,2-a] quinoxaline | Cucumber | 63 | 12 | 100 |
| Pyrrole | P00796P | Pyrrole-2-carboxaldehyde | Cucumber | 38 | 100 | 100 |
| Pyrrole | P00396P | Ethyl 3,5-dimethyl-2-pyrrolecarboxylate | Cucumber | 63 | 0 | 100 |
| Pyrrole | P00606P | 3-Ethyl-2-methyl-4,5,6,7-tetrahydroindol | Cucumber | 81 | 0 | 100 |
| Pyrrole | P00664P | 1-Methyl-2-pyrrolecarboxylic acid | Cucumber | 63 | 0 | 88 |
| Pyrrole | P00663P | 1-Methyl-2-pyrrolecarboxaldehyde | Cucumber | 50 | 0 | 94 |

(continued)

**Table 17.3** (continued)

| Template | Mod ID | Modulator | Plant | DTH ALA | DTH Mod | DTH A + M |
|---|---|---|---|---|---|---|
| Pyrrole | P00431P | 1-Furfurylpyrrole | Cucumber | 63 | 0 | 94 |
| Pyrrole | P00309P | 1-(Dimethylamino) pyrrole | Cucumber | 38 | 0 | 94 |
| Pyrrole | P00235P | 1-(2-Cyanomethyl)pyrrole | Cucumber | 63 | 0 | 88 |
| Pyrazolene | P00660P | 4-Methyl-2-pyrazolin-5-one | Cucumber | 50 | 0 | 90 |
| Pyrazolene | P00660P | 4-Methyl-2-pyrazolin-5-one | Pigweed | 50 | 0 | 80 |
| Pyrazolene | P00660P | 4-Methyl-2-pyrazolin-5-one | Johnsongrass | 50 | 0 | 10 |
| Pyrazolene | P00660P | 4-Methyl-2-pyrazolin-5-one | Corn | 0 | 0 | 0 |
| Pyrazolene | P00336P | 3,4-Dimethyl-1-phenyl-3-pytazolin-5-one | Cucumber | 50 | 10 | 90 |
| Pyrazolene | P00336P | 3,4-Dimethyl-1-phenyl-3-pytazolin-5-one | Pigweed | 50 | 60 | 50 |
| Pyrazolene | P00336P | 3,4-Dimethyl-1-phenyl-3-pytazolin-5-one | Johnsongrass | 50 | 0 | 100 |
| Pyrazolene | P00336P | 3,4-Dimethyl-1-phenyl-3-pytazolin-5-one | Corn | 0 | 0 | 0 |
| Pyrazole | P00784P | Pseudothiohydrantoin | Cucumber | 0 | 0 | 70 |
| Pyrazole | P00784P | Pseudothiohydrantoin | Pigweed | 0 | 0 | 100 |
| Pyrazole | P00784P | Pseudothiohydrantoin | Johnsongrass | 0 | 0 | 90 |
| Pyrazole | P00784P | Pseudothiohydrantoin | Corn | 0 | 0 | 0 |
| Oxazole | P00361P | 3,3′-Dipropyloxacarbocyanine iodide | Cucumber | 50 | 50 | 95 |
| Oxazole | P00361P | 3,3′-Dipropyloxacarbocyanine iodide | Pigweed | 20 | 80 | 80 |
| Oxazole | P00361P | 3,3′-Dipropyloxacarbocyanine iodide | Johnsongrass | 50 | 20 | 10 |
| Oxazole | P00361P | 3,3′-Dipropyloxacarbocyanine iodide | Corn | 0 | 5 | 5 |
| Oxazole | P00331P | 3,3′-Dipropyloxacarbocyanine iodide | Cucumber | 50 | 100 | 100 |
| Oxazole | P00331P | 3,3′-Dipropyloxacarbocyanine iodide | Pigweed | 20 | 100 | 100 |
| Oxazole | P00331P | 3,3′-Dipropyloxacarbocyanine iodide | Johnsongrass | 50 | 20 | 80 |
| Oxazole | P00331P | 3,3′-Dipropyloxacarbocyanine iodide | Corn | 0 | 10 | 5 |
| Oxazole | P00358P | 2,5-Diphenyloxazole | Cucumber | 50 | 10 | 95 |
| Oxazole | P00358P | 2,5-Diphenyloxazole | Pigweed | 20 | 20 | 80 |
| Oxazole | P00358P | 2,5-Diphenyloxazole | Johnsongrass | 50 | 60 | 50 |
| Oxazole | P00358P | 2,5-Diphenyloxazole | Corn | 0 | 0 | 0 |
| Oxazole | P00567P | 2-Mercaptobenzoxazole | Cucumber | 50 | 50 | 90 |
| Oxazole | P00567P | 2-Mercaptobenzoxazole | Pigweed | 20 | 30 | 30 |
| Oxazole | P00567P | 2-Mercaptobenzoxazole | Johnsongrass | 50 | 90 | 100 |
| Oxazole | P00567P | 2-Mercaptobenzoxazole | Corn | 0 | 0 | 0 |
| Oxazole | P00641P | 3-Methyl-2-oxazolidinone | Cucumber | 50 | 0 | 20 |
| Oxazole | P00641P | 3-Methyl-2-oxazolidinone | Pigweed | 15 | 0 | 100 |
| Oxazole | P00641P | 3-Methyl-2-oxazolidinone | Johnsongrass | 60 | 0 | 30 |

(continued)

**Table 17.3** (continued)

| Template | Mod ID | Modulator | Plant | DTH ALA | DTH Mod | DTH A + M |
|---|---|---|---|---|---|---|
| Oxazole | P00641P | 3-Methyl-2-oxazolidinone | Corn | 0 | 0 | 0 |
| Oxazole | P00201P | 2-Chlorobenzoxazole | Cucumber | 90 | 0 | 100 |
| Oxazole | P00201P | 2-Chlorobenzoxazole | Pigweed | 100 | 20 | 100 |
| Oxazole | P00201P | 2-Chlorobenzoxazole | Johnsongrass | 50 | 0 | 90 |
| Oxazole | P00201P | 2-Chlorobenzoxazole | Corn | 0 | 0 | 0 |
| Oxazole | P00140P | 2-(4-Biphenylyl)-5-phenyl-oxazole | Cucumber | 90 | 0 | 100 |
| Oxazole | P00140P | 2-(4-Biphenylyl)-5-phenyl-oxazole | Pigweed | 100 | 20 | 100 |
| Oxazole | P00140P | 2-(4-Biphenylyl)-5-phenyl-oxazole | Johnsongrass | 50 | 0 | 90 |
| Oxazole | P00140P | 2-(4-Biphenylyl)-5-phenyl-oxazole | Corn | 0 | 0 | 0 |
| Oxazole | P00114P | 2-Benzoxazolinone | Cucumber | 90 | 0 | 80 |
| Oxazole | P00114P | 2-Benzoxazolinone | Pigweed | 100 | 20 | 50 |
| Oxazole | P00114P | 2-Benzoxazolinone | Johnsongrass | 50 | 0 | 90 |
| Oxazole | P00114P | 2-Benzoxazolinone | Corn | 0 | 0 | 0 |
| Oxazole | P00142P | 2,5-Bis(4-biphenyly)oxazole | Cucumber | 90 | 20 | 100 |
| Oxazole | P00142P | 2,5-Bis(4-biphenyly)oxazole | Pigweed | 10 | 0 | 10 |
| Oxazole | P00142P | 2,5-Bis(4-biphenyly)oxazole | Johnsongrass | 50 | 50 | 80 |
| Oxazole | P00142P | 2,5-Bis(4-biphenyly)oxazole | Corn | 0 | 0 | 0 |
| Oxazole | P00289P | 3,3′-Dihexyloxacarbocyanine iodide | Cucumber | 30 | 5 | 10 |
| Oxazole | P00289P | 3,3′-Dihexyloxacarbocyanine iodide | Pigweed | 50 | 0 | 80 |
| Oxazole | P00289P | 3,3′-Dihexyloxacarbocyanine iodide | Johnsongrass | 50 | 20 | 100 |
| Oxazole | P00289P | 3,3′-Dihexyloxacarbocyanine iodide | Corn | 0 | 0 | 5 |
| Oxazole | P00285P | 3,3′-Diethyloxacarbocyanine iodide | Cucumber | 30 | 100 | 95 |
| Oxazole | P00285P | 3,3′-Diethyloxacarbocyanine iodide | Pigweed | 50 | 20 | 0 |
| Oxazole | P00285P | 3,3′-Diethyloxacarbocyanine iodide | Johnsongrass | 50 | 10 | 50 |
| Oxazole | P00285P | 3,3′-Diethyloxacarbocyanine iodide | Corn | 0 | 5 | 5 |
| Oxazole | P00314P | 2,5-Dimethyl-benzoxazole | Cucumber | 30 | 0 | 50 |
| Oxazole | P00314P | 2,5-Dimethyl-benzoxazole | Pigweed | 50 | 0 | 20 |
| Oxazole | P00314P | 2,5-Dimethyl-benzoxazole | Johnsongrass | 50 | 0 | 95 |
| Oxazole | P00314P | 2,5-Dimethyl-benzoxazole | Corn | 0 | 0 | 0 |
| Imidazole | P00568P | 2-Mercaptoimidazole | Cucumber | 50 | 95 | 95 |
| Imidazole | P00568P | 2-Mercaptoimidazole | Pigweed | 40 | 50 | 80 |
| Imidazole | P00568P | 2-Mercaptoimidazole | Johnsongrass | 10 | 10 | 50 |
| Imidazole | P00568P | 2-Mercaptoimidazole | Corn | 0 | 0 | 5 |
| Imidazole | P00570P | 2-Mercapto-1-methylimidazole | Cucumber | 50 | 0 | 50 |
| Imidazole | P00570P | 2-Mercapto-1-methylimidazole | Pigweed | 40 | 0 | 90 |
| Imidazole | P00570P | 2-Mercapto-1-methylimidazole | Johnsongrass | 10 | 0 | 15 |
| Imidazole | P00570P | 2-Mercapto-1-methylimidazole | Corn | 0 | 0 | 0 |
| Imidazole | P00868P | 6-Thioxanthine | Cucumber | 60 | 0 | 80 |
| Imidazole | P00868P | 6-Thioxanthine | Pigweed | 100 | 70 | 90 |
| Imidazole | P00868P | 6-Thioxanthine | Johnsongrass | 95 | 0 | 0 |
| Imidazole | P00868P | 6-Thioxanthine | Corn | 0 | 0 | 0 |

(continued)

**Table 17.3** (continued)

| Template | Mod ID | Modulator | Plant | DTH ALA | DTH Mod | DTH A + M |
|---|---|---|---|---|---|---|
| Imidazole | P00900P | 2,4,5-Triphenylimidazole | Cucumber | 60 | 0 | 95 |
| Imidazole | P00900P | 2,4,5-Triphenylimidazole | Pigweed | 100 | 50 | 100 |
| Imidazole | P00900P | 2,4,5-Triphenylimidazole | Johnsongrass | 95 | 0 | 100 |
| Imidazole | P00900P | 2,4,5-Triphenylimidazole | Corn | 0 | 0 | 0 |
| Imidazole | P00354P | 4,5-Diphenylimidazole | Cucumber | 70 | 0 | 100 |
| Imidazole | P00354P | 4,5-Diphenylimidazole | Pigweed | 50 | 10 | 15 |
| Imidazole | P00354P | 4,5-Diphenylimidazole | Johnsongrass | 50 | 0 | 20 |
| Imidazole | P00354P | 4,5-Diphenylimidazole | Corn | 0 | 0 | 0 |
| Thiazole | P00083P | 2-Amino-2-thiazoline | Corn | 0 | 0 | 0 |
| Thiazole | P00083P | 2-Amino-2-thiazoline | Johnsongrass | 5 | 0 | 0 |
| Thiazole | P00083P | 2-Amino-2-thiazoline | Cucumber | 40 | 0 | 90 |
| Thiazole | P00083P | 2-Amino-2-thiazoline | Johnsongrass | 50 | 0 | 20 |
| Imidazole | P00438P | Guanosine hydrate | Cucumber | 50 | 5 | 100 |
| Imidazole | P00438P | Guanosine hydrate | Pigweed | 50 | 0 | 90 |
| Imidazole | P00438P | Guanosine hydrate | Johnsongrass | 50 | 0 | 80 |
| Imidazole | P00438P | Guanosine hydrate | Corn | 0 | 0 | 0 |
| Imidazole | P00404P | 2-Ethyl-4-methyl-imidazole | Cucumber | 50 | 0 | 10 |
| Imidazole | P00404P | 2-Ethyl-4-methyl-imidazole | Pigweed | 50 | 0 | 100 |
| Imidazole | P00404P | 2-Ethyl-4-methyl-imidazole | Johnsongrass | 30 | 5 | 100 |
| Imidazole | P00404P | 2-Ethyl-4-methyl-imidazole | Corn | 0 | 0 | 0 |
| Imidazole | P00278P | 4,5-Dicyanoimidazole | Cucumber | 50 | 50 | 90 |
| Imidazole | P00278P | 4,5-Dicyanoimidazole | Pigweed | 50 | 100 | 100 |
| Imidazole | P00278P | 4,5-Dicyanoimidazole | Johnsongrass | 30 | 80 | 100 |
| Imidazole | P00278P | 4,5-Dicyanoimidazole | Corn | 0 | 5 | 5 |
| Imidazole | P00578P | 1-(Mesitylenesulfonyl)-imidazole | Cucumber | 50 | 0 | 95 |
| Imidazole | P00578P | 1-(Mesitylenesulfonyl)-imidazole | Pigweed | 50 | 90 | 100 |
| Imidazole | P00578P | 1-(Mesitylenesulfonyl)-imidazole | Johnsongrass | 30 | 0 | 20 |
| Imidazole | P00578P | 1-(Mesitylenesulfonyl)-imidazole | Corn | 0 | 0 | 0 |
| Imidazole | P00369P | 2,2′-Dithiobis(4-tert-butyl-1-isopropylimidazole) | Cucumber | 50 | 50 | 90 |
| Imidazole | P00369P | 2,2′-Dithiobis(4-tert-butyl-1-isopropylimidazole) | Pigweed | 50 | 40 | 20 |
| Imidazole | P00369P | 2,2′-Dithiobis(4-tert-butyl-1-isopropylimidazole) | Johnsongrass | 30 | 0 | 10 |
| Imidazole | P00369P | 2,2′-Dithiobis(4-tert-butyl-1-isopropylimidazole) | Corn | 0 | 0 | 0 |
| Imidazole | P00524P | Inosine-5′-triphosphate, disodium salt dihydrate | Cucumber | 50 | 0 | 95 |
| Imidazole | P00524P | Inosine-5′-triphosphate, disodium salt dihydrate | Pigweed | 50 | 0 | 20 |
| Imidazole | P00524P | Inosine-5′-triphosphate, disodium salt dihydrate | Johnsongrass | 50 | 30 | 10 |
| Imidazole | P00524P | Inosine-5′-triphosphate, disodium salt dihydrate | Corn | 0 | 0 | 10 |
| Imidazole | P00887P | 1-(2,4,6-Triisopropylbenzene-sulfonyl)imidazole | Cucumber | 50 | 0 | 50 |

(continued)

**Table 17.3** (continued)

| Template | Mod ID | Modulator | Plant | DTH ALA | DTH Mod | DTH A + M |
|---|---|---|---|---|---|---|
| Imidazole | P00887P | 1-(2,4,6-Triisopropylbenzene-sulfonyl)imidazole | Pigweed | 30 | 15 | 100 |
| Imidazole | P00887P | 1-(2,4,6-Triisopropylbenzene-sulfonyl)imidazole | Johnsongrass | 30 | 0 | 20 |
| Imidazole | P00887P | 1-(2,4,6-Triisopropylbenzene-sulfonyl)imidazole | Corn | 0 | 0 | 0 |
| Furfural | P00712P | Nitrofurantoin | Pigweed | 50 | 0 | 90 |
| Furfural | P00712P | Nitrofurantoin | Cucumber | 50 | 0 | 50 |
| Furfural | P00712P | Nitrofurantoin | Corn | 0 | 0 | 0 |
| Furfural | P00552P | Kinetin | Pigweed | 50 | 80 | 90 |
| Furfural | P00552P | Kinetin | Cucumber | 50 | 0 | 50 |
| Furfural | P00552P | Kinetin | Corn | 0 | 0 | 0 |

Only modulators that exhibited a negligible effect on corn or a rate of kill of 88 % or better on the other test plants are reported, except for Thiaflavin T. Percent death was monitored 10 days after spraying. *Mod ID* modulator identification number in the database, *DTH ALA* death due to 5 m*M* ALA treatment, *DTH Mod* death due to 20 m*M* modulator treatment, *DTH A + M* death due to 5 m*M* modulator treatment. Modulators that are effective by themselves without ALA are probable inducers

ALA and TDPH modulators accumulated massive amounts of tetrapyrroles in darkness (Rebeiz et al. 1991). Although Proto was the main tetrapyrrole that accumulated, significant amounts of MPE and Pchlide were also formed. In the light, the excised roots that accumulated tetrapyrroles, exhibited significant phytotoxicity (Rebeiz et al. 1991).

### 17.7.2 *Effects of TDPH on Attached Cucumber Roots*

To determine the effects of TDPH treatments on attached roots, cucumber seedlings were watered once with a solution consisting of 4 mM ALA and 3 mM modulator (Chung and Rebeiz, unpublished). The treated seedlings were kept in darkness for various periods of time prior to tetrapyrrole analysis and exposure to light. As was observed with excised roots, the roots of intact seedlings watered with a solution of ALA + 10-phenanthroline accumulated massive amounts of tetrapyrroles (Rebeiz et al. 1991). In this case too the major tetrapyrrole pool that accumulated in darkness consisted of Proto. However, this tetrapyrrole accumulation was not toxic to the root cells in darkness. Even after 2 days in darkness no apparent damage to the root system was observed. These results indicated that although plant roots do react to treatment with ALA and TDPH modulators by accumulating tetrapyrroles, they do not exhibit the phenomenon of dark tetrapyrrole-dependent death which was observed in some insects (Rebeiz et al. 1991).

## 17.8 Translocation of TDPH in Intact Plant Seedlings

The translocation of TDPH in intact plant seedlings was investigated by Chung and Rebeiz (unpublished).

### *17.8.1 Acropetal (Upward) Translocation*

Acropetal (upward) translocation, presumably via the xylem, was monitored by following the dark-accumulation of tetrapyrroles in the vegetative parts of the cucumber seedlings upon watering with solutions of ALA and modulators. Translocation of ALA and modulator from the roots to the cotyledons of cucumber seedlings appeared to be very extensive as evidenced by the massive accumulation of tetrapyrroles in the cotyledons (Rebeiz et al. 1991). It is very unlikely that the build up of tetrapyrroles in the cotyledons was caused by translocation of protoporphyrinogen from the roots.

It is also worth noting that in contrast to excised or intact roots, the cotyledons accumulated mainly protochlorophyllide instead of protoporphyrin. Furthermore, it was suspected that the acropetal translocation of ALA plus modulator resulted in the accumulation of significant amounts of tetrapyrroles in the cucumber hypocotyls. This was suggested by the rapid desiccation and death of the hypocotyls, which preceded that of the cotyledons.

### *17.8.2 Basipetal (Downward) Translocation*

Basipetal (downward) translocation, presumably via the phloem, was monitored by following the accumulation of tetrapyrroles in the hypocotyl and roots when the cotyledons of cucumber seedlings were sprayed with ALA and modulators. Tetrapyrrole accumulation, mainly Pchlide, was observed in the cotyledons and in the hypocotyls. None was observed in the roots. This is turn suggested that the basipetal translocation of TDPH was of limited range and did not proceed past the hypocotyl in hypogenous seedlings (Smith and Rebeiz 1979).

Finally, it should be pointed out that all the above work was carried out on seedlings growing in vermiculite. When similar experiments were carried out on seedlings grown in soil, less dramatic results were observed, suggesting that the effect of the soil environment on the availability of ALA and modulators to the root system is significant (Rebeiz et al. 1991).

## 17.9 Is a Postspray Dark Incubation Period Needed for Effective TDPH Activity?

Since in addition to light, a certain level of tetrapyrrole accumulation in the dark is essential for the expression of photodynamic damage, it was natural to wonder whether a post-spray incubation period is essential for effective TDPH treatments. Preliminary experiments reported by Rebeiz et al. (1984a, b) indicated that this was apparently the case. Further experimentation revealed, however, that the reduced photodynamic injury observed in plants treated with ALA and Dpy, and exposed to light without a post-spray dark incubation period, was more probably related to the penetration of the active ingredient to target sites, than to tetrapyrrole accumulation per se. For example, with appropriate formulation, it was possible to achieve the destruction of broadleaf weeds in a Kentucky bluegrass lawn without a post-spray dark incubation period (Rebeiz et al. 1988b). These studies were extended by Mayasich and Rebeiz to ten common weed species (Rebeiz et al. 1991). The weed species were sprayed with ALA and one of four different TDPH modulators using a solvent system developed for experimental field applications (Rebeiz et al. 1988b). Some treatments involved a postspray dark incubation period while others did not. Tetrapyrrole accumulation and photodynamic injury were then evaluated. Seedlings that were subjected to a post-spray dark incubation period accumulated large amounts of tetrapyrroles in darkness. The steady state formation of tetrapyrroles in plants that were not exposed to a post-spray dark incubation period was monitored 1 h after spraying and exposing the plants to low light intensity.

The low light treatment was meant to allow sampling the steady state formation of tetrapyrroles in the light without excessive tetrapyrrole destruction (Rebeiz et al. 1991). In response to ALA plus modulator treatment, tetrapyrrole biosynthesis and accumulation was observable under the low light conditions used in these experiment Under these conditions, however, tetrapyrrole accumulation was much lower than in darkness, probably due to photodestruction and metabolism. Nevertheless, for all practical purposes photodynamic damage in weeds that did not receive a post-spray dark-treatment was essentially as good as in weeds that did (Rebeiz et al. 1991). Minor differences in response between the dark and light treatments was attributed to a combination of factors, that included greening group affiliation of the weed species, and type of accumulated tetrapyrrole in response to the ALA plus modulator treatment. It is obvious that the pre-accumulation of massive amounts of tetrapyrroles is not essential for effective photodynamic injury to take place. What appears to be needed is a steady state supply of tetrapyrroles at a rate sufficiently large to initiate and sustain damaging free radical reactions. The same phenomenon was observed with insects treated with ALA and Dpy in the light (Rebeiz et al. 1988a).

## 17.10 Discrepancy Between the Effects of ALA With and Without TDPH Modulators on Greenhouse-Grown Plants and Field-Grown Plants

The ultimate interest in any herbicide resides in its eventual use to eliminate undesirable weeds under field conditions. Yet as reported in Tables 17.4 and 17.5, a serious discrepancy existed between the effects of ALA with and without TDPH modulators on greenhouse-grown plants and field-grown plants.

Due to climatic conditions in Illinois, it was not possible to investigate the molecular basis of this greenhouse-field discrepancy year round. In order to circumvent this problem, it was decided to develop a greenhouse model plant system that simulates the field effects of ALA treatments, and then use this model to determine whether the greenhouse-field discrepancy is due to poor ALA penetration, poor ALA metabolism, or both. From tetrapyrrole accumulation profiles, it was decided that older greenhouse-grown plants were a good model for young field-grown plants that had already developed a thick cuticle.

The experiments described below were aimed at understanding the molecular basis of this discrepancy (Kulur 1996).

### *17.10.1 Tetrapyrrole and ALA Accumulation and Photodynamic Damage in Morningglory Seedlings of Various Ages, Using Whole Leaves for Analysis*

To determine whether age-dependency of ALA-dependent photodynamic damage was caused by slow metabolism, poor ALA translocation, or both, treated and control morningglory leaves were analyzed separately for tetrapyrrole and ALA content (Kulur 1996). Tetrapyrrole accumulation was used as a marker of metabolic activity by the treated tissue. ALA content was considered as a crude putative marker of applied ALA availability, which in turn may be related to ALA penetration to inner tissues where conversion of ALA to tetrapyrroles takes place.

Ten, fifteen and twenty-day old morningglory seedlings thinned down to one primary leaf per seedling, were sprayed with solvent alone (control) and solvent + 2 lbs per acre ALA (treatment). After spraying, the plants (six plants per treatment) were placed in black foam rubber buckets, which were covered with aluminum foil and placed in darkness overnight. The following day, two leaves were excised from each container and were weighed. Since it was assumed that during overnight dark-incubation, all ALA might have entered the leaf tissue, one unwashed leaf was used for tetrapyrrole accumulation and the other for ALA analysis. The remaining four seedlings were used for assessment of photodynamic injury after placing them in the growth room under 211 W $m^{-2}$ of metal halide light. 1,000 W metal halide lamps provided light intensity.

The amount of ALA detected in ALA-treated seedlings was significantly higher in treated leaves of all ages in comparison to controls. However the amounts of

**Table 17.4** Effect of δ-aminolevulinic acid (ALA) sprays on greenhouse-grown velvetleaf and tall morningglory

| | | | Percent injury | |
|---|---|---|---|---|
| | | | VL | MG |
| Location | Treatment | Rate (g/acre) | (%) | |
| Greenhouse | ALA | 900 | 98.29 ± 2.9 | 97.00 ± 2.9 |

Plants were germinated in vermiculite in glass containers 7.5-cm deep and 9 cm in diameter under a 14-h light/10 h dark photoperiod. Light intensity (metal halide) was about 211 W $m^{-2}$. After 10 days VL (velvetleaf) or 20 days MG (morningglory) of growth, the seedlings were sprayed with 2 lbs. Per acre of ALA. ALA was dissolved in a solution made up of 2.25 % acetone, 0.25 % Sylgard, 1.0 % Tween -80, 9.00 % polyethylene glycol 600, 6 % soybean oil, 79.50 % water. Photodynamic damage was assessed 10 days after spraying. Values are means of three replications ± standard deviation *VL* velvetleaf, *MG* Morningglory (Reproduced from Kulur 1996)

**Table 17.5** Comparison of the effects of ALA sprays, with and without modulators, and acifluorfen, in a randomized plot design, on soybean and several weed species under field conditions

| | | | Percent injury | | | | | |
|---|---|---|---|---|---|---|---|---|
| | | | Soybean | VL | GF | TMG | JW | WM |
| Location | Treatment | Rate (g/acre) | (%) | | | | | |
| Field | ALA | 900 | 24 | 71 | 56 | 53 | 58 | 50 |
| | Dpy | 300 | 7 | 30 | 25 | 23 | 23 | 18 |
| | ALA + Dpy | 900 + 300 | 21 | 65 | 52 | 43 | 55 | 55 |
| | Oph | 300 | 16 | 51 | 18 | 20 | 28 | 30 |
| | ALA + Oph | 900 + 300 | 30 | 79 | 63 | 66 | 73 | 78 |
| | Acifluorfen | 171 | 47 | 96 | 92 | 97 | 99 | 99 |
| | Control | 0 | 0 | 0 | 0 | 0 | 0 | 0 |
| | LSD (0.05) | – | 7 | 14 | 30 | 19 | 17 | 20 |

Values are means of three replicates. Photodynamic injury was evaluated 17 days after spraying. Application date = 07/08/87. Plot size = 5 × 18 ft. Spray volume = 151 l/acre. *ALA* δ-aminolevulinic acid, *Dpy* 2,2′-dipyridyl, *Oph* 1,10-orthophenanthroline, *VL* velvetleaf, *GF* giant foxtail, *TMG* tall morningglory, *JW* jimsonweed, *WM* wild mustard, *LSD* least significant difference (Adapted from C. A. Rebeiz and R. Liebel, unpublished). Other conditions are as in Table 17.4

ALA detected in treated leaves of all ages were not significantly different from one another (Table 17.6, Fig. 17.4). These results suggested that under our growth conditions, leaf age had no bearings on ALA penetration to active sites of tetrapyrrole metabolism.

Further insight into the metabolic fate of ALA was derived from an examination of tetrapyrrole accumulation. ALA-treated plants accumulated significantly higher amounts of tetrapyrroles than untreated ones (Table 17.6, Fig. 17.4). The accumulated tetrapyrroles consisted mainly of Pchlide *a*. The accumulation of Pchlide *a* was age-dependent and the amount of tetrapyrroles accumulated by 10 and 15-day old plants was higher than in 20-day old plants (Table 17.6, Fig. 17.4). Together with ALA content (see above), these results suggested that

**Table 17.6** Tetrapyrrole accumulation and ALA content of grenhouse-grown whole morningglory leaves of various ages following ALA treatment (2 lbs per acre)

| ALA content and tetrapyrrole accumulation after 15 h dark-incubation | | | | | | |
|---|---|---|---|---|---|---|
| | 10-day old | | 15-day old | | 20-day old | |
| Tetrapyrrole or % injury | C | T | C | T | C | T |
| | nmoles per g fresh weight | | | | | |
| ALA | 0 ± 0 | 1.6 ± 0.9 | 0.1 ± 0 | 1.4 ± 0.2 | 0.1 ± 0.1 | 1.9 ± 1.5 |
| Mp(e) | 0.1 ± 0 | 2.4 ± 2.4 | 0 ± 0 | 4.7 ± 6.3 | 0.1 ± 0.1 | 1.7 ± 2.2 |
| Pchlide *a* | 2.2 ± 0.5 | 24.3 ± 4.6 | 1.6 ± 0.5 | 22.6 ± 13.3 | 0.7 ± 0.6 | 4.5 ± 0.7 |
| % Injury | 0 ± 0 | 96.3 ± 5.4 | 0.0 ± 0 | 83.8 ± 2.1 | 0 ± 0.0 | 72.9 ± 12.7 |
| LSD (0.05) age | ALA = 0.4 | | | | | |
| | Mp(e) = 2.8 | | | | | |
| | Pchlide *a* = 6.6 | | | | | |
| | Injury = 0.5 | | | | | |
| LSD (0.05) treatment | ALA = 0.4 | | | | | |
| | Mp(e) = 2.2 | | | | | |
| | Pchlide *a* = 5.4 | | | | | |
| | Injury = 0.4 | | | | | |

Values are means of six replicates. Photoperiodic damage was assessed 10 days after treatment (Reproduced from Kulur 1996)

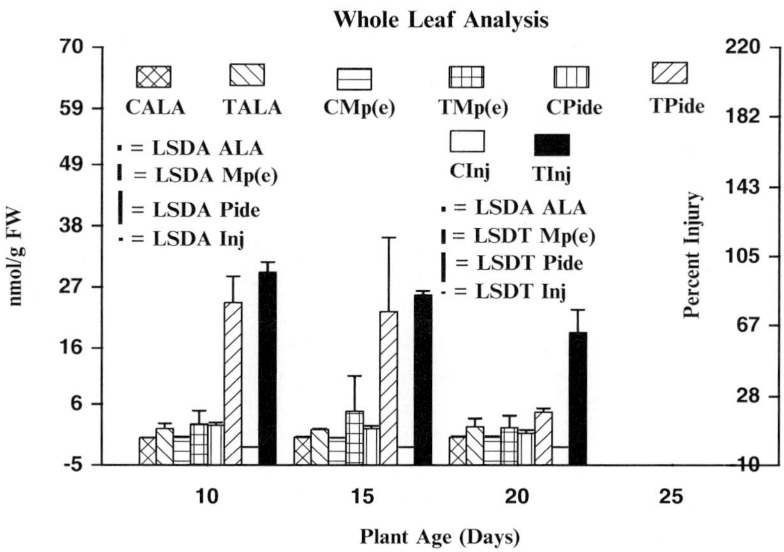


**Fig. 17.4** ALA content, tetrapyrrole accumulation and photodynamic injury in unwashed whole morningglory leaves of different ages. The data was adapted from Table 17.4. C control, T treated, *ALA* δ-aminolevulinic acid, *Mp(e)* Mg-Proto and/or its methyl ester, *Pide* Protochlorophyllide *a*, *Inj* photodynamic injury, *LSD* least significant difference at the 5 % level. *Vertical* graph bars refer to standard deviation. Text bars refer to LSD magnitudes (Reproduced from Kulur 1996)

the higher tetrapyrrole content of 10 and 15-day old leaves was due to more active tetrapyrrole metabolism in younger leaves rather than ALA availability.

Photodynamic damage appeared to parallel tetrapyrrole accumulation as 10-day old seedlings exhibited higher photodynamic damage (96.28 %), than 15-day old (83.8 %) and 20-day old seedlings (62.86 %).

### 17.10.2 ALA Content, Tetrapyrrole Accumulation and Photodynamic Damage in Unwashed Morningglory Primary Leaf Sections

To confirm that differences in tetrapyrrole accumulation in morningglory seedlings of different ages were due to differences in active tetrapyrrole metabolism rather than ALA availability, the ALA treatment experiment (see above) was repeated with the following modification. Instead of monitoring tetrapyrrole accumulation and ALA content on one leaf and photodynamic damage on other leaves, tetrapyrrole and ALA analysis were performed on a small leaf section and photodynamic damage was evaluated on the remaining leaf parts. It was conjectured that such an approach would minimize sampling errors and give a more accurate picture of what is going on.

Tetrapyrrole and ALA analyses were similar to those described earlier for whole leaves, except that in this case unwashed leaf sections were used for analysis. After the plants were sprayed at a rate of 2lbs ALA per acre, and placed in darkness overnight, two small sections of the primary leaf were cut from each of the six plants. The tissue sections were pooled, weighed and, frozen in liquid nitrogen for tetrapyrrole analysis. Next, a second leaf section was removed from each plant. The leaf sections were pooled and used for ALA determination. In this manner, about two third of the primary leaf from each plant was removed for analysis. Photodynamic damage was assessed on the remaining one-third leaf sections.

As was observed with unwashed whole leaves, the amount of ALA detected in ALA-treated seedlings was significantly higher in treated leaves of all ages in comparison to controls. Again, the amounts of ALA detected in treated leaves of all ages were not significantly different from one another (Table 17.7, Fig. 17.5). These results confirmed that under our growth conditions, leaf age had no bearings on ALA penetration to active sites of tetrapyrrole metabolism.

As was observed with whole leaves, ALA-treated plants accumulated significantly higher amounts of tetrapyrroles than untreated ones (Table 17.7, Fig. 17.5). Here again, the accumulated tetrapyrroles consisted mainly of Pchlide *a*. In this case too the accumulation of Pchlide *a* was age-dependent and the amount of tetrapyrroles accumulated by 10 and 15-day old plants was higher than in 20-day old plants (Table 17.7, Fig. 17.5). Together with the ALA content data (see above) these results confirmed that the higher tetrapyrrole content in 10 and 15-day old

**Table 17.7** Tetrapyrrole accumulation and ALA content of unwashed morningglory leaf sections of various ages following ALA treatment (2lbs ALA per acre)

| ALA content and tetrapyrrole accumulation after 15 h dark-incubation | | | | | | |
|---|---|---|---|---|---|---|
| | 10-day old | | 15-day old | | 20-day old | |
| Tetrapyrrole or % injury | C | T | C | T | C | T |
| | nmoles per g fresh weight | | | | | |
| ALA | 0.4 ± 0.2 | 2.3 ± 0.4 | 0.1 ± 0.0 | 1.8 ± 0.2 | 0.1 ± 0.1 | 2.0 ± 0.6 |
| Mp(e) | 0.0 ± 0.0 | 12.9 ± 13.3 | 0.0 ± 0.0 | 2.7 ± 3.4 | 0.0 ± 0.0 | 0.9 ± 0.9 |
| Pchlide *a* | 2.2 ± 1.0 | 23.1 ± 11.4 | 0.8 ± 0.3 | 8.7 ± 4.8 | 1.2 ± 1.0 | 3.3 ± 1.1 |
| % Injury | 0 0 ± 0.0 | 100.0 ± 0.0 | 0.0 ± 0.0 | 89.7 ±15.2 | 0.0 ± 0. 0 | 64.1 ± 17.6 |
| LSD (0.05) age | ALA = 0.3<br>Mp(e) = 3.4<br>Pchlide *a* = 3.6<br>Injury = 0.8 | | | | | |
| LSD (0.05) treatment | ALA = 0.2<br>Mp(e) = 2.8<br>Pchlide *a* = 3.0<br>Injury = 0. 5 | | | | | |

Values are means of six replicates. Photoperiodic damage was assessed 10 days after treatment (Reproduced from Kulur 1996)

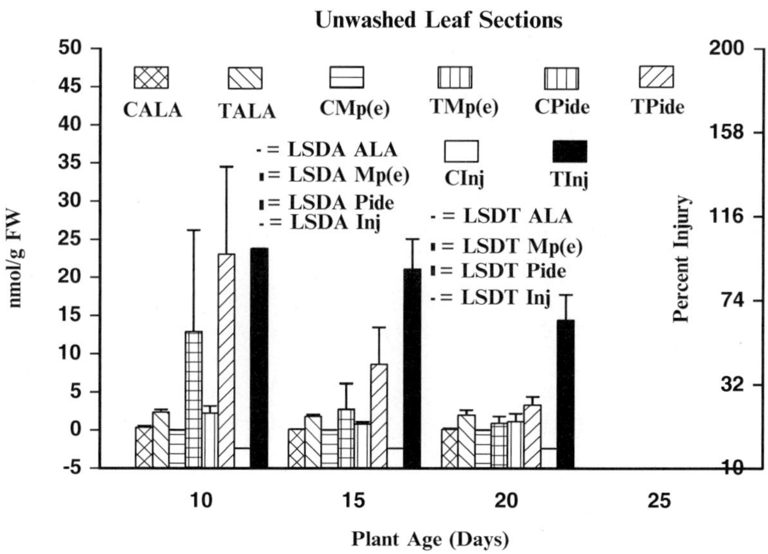


**Fig. 17.5** ALA content, tetrapyrrole accumulation and photodynamic injury in unwashed morningglory leaf sections of different ages. The data was adapted from Table 17.5. All abbreviations are as in Fig. 17.4 (Reproduced from Kulur 1996)

leaves was more likely due to more active tetrapyrrole metabolism in younger leaves than to ALA availability.

Also, as in whole, unwashed leaves, photodynamic damage appeared to parallel tetrapyrrole accumulation as 10-day old seedlings exhibited higher photodynamic damage (100 %), than 15-day old (89.7 %) and 20-day old plants (64.1 %). Altogether the above results confirmed the conclusions drawn from the analysis of unwashed whole leaves.

### *17.10.3 ALA Content, Tetrapyrrole Accumulation and Photodynamic Damage in Washed Morningglory Primary Leaf Sections*

To further confirm that differences in tetrapyrrole accumulation in morningglory seedlings of different ages was due to differences in active tetrapyrrole metabolism rather than ALA availability, the above experiment was repeated with the following modification. Instead of monitoring tetrapyrrole accumulation and ALA content on unwashed leaf sections, tetrapyrrole and ALA analyses were performed on washed leaf sections, and photodynamic damage was evaluated on the remaining leaf parts. It was conjectured that such an approach would give a more accurate picture of the size of the active ALA pool inside the leaf tissue by eliminating the contribution of metabolically inactive surface ALA that did not penetrate the tissue. As a consequence a better evaluation of the relationship between ALA metabolic availability and photodynamic damage may be obtained.

Tetrapyrrole and ALA analyses were similar to those described earlier for unwashed leaf sections, except that in this case washed leaf sections were used for analysis. In other words, after the plants were sprayed and placed in darkness overnight, the leaves were washed with distilled water before excising the leaf sections.

Contrary to what was observed with unwashed leaf sections, the difference in ALA content between control and ALA-treated seedlings although statistically significant was extremely small (Table 17.8, Fig. 17.6). This in turn indicated that the higher ALA concentrations observed in ALA-treated plants in comparison to controls, in unwashed leaves and leaf sections were probably due to traces of metabolically inactive ALA that remained on the unwashed leaf surface. As pointed out below, it also shed additional light about the interaction of ALA availability, tetrapyrrole biosynthesis and leaf age.

As was observed with unwashed leaf material, ALA-treated plants accumulated significantly higher amounts of tetrapyrroles than untreated ones (Table 17.8, Fig. 17.6). Most of the accumulated tetrapyrroles consisted of Pchlide *a*. In this case too the accumulation of Pchlide *a* was age-dependent and the amount of tetrapyrroles accumulated by 10 and 15-day old plants was higher than in 20-day

**Table 17.8** Tetrapyrrole accumulation and ALA content of washed morningglory leaf sections of various ages following ALA treatment (2lbs per acre)

| ALA content and tetrapyrrole accumulation after 15 h dark-incubation | | | | | | |
|---|---|---|---|---|---|---|
| | 10-day old | | 15-day old | | 20-day old | |
| Tetrapyrrole or % injury | C | T | C | T | C | T |
| | nmoles per g fresh weight | | | | | |
| ALA | 0.1 ± 0.3 | 0.2 ± 0.1 | 0.1 ± 0.0 | 0.2 ± 0.0 | 0.1 ± 0.0 | 0.2 ± 0.0 |
| Mp(e) | 0.0 ± 0.0 | 0 7 ± 0.3 | 0.0 ± 0.0 | 0.0 ± 0.0 | 0.0 ± 0.0 | 0.1 ± 0.0 |
| Pchlide *a* | 1.6 ± 0.2 | 9.0 ± 4.2 | 0.5 ± 0.1 | 2.9 ± 1.6 | 0.2 ± 0.0 | 1.0 ± 0.0 |
| % Injury | 0 0 ± 0.0 | 83.3 ± 16.7 | 0.0 ± 0.0 | 61.1 ±19.2 | 0.0 ± 0. 0 | 55.6 ± 14.0 |
| LSD (0.05) age | ALA = 0.04<br>Mp(e) = 0.15<br>Pchlide *a* = 2.2<br>Injury = 3.5 | | | | | |
| LSD (0.05) treatment | ALA = 0.03<br>Mp(e) = 0.12<br>Pchlide *a* = 1.8<br>Injury = 2.5 | | | | | |

Values are means of six replicates. Photoperiodic damage was assessed 10 days after treatment (Reproduced from Kulur 1996)

old plants (Table 17.8, Fig. 17.6). Also, as in whole, unwashed leaves, photodynamic damage appeared to parallel tetrapyrrole accumulation as 10-day old seedlings exhibited higher photodynamic damage (83.3 %), than 15-day old (61.1 %) and 20-day old plants (55.6 %).

Thus the key experiment that gave insight about the interaction of ALA availability, tetrapyrrole metabolism and tissue age was the experiment just described, involving washed leaf sections. The small amounts of ALA detected in treated leaves of all ages were not significantly different from one another (Table 17.8, Fig. 17.6). Yet, ALA-treated younger leaves accumulated significantly higher amounts of tetrapyrroles than older ones (Table 17.8, Fig. 17.6). In other words, the same amount of metabolically active ALA was detected in young and old washed leaf sections, but younger leaves accumulated more tetrapyrroles than older ones. This lead to several conclusions that are discussed below.

The detection of very small amounts of ALA, and much larger amounts of tetrapyrroles in young leaves is compatible with the notion that exogenous ALA translocates rapidly to inner tissues, where it is rapidly converted to tetrapyrroles. The rapid translocation of ALA to inner tissues is mandatory for the support of the observed high rates of tetrapyrrole biosynthesis and accumulation. In this case, the detection of very small amounts of ALA accumulation in the inner tissues can best be explained by its rapid conversion to tetrapyrroles.

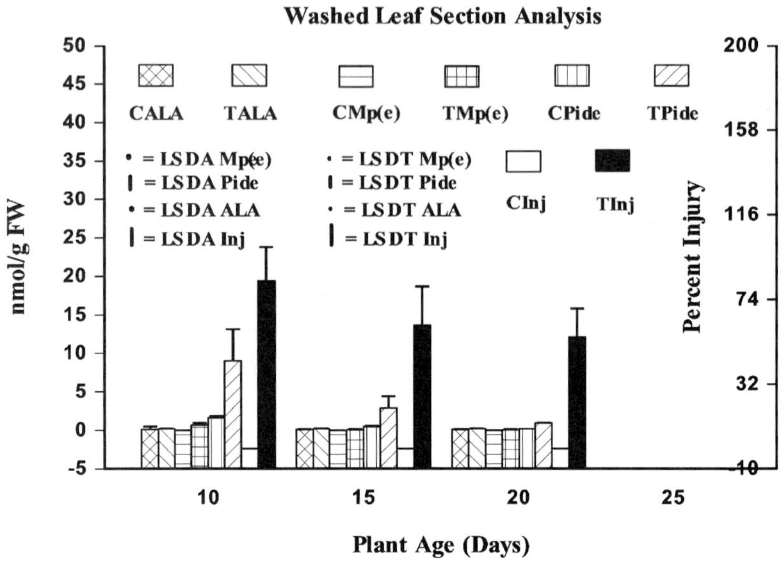


**Fig. 17.6** ALA content, tetrapyrrole accumulation and photodynamic injury in washed morningglory leaf sections of different ages. The data was adapted from Table 17.6. All abbreviations are as in Fig. 17.4 (Reproduced from Kulur 1996)

The detection of equally small amounts of ALA and lesser amounts of tetrapyrrole accumulation in older leaves can be explained by several scenarios, none of which is compatible with a rapid translocation of exogenous ALA to inner tissue. If rapid translocation of exogenous ALA to inner tissue took place in older leaves, one should observe one of two phenomena: (a) larger amounts of ALA accumulation, if ALA conversion to tetrapyrrole is sluggish or (b) larger amount of tetrapyrrole accumulation if ALA conversion to tetrapyrroles is rapid.

Since none of these phenomena were observed in older tissues, two other scenarios involving a slow translocation of exogenous ALA to inner tissues may be involved: (a) In one, slow translocation of exogenous ALA to inner tissues is accompanied also by slow conversion of ALA to tetrapyrroles, or (b) although older tissues have very active tetrapyrrole biosynthetic capabilities, the very slow translocation of exogenous ALA to inner tissue is a limiting factor, that results in poor tetrapyrrole accumulation.

Distinction between the last two hypotheses is very important since it dictates how to formulate ALA for field use. In case older tissues exhibit low translocation rates of exogenous ALA to inner tissues and low conversion rates of ALA to tetrapyrroles, then ALA formulations should include a TDPH modulator to activate tetrapyrrole anabolism. On the other hand, in case older tissues exhibit low translocation rates of exogenous ALA to inner tissues, and high conversion rates of ALA to tetrapyrroles, then ALA formulations that improve the translocation of ALA to inner tissues should be a focal point. Distinction between these two hypotheses is carried out in the next experiments described below.

## 17.11 Effects of Two Different Treatments on the Availability of Metabolically Active ALA and Concomitant Photodynamic Damage in Morningglory

As a result of the previous experiments, it was proposed that in older greenhouse-grown morningglory seedlings which simulate the response of younger field-grown seedlings, the reduction in photodynamic damage in comparison to younger, greenhouse-grown seedlings maybe caused by one of two factors: (a) Slow translocation of exogenous ALA to inner tissues, coupled with slow conversion of ALA to tetrapyrroles or (b) slow translocation of exogenous ALA to inner tissues coupled with fast conversion of ALA to tetrapyrroles. As discussed in above, distinction between these two possibilities is essential for effective ALA field formulations. In an effort to determine whether the slow translocation of exogenous ALA to inner tissues is accompanied by sluggish or active tetrapyrrole metabolism, two sets of experiments were designed.

In a first set of experiments, thioflavin T, a desiccant (Rebeiz et al. 1994) was used jointly with ALA in an effort to improve ALA translocation to inner tissues, since at low concentrations, thioflavin T is supposed to create holes in the cuticle and facilitate herbicide penetration.

It was conjectured that, if ALA penetration was improved by that treatment and if tetrapyrrole metabolism was highly active in older leaves, then improved ALA accumulation and photodynamic damage should be observed.

In a second set of experiments, an attempt was made to incubate older leaves under conditions of unlimited ALA supply. It was conjectured that if active ALA conversion to tetrapyrroles were limiting in older leaves, then high levels of tetrapyrrole accumulation and photodynamic death would be observed.

### *17.11.1 Response of Various Age Groups of Morningglory Seedlings to ALA Treatments With and Without Thioflavin T*

As was observed in previous experiments, the amount of ALA detected in washed leaf sections in ALA-treated seedlings was slightly, but significantly higher in treated leaves of all ages in comparison to controls (Table 17.7, Fig. 17.7b, c). ALA-treated plants accumulated significantly higher amounts of tetrapyrroles than untreated ones. Most of the accumulated tetrapyrroles consisted of Pchlide *a*. Also, the accumulation of Pchlide *a* was age-dependent and the amount of tetrapyrroles accumulated by 10 and 15-day old plants was higher than in 20-day old plants.

**Table 17.9** Tetrapyrrole accumulation and ALA content of washed morningglory leaf sections of various ages following plant treatment with ALA and thioflavin T (0.5 lbs of thioflavin T and/or 2lbs per acre of ALA)

| | | Accumulation after 15 h-incubation | | | |
|---|---|---|---|---|---|
| | | ALA | Mp(e) | Pchlide *a* | |
| Age (days) | Treatment | nmoles per g fresh weight | | | Injury (%) |
| 10 | Control | 0.2 ± 0.1 | 0.0 ±0.0 | 2.7 ±1.2 | 0.0 ± 0.0 |
| | 2 ALA | 0.3 ± 0.1 | 2.2 ± 2.1 | 17.8 ± 7.1 | 95.6 ± 8.7 |
| | 0.5 TFT | 0.2 ± 0.1 | 0.0 ± 0.0 | 2.3 ± 1.1 | 0.0 ± 0.0 |
| | 2 ALA + 0.5 TFT | 0.3 ± 0.1 | 1.0 ± 0.7 | 16.7 ± 5.1 | 100 ± 0.0 |
| | 2 TFT | 0.1 ± 0.0 | 0.0 ± 0.0 | 3.1 ± 0.9 | 4.2 ± 0.2 |
| | 2 ALA + 2 TFT | 0.2 ± 0.1 | 3.3 ± 4.2 | 22.0 ± 19.5 | 87.5 ± 16.0 |
| 15 | Control | 0.1 ± 0.0 | 0.0 ± 0.0 | 0.9 ± 0.4 | 0.0 ± 0.0 |
| | 2 ALA | 0.2 ± 0.1 | 0.9 ± 0.8 | 10.3 ± 7.7 | 75.0 ± 21.5 |
| | 0.5 TFT | 0.2 ± 0.0 | 0.0 ± 0.0 | 1.3 ± 1.5 | 0.0 ± 0.0 |
| | 2 ALA + 0.5 TFT | 0.2 ± 0.0 | 1.0 ±0.8 | 10.6 ± 3.0 | 87.5 ± 8.3 |
| | 2 TFT | 0.1 ± 0.1 | 0.0 ± 0.0 | 1.7 ± 0.4 | 20.9 ± 25.1 |
| | 2 ALA + 2 TFT | 0.2 ± 0.0 | 0.5 ± 0.2 | 9.8 ± 3.4 | 27.5 ± 8.3 |
| 20 | Control | 0.1 ± 0.0 | 0.0 ± 0.0 | 0.5 ± 0.3 | 0.0 ± 0.0 |
| | 2 ALA | 0.2 ± 0.1 | 0.1 ± 0.1 | 3.2 ± 2.7 | 52.8 ± 5.7 |
| | 0.5 TFT | 0.1 ± 0.1 | 0.0 ± 0.0 | 0.8 ± 0.6 | 12.6 ± 16.1 |
| | 2 ALA + 0.5 TFT | 0.3 ± 0.1 | 0.2 ± 0.1 | 5.1 ± 0.9 | 55.7 ± 6.6 |
| | 2 TFT | 0.1 ± 0.0 | 0.1 ± 0.0 | 1.1 ± 2.7 | 21.0 ± 22.4 |
| | 2 ALA + 2 TFT | 0.2 ± 0.0 | 0.1 ± 0.1 | 3.8 ± 2.4 | 50.2 ± 22.4 |
| LSD | ALA = 0.0 | Pchlide *a* = 3.2 | | | |
| Age | Mp(e) = 0.7 | Injury = 1.0 | | | |
| LSD | ALA = 0.1 | Pchlide *a* = 4.6 | | | |
| Treatment | Mp(e) = 0.9 | Injury = 0.1 | | | |

Values are means of four replicates. Photoperiodic damage was assessed 2 days after treatment. *TFT* Thioflavin T

**Fig. 17.7** Thioflavin T

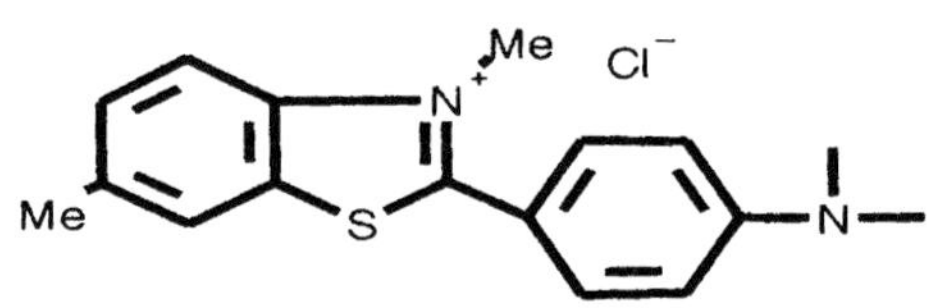


The results reported in Table 17.9, and Fig. 17.8, confirmed that thioflavin T was not an inducer of tetrapyrrole biosynthesis since in the absence of added ALA it did not trigger an enhancement of tetrapyrrole biosynthesis (Rebeiz et al. 1990). Although thioflavin T slightly improved the penetration of ALA to inner tissue as evidenced by detection of higher amounts of ALA in thioflavin T + ALA treated tissues, the effect was marginal and for all practical purposes insufficient. So were the effects on tetrapyrrole accumulation as evidenced by nearly equal amounts of Pchlide *a* accumulation in the presence and absence of thioflavin T (Table 17.9, Fig. 17.8). Photodynamic damage appeared to parallel tetrapyrrole accumulation as

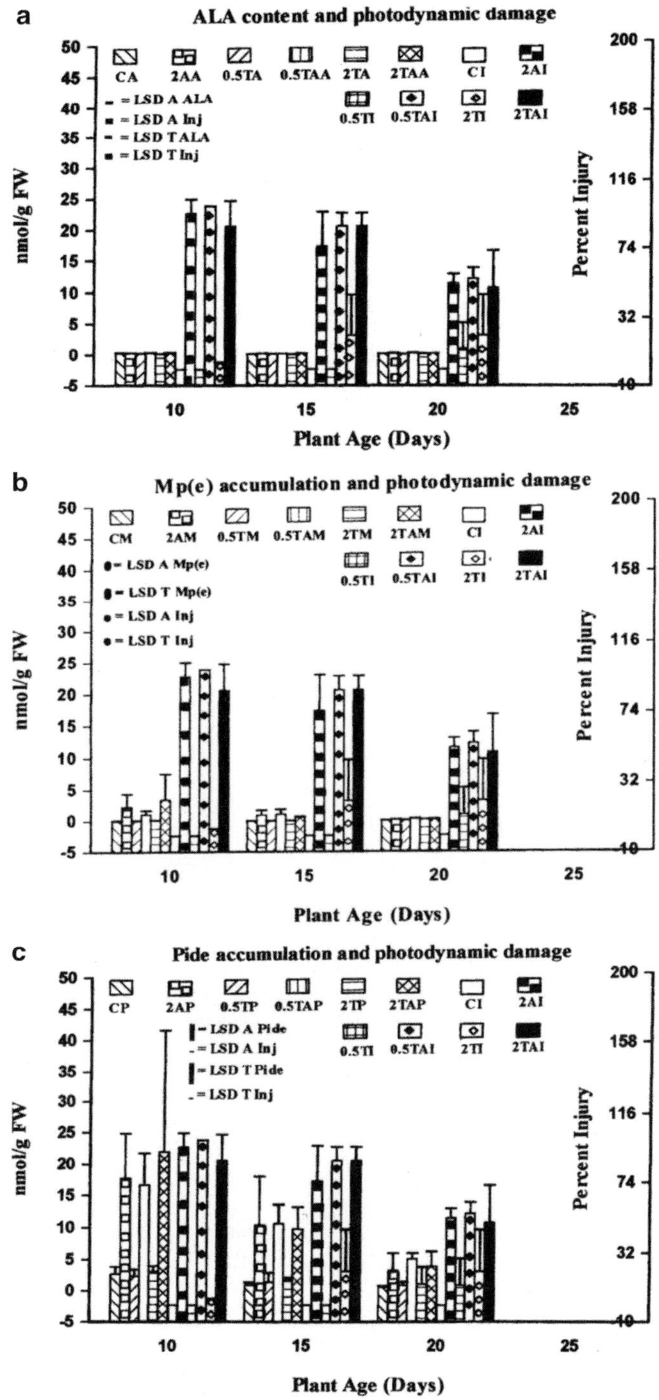


**Fig. 17.8** Tetrapyrrole accumulation in washed leaves excised from ALA and thioflavin T-treated seedlings. The data was adapted from Table 17.7. All results refer to ALA or tetrapyrrole content

10 day-old seedlings exhibited higher photodynamic damage (100–87.5 %), than 15-day old (87.5–27.5 %) and 20-day old plants (55.7–50.2 %). In 10 and 15-day old seedlings, higher concentrations of thioflavin T appeared to result in lower levels of photodynamic injury. All in all, the enhancement of ALA translocation by thioflavin T was not large enough to allow determination of the activity level of tetrapyrrole biosynthesis in older leaf tissues.

### *17.11.2 Response of 20-Day Old Morningglory Leaves to Conditions That Simulate Improved ALA Penetration to Inner Tissues*

Since the use of a desiccant failed to substantially improve ALA translocation to inner tissue, and to shed additional light upon the activity level of tetrapyrrole biosynthesis in older leaves, an alternative strategy was explored. It was conjectured that if the older plant tissue was left in contact with a solution of ALA overnight, enough ALA may translocate to the inner tissues and would result in improved tetrapyrrole accumulation and concomitant photodynamic injury. This would be true if the Chl biosynthetic pathway is as active in older leaves as in younger ones (Table 17.10).

The amount of ALA detected in leaves incubated with ALA was significantly higher than in control leaves (Table 17.10). It was also noticed that ALA content in the treated leaves was much higher than that accumulated by leaves sprayed

**Fig. 17.8** (continued) of washed leaf sections excised from control or treated seedlings. (**a**) ALA content and photodynamic injury: CA ALA content of control, 2AA ALA content after treatment with 2 lbs per acre of ALA, 0.5TAA ALA content after treatment with 0.5 lbs. per acre of thioflavin T, 0.5TA ALA content after treatment with 0.5 lbs. per acre of thioflavin T and 2 lbs. per acre of ALA, 2TAA ALA content after treatment with 2 lbs. per acre of thioflavin T, 2TA ALA content after treatment with 2 lbs. per acre of thioflavin T and 2 lbs. per acre of ALA. (**b**) Mp (e) accumulation and photodynamic injury, CM Mp(e) content of control, 2 AM Mp(e) content after treatment with 2 lbs. per acre of ALA, 0.5TM Mp(e) content after treatment with 0.5 lbs. per acre of thioflavin T, 0.5TAM Mp(e) content after treatment with 0.5 lbs. per acre of thioflavin T and 2 lbs. per acre of ALA, 2TM Mp(e) content after treatment with 2 lbs. per acre of thioflavin T, 2TAM Mp(e) content after treatment with 2 lbs. per acre of thioflavin T and 2 lbs. per acre of ALA, (**c**) Pchlide *a* accumulation and photodynamic injury, CP Pchlide *a* content of control, 2AP Pchlide *a* content after treatment with 2 lbs. per acre of ALA, 0.5TP Pchlide *a* content after treatment with 0.5 lbs. per acre of thioflavin T, 0.5TAP Pchlide *a* content after treatment with 0.5 lbs. per acre of thioflavin T and 2 lbs. per acre of ALA, 2TP Pchlide *a* content after treatment with 2 lbs. per acre of thioflavin T, 2TAP Pchlide *a* content after treatment with 2 lbs. per acre of thioflavin T and 2 lbs. per acre of ALA, CI Photodynamic damage in control leaves, 2AI Photodynamic damage in leaves treated with 2 lbs. per acre ALA, 0.5TI Photodynamic damage in leaves treated with 0.5 lbs. per acre thioflavin, 0.5TAI Photodynamic damage in leaves treated with 0.5 lbs. per acre thioflavin T and 2 lbs. per acre ALA, 2TI Photodynamic damage in leaves treated with 2 lbs. per acre thioflavin T, 2TAI Photodynamic damage in leaves treated with 2 lbs. per acre thioflavin T and 2 lbs. per acre ALA. All other abbreviations are as in Fig. 17.4

**Table 17.10** Tetrapyrrole accumulation and ALA content of 20-day old morningglory leaves incubated in ALA solution (10 ml of 20 mM ALA solution)

| ALA content and tetrapyrrole accumulation after 15 h dark-incubation | | |
|---|---|---|
| | Control | Treated |
| Tetrapyrrole or % injury | nmoles per g fresh weight | |
| ALA | 0.319 ± 0.09 | 3.19 ± 1.03 |
| Proto | 0.069 ± 0.07 | 41.03 ± 29.79 |
| Mp(e) | 0.00 ± 0.00 | 13.69 ± 9.88 |
| Pchlide a | 1.14 ± 1.08 | 31.22 ± 16.16 |
| % Injury | 0.00 ± 0.00 | 100 ± 0.00 |
| | ALA = 0.94 | |
| | Proto = 27.10 | |
| | Mp(e) = 8.99 | |
| | Pchlide a = 14.73 | |
| | Injury = 0.00 | |

Values are means of six replicates. Photoperiodic damage was assessed 2 days after treatment

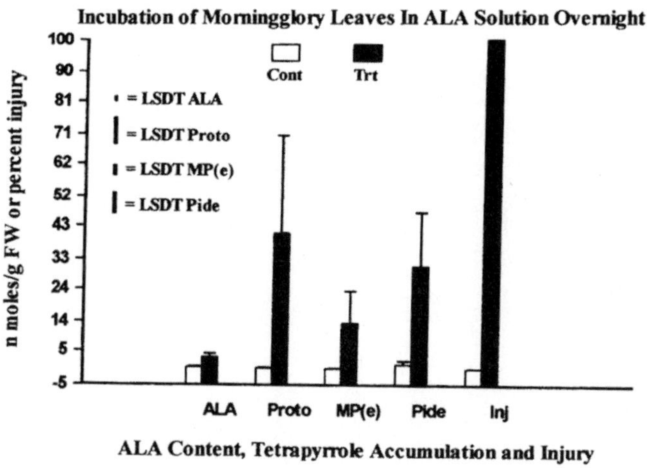


**Fig. 17.9** Tetrapyrrole accumulation and ALA content in 20-day old morningglory leaves incubated in ALA solution. The data was adapted from Table 17.8. All results refer to ALA or tetrapyrrole content in washed whole leaves of 20-day old morningglory plants, incubated in 10 ml ALA solution. Cont = Tissue incubated in 10 ml water. Trt = Tissue incubated in 10 ml water + 29 mM ALA. All other abbreviations are as in Fig. 17.7

with ALA solutions, as in any of the previous experiments (Tables 17.4, 17.5, 17.6 and 17.7 and Figs. 17.4, 17.5, 17.6 and 17.7). Leaves incubated with ALA also accumulated significantly higher amounts of tetrapyrroles than leaves that were incubated in water only. The highest level of accumulated tetrapyrroles consisted of Proto. Treated leaves also accumulated significantly higher amounts of Mp(e) and

Pchlide *a*. This pattern of tetrapyrrole accumulation was different from any of the previous experiments, where the predominant tetrapyrrole accumulated by treated leaves was Pchlide *a* and only minimal amounts of Proto and Mp(e) accumulation were observed.

Thus the results of the above investigations confirm that older greenhouse-grown leaf tissues that simulate the behavior of younger field-grown tissues toward ALA treatment have active tetrapyrrole biosynthetic capabilities. When the translocation barrier to ALA was overcome by prolonged contact of treated tissue with ALA, the older tissue accumulated as much tetrapyrroles as younger tissues and was subjected to the same extent of photodynamic injury (Tables 17.5, 17.6, 17.7 and 17.8 and Figs. 17.5, 17.6, 17.7, 17.8 and 17.9). It indicated that older tissues were capable of high rates of tetrapyrrole biosynthesis when ALA supply is not limiting. The implications of this finding to the development of ALA field formulations are very significant. It implies that if ALA formulations that overcome the barrier to ALA translocation under field conditions were developed, then ALA may become a potent photodynamic field herbicide.

## 17.12 Epilogue

It is hoped that based upon the information provided in this chapter younger investigators may develop efficient field-formulation for safe ALA + TDPH modulators. I took Monsanto 6 year of experimentation to develop a formulation for the efficient penetration of Glyphosate (Round up) into field-grown plants. It may take a shorter time to develop one for photodynamic herbicides. It should me mentioned that once singlet oxygen is produced inside plant tissues, there is no known mechanism that the plant tissue can use to detoxify it.

## References

Abd-El-Mageed HA, El Sahhar KF, Robertson KR et al (1997) Chloroplast biogenesis 77. Two novel monovinyl and divinyl light-dark greening groups of plants and their relationship to the chlorophyll a biosynthetic heterogeneity of green plants. Photochem Photobiol 66:89–96

Amindari SM, Splittstoesser WE, Rebeiz CA (1995) Photodynamic effects of several metabolic tetrapyrroles on isolated chloroplasts. In: Heitz JR, Downum KR (eds) Light-activated pest control. American Chemical Society, Washington, DC, pp 217–246

Bassi R, Rigoni F, Giacometti GM (1990) Chlorophyll binding proteins with antenna function in higher plants and green algae. Photochem Photobiol 52:1187–1206

Bazzaz MB, Rebeiz CA (1978) Chloroplast culture: the chlorophyll repair potential of mature chloroplasts incubated in a simple medium. Biochim Biophys Acta 504:310–323

Bednarick DP, Hoober JK (1985) Biosynthesis of a chlorophyllide b-like pigment in phenanthroline-treated Clamydomonas reinhardtii. Arch Biochem Biophys 240:269–275

Belanger FC, Rebeiz CA (1984) Chloroplast biogenesis 47: spectroscopic study of net spectral shifts induced by ligand coordination in metalated tetrapyrroles. Spectrochim Acta 40A:807–827

Butler WL, Kilajima M (1975) Florescence quenching in photosystem II of chloroplasts. Biochim Biophys Acta 396:72–85

Castelfranco PA, Rich PM, Beale SI (1974) The abolition of the lag phase in greening cucumber cotyledons by exogenous δ-aminolevulinic acid. Plant Physiol 53:615–618

Cohen CE, Rebeiz CA (1978) Chloroplast biogenesis 22. Contribution of short wavelength and long wavelength protochlorophyll species to the greening of higher plants. Plant Physiol 61:824–829

Duggan JX, Gassman M (1974) Induction of porphyrin biosynthesis in etiolated bean leaves by chelators of iron. Plant Physiol 53:206–215

Duggan JX, Rebeiz CA (1982) Chloroplast biogenesis 38. Quantitative detection of a chlorophyllide b pool in higher plants. Biochim Biophys Acta 714:248–260

Duke SO, Rebeiz CA (1994) Porphyrinogenesis as a tool in pest management. In: Duke SO, Rebeiz CA (eds) Porphyric pesticides: chemistry, toxicology, and pharmaceutical applications, vol 559. American Chemical Society, Washington, DC, pp 1–16

Granick S (1961) Magnesium protoporphyrin monoester and protoporphyrin monomethyl ester in chlorophyll biosynthesis. J Biol Chem 236:1168–1172

Hinchigeri SB, Nelson DW, Richards WR (1984) The purification and reaction mechanism of S-adenosyl-L-methionine: magnesium protoporphyrin methyltransferase from hodopseudomonas spheroides. Photosynthetica 18:168–178

Hipkins MF (1986) Introduction to photosynthetic energy transduction. In: Hipkins MF, Baker NR (eds) Photosynthesis energy transduction, a practical approach. IRL Press, Washington, DC, pp 1–7

Kulur S (1996) Study of the cause of greenhouse simulated field discrepancy of the delta-sminolevulinic acid-dependent photodynamic herbicide. In: Natural resources and environmental sciences. University of Illinois, Urbana-Champaign, p 61

Mayasich JM, Mayasich SA, Rebeiz CA (1990) Photodynamic herbicides. Response of corn (Zea mays) soybean (Glycine max) and several weed species to dark- applied photodynamic herbicide modulators. Weed Sci 36:10–15

Perkins HJ, Roberts DWA (1960) Chlorophyll biosynthesis in wheat leaves. Biochem Biophys Acta 45:613–620

Rebeiz CA (1967) Studies on chlorophyll biosynthesis in etiolated excised cotyledons of germinating cucumber at different stages of seedling development. Magon Serie Scientifique 13:1–21

Rebeiz CA (1991) Tetrapyrrole-dependent photodynamic herbicides and the chlorophyll biosynthetic pathway. In: Pell E, Steffen K (eds) Active oxygen/oxidative stress and plant metabolism. American Society Plant Physiologists, Rockville, pp 193–203

Rebeiz CA (2002) Analysis of intermediates and end products of the chlorophyll biosynthetic pathway. In: Smith A, Witty M (eds) Heme chlorophyll and bilins, methods and protocols. Humana Press, Totowa, pp 111–155

Rebeiz CA, Bazzaz MB (1978) Cell-free agriculture: the concept and its initial implementation. In: Charles DS (ed) Biotechnology in energy production and conservation. Wiley, New York, pp 453–471

Rebeiz CA, Belanger FC (1984) Chloroplast biogenesis 46: calculation of net spectral shifts induced by axial ligand coordination in metalated tetrapyrroles. Spectrochim Acta 40A:793–806

Rebeiz CA, Abou Haidar M, Yaghi M et al (1970) Porphyrin biosynthesis in cell-free homogenates from higher plants. Plant Physiol 46:543–549

Rebeiz CA, Mattheis JR, Smith BB et al (1975) Chloroplast biogenesis. Biosynthesis and accumulation of Mg-protoporphyrin IX monoester and longer wavelength metalloporphyrins by greening cotyledons. Arch Biochem Biophys 166:446–465

Rebeiz CA, Montazer-Zouhoor A, Daniell H (1984a) Chloroplast culture X: thylakoid assembly in vitro. Isr J Bot 33:225–235

Rebeiz CA, Montazer-Zouhoor A, Hopen HJ et al (1984b) Photodynamic herbicides: 1. Concept and phenomenology. Enzyme Microb Technol 6:390–401
Rebeiz CA, Montazer-Zouhoor A, Mayasich JM et al (1987) Photodynamic herbicides and chlorophyll biosynthesis modulators. In: Heitz JR, Downum KR (eds) Light activated I pesticides, vol 339, ACS symposium series. American Chemical Society, Washington, DC, pp 295–328
Rebeiz CA, Juvik JA, Rebeiz CC (1988a) Porphyric insecticides 1. Concept and phenomenology. Pestic Biochem Physiol 30:11–27
Rebeiz CA, Montazer-Zouhoor A, Mayasich JM et al (1988b) Photodynamic herbicides. Recent developments and molecular basis of selectivity. Crit Rev Plant Sci 6:385–434
Rebeiz CA, Reddy KN, Nandihalli UB et al (1990) Tetrapyrrole-dependent photodynamic herbicides. Photochem Photobiol 52:1099–1117
Rebeiz CA, Nandihalli UB, Reddy K (1991) Photodynamic herbicides and chlorophyll biosynthesis modulators. In: Baker NR, Percival M (eds) Herbicides. Elsevier, Amsterdam, pp 173–208
Rebeiz CA, Amindari S, Reddy KN et al (1994) Delta-aminolevulinic acid based herbicides and tetrapyrrole biosynthesis modulators. In: Duke SO, Rebeiz CA (eds) Porphyric pesticides: chemistry, toxicology, and pharmaceutical applications, vol 559. American Chemical Society, Washington, DC, pp 48–64
Reddy KN, Rebeiz CA (1994) Modulators of the porphyrin pathway beyond protox. In: Duke SO, Rebeiz CA (eds) Porphyric pesticides: chemistry, toxicology, and pharmaceutical applications, 559th edn. American Chemical Society, Washington, DC, pp 161–190
Sisler EC, Klein W (1963) The effect of age and various chemicals on the lag phase of chlorophyll synthesis in dark grown bean seedlings. Physiol Plant 16:315–322
Smith BB, Rebeiz CA (1979) Chloroplast biogenesis XXIV. Intrachloroplastic localization of the biosynthesis and accumulation of protoporphyrin IX, magnesium protoporphyrin IX, Magnesium-protoporphyrin monoester and longer wavelength metalloporphyrins during greening. Plant Physiol 63:227–231
Virgin HI (1961) On the formation of protochlorophyll in normal green leaves of varying ages. Physiol Plant 14:384–392
Weiss CA (1975) The molecular orbital theory of chlorophyll. N Y Acad Sci 244:204–213
Weiss CA (1978) Electronic absorption spectra of chlorophylls. In: Dolphin D (ed) The porphyrins, vol III. Academic, New York, pp 211–223
Wickliff JL, Aronoff S (1963) Turnover of chlorophyll a in mature soybean leaves. Plant and Cell Physiology, Tokyo, pp 441–449

# Chapter 18
# Porphyric Insecticides

> *A well-cultivated mind is made up of all the minds of preceding ages; it represents only the one single mind educated by all previous ones. (Adapted from Fontenelle)*

## 18.1 Prologue

The discovery of porphyric insecticides (Gut et al. 1993, 1994a, b; Rebeiz et al. 1988a, 1990a, 1995; Rebeiz 1993) was a direct fallout of the discovery and development of photodynamic herbicides (see Chap. 17). Since plant and animal cells share the same tetrapyrrole biosynthetic pathway from ALA to protoporphyrin IX (Proto), it was conjectured that it should be possible to adapt the tetrapyrrole-dependent photodynamic herbicide (TDPH) phenomenon to the photodynamic control of insects.

## 18.2 Porphyric Insecticides

The concept and phenomenology of porphyric insecticides were described in 1988 (Rebeiz et al. 1988a). Since then significant progress has been achieved in expanding the scope of this insecticidal system, in understanding its mode of action and in the development of specific insecticidal applications. The description of these developments constitutes the subject of this chapter.

C.A. Rebeiz, *Chlorophyll Biosynthesis and Technological Applications*, 
DOI 10.1007/978-94-007-7134-5_18, 

### *18.2.1 Principle*

Porphyric insecticides are made of compounds which force insects to accumulate undesirable amounts of metabolic intermediates of the heme metabolic pathway, namely protoporphyrin IX (Proto). In the light, the accumulated Proto photosensitize the formation of singlet oxygen which kills treated insects by oxidation of their cellular membranes. Photodynamic (porphyric) insecticides usually consist of a 5-carbon amino acid, δ-aminolevulinic acid (ALA), the precursor of all tetrapyrroles in plant and animal cells, and one of several chemicals referred to as modulators. ALA and the modulators act in concert. The amino acid serves as a building block of Proto accumulation, while the modulator alters quantitatively and qualitatively the pattern of Proto accumulation.

### *18.2.2 Demonstration of Protoporphyrin IX Accumulation in* T. ni *Treated with ALA and 2,2′-Dipyridyl (Dpy)*

Demonstration of the potential for Proto accumulation in treated insects was initially achieved by spraying Trichoplusia *ni* (*T. ni*) larvae with 40 mM ALA + 30 mM 2,2,-dipyridyl (Dpy) (Rebeiz et al. 1988a). Treated larvae were placed overnight in darkness at 28 °C in order to allow tetrapyrrole accumulation. Extraction of treated, dark-incubated larvae with ammoniacal acetone, followed by spectrofluorometric examination of the larval extract, revealed the accumulation of massive amounts of a fluorescent compound which was not present in control larvae sprayed with solvent only (Fig. 18.1). Following chemical derivatization coupled to spectrofluorometric analysis, the accumulated compound was identified as a tetrapyrrole, specifically Proto (Rebeiz et al. 1988a).

### *18.2.3 Insecticidal Effects of the ALA + Dpy Treatment*

Insecticidal effects of ALA + Dpy-dependent accumulation of Proto was demonstrated by (a) inducing the accumulation of Proto in third instar *T. ni.*, (b) initiating photodynamic death by exposing treated larvae to light and (c) establishing correlations between photodynamic death and Proto accumulation, photodynamic death and length of larval exposure to light, and Proto accumulation and larval growth.

In initial experiments, third instar *T. ni* larvae were sprayed with ALA + Dpy at a pH of 3.5 and placed in darkness overnight to allow Proto accumulation to take place. At the end of dark incubation, some larvae were analyzed for Proto accumulation while others were exposed to 14-h light/10-h dark regime to trigger photodynamic damage. The ALA + Dpy treatment resulted in massive Proto accumulation

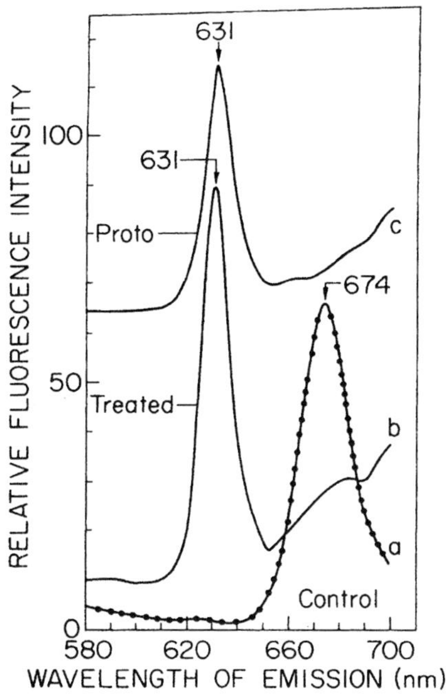


**Fig. 18.1** Room temperature emission spectra in hexane-extracted acetone of *T. ni* larvae treated with ALA + Dpy or with solvent only (control). Third instar larvae were treated with 40 mM ALA + 30 mM Dpy, or solvent only and placed in darkness at 28 °C for 17 h. Control and ALA + modulator-treated larvae were homogenized in ammoniacal acetone. Excitation was at 400 nm. The spectra were recorded at emission and excitation slits of 4 nm. (*a*) Hexane-extracted acetone extract of control larvae; the emission peak at 674 nm is that of pheophorbide a; (*b*) extract of treated larvae; (*c*) authentic Proto in hexane- extracted acetone (Reproduced from Rebeiz et al. 1988a)

and significant larval mortality after three photoperiods (Table 18.1). A high degree of correlation was observed between Proto accumulation in darkness and larval death in the light (Table 18.1). A few hours after exposure to light, the larvae became sluggish and flaccid due to loss of body fluids. Death was accompanied by extensive desiccation (Fig. 18.2).

In follow-up experiments, third instar *T. ni* were sprayed with ALA + Dpy, and the treated larvae were placed in darkness overnight to allow for Proto accumulation. While some larvae were exposed to light to trigger photodynamic death, others were left in darkness for an equal period of time. It was observed that some larval death occurred during the overnight dark incubation before exposure to light (Table 18.2, A2, B2).

At that time the cause of this dark-death phenomenon was not understood. Later on, a hypothesis was proposed explaining the molecular basis of this dark-death. The growth of larvae that survived the initial 17-h dark-incubation was not inhibited by further dark incubation (Table 18.2, A1, A2 and B1, B2). The bulk of larval death occurred during illumination of treated larvae (Table 18.2, A2–A4 and B2–B4). During illumination the accumulated Proto disappeared probably as a

**Table 18.1** Effect of ALA + Dpy spray on the biosynthesis and accumulation of Proto on the extent of larval death in *T. ni*

| Experiment | Treatment | Proto content after 17 h of dark incubation (nmol per 100 mg protein) | Larval death (%) |
|---|---|---|---|
| A | Control | 0.1 | 3 |
| | Treated | 85.0 | 80 |
| | Δ Change | 84.9 | 77 |
| B | Control | 0.0 | 1 |
| | Treated | 96.7 | 75 |
| | Δ Change | 96.7 | 74 |
| C | Control | 0.1 | 13 |
| | Treated | 12.8 | 47 |
| | Δ Change | 12.7 | 34 |
| D | Control | 0.1 | 6 |
| | Treated | 89.7 | 89 |
| | Δ Change | 89.6 | 83 |
| Correlation coefficient | | 0.945 | |
| Level of significance | | 0.1 % | |

Larvae were in the third instar. Larval death refers to percent death at the beginning of the fourth photoperiod, i.e. after 3 days in the growth chamber. Δ Change refers to the difference in Proto content between the ALA + Dpy-treated larvae and the control larvae which were sprayed with solvent only, after 17-h post-spray dark incubation period (Adapted from Rebeiz et al. 1988a)

**Fig. 18.2** Larval death in control (*C*) and treated (40 mM ALA + 30 mM Dpy) third instar *T. ni* larvae (*D*) 24 h after spraying. *C* control, *D* treated (Adapted from Rebeiz et al. 1988a)

result of photo destruction, a well-known tetrapyrrole phenomenon. Significant correlations were observed between larval death and the extent of post-spray exposure to light and between the extent of post-spray exposure to light and inhibition in body weight gain per surviving larvae (Table 18.2).

### 18.2.4 *Synergistic Effects of ALA and Dpy on Proto Accumulation and Larval Death in* T. ni

To determine the relative effects of each component of the ALA + Dpy spray on the insect, third instar *T. ni* larvae were sprayed with ALA alone, Dpy alone and

**Table 18.2** Effect of Proto accumulation on *T. ni* larval death and body weight change in darkness and in the light

| Light-dark regime after illumination treatment | | | Proto content after the 17-h post-spray treatment and 0–6 h illumination | Post-spray larval death after treatment | | Post-dark incubation body weight |
|---|---|---|---|---|---|---|
| | Entry | Incubation | (protein) | 17 h dark (%) | Total 65-h (%) | per live larva (mg) |
| A | 1 | Control, two 14-h light-10-h dark photoperiods | 0.2[b] | 1 | 12 | 45.1 |
| | 2 | Treated[a], 0-h light + 48-h dark | 62.0[b] | 28 | 31 | 44.5 |
| | 3 | Treated, 3-h light + 45-h dark | 0.4[c] | 21 | 60 | 43.4 |
| | 4 | Treated, 6-h light + 42-h dark | 1.1[d] | 22 | 73 | 26.6 |
| | 5 | Treated, after two 14-h light-10-h dark photoperiod | 62.0[b] | 19 | 80 | 16.6 |
| B | 1 | Control, two 14-h light-10-h dark photoperiods | 0.2[b] | 0 | 22 | 44.1 |
| | 2 | Treated[a], 0-h light + 48-h dark | 40.1[b] | 19 | 48 | 45.7 |
| | 3 | Treated, 3-h light + 45-h dark | 1.2[c] | 20 | 65 | 32.1 |
| | 4 | Treated, 6-h light + 42-h dark | 0.6[c] | 21 | 77 | 20.5 |
| | 5 | Treated, after two 14-h light-10-h dark photoperiod | 40.1[b] | 22 | 95 | 3.6 |
| Correlation coefficient | | | 0.808[e] | | −0.815[f] | |
| Levels of significance | | | 5 % | | 1 % | |

Adapted from Rebeiz et al. (1988a)
[a]Treatment consisted in spraying larvae with 40 mM ALA + 30 mM Dpy
[b]Proto content after the post-spray 17-h dark incubation
[c]Proto content after the post-spray 17-h dark incubation and 3 h of illumination
[d]Proto content after the post-spray 17-h dark incubation and 6 h of illumination
[e]Correlation between larval death and the extent of post-spray exposure to light. In computing the correlation coefficient, the larval death of the control was subtracted from that of the treated
[f]Correlation between the extent of post-spray exposure to light and average body weight change per surviving larva. In calculating the correlation coefficient, the average body weight of the control larvae was subtracted from that of the treated larvae

ALA + Dpy. After dark incubation the larvae were analyzed for Proto accumulation, while duplicate sets of larvae were exposed to light. Treatment with ALA alone or Dpy alone, resulted in Proto accumulation and concomitant photodynamic death (Table 18.3, A). However, treatment with the ALA + Dpy mixture exhibited definite

**Table 18.3** Synergistic effects of ALA and Dpy on Proto accumulation and larval death in *T. ni*[a]

| Experiment | Entry | Treatment | Proto content (nmol/100 mg protein) | Larval death after 3 days in the greenhouse (%) |
|---|---|---|---|---|
| A | 1 | Control | 0 | 6 |
| | 2 | 40 mM ALA | 2 | 26 |
| | 3 | 30 mM Dpy | 15 | 41 |
| | 4 | 40 mM ALA + 30 mM Dpy | 80 | 90 |
| B | 1 | Control | 0 | 2 |
| | 2 | 30 mM Dpy | 11 | 61 |
| | 3 | 30 mM Dpy + 10 mM ALA | 75 | 86 |
| | 4 | 30 mM Dpy + 20 mM ALA | 89 | 76 |
| | 5 | 30 mM Dpy + 40 mM ALA | 73 | 92 |
| C | 1 | Control | 0 | 7 |
| | 2 | 15 mM Dpy | 1 | 22 |
| | 3 | 15 mM Dpy + 10 mM ALA | 8 | 42 |
| | 4 | 15 mM Dpy + 20 mM ALA | 34 | 40 |
| | 5 | 15 mM Dpy + 40 mM ALA | 27 | 43 |
| D | 1 | Control | 0 | 5 |
| | 2 | 40 mM ALA | 1 | 7 |
| | 3 | 40 mM ALA + 5 mM Dpy | 3 | 4 |
| | 4 | 40 mM ALA + 10 mM Dpy | 7[c] | 18 |
| | 5 | 40 mM ALA + 20 mM Dpy | 12 | 34 |
| | 6 | 40 mM ALA + 40 mM Dpy | 15 | 71 |
| Correlation between Proto content and larval death | | | 0.857 | |
| Level of significance | | | 0.1 % | |

Adapted from Duggan and Gassman (1974)
[a]Larvae were in the third instar

synergistic effects, with Proto accumulation (80.4 nmol) and larval death (90 %) far exceeding the sum of Proto accumulation (2.5 + 15.5 = 18 nmol) and larval death (26 + 41 = 67 %) caused by separate ALA and Dpy treatments (Table 18.3, A). The observed levels of Proto accumulation indicated that Dpy was both an inducer and an enhancer of Proto accumulation. In the absence of added ALA, Dpy caused the accumulation of Proto over and beyond the level in control larvae, sprayed with solvent only (Table 18.3, A3, B2). Under these circumstances, Dpy acted as an inducer of Proto formation (Rebeiz et al. 1988a). In the presence of ALA, Dpy enhanced the conversion of ALA to Proto. This was evidenced by the dramatic increase in Proto

**Table 18.4** Susceptibility of various instars of *T. ni* to treatment with ALA + Dpy

| Experiment | Entry | Instar | | Larval death over and beyond the controls after 3 days in the greenhouse (%) |
|---|---|---|---|---|
| A | 1 | 1st | | 76 |
| | 2 | 2nd | | 42 |
| | 3 | 3rd | | 43 |
| B | 1 | 1st | | 64 |
| | 2 | 2nd | | 63 |
| | 3 | 3rd | | 27 |
| C | 1 | 1st | | 75 |
| | 2 | 2nd | | 28 |
| | 3 | 3rd | | 11 |
| D | 1 | 1st | | 70 |
| | 2 | 2nd | | 23 |
| | 3 | 3rd | | 11 |
| Correlation coefficient between the rank of the instar and larval death | | | −0.897 | |
| Level of significance | | | 0.1 % | |

Adapted from Rebeiz et al. (1988a)

accumulation and larval death in ALA + Dpy treatments, as a result of the greatly improved conversion of exogenous ALA to Proto (Table 18.3). There also appeared to be a positive relationship between Dpy concentration and enhanced Proto formation (Table 18.3, D).

### 18.2.5 *Effect of Age on* T. ni *Herbicidal Susceptibility*

The possible effect of insect age on susceptibility toward porphyric insecticides was investigated by comparing the mortality of first, second, third and fourth *T. ni* instars, sprayed with ALA (40 mM) + Dpy (30 mM) at pH 3.5. After 17-h dark incubation, the treated larvae were exposed to light. As shown in Table 18.4, the susceptibility of *T. ni* to the ALA + Dpy treatment was inversely related to the rank of the instar, with young, first instars being most susceptible and older fourth instars being least susceptible (Rebeiz et al. 1988a).

The relative susceptibility of the stage of development within a particular instar was also investigated. The mid, early, and late stages of every instar were assigned a rank of 1, 2, or 3, one denoting the least susceptible and three denoting the most susceptible stages. To test the relationship between larval death and the early, mid and late stages of a particular instar, irrespective of stage-dependent susceptibility, percent death values within an experiment (one for each instar) were normalized to a value of 100 %. The latter represented the percent death for the late stage within each instar. As shown in Table 18.5, the relationship between age within an instar

**Table 18.5** Susceptibility of *T. ni* larvae of various ages within an instar to treatment with ALA + Dpy

| | | | | Percentage larval death over and beyond the controls after 3 days in the greenhouse | |
|---|---|---|---|---|---|
| Experiment | Entry | Stage of the instar | Susceptibility ranking | Before normalization (%) | After normalization (%) |
| A | 1 | Mid third | 1 | 47 | 52 |
| | 2 | Early third | 2 | 60 | 66 |
| | 3 | Late third | 3 | 91 | 100 |
| B | 1 | Mid third | 1 | 47 | 92 |
| | 2 | Early third | 2 | 39 | 76 |
| | 3 | Late third | 3 | 51 | 100 |
| C | 1 | Mid third | 1 | 21 | 57 |
| | 2 | Early third | 2 | 18 | 49 |
| | 3 | Late third | 3 | 27 | 100 |
| D | 1 | Mid third | 1 | 26 | 54 |
| | 2 | Early third | 2 | 35 | 73 |
| | 3 | Late third | 3 | 48 | 100 |
| Correlation between the stage of development within an instar and larval death | | | | 0.739 | |
| Level of significance | | | | 1 % | |

Adapted from Rebeiz et al. (1988a)

and mortality was highly significant, with the late stage of the instar being more susceptible to the ALA + Dpy treatment than the early and mid stages (Rebeiz et al. 1988a). This period of maximum susceptibility corresponded to the period when the larvae were quiescent and the new cuticle for the next instar was being actively synthesized beneath the old cuticle.

### 18.2.6 *Effectiveness of the ALA + Dpy Treatment in the Absence of a Post-spray Dark Incubation Period*

To determine if a post-spray dark incubation period, was required for expression of insecticidal activity, the mortality of second instar *T. ni* treated in the dark or light was compared. In the dark treatment, treated larvae (ALA + Dpy) were subjected to a 17-h dark incubation prior to exposure to light (dark sprays). Light treatment consisted of spraying larvae at the beginning of the light phase of a 14-h light/10-h dark photoperiod prior to exposure to light (light sprays). As shown in Table 18.6, the light sprays were as effective as the dark sprays in causing larval death. This indicated that, although in the light Proto is destroyed as rapidly as it is formed, the steady state formation of Proto in the light was enough to cause extensive photodynamic damage.

**Table 18.6** Comparison of the effectiveness of ALA + Dpy light sprays with dark sprays in *T. ni*

| Experiment | Treatment[a] | Larval death after 3 days in the greenhouse (%) |
|---|---|---|
| A | DSP[b] control | 21 |
| | DSP treated | 90 |
| | DSP Δ change[c] | 69 |
| B | LSP[d] control | 25 |
| | LSP treated | 94 |
| | LSP Δ change | 69 |
| C | DSP[b] control | 14 |
| | DSP treated | 93 |
| | DSP Δ change | 79 |
| D | LSP control | 5 |
| | LSP treated | 83 |
| | LSP Δ change | 78 |

Adapted from Rebeiz et al. 1988a
[a]Treatment was either by spraying with 20 mM ALA + 15 mM Dpy (A and B) or by spraying with 40 mM ALA + 30 mM Dpy (C and D) at a pH of 3.5 and a rate of spray equivalent to 40 gal per acre. First instar larvae were used
[b]DSP = dark spray
[c]Δ change = control – treated
[d]LSP = light spray

## 18.3 Use of 1,10-Orthophenanthroline (Oph) as a Porphyric Insecticide Modulator

During preliminary evaluation of the mode of action of Putative photodynamic herbicide modulators (Rebeiz et al. 1988b) it was observed that all modulators that exhibited significant herbicidal activity fell in two major groups: (a) a group that enhanced the conversion of ALA to Proto, hence its designation as a "primitive modulator group", since Proto is an early precursor of chlorophyll (Chl) in plants, and (b) a group that enhanced the conversion of ALA to Pchlide, one of the terminal precursors of Chl, hence its designation as an "advanced modulator group". Since insects and plants share the same porphyrin-heme biosynthetic pathway between ALA and Proto it was conjectured that primitive photodynamic herbicide modulators may also exhibit significant porphyric insecticidal properties.

It was also observed elsewhere that 1,10-orthophenanthroline (Oph) is an especially potent primitive photodynamic herbicide modulator, which induced the formation of massive amounts of Proto and Mg-Proto in a variety of weed species (Rebeiz et al. 1991). It was therefore a good candidate for additional investigation of its potential porphyric insecticidal properties.

### *18.3.1 Porphyric Insecticidal Properties of 1,10-Phenanthroline (Oph)*

In initial experiments, third instar *T. ni* were sprayed with 40 mM ALA with and without 30 mM Oph. After 17 h of dark-incubation some of the treated larvae were analyzed for tetrapyrrole accumulation, while others were exposed to light. Oph exhibited very potent porphyric insecticidal properties (Table 18.7). In the absence of added ALA it induced the massive accumulation of Proto. In other words in the absence of added ALA it behaved as an inducer of Proto accumulation. In the presence of added ALA the inducing properties of Oph were obscured by the massive enhancement of ALA conversion to Proto. To put it differently, Oph behaved as an inducer-enhancer of Proto accumulation in *T. ni* (Rebeiz et al. 1990a) The correlation between photodynamic death and Proto accumulation was highly significant (Table 18.7).

### *18.3.2 Zn-Proto Accumulation in* T. ni *Larvae Treated with ALA and Oph*

It has been our experience that dicarboxylic and monocarboxylic tetrapyrroles of plant, insect, and animal tissues are found in the hexane-extracted acetone fraction of extracted tissues (Rebeiz 2002). Thus in the hexane-extracted acetone fraction of the ALA + Oph-treated insects, in addition to fluorescence emission originating in the Proto pool, another fluorescence emission band of smaller amplitude was observed. It exhibited an emission maximum at 590 nm at room temperature and at 587 nm at 77 K in ether. Since the band exhibited fluorescence properties at room temperature and at 77 K that were identical to those of Zn-Proto (Fig. 18.3), it was assigned to the biosynthesis and accumulation of Zn-Proto in the treated larvae. The amount of Zn-Proto formed following treatment with Oph is shown in (Table 18.7). It too correlated positively with insect death. After the same period of dark-incubation, no significant amounts of Proto or Zn-Proto were detected in control insects.

### *18.3.3 Proposal of a Dark-Death Hypothesis*

The discovery of Zn-Proto accumulation suggested an explanation for insect mortality observed during dark incubation. In addition to damage via singlet oxygen, it is conceivable, that ALA + modulator-dependent larval death may also be caused by the induction of a premature release of $O_2^-$ and ·OH radicals from the active site of a damaged cytochrome *c* oxidase. Indeed cytochrome *c* oxidase is the major consumer of $O_2$ in eukaryotic cells. Because of spin restrictions, $O_2$ cannot accept four electrons at once. During cytochrome *c*-mediated electron transport it therefore accepts electrons, one at a time. In the process, $O_2$ passes through a series of partially-reduced intermediates including the highly reactive superoxide radical ($O_2^-$) and the hydroxy radical (˙OH) (Halliwell 1984). In mitochondria, these highly reactive oxygen radicals are kept tightly bound to the active site of cytochrome

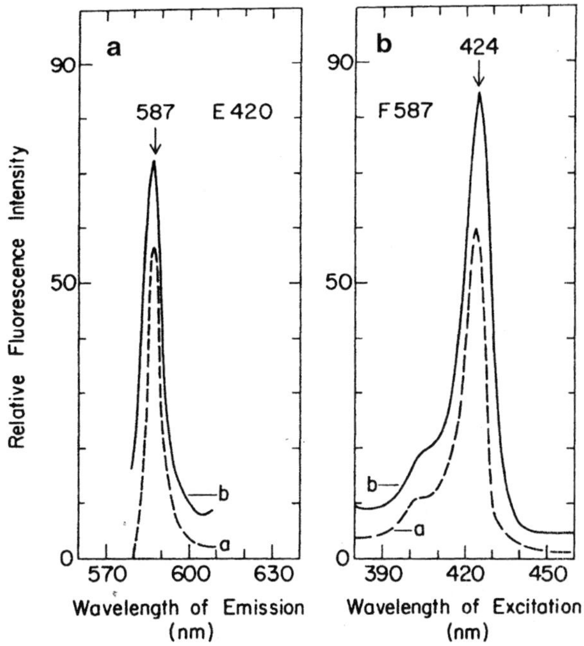


**Fig. 18.3** Fluorescence spectra of Zn-Proto in ether at 77 K. (**a**) Fluorescence emission and (**b**) fluorescence excitation spectra in ether at 77 K of (*a*) authentic Zn-Proto and (*b*) of the ether extract of third instar *T. ni* larvae. The larvae were sprayed with 40 mM ALA + 30 mM Oph and were incubated in darkness for 17 h prior to extraction. The emission and excitation spectra were recorded at the emission (*F*) and excitation (*E*) wavelengths indicated on the figure, at 4 nm emission and excitation slit widths. *Arrows* point to wavelengths of interest (Adapted from Rebeiz et al. 1990a)

*c* oxidase, and under normal conditions are only released when they are fully reduced to $H_2O$ (Halliwell 1984). It is conceivable therefore that premature release of these radicals in the intracellular environment may trigger peroxidation of the membrane lipoprotein, causing the same type of damage as singlet oxygen-mediated photodynamic damage. This explanation is compatible with the observed accumulation of Zn-Proto in treated insects. Indeed, Zn-Proto is not a natural metabolic intermediate of the porphyrin-heme pathway. Its occurrence in living cells and tissues usually denotes a poisoned porphyrin-heme metabolism (Lamola and Yamane 1974). Most ferrochelatases (the enzymes that insert ferrous iron into Proto to form heme) can insert Zn instead of iron into Proto to yield Zn-Proto, particularly under unfavorable reaction conditions (Lamola and Yamane 1974). Thus it is possible that the accumulation of Zn-Proto as a result of treatments containing Dpy or Oph may be caused by damage to the ferrochelatase system causing the enzyme to insert Zn instead of ferrous Fe into some of the Proto. If it ensues that some of the cytochrome *c* prosthetic groups consist of Zn-Proto instead of heme in treated insects, then those cytochrome *c* oxidase molecules containing Zn-Proto instead of heme may no longer be able to prevent the premature release of oxygen superoxide and hydroxy free radicals, by holding them tight to the reaction centers until they are fully reduced. The intracellular release of these destructive free radicals in the biological membrane environment could then contribute to the free radical damage that results in insect death.

**Table 18.7** Effect of ingested ALA, Oph AND Dpy on the biosynthesis and accumulation of Proto and Zn-Proto and on the extent of larval death in *T. ni*[a]

| Experiment | Treatment | Proto content (nmol/100 mg protein) | Zn-proto content (nmol/100 mg protein) | Larval death[b] (%) |
|---|---|---|---|---|
| A | Control | 0.0 | 0.0 | 14.5 |
| | 16 mM ALA | 1.5 | 0.0 | 5.2 |
| | 12 mM Oph | 88.1 | 1.3 | 40.4 |
| | 16 mM ALA + 12 mM Oph | 224.6 | 1.5 | 94.4 |
| B | Control | 0.0 | 0.0 | 4.5 |
| | 16 mM ALA | 2.6 | 0.0 | 4.7 |
| | 12 mM Oph | 14.3 | 1.9 | 51.4 |
| | 16 mM ALA + 12 mM Oph | 160.2 | 2.0 | 95.4 |
| C | Control | 0.4 | 0.0 | 18.2 |
| | 16 mM ALA | 3.1 | 0.0 | 49.3 |
| | 12 mM 2,2′-Dpy | 20.3 | 8.8 | 11.4 |
| | 16 mM ALA + 12 mM 2,2′-Dpy | 55.2 | 11.4 | 100.0 |
| Correlation between pigment content and larval death | | 0.703 | 0.721 | |
| Level of significance | | 2.1 % | 0.9 % | |

Adapted from Rebeiz et al. (1990a)

[a]Third instar larvae were placed on control and baited diets and held for 17 h in darkness. The larvae were then sampled for tetrapyrrole content, and placed in the light for observation of photodynamic injury

[b]Refers to larval death at the beginning of the fourth photoperiod, *i.e.* after 3 days in the growth chamber

## 18.3.4 Insecticidal Effectiveness of Ingested ALA and Oph or Dpy

Since control of insects by ingestion is as viable an option as control by spraying, and offers certain advantages under household conditions, studies were conducted to determine whether combinations of ALA and porphyric insecticide modulators would be effective if ingested with the food. Initially the effect of ALA (16 mM final concentration) and Oph (12 mM final concentration) were determined by incorporating them into the diet of *T. ni* larvae. Upon exposure to light, following 17 h of dark incubation, larvae underwent violent convulsions and vomiting and died within 20–40 s. Tetrapyrrole analysis of the treated larvae immediately after dark incubation revealed significant amounts of Proto and Zn-Proto accumulation. Correlation between tetrapyrrole accumulation and larval death was significant (Table 18.8, A). Similar results were obtained when ALA and Dpy were administered to the larvae with the diet (Table 18.8, C). The above results indicated that in addition to contact via spraying, porphyric insecticides had the potential to be very potent when ingested.

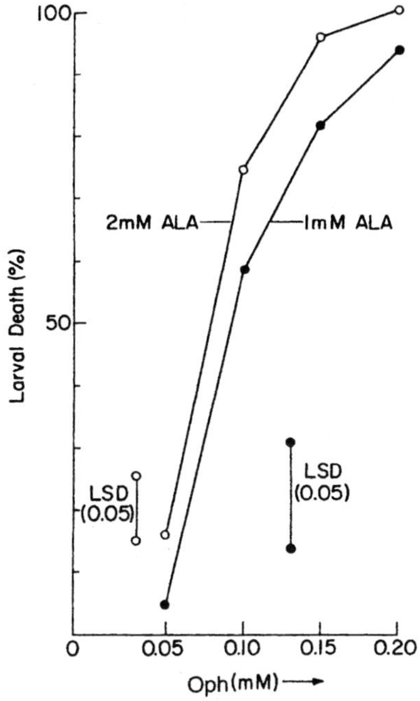


**Fig. 18.4** Response of *T. ni* larvae to the ingestion of two concentrations of ALA and variable concentrations of Oph. Fourth instar larvae were placed for 17 h in darkness on diet containing 1 or 2 mM ALA and O-Ph concentrations ranging from 0.05 to 0.2 mM. Every treatment was replicated three times. At the end of dark-incubation a subsample of larvae were analyzed for tetrapyrrole content while those remaining were placed in the light for observation of photodynamic damage. The data was first analyzed as a factorial design with two levels of ALA and four levels of Oph. Since the ALA × Oph interaction was not significant, the data was reanalyzed separately for every level of ALA, as a randomized complete block in order to generate LSD values for the two graphs (Adapted from Gut et al. 1993)

### *18.3.5 Concentrations of Dietary ALA and 1,10-Phenanthroline Needed to Achieve 50 and 100 % Larval Kill in* T. ni

Fifty percent larval kill in *T. ni* was achieved with diet containing as little as 1 mM ALA and 0.1 mM Oph (Fig. 18.4). When the concentration of ALA and Oph were raised to 2 and 0.2 mM respectively, 100 % mortality was achieved (Fig. 18.4). Typical tetrapyrrole accumulation profiles that accompanied dark-treatments at such low concentrations of ALA and Oph mixed with the diet are reported in Table 18.9. At these low concentrations of ALA and modulator, larval death occurred after an extremely small amount of food consumption. The effective concentrations of ALA and modulator given in Table 18.8, are orders of magnitudes lower than those previously reported (Rebeiz et al. 1988a).

**Table 18.8** Tetrapyrrole accumulation accompanying the ingestion of diet supplemented with 2 mM ALA and varying Oph concentrations[a]

| Treatment | Proto content (nmol/100 mg protein) | Zn-proto content (nmol/100 mg protein) | Larval[b] death (%) |
|---|---|---|---|
| Control | 0.6 | 0.0 | 16.0 |
| 2 mM ALA | 3.6 | 0.0 | 54.0 |
| 2 mM ALA + 0.1 mM Oph | 10.4 | 0.0 | 84.0 |
| 2 mM ALA + 0.4 mM Oph | 46.4 | 2.7 | 100.0 |
| 2 mM ALA + 1.6 mM Oph | 67.0 | 4.1 | 100.0 |

Adapted from Rebeiz et al. (1990a)

[a]Fourth instar larvae were placed on control and baited diets. After 17 h in darkness the larvae were sampled for tetrapyrrole analysis. The remaining larvae were placed in the light for observation of photodynamic injury

[b]Refers to larval death at the beginning of the fourth photoperiod, *i.e.* after 3 days in the growth chamber

## *18.3.6 Phenomenology of Baited Food Consumption and Photodynamic Damage in* T. ni

To further the development of ALA-dependent insecticidal field strategies, various studies were performed to determine the relationship between food intake, Proto accumulation and photodynamic death in fourth instar *T. ni* larvae. The following was observed: (a) With diet baited with 1 mM ALA and 0.5 mM Oph, at least 4 h of feeding on the treated diet was required before exposure to light in order to achieve photodynamic mortality rates of 90 % or better, (b) it did not matter whether food consumption took place in the light or in darkness, (c) ingested baited food was detoxified if the larvae were taken off the baited food and placed on untreated diet in darkness for 4-h or longer, and (d) Proto accumulation in the body of the larvae increased exponentially, as the concentration of ALA and Oph in the diet increased to 1 and 0.75 mM respectively. Beyond these concentrations, the increase in Proto accumulation slowed down considerably.

## *18.3.7 Inhibition by Metal Cations of the Insecticidal Properties of Oph*

It has been reported that the tetrapyrrole-inducing properties of bidentate metal chelators, such as Dpy in plants, were not expressed in the presence of metallic cations such as $Fe^{++}$ and $Zn^{++}$ (Duggan and Gassman 1974). This phenomenon was a convenient tool to further determine whether the insecticidal properties of Oph, another bidentate chelator, were obligatorily linked to Proto accumulation in insects. To this effect third instar *T. ni* larvae were placed overnight on treated food which in addition to ALA (1.0 mM) and Oph (0.5 mM) contained various concentrations of $FeCl_2$, $FeCl_3$ and $ZnCl_2$. After 17 h incubation, some larvae were monitored for Proto accumulation while others were exposed to light to trigger photodynamic damage.

**Table 18.9** Effect of various concentrations of added $FeCl_3$ on Proto accumulation and photodynamic damage in third instar *T. ni* larvae

| Treatment | Proto content (nmol/100 mg protein) | Larval death (%) |
|---|---|---|
| 1 mM ALA | 1.1 | 0.0 |
| 1 mM ALA 0.5 mM Oph | 46.7 | 74.0 |
| 1 mM ALA + 0.5 mM Oph + 0.5 mM $FeCl_3$ | 11.2 | 14.7 |
| 1 mM ALA + 0.5 mM Oph + 1.0 mM $FeCl_3$ | 2.1 | 6.7 |
| 1 mM ALA + 0.5 mM Oph + 2.0 mM $FeCl_3$ | 0.8 | 1.3 |

Adapted from Rebeiz et al. (1990a)

All three cations were effective in blocking porphyric insecticidal damage in the following order $Fe^{+++} > Fe^{++} > Zn^{++}$ (Gut et al. 1993). Table 18.9 summarizes the effects of various concentrations of $FeCl_3$ on Proto accumulation and photodynamic damage in third instar *T. ni*. $FeCl_3$ strongly inhibited Proto accumulation and photodynamic damage. The inhibitory action of this metallic cation is further proof that insecticidal photodynamic damage is a Proto-dependent phenomenon.

## 18.4 Tissue Cellular and Subcellular Sites of Tetrapyrrole Accumulation in Various Insect Tissues

For a more thorough understanding of the mode of action of porphyric insecticides, the phenomenology of tissue, cellular and subcellular sites of tetrapyrrole accumulation in representative insect species was investigated.

### *18.4.1 Site of Tetrapyrrole Accumulation in Sprayed* T. ni *Larvae*

To determine the site of tetrapyrrole accumulation in *T. ni* larvae sprayed with ALA (40 mM) + Dpy (30 mM), the integument, hemolymph and gut of sprayed early fifth instar larvae were separated and analyzed for pigment content. On a unit protein basis, about 59 % of the accumulated Proto was observed in the hemolymph, 35 % in the gut and 6 % in the integument (Lee and Rebeiz 1995).

### *18.4.2 Tissue and Organ Response to Porphyric Insecticides in Several Insect Species*

Further understanding of the response of insect organs and tissues to porphyric insecticide treatment was obtained by investigating the response of isolated organs

**Table 18.10** Response of isolated organs and tissues to incubation with ALA and Dpy or Oph[a]

| | | Proto content after 5 h incubation in darkness the media listed below (nmol/per 100 mg protein) | | | | | |
|---|---|---|---|---|---|---|---|
| Insect Oph | Tissue | Buffer | ALA | Dpy | ALA + Dpy | Oph | ALA+ |
| *T. ni* | Midgut | 0.6 | 0.6 | 4.4 | 3.0 | 8.6 | 6.5 |
| | Integument + fat body | 0.2 | 0.3 | 1.7 | 1.5 | –[b] | – |
| | Integument | 0.2 | 0.1 | – | – | – | – |
| | Fat body | 0.6 | 0.6 | 1.9 | 5.1 | 4.8 | 2.1 |
| *H. zea* | Midgut | 1.8 | 0.7 | 15.9 | 20.5 | 23.2 | 29.2 |
| | Integument + fat body | 0.5 | 0.7 | 1.9 | 2.1 | – | – |
| | Integument | 0.2 | 0.2 | 0.6 | 0.6 | 0.7 | 0.8 |
| | Fat body | 0.1 | 0.8 | 10.3 | 11.2 | 8.6 | 8.3 |
| *B. germanica* | Male abdomen + gut | 0.7 | 0.8 | 6.4 | 6.2 | – | – |
| | Male gut | 3.0 | 2.0 | – | – | 4.8 | 2.3 |
| | Female gut | 2.7 | 4.8 | – | – | 6.1 | 6.9 |
| | Male abdomen – gut | 0.2 | 0.3 | – | – | 2.9 | 4.5 |
| | Female abdomen – gut | 0.4 | 0.3 | – | – | 1.5 | 1.3 |
| *A. grandis* | Abdomen + gut | 0.5 | 0.8 | 1.9 | 1.7 | – | – |

Adapted from Lee and Rebeiz (1995)

[a]Incubation was in 5 mM ALA, 1.5 mM Dpy orin ALA + Dpy or Oph at pH 5.5 for 5 h in darkness

[b]Not determined

and tissues to incubation with ALA + Dpy or ALA + Oph. In these experiments, the following insects were used: Adult *Blattella germanica* (German cockroach), Adult *Anthonomus grandis* (cotton boll weevil), fifth instar larvae of *Heliothus zea* (corn earworm) and fifth instar larvae of *T. ni* (cabbage looper).

One week-old cockroaches, were starved overnight, and anesthesized with $CO_2$ prior to dissection. Abdomens of four cockroaches were removed with small surgical scissors, rinsed in cold buffer (0.1 M potassium phosphate, pH 7.0), cut into small pieces, and transferred to a small plastic petri dish (5 cm in diameter), containing 3 ml of incubation buffer as well as ALA and a modulator at a specific PH. Abdomens of 3 day old, adult cotton boll weevils were excised and processed in a similar manner. For larvae of *H. zea* an *T. ni*, midguts were dissected and placed in ice-chilled phosphate buffer pH 7.0, cleared of tracheal branches and Malpighian tubules under a stereoscopic microscope, slit open and cleared of residual food remains, and rinsed in fresh phosphate buffer. Fat bodies were also collected from dissected larvae with a small spatula, washed twice in phosphate buffer and stored in cold fresh buffer. Integuments were removed from the mid-section of the body, cut into small pieces and placed in fresh cold phosphate buffer along with the fat bodies and tracheal branches. Organs and tissues prepared as just described were placed in incubation buffer containing ALA, a modulator or ALA + modulator (Table 18.10), and were incubated in darkness for 5 h. After incubation the organ and tissue pieces were homogenized in acetone: 0.1N $NH_4OH$ (9:1 v/v) and after centrifugation the acetone extract was used for tetrapyrrole determination by spectrofluorometry (Rebeiz 2002). Concomitant photodynamic damage was assessed by monitoring the decrease in oxygen consumption of the incubated tissue in the light. Oxygen consumption was determined polarographically, using a Clark oxygen electrode.

The results of isolated organ and tissue investigations for each of the aforementioned four insect species are summarized Table 18.10. In *T. ni*, and H. zea, significant Proto accumulation was observed in the midgut, and fat bodies. Proto accumulation occurred when tissues were incubated with Dpy, ALA + Dpy, Oph, and ALA + Oph (Table 18.10). No response to treatment with ALA alone was observed. In cockroaches more of the Proto appeared to accumulate in the male and female guts than in their abdomen. As in *T. ni* and H. zea, the response was elicited by each of the treatments that included Dpy or Oph. Cotton boll weevil abdomens appeared to be less responsive than the abdomens of the other three species.

To determine whether Proto accumulation resulted in photodynamic damage in incubated tissues, *T. ni* midguts were incubated in darkness either in buffer, with ALA, or with Oph + ALA. Oxygen consumption of the tissue was then monitored before and after exposure to 2 h of illumination. It was assumed that decrease in $O_2$ consumption indicated photodynamic damage followed by cell death. A 30 % decrease in $O_2$ consumption was observed in mid guts treated with Oph or with ALA + Oph after 2 h in the light (Lee and Rebeiz 1995).

### *18.4.3 Subcellular Localization of Proto Accumulation in* T. ni

The decrease in oxygen consumption observed in isolated *T. ni* midguts (Table 18.10), suggested that toxicity of porphyric insecticides may result, among other things, from photodynamic damage to mitochondria. This issue was investigated by Lee and Rebeiz as described below (Lee and Rebeiz 1995).

Fifth-instar *T. ni* larvae were placed on diets containing ALA (4 mM) and Oph (3 mM) in darkness for 17 h. After dark-incubation, the site of Proto accumulation in various subcellular components of the larvae was determined. Most of the Proto was found in the mitochondrial (37 %) and microsomal (35 %) fractions, while the balance (28 %) was found in the cytosol. In order to ascertain that the mitochondrial Proto was not due to contamination by microsomal and cytosolic Proto, The Proto content of mitochondria purified on Percoll gradients was also determined. Percoll-purified mitochondria were highly active as evidenced by their succinate cytochrome *c* reductase activity, and contained 534 nmol Proto per 100 mg mitochondrial protein. These results suggested that Proto formation may take place in the mitochondria and microtomes both of which need Proto for heme formation, while the presence of Proto in the cytosol may be due to leakage from the mitochondrial and microsomal compartments.

### 18.4.4 *Photodynamic Effects of Proto Accumulation on Mitochondrial Function in* T. ni

To determine the possible photodynamic effects of mitochondrial Proto accumulation upon mitochondrial function, mitochondria were isolated from fifth-instar *T. ni* larvae which were dark-treated for 17 h with ALA (4 mM) and Oph (3 mM). The isolated mitochondrial suspension was exposed to 900 W/m$^2$ of white fluorescent light for 30 min at 25° C before monitoring the activity of various mitochondrial marker enzymes, namely: succinate oxidase, NADH dehydrogenase and NADH-cytochrome *c* reductase. All three-enzyme activities decreased significantly in a time-dependent manner in comparison to dark mitochondrial controls. These results strongly indicated that Proto accumulation in mitochondria triggers mitochondrial damage in the light and may contribute significantly to photodynamic damage in treated insects (Lee and Rebeiz 1995).

## 18.5 Screening of Other Porphyric Insecticide Modulators and Their Effects on Four Different Insect Species

Earlier photodynamic herbicide structure-function studies described elsewhere led to the assembly of two databases of commercially available compounds with potential photodynamic herbicidal properties (Rebeiz et al. 1990b, 1991). These databases consisted of a set of 6-membered N-heterocyclic compounds (Rebeiz et al. 1991) and a set of 5-membered N-heterocyclics (Rebeiz et al. 1990b). A substructure computer search of these databases identified 322 putative photodynamic herbicide modulators (see Chap. 17). Extensive testing of these modulators on a variety of plant species led to the identification of about 150 modulators with excellent photodynamic herbicidal properties (Rebeiz et al. 1990b). Encouraged by these results, a screening effort was undertaken to determine whether these 150 modulators exhibited porphyric insecticidal properties. Screening by food ingestion was performed on the German cockroach, cotton boll weevil, corn earworm and cabbage looper as described below.

For *T. ni* and H. zea, ALA (4 mM) and a modulator (3 mM) were added to liquefied Waldbauer's medium (Waldbauer et al. 1984) at 55–60 °C. The mixture was blended for 2 min in a Sorval Omnimixer. The treated and control diets (the latter lacking ALA and modulators) were poured into 12-ml plastic molds and were allowed to cool down and to solidify before storage in a refrigerator at 4 °C. The food was generally used within 2 days, and was never stored for more than 2 weeks. Treatment of *T. ni* consisted of placing 15–20 third-instar larvae with one block of control or treated food, in a cardboard cup (about 9 cm h × 10 cm diameter) sealed with a plastic lid. Each treatment was replicated three times. Treatment of *H. zea* entailed placing a 3 ml control or treated diet cube and a single third instar *H. zea* larvae in each cell of a 20 cell plastic tray. The tray was sealed with a glass plate to prevent the escape and/or desiccation of the larvae. For both species, treatment was replicated three times. After 17 h of dark incubation at room temperature, the control and treated insects were placed in the growth chamber at 28 °C, [about 21.1 mW·cm$^{-2}$ of white light (metal halide)] under a 14 h light-10 h dark photoperiod. After 6 h in the light, untreated diet was added to each treatment. The untreated food was replenished daily. Insect death was monitored over a 6-day period.

*Anthonomus grandis* were obtained from a colony maintained at the Boll Weevil Research Laboratory, USDA, SEA, Mississippi State, Mississippi. A single tray of eggs dispersed on boll weevil diet was received weekly and held at room temperature until adults emerged. ALA and a modulator were added to liquefied, warm, boll weevil diet (BioServe Inc.) to a final concentration of 8 mM of ALA and 6 mM modulator. The mixture was blended for 2 min in a Sorval Omnimixer. Treated and control diets (the latter lacking ALA and modulators) were poured into 12 ml plastic molds and allowed to cool down and solidify before storage in a refrigerator at 4° C. Treatment consisted of placing 15 adults and a block of control or baited food in a cardboard carton (9 cm h × 10 cm diameter) sealed with a plastic lid. In order to minimize desiccation, a small petri dish (about 2.5 cm in diameter) containing cotton moistened with water was placed in the bottom of each carton. The baited food was replenished daily for the duration of the experiment. Dark incubation, exposure to light and evaluation of mortality was as described for *T. ni* and *H. zea*.

*Blattella germanica* were obtained from a colony maintained in a Biotronette Mark III Environmental Chamber at 28 °C, and 50 % relative humidity. The chamber was set for an 8-h light-16-h dark photoperiod. The colony was initiated from egg cases purchased from Carolina Biological Supply and was maintained on a diet of dog food and water. Sub-colonies were established weekly in separate 42-cm × 27-cm plastic animal cages. Waldbauer's medium was prepared exactly as described for *T. ni*. However concentrations of ALA and modulator were raised to 24 and 16 mM, respectively. Treatment consisted in placing a 12 ml baited block of food and 15 adults in a cardboard carton sealed with a plastic lid. As was described for *A. grandis*, a small petri dish containing moistened cotton was placed in the bottom of the carton. Control and treated containers were then placed in darkness at room temperature for 40 h. After dark-incubation the cartons were placed under subdued light (20–40 ft. candles) for the duration of the experiment. The baited diet, and water in the cotton dish, were replenished daily. Mortality was recorded over a 6-day period.

The results of the screening effort are described in Table 18.11. Thirty six compounds belonging to ten different chemical families (templates) were effective (>70 % mortality) against at least one insect species. *T. ni* was generally more susceptible than the other species. Structure-activity studies of some of these compounds are described below.

**Table 18.11** Porphyric Insecticide Modulators with 70 % or Better Activity Against Third Instar Larvae of *T. ni*, *H. zea*, and Against Adult *A. grandis* and *B. germanica*

| | | Mortality after 6 days (%) | | | |
|---|---|---|---|---|---|
| Template | Modulator | *T. ni* | *H. zea* | *A. grandis* | *B. germanica* |
| 1,10-Phenanthroline | 1,10-Phenanthroline | 100 | 100 | 90 | 100 |
| | 4-Methyl-1,10-Phenanthroline | 99 | 95 | 76 | 87 |
| | 5-Methyl-1,10-Phenanthroline | 97 | 55 | 64 | 30 |
| | 4,7-Dimethyl-1,10-Phenanthroline | 100 | 90 | 85 | 96 |
| | 5,6-Dimethyl-1,10-Phenanthroline | 97 | 50 | 51 | 93 |
| | 3,4,7,8-Tetramethyl-1,10-Phenanthroline | 85 | 79 | 44 | 27 |
| | 5-Nitro-1,10-Phenanthroline | 99 | 63 | 58 | 38 |
| | 5-Chloro-1,10-Phenanthroline | 96 | 84 | 40 | 82 |
| | 4,7-Diphenyl-1,10-Phenanthroline | 96 | 0 | 0 | 0 |
| (Di)pyridyls | 2,2′-Dipyridyl | 100 | 100 | 100 | 100 |
| | 4,4′-Dimethyl-2,2′-Dipyridyl | 71 | 25 | 20 | 7 |
| | Phenyl-2-Pyridyl ketoxime | 71 | 5 | 7 | 0 |
| | 2,2′-Dithiobis (pyridine N-oxide) | 73 | 0 | 67 | 67 |
| | Benzyl viologen dichloride monohydrate | 0 | 47 | 100 | 53 |
| Quinoxaline | Neutral red | 81 | 20 | 11 | 7 |
| 8-Hydroxyquinoline | 8-Hydroxy-7-(4-sulfo-1-naphtylazo)-5-quinoline | 74 | 35 | 2 | 0 |
| | 8-Hydroxy-5-sulfonic acid monohydrate | 86 | 10 | 0 | 0 |
| | 8-Hydroxy-7-iodo-5-quinoline | 78 | 10 | 0 | 0 |
| 2-Oxypyridine | 2-Metoxy-5-nitropyridine | 56 | 0 | 64 | 100 |
| | Isocarbostyril | 84 | 0 | 60 | 80 |
| Pyridiniums | 1-1-Diethyl-4,4-carbocyanine iodide | 67 | 95 | 100 | 0 |
| | 1-1-Diethyl-2,4-cyanine iodide | 36 | 11 | 53 | 93 |
| | Cetylpyridinium chloride monohydrate iodide | 49 | 0 | 78 | 40 |
| | 2-Chloro-1-methylpyridinium iodide | 49 | 5 | 0 | 83 |
| | 2-(4-(Dimethylamino)-styril)-ethylpyridinium | 30 | 10 | 89 | 13 |
| | 2-(4-(Dimethylamino)-styril)-methylpyridinium | 11 | 0 | 73 | 80 |
| | Poly (4-vinylpyridinium) dichromate | 85 | 5 | 4 | 3 |
| | 5-Phenyl-2-(4-pyridyl) oxazole | 76 | 20 | 0 | 0 |
| | Bis-N-methyl acidinium nitrate | 75 | 0 | 56 | 27 |
| Pyrrole | 3-Ethyl-2-methyl-4,5,6,7-tetrahydroindol-4-one | 93 | 25 | 11 | 0 |
| | Tert-butyl-4-acetyl-3,5-dimethyl-2-pyrrole-carboxylate | 71 | 0 | 9 | 3 |
| | 1-Phenylpyrrole | 71 | 0 | 9 | 3 |
| Pyrrolidine | 4-Pyrollidinopyridine | –[b] | – | – | 100 |
| Tetrazole | 3-3′-(4,4′-Biphenylene) bis | – | – | – | 100 |
| Thiazole | 3,6-Dimethyl-2-(4-dimethylaminophenyl) | – | – | – | 100 |
| | Thioflavin T | – | – | – | 100 |

Adapted from Gut and Rebeiz, unpublished

[a]Procedural details are given in the text

[b]Not determined

## 18.6 Structure-Activity Studies of Porphyric Insecticide Modulators

The identification of a significant number of chemicals exhibiting insecticidal activity encouraged detailed structure-function analyses of these modulators. Accordingly structure-function studies of modulators belonging to the Oph, Dpy, pyridinium, oxypyridine, pyrrole and 8-hydroxyquinoline templates were investigated. The chemical structures of the selected modulators and their template affiliation are depicted below in Figs. 18.5, 18.6 and 18.7.

Quantitative structure activity relationship (QSAR) analysis was performed on a Digital Equipment Co. workstation Model 3520, operating on a VMS platform. Chem-X software, (Chemical Design Limited, Oxford, England) was used to build 3-dimensional chemical structures and to carry out QSAR analysis. The latter involved, among other things: (a) optimization of the chemical structures via Mopac (QCPE, version 5, 1989) using MNDO or PM1 Hamiltonians, (b) writing the optimized structures of Oph and its analog to a database, (c) calculation of 24 different electronic and physical organic properties for each structure, i.e. descriptors, (d) correlation analysis between the 24 descriptors and biological activity, and (e) stepwise multiple regression analysis to determine the nature of relationships between biological activity and various descriptors.

Calculation of electrostatic potential energy levels at various sites of a given molecule was performed using Chem-X software. Chem-X treats the charge on each atom in a molecule as a point charge positioned at the center of the atom. A positive unit charge equivalent to that of a proton is placed at each grid point and the electrostatic interaction between groups of atoms and the unit charge is calculated. The number of grid points used in the calculation, one point per Angstrom in this case, is usually set by the operator. After calculations were completed, electrostatic isopotential contour lines were drawn. The level of potential energy in kcal per mole was also selected by the operator. We chose values of 10 kcal/mol for positive potential energy levels and −10 kcal/mol for negative potential energy levels. Since the interaction of a positive probe with a positive region of the molecule generated positive energy levels they were interpreted as repelling. Likewise since the interaction of a positive charge probe with a negative region of the molecule generated negative energy levels they were interpreted as binding or attracting energy levels. In this manner the attraction or repulsion at various loci of a particular molecule toward a positive charge was well defined by the negative and positive potential energy contour lines respectively, which in fact delineated positive charge binding or repelling electrostatic volumes surrounding various sections of a molecule.

Quantitative and positional differences between the electrostatic fields of various modulators were calculated by determining the exclusive positive charge binding and repelling volumes for each analog in comparison to a reference molecule. The calculation of exclusive volumes (i.e. non overlapping volumes between any two molecules) was achieved via a Chem-X software module that calculated exclusive electrostatic field volumes for pairs of molecules from the electrostatic volumes of each individual molecule. In so doing it became possible to compare quantitative and qualitative positional differences between various analogs within each template.

| Modulator | Chemical Structure |
|---|---|
| Phenanthrene | 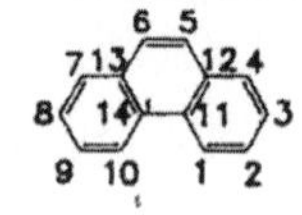 |
| 1,10–Phenanthroline (Oph) | 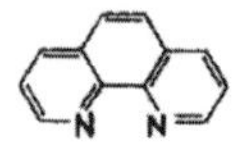 |
| 4–Methyl–Oph | 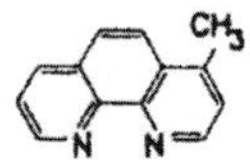 |
| 5–Methyl–Oph | 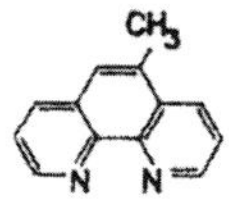 |
| 5,6–Dimethyl–Oph | 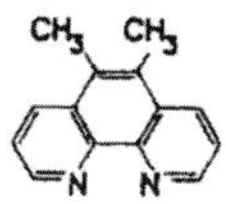 |
| 4,7–Dimethyl–Oph | 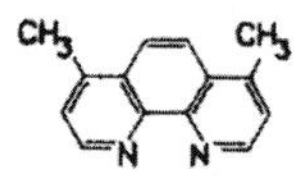 |
| 3,4,7,8–Tetramethyl–Oph | 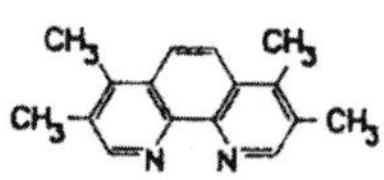 |
| 4,7–Dihydroxy–Oph | 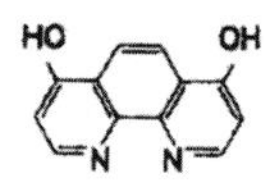 |
| 4,7–Diphenyl–Oph | 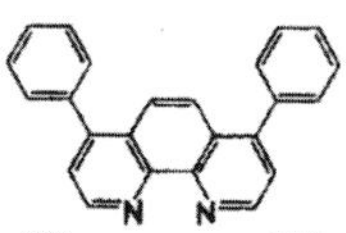 |
| 2,9–Dimethyl–4,7–diphenyl–Oph | 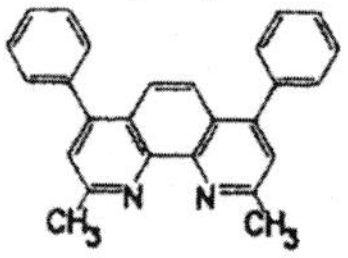 |
| 4,7–Dimethyl–5,6 dihydro–dibenzo–Oph | 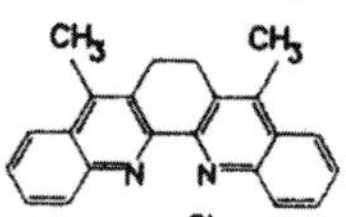 |
| 5–Chloro–Oph | 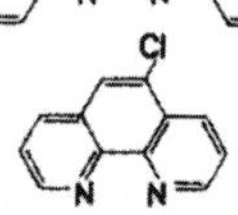 |
| 5–Nitro–Oph | 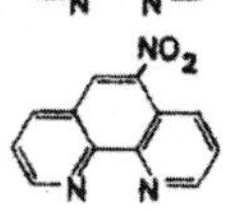 |

**Fig. 18.5** Chemical names and structures of the substituted phenanthrolines used in the structure-activity investigations

| Modulator | Chemical Structure |
|---|---|
| 1,10-Phenanthroline (Oph) | 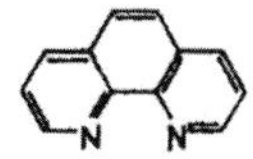 |
| 2,2'-Dipyridyl |  |
| 2,2':6'2"-Terpyridine | 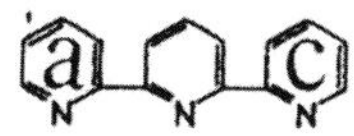 |
| 4,4'-Dimethyl-2,2'-dipyridyl | 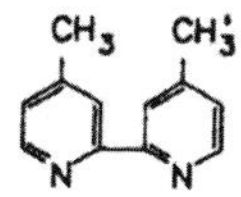 |
| 4,4'-Diphenyl-2,2'-dipyridyl | 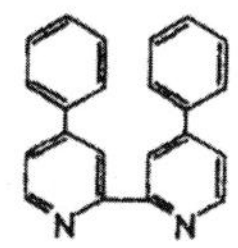 |
| 2,2'-Biquinoline | 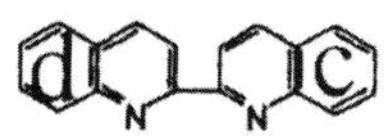 |
| 2,2'-Dipyridyl disulfide | 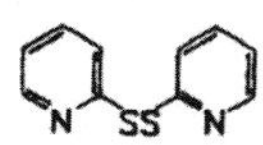 |
| 2,2'-Dithiobis (5-nitropyridine) | 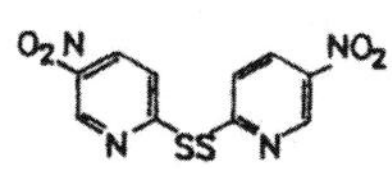 |
| 2,2'-Dithiobis (pyridine N-oxide) | 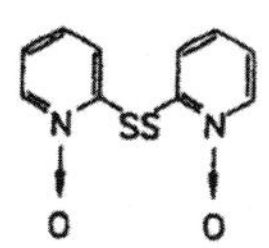 |
| Di-2-pyridyl ketone oxime | 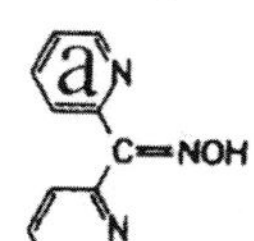 |
| Phenyl 2-pyridyl ketoxime | 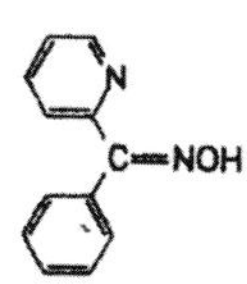 |

**Fig. 18.6** Chemical names and structures of the substituted pyridyls used in the structure-activity investigations

| Template | Modulator | Chemical Structure |
| --- | --- | --- |
| Pyridinium | 1–(3–Sulfopropyl) pyridinium hydroxide | 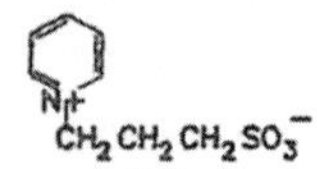 |
| Pyridinium | 1,1'–Diethyl–2,4'–cyanine iodide | 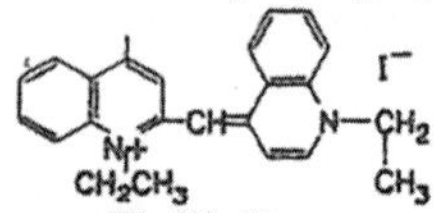 |
| Pyridinium | 1,1'–Diethyl–4,4'–carbocyanine iodide | 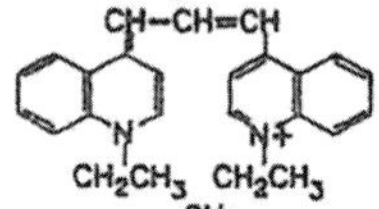 |
| Pyridinium | Bis–N–methylacridinium nitrate | 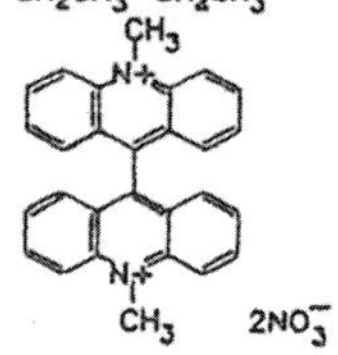 |
| 8–Hydroxyquinoline | 8–Hydroxyquinoline | 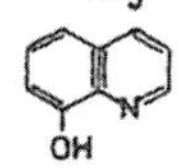 |
| 8–Hydroxyquinoline | 8–Hydroxy–5–nitroquinoline | 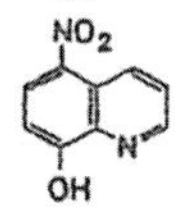 |
| 8–Hydroxyquinoline | 8–Hydroxyquinoline–5–sulfonic acid | 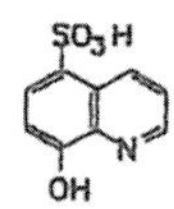 |
| 8–Hydroxyquinoline | 8–Hydroxy–7–(4–sulfo–1–napthyl–azo)–5–quinoline–sulfonic acid | 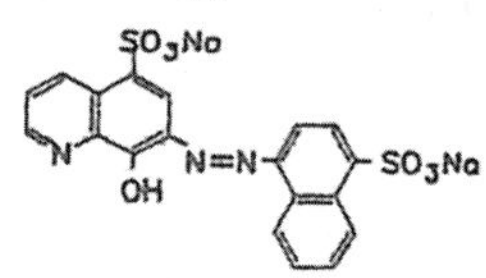 |
| 2–Oxypyridine | Isocarbostyril | 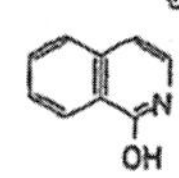 |
| 2–Oxypyridine | 1,5–Isoquinolinediol | 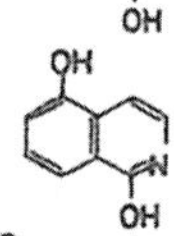 |
| Pyrrole | tert–Butyl 4–acetyl–3,5–dimethyl–2–pyrrole–carboxylate | 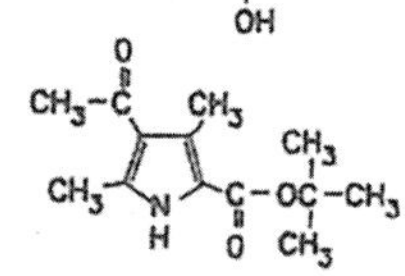 |
| Pyrrole | 4,5,6,7–Tetrahydroindole | 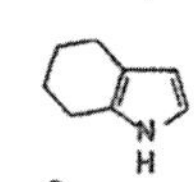 |
| Pyrrole | 3–Ethyl–2–methyl–4,5,6,7–tetrahydroindole–4–one | 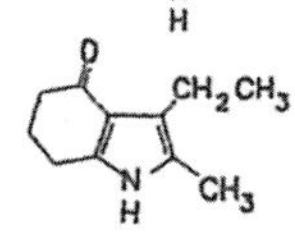 |

**Fig. 18.7** Chemical names and structures of substituted pyridiniums, oxypyridines, quinolines and pyrroles used in the structure-activity investigations

In the study of structure/function correlations the emphasis was placed on the analysis of relationships between tetrapyrrole modulating activity, photodynamic damage and alterations in the electrostatic field of various modulators. With this approach the underlying hypothesis was structural complementarity between enzymatic receptor sites and modulators that favored electrostatic binding of various modulators to or close to the receptor sites of enzymes that catalyze various reactions of the heme biosynthetic pathway. It was conjectured that the detection of exclusive positive or negative charge binding volumes in groups of analogs that exhibited similar tetrapyrrole biosynthesis modulating activity may be an indication that these analogs may exert their effects by binding to the same enzymatic receptor site. In all cases the modulation of tetrapyrrole biosynthetic activity by various analogs was related to the presence of unique positive charge binding or repelling volumes. This in turn suggested that various modulators exerted their effects by binding to specific negatively charged areas that may well be close to tetrapyrrole biosynthetic enzyme reaction sites.

Early fourth-instar *T. ni* larvae were used in evaluating modulator effects on Proto accumulation and larval mortality. Chemicals were incorporated into liquefied Waldbauer's medium as described in section above in "IV". Twenty six larvae were placed in each diet cup and allowed to feed in darkness for 17 h. Following dark incubation, six larvae were assayed for tetrapyrrole accumulation. Mortality of the remaining 20 larvae was monitored prior to light exposure and daily thereafter, for a period of 6 days. Modulators were evaluated alone and in combination with ALA in a randomized complete block design with three replicates. Percent larval mortality was arcsine transformed prior to statistical analysis. Structure-Activity relationship of five selected modulators are reported below.

### *18.6.1 Structure-Activity Relationship of Substituted Phenanthrolines*

The porphyric insecticide modulating activity of Oph and eight of its analogs are depicted in Table 18.12. The insecticidal efficacy of Oph and its analogs was closely associated with their ability to enhance the conversion of exogenous ALA to Proto (Table 18.13). As was observed for photodynamic herbicidal effects in plants, (Rebeiz et al. 1990b) the presence of N atoms at position 1 and 10 of the macrocycle was essential for porphyric insecticidal activity. This was evidenced by the very limited activity of phenanthrene, in which position 1 and 10 are occupied by carbon instead of N atoms (Fig. 18.5, Table 18.12, # 7, 8). On the other hand, enhancement of Proto formation and porphyric insecticidal activity were maintained following peripheral methyl, chloro and nitro group substitution of Oph (Table 18.12, # 11–18). In contrast, enhancement of Proto formation and photodynamic toxicity was diminished by phenyl or benzyl (data not shown) substitution (Table 18.12, # 5, 6, 9, 10).

**Table 18.12** Effects of ALA, phenanthrene and 1,10-phenanthroline (Oph) on Proto accumulation and larval death in *T. ni*

| Entry | Treatment[a] | Proto content (nmol/100 mg protein) | Larval mortality after 6 days in the greenhouse (%) |
|---|---|---|---|
| 1 | Control | 1.5a[b] | 0.0a |
| 2 | ALA | 3.0a | 15.0abcd |
| 3 | 2,9-Dimethyl-4,7-diphenyl Oph | 0.8a | 0.0a |
| 4 | Above + ALA | 1.6a | 10.0abc |
| 5 | 2,9-Dimethyl-4,7-diphenyl Oph | 1.2a | 0.0a |
| 6 | Above + ALA | 1.6a | 18.3abcd |
| 7 | Phenanthrene | 1.0a | 0.0a |
| 8 | Above + ALA | 3.9a | 23.3bcd |
| 9 | 4,7-Diphenyl Oph | 1.3a | 0.0a |
| 10 | Above + ALA | 8.6a | 36.7 cd |
| 11 | 3,4,7,8-Tetramethyl Oph | 5.1a | 11.7abcd |
| 12 | Above + ALA | 117.0c | 95.0e |
| 13 | 4,7-Dimethyl Oph | 15.9a | 8.3ab |
| 14 | Above + ALA | 131.5c | 96.7e |
| 15 | 5-Cl-Oph | 2.0a | 10.0abc |
| 16 | Above + ALA | 133.9c | 96.7e |
| 17 | 5-Nitro-Oph | 50.4b | 41.7d |
| 18 | Above + ALA | 190.1d | 98.3e |
| 19 | Oph | 8.6a | 5.0a |
| 20 | Above + ALA | 207.0d | 100.0e |
| Correlation coefficient | | 0.88 | |
| Level of significance | | 0.1 % | |

Adapted from Gut et al. (1993)

[a]Treatment consisted of control diet lacking any added ALA or modulator, diet containing 4 mM ALA, 3 mM modulator or 4 mM ALA + 3 mM modulator, in a randomized complete block design with three replicates. Percent larval death was arcsine transformed prior to statistical analysis

[b]Means followed by the same letter within a column are not significantly different at the 5 % level of significance

Quantitative structure activity calculations suggested a relationship between peripheral group substitution and some physico-chemical properties of the substituted compounds. Electron density changes in Oph and its analogs that appeared to be related to reduced efficacy included (Gut et al. 1993) (a) Appearance of positive charge binding volumes at position 4 and 7 of the 1,10-phenanthroline macrocycle, which flanks positive charge repelling volumes, (b) a dramatic increase in super-delocalisability (i.e. electron density) over some unoccupied molecular orbitals, and (c) electronic charge at position 1 and 10 of the macrocycle. Large increases in Van der Waals volumes also exerted negative effects on insecticidal efficacy (Gut et al. 1993).

### *18.6.2 Structure-Activity Relationship of Substituted Pyridyls*

The porphyric insecticide modulating activity of Dpy and eight of its analogs are depicted below in Table 18.13. The insecticidal efficacy of Dpy and its analogs was

**Table 18.13** Effects of ALA and substituted dipyridyls, on Proto accumulation and larval mortality in *T. ni*

| Entry | Treatment[a] | Proto content (nmol/100 mg protein) | Larval mortality after 6 days in the greenhouse (%) |
|---|---|---|---|
| 1 | Control | 0.9a[b] | 0.0a |
| 2 | ALA | 2.3a | 1.7a |
| 3 | 2,2′-Dipyridyl disulfide | 1.0a | 0.0a |
| 4 | Above + ALA | 1.6a | 0.0a |
| 5 | 2,2′-Biquinoline | 0.9a | 0.0a |
| 6 | Above + ALA | 2.0a | 0.0a |
| 7 | 4,4′-Diphenyl-2,2′-dipyridyl | 1.2a | 1.7a |
| 8 | Above + ALA | 2.3a | 1.7a |
| 9 | 2,2′-Dithiobis (5-nitropyridine) | 0.7a | 0.0a |
| 10 | Above + ALA | 1.8a | 5.0a |
| 11 | 2,2′-Dithiobis (pyridine) N-oxide | 0.7a | 0.0a |
| 12 | Above + ALA | 10.46a | 56.7c |
| 13 | 4,4′-Dimethyl-2,2′-dipyridyl | 1.9a | 0.0a |
| 14 | Above + ALA | 18.1a | 28.3b |
| 15 | Phenyl 2-pyridyl ketoxime | 1.5a | 1.7a |
| 16 | Above + ALA | 21.9a | 33.3b |
| 17 | 2,2′:6′,2″-Terpyridine | 6.2a | 11.7a |
| 18 | Above + ALA | 181.1b | 98.2d |
| 19 | 2,2′-Dipyridyl | 15.0a | 48.3c |
| 20 | Above + ALA | 176.3b | 100.0d |
| Correlation coefficient | | 0.91 | |
| Level of significance | | 0.1 % | |

Adapted from Gut et al. (1993)

[a]Treatment consisted of control diet lacking any added ALA or modulator, diet containing 4 mM ALA, 3 mM modulator or 4 mM ALA + 3 mM modulator, in a randomized complete block design with three replicates. Percent larval death was arcsine transformed prior to statistical analysis

[b]Means followed by the same letter within a column are not significantly different at the 5 % level of significance

closely associated with their ability to enhance the conversion of exogenous ALA to Proto (Table 18.13). As was observed with Oph, the presence of N atoms at position 1 and 10 of the macrocycle appeared to be essential for porphyric insecticide activity. This was evidenced by a 30 % reduction in activity in phenyl 2-pyridyl ketoxime with a N atom at position 1 and a carbon atom at position 10 (Fig. 18.7) as compared to dipyridyl ketoxime with 2 N atoms at positions 1 and 10 of the macrocycle (data not shown). The presence of a third 6-membered N heterocyclic ring as in 2,2′:6′2″-terpyridine did not decrease insecticidal activity (Fig. 18.6, Table 18.14, # 17). Ring substitution at the periphery of the Dpy macrocycle, however, had a highly negative effect on insecticidal performance. For example the addition of phenyl, benzyl, or methyl groups as in 4,4′-diphenyl-Dpy, biquinoline and 4,4′-dimethyl-Dpy, dramatically reduced the insecticidal activity (Table 18.13, # 5, 6, 7, 8, 13, 14).

Quantitative structure activity calculations again revealed a potential relationship between some physico-chemical properties of various analogs and photodynamic insecticidal activity. For example, substantial increase in the distance between N1 and N10 which ensued following the insertion of a disulfide bridge between the two rings, as in 2,2′-dipyridyl disulfide, eliminated completely the insecticidal activity (Fig. 18.6, Table 18.14, # 3, 4). As was observed for Oph analogs, electron density changes in Dpy and its analogs that appeared to be related to reduced efficacy included, (a) appearance of positive charge binding volumes at position 4 and 4′ of the dipyridyl macrocycle, which flanked positive charge repelling volumes, (b) the acquisition of a fourth phenyl or pyridyl ring which generated alternating positive charge binding and repelling electrostatic fields between rings C and D, and (c) a change in sign and magnitude of the 1, 2, 2′ 1′ torsion angles which was accompanied by the appearance of positive charge repelling volumes at the C5′-C6′, or C6′ positions (Gut et al. 1993).

### *18.6.3 Structure-Activity Relationship of Substituted Pyridiniums*

The structure-activity relationships of four substituted Pyridiniums was investigated (Fig. 18.4). Porphyric insecticide damage was positively correlated with Proto accumulation except for 1,1′-diethyl-4,4′-carbocyanine iodide (Table 18.14, # 9,10). Although light was absolutely required for the expression of activity, enhancement of Proto accumulation appeared to be less pronounced than was indicated by the high insect mortality. The only total loss of insecticidal activity was observed in 1-(3-sulfopropyl) pyridinium hydroxide (Fig. 18.7, Table 18.14, # 3, 4). QSAR investigations of this compound revealed a 3.3 fold increase in the value of its dipole moment (17.03 D) in comparison to 1,1′-diethyl-2,4′-cyanine iodide (5.11 D) (Table 18.14, # 7, 8) that lacked a sulfonyl group at the $N^+$ position (Fig 18.7). This was accompanied by a substantial increase in positive charge binding volumes and a substantial decrease in positive charge repelling volumes around the molecule. In other words, it appeared that the presence of an unbalanced positive charge at the $N^+$ position was essential for pyridinium activity. This was also corroborated by the reduced insecticidal activity of Bis-N-methylacridinium nitrate that has an $NO_3^-$ counterion (Table 18.14, # 5, 6).

### *18.6.4 Structure-Activity Relationship of Substituted Quinolines and Oxypyridines*

Structure-activity relationships of four substituted 8-hydroxyquinolines, and two substituted oxypyridines were also investigated (Fig. 18.7). Porphyric insecticidal effects were positively correlated with Proto accumulation, and the presence of both

**Table 18.14** Effects of ALA, and substituted pyridiniums, 2-oxypyridines and pyrroles on Proto accumulation and larval death in *T. ni*

| Entry | Template | Treatment[a] | Proto content (nmol/100 mg protein) | Larval mortality after 6 days in the greenhouse (%) |
|---|---|---|---|---|
| 1 | Pyridinium | Control | 1.4abc[b] | 1.7a |
| 2 | | ALA | 2.2bcd | 3.3a |
| 3 | | 1-(3-Sulfopropyl) pyridinium hydroxide | 2.4 cd | 0.0a |
| 4 | | Above + ALA | 4.4de | 0.0a |
| 5 | | Bis-N-methylacridinium nitrate | 5.6e | 21.7b |
| 6 | | Above + ALA | 10.9f | 65.0c |
| 7 | | 1,1′-Diethyl-2,4′-cyanine iodide | 58.1 h | 75.0c |
| 8 | | Above + ALA | 87.0 h | 85.0 cd |
| 9 | | 1,1′-Diethyl-4,-4′-carbocyanine iodide | 3.5a | 96.7d |
| 10 | | Above + ALA | 21.9 g | 93.3d |
| 11 | Oxypyridine | 1,5-Isoquinolinediol | 1.0a | 1.7a |
| 12 | | Above + ALA | 3.7de | 33.0bb |
| 13 | | Isocarbostyril | 1.2ab | 1.7a |
| 14 | | Above + ALA | 16.3 fg | 86.7 cd |
| 15 | Pyrrole | 4,5,6,7-Tetrahydroindole | 1.3abc | 1.7a |
| 16 | | Above + ALA | 2.6 cd | 1.7a |
| 17 | | ter-Butyl 4-acetyl-3,5-dimethyl-2-pyrrole-carboxylate | 1.3abc | 0.0a |
| 18 | | Above + ALA | 2.7d | 20.4b |
| 19 | | 3-Ethyl-2-methyl-4,5,6,7-tetrahydroindole-4-one | 0.9a | 0.0a |
| 20 | | Above + ALA | 10.0f | 83.3 cd |
| Correlation coefficient | | | 0.76 | |
| Level of significance | | | 0.1 % | |

Adapted from Gut and Rebeiz Gut et al. (1993)

[a]Treatment consisted of control diet lacking any added ALA or modulator, diet containing 4 mM ALA, 3 mM modulator or 4 mM ALA + 3 mM modulator, in a randomized complete block design with three replicates. Percent larval death was arcsine transformed prior to statistical analysis

[b]Means followed by the same letter within a column are not significantly different at the 5 % level of significance

ALA and modulator were required for activity (Tables 18.14 and 18.15). 8-Hydroxyquinoline and the $NO_2^-$ substituted analog, 8-hydroxy-5-nitroquinoline in which the hydrogen atom at position 5 of the macrocycle was replaced by an $NO_2$– group was inactive (Fig. 18.7, Table 18.15, # 3–6). When the nitro group was replaced by a sulfonyl group, as in 8-hydroxyquinoline-5-sulfonic acid, substantial insecticidal activity was observed (Fig. 18.7; Table 18.15, # 9, 10). Further substitution with a 4-sulfo-1-naphtylazo group at position 7 of the macrocycle as in 8-hydroxy-7-(4-sulfo-1-naphthylazo)-5-quinoline-sulfonic acid resulted in a minor

**Table 18.15** Effects of ALA, and substituted quinolines on Proto accumulation and larval mortality in *T. ni*

| Entry | Treatment[a] | Proto content (nmol/100 mg protein) | Larval mortality after 6 days in the greenhouse (%) |
|---|---|---|---|
| 1 | Control | 1.0b[b] | 0.0a |
| 2 | ALA | 1.7c | 0.0a |
| 3 | 8-Hydroxyquinoline | 1.0a | 0.0a |
| 4 | Above + ALA | 1.9c | 0.0a |
| 5 | 8-Hydroxy-5-nitroquinoline | 0.7a | 0.0a |
| 6 | Above + ALA | 3.0d | 0.0a |
| 7 | 8-Hydroxy-7-(4-sulfo-1-naphtylazo)-5-sulfonic acid | 1.8c | 1.7a |
| 8 | Above + ALA | 41.8 g | 76.7b |
| 9 | 8-Hydroxyquinoline-5-sulfonic acid | 1.8c | 0.0a |
| 10 | Above + ALA | 32.1f | 81.7b |
| Correlation coefficient | | 0.85 | |
| Level of significance | | 1 % | |

Adapted from Gut et al. (1993)
[a]Treatment consisted of control diet lacking any added ALA or modulator, diet containing 4 mM ALA, 3 mM modulator or 4 mM ALA + 3 mM modulator, in a randomized complete block design with three replicates. Percent larval death was arcsine transformed prior to statistical analysis
[b]Means followed by the same letter within a column are not significantly different at the 5 % level of significance

reduction in activity (Table 18.15, # 7, 8). The presence of a sulfonyl group at position 5 of the macrocycle was an essential requirement for activity. The main effect of this substitution was a significant increase in the electrostatic positive charge binding volumes around the sulfonic acid group.

Moving the exocyclic OH group from position 8 to position 1 and moving the endocyclic N atom from position 1 to position 2, as in isocarbostyril, an oxypyridine compound, also resulted in substantial insecticidal activity in comparison to 8-hydroxyquinoline (Table 18.14, # 13, 14; Table 18.15 #, 3, 4). This change resulted in an increase in the magnitude of the dipole moment from 2.33 D to 3.53 D, as well as an increase in electron density (from 0.16 to 0.23 kcal/mol) in the unoccupied molecular orbitals in isocarbostyril. Addition of a hydroxyl group at position 5, as in 1,5-isoquinolinediol, diminished considerably the observed insecticidal activity (Fig. 18.7; Table 18.14, # 11, 12). This was accompanied by a decrease in dipole moment and further increase in electron density of the unoccupied molecular orbitals.

## *18.6.5 Structure-Activity Relationship of Substituted Pyrroles*

Structure-activity relationships of three substituted pyrroles were investigated (Fig. 18.7). As was generally observed for the other templates, porphyric insecticidal effects were positively correlated with Proto accumulation and the presence of both ALA and modulator were required for activity (Table 18.14, # 19, 20). Highly substituted pyrroles such as tert-butyl 4-acetyl-3,5-dimethyl-2-pyrrole-carboxylate exhibited low insecticidal activity (Fig. 18.7; Table 18.14, # 17, 18). However tetrahydroindoles with an isolated keto group at position 4, as in 3-ethyl-2-methyl-4,5,6,7-tetrahydroindol-4-one, were active (Fig. 18.7; Table 18.14, # 19, 20). Removal of the ketone group resulted in complete loss of activity as in 4,5,6,7-tetrahydroindole (Fig. 18.7; Table 18.14, # 15, 16). This was accompanied by a substantial decrease in electron density (from 1.03 to 0.52 kcal/mol) of the unoccupied molecular orbitals, and a significant decrease (from 132.3 to 61.8 A°3) in volumes of the positive charge binding electrostatic field.

## 18.7 Epilogue

It is hoped that based on the reported investigations it will be possible to develop safe formulation for the control of household pests. It will be more difficult to develop formulations for the control of insects on living plants as the ALA and modulators so far described appear to destroy the plant as well as the insect.

## References

Duggan JX, Gassman M (1974) Induction of porphyrin biosynthesis in etiolated bean leaves by chelators of iron. Plant Physiol 53:206–215

Gut L, Lee K, Juvik JA et al (1993) Porphyric insecticides. IV: structure-activity study of substituted phenanthrolines. Pestic Sci 39:19–30

Gut L, Lee K, Juvik JA et al (1994a) Porphyric insecticides 6. Structure activity study of substituted pyridyls. Pestic Biochem Physiol 50:1–14

Gut LJ, Juvik JL, Rebeiz CA (1994b) Porphyric insecticides. In: Duke SO, Rebeiz CA (eds) Porphyric pesticides, chemistry, toxicology and pharmaceutical applications, vol 559. American Chemical Society, Washington, DC, pp 206–232

Halliwell B (1984) Oxygen-derived species and herbicide action. What's New in Plant Physiol 15:21–24

Lamola AA, Yamane T (1974) Zinc protoporphyrin in the erythrocytes of patients with lead intoxication and iron deficiency anemia. Science 186:936–938

Lee K, Rebeiz CA (1995) Subcellular localization of protoporphyrin IX and its photodynamic effects on mitochondrial function of the cabbage looper (*Trichoplusia ni*). In: Heitz JR, Downum KR (eds) Light activated pesticides, vol 616. American Chemical Society, Washington, DC, pp 152–164

Rebeiz CA (1993) Porphyric insecticides. J Photochem Photobiol B Biol 18:97–99

Rebeiz CA (2002) Analysis of intermediates and end products of the chlorophyll biosynthetic pathway. In: Smith A, Witty M (eds) Heme chlorophyll and bilins, *methods and protocols*. Humana Press, Totowa, pp 111–155

Rebeiz CA, Juvik JA, Rebeiz CC (1988a) Porphyric insecticides 1. Concept and phenomenology. Pestic Biochem Physiol 30:11–27

Rebeiz CA, Montazer-Zouhoor A, Mayasich JM et al (1988b) Photodynamic herbicides. Recent developments and molecular basis of selectivity. Crit Rev Plant Sci 6:385–434
Rebeiz CA, Juvik JA, Rebeiz CC et al (1990a) Porphyric insecticides 2. 1,10-Phenanthroline, a potent porphyric insecticide modulator. Pestic Biochem Physiol 36:201–207
Rebeiz CA, Reddy KN, Nandihalli UB et al (1990b) Tetrapyrrole-dependent photodynamic herbicides. Photochem Photobiol 52:1099–1117
Rebeiz CA, Nandihalli UB, Reddy K (1991) Photodynamic herbicides and chlorophyll biosynthesis modulators. In: Baker NR, Percival M (eds) Herbicides. Elsevier, Amsterdam, pp 173–208
Rebeiz CA, Gut LJ, K. L et al (1995) Photodynamics of porphyric insecticides. Crit Rev Plant Sci 14:329–366
Waldbauer GA, Cohen RW, Friedman S (1984) An improved procedure for laboratory rearing of the corn earworm heliothus zea (Lepidoptera:Noctuidea). Gt Lakes Entomol 17:113–120

# Chapter 19
# ALA-Dependent Cancericides

*I came to realize that life lived to help others is the only one that matters and that it is my duty and my highest and best use as a human. (Adapted from Ben Stein)*

## 19.1 Prologue

In 1978, Dougherty and coworkers observed that upon injection of hematoporphyrin derivative (HpD), a chemically synthesized porphyrin molecule, into cancerous tissues the compound accumulated to higher concentrations in malignant tissue than in normal tissues. This finding constituted the basis for use of porphyrins, specifically HpD, in photoradiation cancer therapy. The work was summarized later by Dougherty (Dougherty 1987). Photodynamic Therapy (PDT), has concentrated primarily on the use of hematoporphyrin (Hp) and its derivatives, such as dihydrohematoporphyrin and commercially available Photofrin I and II, which are patented derivatives of Hp.

When the photodynamic herbicides technology (see Chap. 17) was described in 1984 (Rebeiz et al. 1984). I received several inquiries from the medical community about the possibility of developing the technology to kill cancer cells. I responded by the affirmative since plants (Rebeiz et al. 1984), insects (Rebeiz et al. 1988) and animals shared the same heme pathway up to protoporphyrin IX (Proto). However due to involvement with photodynamic herbicide and insecticide development I could not immediately work on developing that technology.

Then in 1989, Malik et al. investigated the possibility of stimulating endogenous protoporphyrin production by supplementing the medium of transformed cells with ALA (Malik et al. 1989). The K562 cell line did not synthesize Proto during 4 days of incubation with ALA in darkness. However, FELC (Friend erythroleukemia cells) and Bsb (a highly metastic cell line) were sensitive to ALA treatment and synthesized porphyrins in the dark during 4 days of incubation. Trypan blue


DOI 10.1007/978-94-007-7134-5_19, 

exclusion and $^{3}$H-blue thymidine incorporation were used to determine cell viability and proliferation, respectively. Chromatography was used to determine the type of porphyrin species that accumulated. Flow cytometry was used to determine porphyrin content by monitoring the red fluorescence (longer than 630 nm) emitted from individual cells. The fluorescence was proportional to porphyrin content. The cells that synthesized Proto became sensitive to light and exhibited 99 % inactivation (cell death) after 4 days of dark incubation at the optimal Proto concentration.

Then Kennedy and coworkers (Kennedy et al. 1990; Kennedy and Pottier 1992, 1994) used ALA to treat different types of skin cancers. In their protocol, ALA was mixed into a paste (10 % glaxal base 35–100 mg ALA/lesion), then applied topically to the skin. The skin was then photoirradiated with laser light. The results appeared to be promising. However highly intense light (52 mW/h/cm$^{2}$) and high concentrations of ALA were required. As a consequence, these conditions caused some damage to healthy tissues.

Then it was reported that better ALA-dependent photoradiation therapy results were obtained by using long chain-fatty acid esters of ALA. That in turn reduced the hydrophilic properties of ALA and improved its penetration into treated tissues (Kloek et al. 1996). The ALA esterases that converted the ALA esters into ALA before conversion to Proto were very active in animal tissues and less active in insect and plant tissues (Kolossov and Rebeiz 2004).

The rest of this chapter is devoted to the discussion of using ALA with porphyrin modulators for the photodynamic destruction of cancer cells.

## 19.2 Photodestruction of Tumor Cells by Induction of Protoporphyrin IX Accumulation by ALA and 1,10-Orthophenanthroline

Photodynamic herbicides (Chap. 17) and porphyric insecticides (Chap. 18) are two novel technologies that manipulate the photosensitizing capability of metabolic porphyrins. These two novel technologies destroy undesirable plants and insects following co-treatment with δ-aminolevulinic acid (ALA), a naturally occurring 5-carbon amino acid, and one of a number of tetrapyrrole biosynthesis modulators (Rebeiz et al. 1984, 1988). The amino acid and the modulator act in concert. The amino acid serves as a building block for intracellular tetrapyrrole accumulation, while the modulator amplifies the accumulation of harmful tetrapyrroles. In the light, the accumulated tetrapyrroles photosensitize the formation of singlet oxygen which kills treated plants or insects by oxidation of their cellular membranes.

In the next four sections it is shown that treatment of rapidly multiplying immortalized cells, with ALA and 1–8-orthophenanthroline (Oph) caused the cells to accumulate much larger amounts of Proto than untreated cells. This endogenous Proto accumulation caused in turn rapid cell death in the light. Slower growing cells responded to such treatments by accumulating much lower levels of Proto.

### 19.2.1 Identification of the Porphyrin That Accumulated in MLA 144 Cells After Treatment with δ–Aminolevulinic Acid and 1,10-Phenanthroline as Protoporphyrin IX

Since herbicide and insecticide investigations revealed that Oph induced tetrapyrrole accumulation in plants (Chap. 17) and Proto in insects (Chap. 15) it was conjectured that it may do the same in cancer cells.

Therefore Initial investigations attempted to determine whether significant Proto biosynthesis and accumulation could be induced in proliferating MLA 144 gibbon lymphoma cells after incubation in total darkness with 1 mM ALA and 0.75 mM Oph. It was observed that The ALA + Oph treated cells accumulated large amounts of a pigment that exhibited emission and Soret excitation maxima at 298 K and 77 K identical to authentic Proto (Rebeiz et al. 1992). After Mg insertion, the pigment exhibited identical fluorescence properties as Mg-Proto (Rebeiz et al. 1992).

To further confirm the identity of the accumulated product, its HPLC mobility was compared to authentic Proto on a Waters Bondapack reverse phase C18-bonded column. Authentic Proto, dissolved in methanol, and the pigment extracted from MLA 144 cells exhibited similar retention times of 5.4–5.6 min (Rebeiz et al. 1992). Since the fluorescence properties of this pigment at 298 K and 77 K were identical to authentic Proto before and after Mg insertion, and since its HPLC mobility was similar to that of authentic Proto, the observed fluorescence was ascribed to the biosynthesis and accumulation of Proto by the treated cells. Protoporphyrin IX accumulation in the treated cells amounted to 29.4 nmol/100 mg of cell protein (Rebeiz et al. 1992). After the same period of dark-incubation Proto was not detected in untreated cells. It was therefore concluded that after treatment with ALA and Oph, MLA 144 cells accumulated large amounts of Proto.

### 19.2.2 Induction of Cell Lysis of MLA 144 Cells Treated with ALA and Oph

Next, the extent of cell destruction in the light following the induction of tetrapyrrole accumulation by MLA 144 cells was investigated. The cells were labeled with $Na^{51}Cr$ and then induced to accumulate Proto by treatment with 1 mM ALA and 0.75 mM Oph for 3.0 h in darkness. After exposure to white light (sodium halide, 2.11 mW $cm^{-2}$) for an additional 30 min, cell destruction was evaluated by monitoring the release of $^{51}Cr$ into the incubation medium. ALA + Oph-treated cells that had accumulated Proto exhibited severe photodynamic damage in the light as compared to control cells which had not accumulated any tetrapyrroles (Rebeiz et al. 1992).

The relative merits of ALA and Oph in inducing Proto accumulation and cell lysis were also investigated by treatment of MLA 144 cells in darkness either with ALA, with Oph, or with a combination of the two compounds. Although ALA alone induced significant levels of Proto accumulation and photodynamic cell injury, the most dramatic tetrapyrrole accumulation and cell destruction were observed when cells were treated jointly with ALA + Oph (Rebeiz et al. 1992). The addition of Oph doubled Proto accumulation and significantly increased cell lysis in the light. 1,10-Phenanthroline, alone, did not induce Proto accumulation.

### *19.2.3 Proto-Dependent Photodestruction of MLA and WEHI 164-Clone13 Cells Following ALA and Oph Treatments*

The susceptibility of other transformed cell lines to ALA + Oph treatment was also investigated. The response of a human granulocytic leukemia cell line (K562), and a murine fibrosarcoma cell line (WEHI 164 clone 13), was compared to that of MLA 144 cells (Rebeiz et al. 1992). Both transformed cell lines exhibited photodynamic injury similar to the MLA 144 cells.

However, the human granulocytic leukemia cells were extremely sensitive to treatment as evidenced by the lower concentrations of ALA and Oph that induced significant photodynamic cell lysis (Rebeiz et al. 1992).

### *19.2.4 Enhancement of Proto Accumulation by Murine Splenocyte Treatment*

Since increased tetrapyrrole biosynthesis and accumulation is more likely to take place in rapidly growing and multiplying cells in need of heme for cytochrome formation, it was conjectured that slowly proliferating cells might be less prone to accumulate tetrapyrroles than rapidly multiplying cells. To test this hypothesis, splenocyte suspensions from BALB/c mice were prepared and, following incubation in the presence or absence of the mitogenic lectin Con A (1.25 pLg/mL for 40 h), were untreated or treated with 1 mM ALA + 0.75 mM Oph for 3.5 h (Rebeiz et al. 1992).

The amount of Proto accumulation in these two cell populations was compared to ALA + Oph treated MLA 144 cells. Resting splenocytes treated with ALA + Oph accumulated much less Proto compared to similarly treated Con A-activated splenocytes, which exhibited a 50-fold increase in cell proliferation as indicated by tritiated thymidine incorporation.

Concanavalin A-activated splenocytes accumulated levels of Proto comparable to those of ALA + Oph treated MLA 144 lymphoma cells.

However resting and Con A-activated splenocytes, as well as MLA 144 cells, treated with medium alone, did not accumulate any detectable amounts of Proto (Rebeiz et al. 1992).

## 19.3 Intracellular Localization of Heme Biosynthesis in Animal Cells

Since Proto accumulation is the essential factor in causing photodynamic damage in animal cells, the next three sections investigated its intracellular accumulation in mammalian cells.

The intracellular site of heme biosynthesis in mammalian cells has been investigated by (Granick 1967). Using ox liver mitochondria, Granick proposed the following scheme: ALA is formed in the mitochondria and then translocates out of the mitochondria into the cytoplasm. In the cytoplasm, ALA is converted to coproporphyrinogen III (Coprogen III). Then Coprogen III translocates back into the mitochondrion where it is converted to Proto and heme. While this hypothesis has been popular for many years, it is beset with problems. Translocations into and out of the mitochondria are energetically wasteful. Secondly, since no evaluation of mitochondrial breakage during mitochondria isolation has been provided, it was difficult to evaluate contamination of the cytoplasm by mitochondrial content. It was therefore decided to reevaluate this hypothesis in neoplastic cells as described below.

### *19.3.1 Purity of the Mitochondrial Preparations*

First the purity of the mitochondrial preparations used in the studies was evaluated. Subcellular organelles were isolated from MLA 144 cells by differential centrifugation in sucrose buffer. To determine the purity of the mitochondrial fraction, the activity of three marker enzymes for mitochondria, cytoplasm and endoplasmic reticulum (ER)/microsomes was determined. Succinate cytochrome c reductase (SCR), a mitochondrial marker located the inner mitochondrial membrane, was used as an indicator of mitochondrial activity. Washed mitochondria exhibited a significant enrichment in SCR activity in comparison to the cell homogenate (Rebeiz et al. 1996a). There was little or no SCR activity in the cytoplasm and ER/microsomes, thus indicating that mitochondria were not present in these fractions (Rebeiz et al. 1996a). In addition, hydroxypyruvate reductase (HPR), an ER/microsomal marker and lactate dehydrogenase, a cytoplasmic marker, were significantly lower in the mitochondrial fraction (Rebeiz et al. 1996a). These results indicated that it was possible to prepare mitochondria from MLA 144 cells with reduced contamination by cytoplasm and ER/microsomal fractions.

Next it was determined whether mitochondria were the site of Proto accumulation as described below.

### *19.3.2 Protoporphyrinogen Accumulation in the Mitochondria of MLA 144 Cells Treated with ALA and Oph*

Homogenates of MLA 144 cells treated with 1 mM ALA and 0.75 mM Oph for 3 h in darkness accumulated significant amounts of protoporphyri(nogen) [Proto(gen)] i.e. a mixture of Proto and its reduced hexahydro analog protoporphyrinogen (Protogen), but did not accumulate other porphyrins. After oxidation of Protogen to Proto, the latter was identified by its fluorescence emission and Soret absorption maxima at 632 nm and 404 nm, respectively, in hexane-extracted acetone at room temperature (Rebeiz et al. 1992). Proto was also identified by chromatography with standard Proto on a Waters Bondapak reverse-phase C18-bonded column (Rebeiz et al. 1992).

Following subcellular fractionation of the treated cells, the Proto(gen) content of each subcellular fraction was determined. The total Proto(gen) content of treated cells averaged $221 \pm 18$ nmol/100 mg protein ($n = 6$). The $1{,}000 \times g$ pellet, which contains mainly nuclei, plasma membranes and unbroken cells accumulated significantly less Proto(gen) ($95 \pm 10$ nmol/100 mg protein). The bulk of the biosynthesized Proto(gen) accumulated in the mitochondria ($235 \pm 45$ nmol/100 mg protein, $p < 0.05$). Some Proto(gen) accumulated in the ER/microsomes ($62 \pm 8.7$ nmol/100 mg protein), but very little was observed in the cytoplasm ($9 \pm 2$ nmol/100 mg protein) (Rebeiz et al. 1992). These results confirmed that in MLA 144 cells mitochondria were the primary site of Proto biosynthesis (Dailey 1990). It was conjectured that the Proto found in the ER/microsomes may be due to contamination by broken mitochondria and/or to transport of Proto(gen).

### *19.3.3 Biosynthetic Origin of Protoporphyrinogen Accumulation in the Mitochondria*

To determine the biosynthetic origin of Proto(gen) accumulation in the mitochondria, the ability of isolated MLA 144 subcellular fractions to synthesize porphyrin(ogen)s from ALA in the presence of Oph was investigated.

In addition to Proto(gen), the formation of other porphyrin(ogen)s was observed. Prior to identification, porphyrinogens were converted to porphyrins as described in (Rebeiz 2002). The oxidized form of coproporphyrinogen (Coprogen), i.e. copropporphyrin (Copro), was identified by its fluorescence emission and Soret excitation at 620 and 394 nm, respectively, in hexane-extracted acetone at room temperature (Rebeiz et al. 1975) and by chromatography with standard Copro on a Waters Bondapak reverse-phase C18-bonded column (Rebeiz et al. 1996a). Other oxidized porphyrinogens such as uroporphyin III (Uro) (RT = 3.4 min) as well as heptaporphyrins (RT = 4.2 min) and hexaporphyrins (RT = 5.8 min) were formed during the conversion of Urogen to Coprogen.

Incubation of MLA 144 whole cell homogenates with ALA and Oph in darkness, resulted in the formation of Proto and lesser amounts of Uro(gen), heptaporphyrin (ogen) and Copro(gen) (Rebeiz et al. 1996a). However, incubation of the cytoplasmic fraction with ALA + Oph resulted mainly in the biosynthesis and accumulation of Uro(gen), heptaporphyrin(ogen) and Copro(gen), and lesser amounts of Proto (gen) and hexaporphyrin(ogen).

On the other hands Isolated mitochondria treated with ALA (1 mM), Oph (0.75 mM) and ATP (15 mM) formed very little Proto(gen) (Rebeiz et al. 1996a). Some Copro(gen) accumulation was observed (10 pmol/5 ml reaction) which might be due to contamination of the mitochondrial fraction by low levels of cytoplasm/ ER. The cytoplasmic/ER fractions accumulated significant amounts of Copro(gen) (25 pmol/5 ml reaction; $p < 0.05$) and much smaller amounts of Proto(gen). More significantly, when the cytoplasmic/ER and mitochondrial fractions were combined, in addition to Copro(gen), significant amounts of Proto were formed (Rebeiz et al. 1996a).

Similar results as those described above were observed when cytoplasm alone, and cytoplasm + mitochondria were incubated with ALA (1 mM), Oph (0.75 mM) and ATP (15 mM), in the absence of added ER. Altogether, these results indicated that Proto biosynthesis and accumulation from ALA in MLA 144 cells required the cooperation of mitochondria and cytoplasm.

### *19.3.4 Cofactor Requirement for the Biosynthesis and Accumulation of Protogen by Mitochondria*

To gain better understanding of the cooperation of mitochondria and cytoplasm during the biosynthesis and accumulation of Proto(gen), in mitochondria, the cofactor requirement of this process was investigated. Preliminary experiments indicated that the conversion of ALA to Protogen by mitochondria + cytoplasm was ATP-dependent (Rebeiz et al. 1996a). Thus the possibility that Coprogen, formed in the cytoplasm, may be actively transported into the mitochondria where it is converted to Proto was therefore investigated. Coprogen (1,250 pmol/ ml) was incubated with isolated mitochondria in the presence or absence of ATP. In the absence of ATP, some Proto biosynthesis was observed (29 ± 11.5 nmol/ 100 mg protein). However, at 100 mM of exogenous ATP, a 62 % increase in Proto formation was observed in comparison to controls ($p < 0.05$), thus indicating that ATP enhances the conversion of exogenous Coprogen to Proto by mitochondria. The high concentration of added ATP may have been mandated by the high concentration of the added Coprogen substrate.

## 19.4 Induction of Apoptosis in Leukemia Cells by Modulators of Heme Biosynthesis

We have shown in preceding sections that the concurrent use of Oph, a tetrapyrrole heme biosynthesis modulator, with ALA, significantly increases the efficacy of ALA treatment of transformed cells by enhancing Proto accumulation. Heme biosynthesis modulators such as Oph enhance Proto accumulation in transformed cells in the presence of ALA, increase cell lysis in vitro and enhance Proto-induced tumor necrosis in vivo with little damage to surrounding normal tissue (see section below).

In insects, modulators have also been associated with the induction of a dark death phenomenon, whereby treated tissues undergo cell death in the absence of light (Chap. 18). The dark cell death, attributed in insects to the formation and accumulation of Zn-Proto has also been observed as well, in cancer cells treated with ALA and Oph. One important possibility in cancer cells is that in addition to enhancing light-induced tumor necrosis by increasing endogenous Proto accumulation, Oph might also cause cell death in darkness by inducing programmed cell death, or apoptosis (Fisher 1994). Apoptosis is a type of cell death that exhibits distinct morphological and biochemical characteristics. It plays a key role in the mode of action of a diverse array of anti-tumor agents (Dive and Hickman 1991; Eastman 1990; Lowe et al. 1993; Strasser et al. 1994) and in oncogenesis (Fannidi et al. 1992; Hart et al. 1987; Symonds et al. 1994). In addition, many of the intermediate signaling molecules involved in apoptosis, such as p53, are known to cause growth arrest of neoplastic cells (Radvanyl et al. 1993; Rebeiz et al. 1994; Yonish-Rouach et al. 1991; Zhu and Anasetti 1995), prior to apoptosis induction.

One important intracellular mechanism occurring during apoptosis induction is the changes in mitochondrial transmembrane potential caused by the opening of large pores, which are part of the permeability transition (PT) phenomenon (Kroemer et al. 1995). The proteins that form these pores have not been identified, but the mitochondrial peripheral-type benzodiazepine receptor (M-PBR) has been implicated during PT (Kinally et al. 1993; Pastorino et al. 1994). The M-PBR may also be involved in porphyrin/porphyrinogen transport since coproporphyrinogen (Taketani et al. 1994), Proto (McEnery et al. 1992; Verma and Snyder 1988) and heme (Taketani et al. 1995) bind to the M-PBR. We have recently shown that coproporphyrinogen transport into mitochondria is enhanced by ATP (Rebeiz et al. 1996a), and the adenine nucleotide is a component of the M-PBR complex (McEnery et al. 1992).

We have therefore investigated the role of apoptosis in the Oph-mediated dark death phenomenon using MLA 144 leukemic T cells. While Oph and Proto, but not ALA, induced growth arrest, only Oph induced apoptosis. The results also suggested that induction of apoptosis by Oph may occur via the M-PBR and that apoptosis contributes to the death of MLA 144 cells in darkness.

### 19.4.1 Inhibition of DNA Synthesis by Oph

Many anti-tumor agents cause cells to pause during the cell cycle and induce growth arrest prior to cell death (Dive and Hickman 1991; Strasser et al. 1994; Yonish-Rouach et al. 1991). We therefore investigated whether ALA, Oph or a combination of the two reagents could reduce cell proliferation. This was done by measuring rates of DNA synthesis. All doses of Oph, but not ALA, caused a significant decrease in DNA synthesis after 3 h of incubation, as evidenced by reduced $^{3}$H-thymidine incorporation into the cells. Proliferation of MLA 144 cells was inhibited 64 % at 0.75 mM Oph, a dose that was previously shown to be non-toxic and to enhance Proto accumulation in the presence of ALA in vitro (Rebeiz et al. 1992). The level of inhibition was similar when ALA was used in combination with Oph, at all doses tested. Treatment of MLA 144 cells with ALA alone at any concentration did not significantly alter the rate of $^{3}$H-thymidine incorporation. DNA synthesis was also inhibited after 6 h and 24 h incubation with Oph (Rebeiz et al. 2001).

### 19.4.2 Reduction of Cell Proliferation by Proto and Non-chelating Isomers of Oph

Ortel et al. (1993) reported that the metal chelator desferrioxamine inhibited cell proliferation by chelating cellular iron, and this chelation suppressed progression through the cell cycle. To test the possibility that inhibition of cell growth by Oph could be due to its iron-chelating properties, two non-chelating positional-isomers of Oph, 1,7-Oph and 4,7-Oph, were used in a DNA synthesis assay. Exogenous Proto, which has been shown to inhibit the proliferation of cells by binding to (M-PBR) in mitochondria (Verma and Snyder 1988), was used as a positive control. Both Oph isomers inhibited $^{3}$H-thymidine incorporation in MLA 144 cells in a dose-dependent manner similar to Oph (Rebeiz et al. 2001). As expected from previous work on exogenous Proto (Verma and Snyder 1988), growth arrest was induced at micromolar concentrations. Surprisingly, exogenous Proto was 10–30 fold more efficient at inhibiting DNA synthesis than Oph or its isomers ($p < 0.05$). The positional isomer 1,7-Oph was two to four fold more effective than Oph in causing growth arrest of MLA 144 cells ($p < 0.05$). These results indicated that the iron-chelating property of Oph was not responsible for inducing cell growth arrest.

### 19.4.3 Cell Viability and Membrane Permeability of MLA 144 Cells Treated with ALA, Oph or Proto

Apoptotic cells still have intact plasma membranes, and may appear viable, while necrotic cells quickly lose membrane intactness (Cohen et al. 1992). It was conjectured that the decreased DNA synthesis observed in MLA 144 cells treated with Oph and Proto could be due to a toxic effect of these chemicals. The plasma membrane integrity of MLA 144 cells subjected to trypan blue exclusion (cell viability) and chromium release (membrane permeability) was therefore assessed. Although treatment with Proto and Oph for 3 h in darkness caused growth arrest, cell viability was not significantly reduced as evidenced by trypan blue exclusion. Indeed, the total cell number did not decrease in Oph or Proto treated cells. Proto-treated (0.05–0.1 mM) cell viability after 3 and 24 h of incubation was similar to controls.

A chromium release assay was also used as a test of membrane permeability. A combination of ALA and Oph did not induce significant chromium release from cells, during 3 h of incubation with ALA (1–4 mM) and Oph (0.75–3 mM) (Rebeiz et al. 2001). Thus, at 0.75–3 mM, Oph did not appear to be toxic to MLA 144 cells and did not induce cell lysis during 3 h of dark of incubation.

However, after 24 h of treatment with Oph (1.5–3 mM), cell membranes no longer excluded trypan blue (15–16 % viability) (Rebeiz et al. 2001). Also, non-chelating isomers of Oph significantly decreased cell viability after 3 and 24 h of incubation.

### *19.4.4 Induction of Apoptosis by Oph*

Internucleosomal cleavage of DNA was first measured to confirm the occurrence of apoptosis. MLA 144 cells that had been serum-deprived for 3 h exhibited slight DNA fragmentation in comparison to cells at time zero (Rebeiz et al. 2001). However, treatment with Oph at either 0.75 mM or 1.5 mM induced a dramatic increase in DNA cleavage after 3 h of dark incubation. A similar DNA laddering pattern was observed when cells were treated with a combination of ALA and Oph. Internucleosomal cleavage of DNA was also evident after 6 h incubation with Oph. Treatment with ALA or Proto alone did not induce DNA fragmentation at either 3 or 6 h. Thus, although Proto did induce growth arrest, unlike Oph, it did not induce apoptosis.

Cells undergoing apoptosis usually exhibit a characteristic hypodiploid peak of DNA when stained with the DNA intercalating dye PI and analyzed by flow cytometry. Thus MLA 144 cells were treated for 3 h with ALA, Oph, Proto, or medium, in a serum-free system prior to ethanol permeabilization and staining with PI. Cell cycle analysis was determined by analysis with MPLUS software. Basal level of apoptosis was $8 \pm 3.7$ % in medium-treated cells (Rebeiz et al. 2001). Treatment with 1.5 mM Oph induced a significant level of apoptosis ($45 \pm 1.8$ %, $p < 0.05$), and a distinct subdiploid peak (apoptotic) became apparent. In addition, the proportion of cells accumulating in early S phase increased in comparison to medium-treated cells ($60 \pm 2.5$ % vs. $48 \pm 2.5$ % respectively, $p < 0.05$). At 0.75 mM Oph, which induced the highest level of internucleosomal DNA cleavage, the majority of the cells became apoptotic ($81 \pm 5.6$ %), and most of the DNA appeared in the hypodiploid peak (Rebeiz et al. 2001). The G0/G1 peak decreased to $9 \pm 0.5$ % ($p < 0.05$) and the G2/M peak became virtually non-existent ($0 \pm 0.1$ %). A significant majority of the cells appeared to be arrested in early S phase ($88 \pm 9.8$ %, $p < 0.05$), a phenomenon often observed in cell cycle dependent apoptosis (Zhu and Anasetti 1995). Neither ALA alone or Proto induced apoptosis, nor did these reagents significantly alter the cell cycle when compared to medium-treated cells (Rebeiz et al. 2001).

### *19.4.5 Abrogation of Induced Apoptosis by Cycloheximide (Rebeiz et al. 2001)*

The classical inductive pathway of apoptosis, as observed in thymocytes during negative selection or following treatment with dexamethasone, is defined by the ability of cycloheximide to inhibit its induction (Cohen et al. 1992). To determine whether Oph causes apoptosis via an inductive pathway, MLA 144 cells were incubated for 18 h with cycloheximide (1.5 μg/ml) and then treated for 3 h with varying amounts of Oph in the continuing presence of cycloheximide. Cells were then double-labeled (Cohen et al. 1992) with Hoechst 33342 and PI. In flow cytometric analysis, apoptotic cells are represented by high Hoechst 33342 and low forward angle light scatter (FALS) with the PI positive (necrotic) cells excluded. Oph at concentrations ranging from 0.375 to 1.5 mM induced a significant increase in the proportion of apoptotic cells, with the highest level of apoptosis occurring at 0.75 mM (85 %, $p < 0.05$). Medium-treated cells, which were serum-deprived for 3 h, were 5 % apoptotic. Cycloheximide alone increased the apoptotic population in medium-treated cells to 15 %, however this was not statistically significant. Co-incubation of cells with cycloheximide and Oph decreased Oph-induced apoptosis. The percentage of apoptotic cells was reduced to background levels at all doses of Oph (Rebeiz et al. 2001). These results indicated that Oph causes apoptosis via the classic induction pathway that requires protein synthesis.

## 19.5 Induction of Tumor Necrosis by ALA and Oph-Dependent Photodynamic Therapy

As was mentioned previously the approved use of ALA and its long chain fatty acid esters to treat skin cancer has met with limited acceptance by the medical community, because of damage to surrounding tissues and limited effectiveness as compared to liquid nitrogen treatments (oral communication from my skin doctor).

In this section we will describe, as an alternative, the use of AlA and Oph as a model research system for treating Meth-A cancerous cells and solid tumors.

The treatment focuses on combined treatment with ALA and Oph in a murine syngeneic solid tumor model using Meth-A cells. It was shown that the combination of ALA and Oph induced the accumulation of porphyrins in Met- A cells and upon exposure to light, significantly reduced the surface area of the tumors and induces their destruction, as determined by their disappearance at both the visual and histological levels. These data supported the possibility that the accumulation of Proto and subsequent photodestruction of tumor cells can be significantly enhanced in vivo by the use of joint AlA an tetrapyrrole modulator

### *19.5.1 Proto Accumulation in ALA and Oph Treated Meth-A Ascites Cell Suspensions*

We determined whether treatment of Meth-A cells in vitro with ALA and Oph would result in Proto accumulation. Meth-A ascites cells were treated with 1 mM ALA and 0.75 mM Oph for 3.5 h in darkness. The combination of ALA and Oph induced very significant increases in Proto accumulation (338 nmol/100 mg protein; $P < 0.05$). Treatment with ALA alone also induced measurable accumulation of Proto (61.73 nmol/100 mg protein) in comparison with control cells, but this was not statistically significant (Rebeiz et al. 1996b). This level of Proto accumulation is approximately six fold lower than the levels that accumulate with ALA + Oph treatment. This finding is also consistent with the specific cell lysis data in which ALA + Oph-induced cell lysis was approximately seven times higher than that arising from treatment with ALA alone (Rebeiz et al. 1996b). It is interesting that the addition of Oph to ALA-treated Meth-A cells induced more Proto accumulation than what was seen previously in MLA 144 cells (Rebeiz et al. 1992). Therefore it was concluded that some cell type specificity was observed because the efficacy of ALA and Oph in stimulating Proto accumulation appeared to vary significantly with cell types.

### *19.5.2 Sensitivity of Meth-A Cells to ALA and Oph Treatment*

Since previous results had shown a positive correlation between the amount of Proto accumulation and specific cell lysis upon illumination in several transformed cell lines treated with ALA and Oph (Rebeiz et al. 1992), Meth-A cells were screened for sensitivity to light activated cell lysis after ALA and Oph treatment.

There was no apparent cell lysis in darkness (3 h) in Meth-A cells in medium, or in cells treated with I mM ALA or 0.75 mM Oph alone (Rebeiz et al. 1996b). However, there was an increase in cell lysis when cells were treated with both compounds in darkness ($P < 0.05$) then were exposed to light. Indeed, upon light activation (30 min, 3.798 J/cm$^2$), 16.8 % of the 1 mM ALA-treated Meth-A cells were lysed ($P < 0.05$). When ALA and Oph were used jointly, specific cell lysis increased over seven fold ($P < 0.05$) compared to the ALA-treated cells.

### *19.5.3 Proto Accumulation in ALA and Oph Treated Meth-A Solid Tumors* In Vivo

Preliminary experiments on Proto accumulation in Meth-A tumors in vivo were conducted using a range of ALA (1–10 mM) and Oph (0.75–7.5 mM) doses. In general, the highest concentration yielded Proto accumulation levels >30 nmol/100 mg protein whereas 1 mM ALA and 0.75 mM Oph yielded Proto levels of 1.1 nmol/100 mg. Therefore, for the next series of experiments, doses ranging from 2.5 to 10 mM ALA were used at different time points (Rebeiz et al. 1996b).

Untreated tumors did not accumulate Proto, whereas significant amounts of Proto accumulated after 3 and 6 h (31 and 17 nmol/100 mg protein, respectively) were observed at the highest dosage of ALA and Oph tested. Twelve hour or more after injection, the levels of accumulated Proto were lower, probably due to degradation by intracellular enzymes (Mattheis and Rebeiz 1977).

A similar pattern of decline in Proto accumulation in single-cell suspensions of a T-cell lymphoma (MLA 144) incubated with 0.5 mM ALA and 0.38 mM Oph for 18 h was also observed. The amount of accumulated Proto fell over tenfold at 18 *versus* 3 h ($1.7 \pm 0.16$ *versus* $19.6 \pm 9.90$ nmol/100 mg protein; $n = 2$). Because substantial Proto had accumulated 3 h after injection with ALA and Oph, this time point was used in subsequent experiments. The next series of experiments focused on the enhancement of ALA-induced Proto accumulation by Oph in Meth-A solid tumors (Rebeiz et al. 1996a) Previous studies with Oph had shown that the optimal Oph concentration to use was about 75 % of the used ALA concentration (Rebeiz et al. 1990). As before, medium-treated tumors or tumors treated with Oph alone at 1.8, 3.75, or 7.5 mM did not accumulate any Proto. Oph at two concentrations, 7.50 and 3.75 mM, enhanced Proto accumulation in the presence of 10 (2.7-fold) and 5 (4.7-fold) mM ALA, respectively ($P < 0.05$). However, at 1.8 mM, Oph did not enhance Proto accumulation when used in conjunction with 1 mM ALA.

### *19.5.4 Effect of ALA and Oph Treatment on the Size and Histopathology of Meth-A Solid Tumors*

Having established that Meth-A tumors accumulate Proto after in vivo injection with ALA and Oph the effects of ALA and Oph phototreatment of the tumors were next investigated by histopathological examination of the tumor and the surrounding normal tissue. Treatment groups consisted of mice with tumors treated with medium or ALA and Oph for 3 h in darkness, followed by illumination for 30 min.

The initial mean tumor sizes in both control and treated mice ($n = 3$) were similar. Mice were allocated randomly to treatment groups. Tumors that received ALA and Oph photodynamic therapy (3 h incubation before a 30-min illumination) were significantly reduced 5.4-fold in size compared to control tumors by day 1 and at all subsequent time points ($P < 0.05$). By days 4 and 5, there were no palpable tumors in 5 of the 6 treated mice. In contrast, the control tumors doubled in size by days 4 and 5 ($P < 0.05$). On day 1, one of the treated mice sacrificed for histopathology appeared to have a palpable tumor, but histopathological examination showed only edema and epidermal necrosis with no evidence of sarcoma tissue present (Rebeiz et al. 1996b). Subsequent histopathology confirmed that in ALA and Oph-phototreated mice, the Meth-A tumor tissue, which is non-metastasizing, was eradicated, except for a small amount of sarcoma tissue in one mouse. In 5 of 6 mice in which the tumor tissue was eradicated, only necrotic sarcoma tissue remained, indicating a high level of tumor necrosis. In addition, one-half of the ALA and Oph-treated mice had panniculitis and epidermal and dermal damage. This is consistent with a local inflammatory response occurring at the site of ALA and Oph injection and is commonly observed in patients treated with ALA photodynamic therapy (Grant et al. 1993; Kennedy et al. 1990; Wolf et al. 1993). Neither panniculitis nor tumor necrosis was evident in control mice.

Histopathological sections of control tumors and lesions 3 days after ALA and Oph photodynamic therapy were examined. Meth-A tumors were undifferentiated, as reported earlier (Chun and Hoffman 1987). In control mice, there was some necrosis of adjacent dermis related to expansion of the tumor, and the tumor showed muscle invasion, which is typical of established tumors (Rebeiz et al. 1996b).

Some panniculitis was apparent, along with necrosis of the epidermis and inflammation of the dermis, sloughing of the skin, and keratinization, which are all indicators of a photodynamic reaction occurring at the site of injection.

## References

Chun M, Hoffman M (1987) Combination immunotherapy of cancer in a mouse model: synergism between tumor necrosis factor and other defense systems. Cancer Res 47:115–118

Cohen JJ, Duke RC, Fadok VAS, Sellins KS (1992) Apoptosis and programmed cell death in immunity. Ann Rev Immunol 10:267–293

Dailey HA (1990) Conversion of coproporphyrinogen to protoheme in higher eukaryotes and bacteria: terminal three enzymes. In: Dailey HA (ed) Biosynthesis of heme and chlorophylls. McGraw-Hill, New York, pp 123–161

Dive C, Hickman JA (1991) Drug-target interactions: only the first step in the commitment to a programmed cell death. Br J Cancer 64:192–196

Dougherty TJ (1987) Photosensitizers: therapy and detection of malignant tumors. Photochem Photobiol 45:879–889

Eastman A (1990) Activation of programmed cell death by anti cancer agents: cisplatin as a model. Cancer Cells 2:275–280

Fannidi A, Harrington EA, Evans GI (1992) Cooperative interaction between c-mye and bel-2 protooncogenes. Nature 359:554–556

Fisher DE (1994) Apoptosis in cancer therapy: crossing the threshold. Cell 78:539–542

Granick S (1967) The heme and chlorophyll biosynthetic chain. In: Goodwin TW (ed) Biochemistry of chloroplasts, vol 2. Academic Press, New York, pp 373–408

Grant WE, Hopper C, McRobert AJ et al (1993) Photodynamic therapy of oral cancer: photosensitization with systemic aminolevulinic acid. Lancet 342:147–148

Hart GJ, Miller AD, Leeper FJ et al (1987) Biosynthesis of the natural porphyrins: proof that hydroxymethylbilane synthase (porphobilinogen deaminase) uses a novel binding group in its catalytic action. J Am Chem Soc Chem Commun 23:1762–1765

Kennedy JC, Pottier RH (1992) Endogenous protoporphyrin IX, a clinically useful photosensitizer for photodynamic therapy. Photochem Photobiol B Biol 14:275–292

Kennedy JC, Pottier RH (1994) Using delta-aminolevulinic acid in cancer therapy. In: Duke SO, Rebeiz CA (eds) Porphyric pesticides, chemistry, toxicology and pharmaceutical applications, vol 559. American Chemical Society, Washington, DC, pp 291–302

Kennedy J, Pottier RH, Pross DC (1990) Photodynamic therapy with endogenous protoporphyrin IX. Basic principles and present clinical experience. J Photochem Photobiol B Biol 6:143–148

Kinally KW, Zorov DB, Antonenko YN et al (1993) Mitochondrial benzodiazepine receptor linked to inner membrane ion channels by nanomolar actions of ligands. Proc Natl Acad Sci USA 90:1374–1378

Kloek J, Gerard MJ, van Henegouwen B (1996) Prodrugs of 5-aminolevulinic acid for photodynamic therapy. Photochem Photobiol 64(6):994–1000

Kolossov v L, Rebeiz CA (2004) Chloroplast biogenesis 91: detection of delta-aminolevulinic acid esterases activity in higher plant and insect tissues. Pestic Biochem Physiol 83:9–20

Kroemer G, Petit P, Zamzami N et al (1995) The biochemistry of programmed cell death. FASEB J 9:1277–1287

Lowe SW, Ruley HE, Jacks T et al (1993) p53-dependent apoptosis modulates the cytotoxicity of anticancer agents. Cell 74:957–967

Malik z, Ehreneberg B, Paraggi A (1989) Inactivation of erythrocytic, lymphocytic and myelocytic leukemic cells by photoexcitation of endogenous porphyrins. J Photochem Photobiol B Biol 4:195–205

Mattheis JR, Rebeiz CA (1977) Chloroplast biogenesis. Net synthesis of protochlorophyllide from protoporphyrin IX by developing chloroplasts. J Biol Chem 252:8347–8349

McEnery MW, Snowman AM, Trifiletti RR et al (1992) Fill in. Proc Natl Acad Sci U S A 89:3170–3174

Ortel B, Tanew A, Honigsman H (1993) Lethal photosensitization by endogenous porphyrins of PAM cells modification by desferrioxamine. J Photochem Photobiol B Biol 17:273–278

Pastorino JG, Simbula G, Gilfor E et al (1994) Protoporphyrin IX, an endogenous ligand of the peripheral benzodiazepine receptor, potentiates induction of the mitochondrial permeability transition and killing of cultured hepatocytes by retenone. J Biol Chem 269:31041–31046

Radvanyl LG, Mills GB, Miller RG (1993) Religation of the T cell receptor after primary activation of mature T cells inhibits proliferation and induces apoptotic cell death. J Immunol 150:5704–5715

Rebeiz CA (2002) Analysis of intermediates and end products of the chlorophyll biosynthetic pathway. In: Smith A, Witty M (eds) Heme chlorophyll and bilins, *methods and protocols*. Humana Press, Totowa, pp 111–155

Rebeiz CA, Mattheis JR, Smith BB et al (1975) Chloroplast biogenesis. Biosynthesis and accumulation of protochlorophyll by isolated etioplasts and developing chloroplasts. Arch Biochem Biophys 171:549–567

Rebeiz CA, Montazer-Zouhoor A, Hopen HJ et al (1984) Photodynamic herbicides: 1. Concept and phenomenology. Enzyme Microb Technol 6:390–401

Rebeiz CA, Juvik JA, Rebeiz CC (1988) Porphyric insecticides 1. Concept and phenomenology. Pestic Biochem Physiol 30:11–27
Rebeiz CA, Juvik JA, Rebeiz CC et al (1990) Porphyric insecticides 2. 1,10-Phenanthroline, a potent porphyric insecticide modulator. Pestic Biochem Physiol 36:201–207
Rebeiz N, Rebeiz CC, Arkins S et al (1992) Photodestruction of tumor cells by induction of endogenous accumulation of protoporphyrin IX: enhancement by 1,10-phenanthroline. Photochem Photobiol 55:431–435
Rebeiz CA, Amindari SM, Reddy NK (1994) Delta-aminolevulinic acid-based herbicides. In: Duke SO, Rebeiz CA et al (eds) Porphyric pesticides: chemistry toxicology and pharmaceutical applications, vol 559. American Chemical Society, Washington, DC, pp 48–64
Rebeiz N, Arkins S, Kelley KW et al (1996a) Enhancement of coproporphyrinogen III transport into isolated leucocyte mitochondria by ATP. Arch Biochem Biophys 333:475–481
Rebeiz N, Arkins S, Rebeiz CA et al (1996b) Induction of tumor necrosis by delta-aminolevulinic acid and 1,10-phenanthroline. Cancer Res 56:339–344
Rebeiz N, Arkins S, Kelley KW et al (2001) Modulator of heme biosynthesis induces apoptosis in leukemia cells. J Clin Laser Med Surg 19:59–67
Strasser A, Harris AW, Jacks T et al (1994) DNA damage can induce apoptosis in proliferating lymphoid cells via p53-independent mechanism inhibitable by bel-2. Cell 79:329–339
Symonds H, Krai L, Remington L et al (1994) p53-dependent apoptosis suppresses tumor growth and progression in vivo. Cell 78:703–711
Taketani S, Kohno H, Okuda M et al (1994) Fill in. J Biol Chem 269:7527–7531
Taketani S, Kohno H, Furukawa T et al (1995) Fill in. J Biochem 117:875–880
Verma A, Snyder SH (1988) Fill in. Mol Pharmacol 34:800–805
Wolf P, Rieger E, Kerl H (1993) Topical photodynamic therapy with endogenous porphyrins after application of 5-aminolevulinc acid. An alternative treatment modality for solar keratoses, superficial squamous cell carcinomas and basal cell carcinomas. J Am Acad Dermatol 28:17–21
Yonish-Rouach E, Resnitzky D, Lotem J et al (1991) Wild type p53 induces apoptosis of myeloid leukaemic cells that is inhibited by interleukin-6. Nature 352:345–347
Zhu L, Anasetti C (1995) Cell cyclic control of apoptosis in human leukemic T. cells. J Immunol 154:192–200

# Index



C.A. Rebeiz, *Chlorophyll Biosynthesis and Technological Applications*,
DOI 10.1007/978-94-007-7134-5, © Springer Science+Business Media Dordrecht 2014

**D**

**E**

## P

**U**



**V**

Printed by Printforce, the Netherlands